**Recent Developments in Pig Nutrition 2**

# Recent Developments in
# Pig Nutrition 2

*Editors*

D J A COLE
W HARESIGN
P C GARNSWORTHY

*Faculty of Agricultural and Food Sciences, University of Nottingham*

Nottingham University Press
Sutton Bonington Campus, Loughborough, Leicestershire LE12 5RD

NOTTINGHAM

First published 1993



**British Library Cataloguing in Publication Data**
Recent Developments in Pig Nutrition – No. 2

I. Cole, D J A
636.4

ISBN 1–897676–41–7

Typeset by The Midlands Book Typesetting Company, Loughborough, Leicestershire
Printed and bound by Quorn Selective Repro Ltd, Loughborough, Leicestershire

# CONTENTS

**Introduction** ix

1 **CHANGES IN CONSUMER PERCEPTIONS OF MEAT QUALITY** 1
A.S. Ambler, Dalehead Foods Ltd, Cambridge Road, Linton, Cambs and J.D. Wood, Department of Meat Animal Science, School of Veterinary Science, University of Bristol, Langford, Bristol, UK

2 **NUTRITIONAL MANIPULATION OF CARCASS QUALITY IN PIGS** 12
C.T. Whittemore, Institute of Ecology and Resource Management, University of Edinburgh, School of Agriculture, West Mains Road, Edinburgh

3 **CONSEQUENCES OF CHANGES IN CARCASS COMPOSITION ON MEAT QUALITY** 20
J.D. Wood, Department of Meat Animal Science, School of Veterinary Science, University of Bristol, Langford, Bristol

4 **ENERGY-PROTEIN INTERACTIONS IN PIGS** 30
A.C. Edwards and R.G. Campbell, Bunge Meat Industries, Corowa, New South Wales, Australia

5 **COMPARISON OF ARC AND NRC RECOMMENDED REQUIREMENTS FOR ENERGY AND PROTEIN IN GROWING PIGS** 47
A.J. Lewis, Department of Animal Science, University of Nebraska, Lincoln, Nebraska, 68506 USA

6 **AMINO ACID NUTRITION OF PIGS AND POULTRY** 60
D.H. Baker, Department of Animal Sciences, University of Illinois, Urbana, Illinois, USA

7 **METHODS OF DETERMINING THE AMINO ACID REQUIREMENTS OF PIGS** 76
H.S. Bayley, Department of Nutritional Sciences, University of Guelph, Guelph, Ontario, N1G 2W1, Canada

8 **ILEAL DIGESTIBILITIES OF AMINO ACIDS IN PIG FEEDS AND THEIR USE IN FORMULATING DIETS** **85**
T.D. Tanksley Jr. and D.A. Knabe, Animal Science Department, Texas A & M University, College Station, Texas 77843, USA

9 **USE OF SYNTHETIC AMINO ACIDS IN PIG AND POULTRY DIETS** **106**
K.E. Bach Knudsen and H. Jorgensen, National Institute of Animal Science, Rolighodsvej 25, Copenhagen V, Denmark

10 **TOWARDS AN IMPROVED UTILISATION OF DIETARY AMINO ACIDS BY THE GROWING PIG** **117**
P.J. Moughan, Department of Animal Science, Massey University, Palmerston North, New Zealand

11 **ROLE OF DIETARY FIBRE IN PIG FEEDS** **137**
A.G. Low, 56 Reigate Road, Brighton, East Sussex, BN1 5AH

12 **PHOSPHORUS AVAILABILITY AND REQUIREMENTS IN PIGS** **163**
A.W. Jøngbloed, H. Everts and P.A. Kemme, IVVO, PO Box 160, 8200 Lelystad, Netherlands

13 **THE WATER REQUIREMENT OF GROWING-FINISHING PIGS – THEORETICAL AND PRACTICAL CONSIDERATIONS** **179**
P.H. Brooks and J.L. Carpenter, Seale-Hayne Faculty of Agriculture, Food and Land Use, Polytechnic South West, Newton Abbot, Devon, UK

14 **WATER FOR PIGLETS AND LACTATING SOWS: QUANTITY, QUALITY AND QUANDARIES** **201**
D. Fraser[1], J.F. Patience[2], P.A. Phillips[1] and J.M. McLeese[2]
[1]Animal Research Centre, Agriculture Canada, Ottawa K1A OC6 and [2]Prairie Swine Centre, Department of Animal and Poultry Science, University of Saskatchewan, Saskatoon S7N OWO, Canada

15 **THE PHYSIOLOGICAL BASIS OF ELECTROLYTES IN ANIMAL NUTRITION** **225**
J.F. Patience, Prairie Swine Centre, Department of Animal and Poultry Science, University of Saskatchewan, Saskatoon S7N OWO, Canada

16 **MANIPULATION OF THE GUT ENVIRONMENT OF PIGS** **243**
T.L.J. Lawrence, Veterinary Field Station, University of Liverpool, Neston, Wirral, Cheshire, L64 7TE

17 **ACIDIFICATION OF DIETS FOR PIGS** **256**
R.A. Easter, Department of Animal Sciences, University of Illinois, Urbana, Illinois, USA

18 **AETIOLOGY OF DIARRHOEA** **267**
J.W. Sissons, Nurish Products, Dairy Food Systems, Protein Technology International, Checkerboard Square, St Louis, MO 63164, USA

19 **IMMUNITY, NUTRITION AND PERFORMANCE IN ANIMAL PRODUCTION** **285**
P. Porter and M.E.J. Barratt, Unilever Colworth Laboratory, Bedford, UK

20 **NOVEL APPROACHES TO GROWTH PROMOTION IN THE PIG** **295**
P.A. Thacker, Department of Animal Science, University of Saskatchewan, Saskatoon, Canada

21 **IMPACT OF SOMATOTROPIN AND BETA-ADRENERGIC AGONISTS ON GROWTH, CARCASS COMPOSITION AND NUTRIENT REQUIREMENTS OF PIGS** **307**
T.S. Stahly, Iowa State University, Department of Animal Science, Ames, Iowa 50011, USA

22 **STRATEGIES FOR SOW NUTRITION: PREDICTING THE RESPONSE OF PREGNANT ANIMALS TO PROTEIN AND ENERGY INTAKE** **317**
I.H. Williams[1], W.H. Close[2] and D.J.A. Cole[3]
[1]University of Western Australia, Nedlands, Perth, Australia
[2]129 Barkham Road, Wokingham Road, Wokingham, Berks, RG11 2RS
[3]University of Nottingham, Faculty of Agricultural and Food Sciences, Sutton Bonington, Loughborough, Leics, LE12 5RD

23 **PREDICTING NUTRIENT RESPONSES OF THE LACTATING SOW** **332**
B.P. Mullan[1], W.H. Close[2] and D.J.A. Cole[3]
[1]Animal Industries, Department of Agriculture, Perth, Western Australia
[2]129 Barkham Road, Wokingham Road, Wokingham, Berks, RG11 2RS
[3]University of Nottingham, Faculty of Agricultural and Food Sciences, Sutton Bonington, Loughborough, Leics, LE12 5RD

24 **NUTRITION OF THE WORKING BOAR** **347**
W.H. Close, 129 Barkham Road, Wokingham Road, Wokingham, Berks, RG11 2RS and F.G. Roberts, Department of Food and Nutritional Sciences, Kings College, Kensington, London W8 7AH

**Index** **371**

# INTRODUCTION

Pig nutrition is a rapidly changing science. The application of the scientific changes is continuous and finds a particular outlet in the compound feed industry. The following chapters are intended to examine relevant developments which have occurred recently. The chapters have previously been published in *Recent Advances in Animal Nutrition* which is produced each year as the proceedings of the University of Nottingham Feed Manufacturers Conference. This is the fifth book in this series and is a successor to the original *Recent Developments in Pig Nutrition*. The series has proved extremely popular and it is hoped that the current book will be valuable to students, teachers, advisory staff, research workers, farmers and many others. The production characteristics of the pig are greatly affected by inherent factors such as genotype and sex. In recent years, intense selection pressures and a shift, in some countries, to the abandoning of castration has resulted in a pig of considerable protein deposition potential. Further there has been a recognition that it is not now satisfactory to produce a diet for a particular class of pig based on liveweight but the genotype particularly, has to be taken into account. Consequently there has developed an attitude of "feeds for breeds".

There has been a clear recognition that the consumers perception of the quality of the end-product is of particular importance and that these perceptions are the subject of change. Particular concerns of the consuming public are cost and influence on human health. In this context there is a demand for carcasses with a low fat content but at the same time the customer still requires aspects of eating quality which are often associated with higher fat levels. In addition there has been an increase in the production of 'natural' products together with an expansion of outdoor rearing. The shift in production systems from more uniform intensive units brings with it the need to consider nutrition under these different conditions. There is, too, a growing awareness of the influence of pig production on the environment. In such a situation it is important to identify clearly the requirements of the animals and to avoid excessive excretion of waste nutrients. Such environmental concern is entirely compatible with efficient production.

Energy and protein form an important basis of the diet. Fundamental to establishing requirements for them is the determination of protein deposition rate. Recent reports in the literature of protein deposition rates well over 200g/day, for highly selected genotypes and animals receiving repartitioning agents, greatly contrast with previously used values of 90–140g/day. There are questions to be raised concerning the techniques used to arrive at such values, but clearly attention needs to be paid to changes which might affect requirement values. Methods of determining the requirements of the pig and the value of the feedstuff continue to attract the interest of the researcher.

The need to reduce pollution and increase efficiency of production stimulates research workers in many ways. Firstly, the increased precision of knowledge will allow the more accurate formulation of diets with less excretion. This is particularly important in the case of nitrogen and phosphorus. The more precise provision of protein has been aided, particularly, by the supply of amino acids as part of the ideal protein and the commercial availability of crystalline amino acids. Enzymes afford an opportunity to increase the utilisation of nutrients and this is particularly well illustrated with phosphorus. Dutch work has given some accurate estimates of requirements for available phosphorus and shown the benefits of dietary microbial phytase. Dietary fibre presents a particular challenge in the production of high quality diets.

An often forgotten aspect of the diet is water but recently there has been a renewed interest in it. Its supply has important influences on performance and doubt has been cast on some nutritional studies where insufficient attention has been paid to water supply. The important interaction between water and food has been clearly identified. A whole complex of aspects relating to the diet probably interact and influence the gut environment and level of performance. Certainly electrolyte balance and acidification have important influences. Diarrhoea is linked to gut environment and continues to be of great interest to the nutritionist.

During the last decade there has been considerable interest in growth promotion and repartition of the deposition of tissues to the benefit of carcass quality. Growth hormone and β-agonists have received particular attention although public opinion has not always favoured them. Interest continues in the further development of probiotics.

Breeding animals have continued to be the focus of considerable research. Much of the work has been directed at establishing nutritional strategies and predicting responses to nutrients in the breeding female. As a focal point, lactation continues to be important. This is because of the consequences which come from undernutrition in lactation and from the problems associated with lack of appetite at this time. There are probably three major factors which influence feed intake in lactation; these are level of feeding in pregnancy, genotype and environmental temperature. Consideration of all these is aimed at achieving the target intake for the suckling period. The working boar suffers considerable neglect from the research worker and author, yet its influence on reproduction is considerable. Nutrition of the boar needs to take account, not only of performance, but also of the influences on behaviour, health and welfare.

The era covered by the chapters in this book has been one of considerable change. The predecessor to this book suggested the need to come to terms with the technologies associated with the application of the microchip. The way ahead, now, will involve coming to terms with the pressures of the consumer, the environment and the need for high technical efficiency. This will coincide with an era of the rapid development of biotechnology in all its meanings. It will demand further advances in the technical and scientific information that underlies pig nutrition.

1

# CHANGES IN CONSUMER PERCEPTIONS OF MEAT QUALITY

A. S. AMBLER
*Dalehead Foods Limited, Cambridge, UK*

*and*

J. D. WOOD
*Department of Meat Animal Science, University of Bristol, UK*

## Introduction

This chapter deals with changing perceptions of quality and with the factors which influence quality, especially those aspects, such as eating quality and appearance, which can be measured and controlled through changes in production and processing. Although much of the discussion applies to meat in general, particular emphasis will be placed on pigmeat.

## Changes in meat consumption and consumer attitudes

Total meat consumption in the UK has remained remarkably constant during the last 20 years at around 65 kg per person per year (carcass weight basis), although the consumption of red meat, especially beef and lamb, has fallen whilst that of poultry has increased (Figure 1.1). Pigmeat consumption has increased slightly, the result of increasing sales of pork offsetting falling sales of bacon and ham.

Underlying and indeed controlling these trends have been several changes in society and in attitudes to food and meat (Woodward, 1988; Baron, 1988; MLC, 1988, 1989a and 1989b; Kempster, 1989). These changes include:

(1) Since 1970, disposable income has greatly increased although a smaller proportion is now spent on meat (3% against 6%). At present, meat accounts for about 25% of food expenditure. People now spend a higher proportion of their income on luxury goods such as video recorders and microwave ovens.
(2) More meals are eaten outside the home and fewer 'formal' meals are eaten inside the home. A large increase in the number of households in which both adults work has meant less time is spent in preparing meals and the 'convenience' attributes of foods are given a higher priority.
    The 'strength of demand' for meat, i.e. after accounting for income and price changes, has fallen for all categories except convenience products since the mid 1970s.
(3) In recent years, consumers have become much more concerned about the health aspects of meat and particularly its fat content. Calls to reduce the consumption of red meat have been central to the advice of many nutritionists

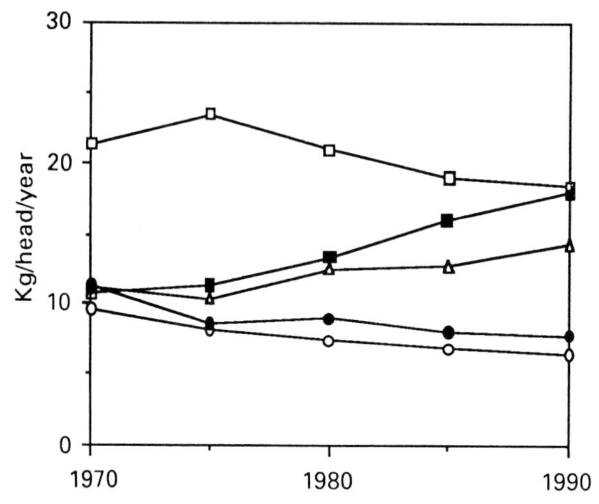


**Figure 1.1** Trends in UK meat consumption. □, beef; ■, poultry; △, pork; ●, bacon; ○, lamb

aiming to reduce the fat content and increase the ratio of polyunsaturated to saturated fatty acids (P:S) in diets. Whereas nutritionists readily distinguish between 'low fat' or skimmed milk and whole milk they less frequently differentiate between high- and low-fat meats. However, in a recent survey, the range in the fat content of pork chops between retail outlets was 12–30% and it was shown that consumers could easily detect variations in fat content by eye (Table 1.1). The more helpful advice to consumers is therefore to continue to eat meat and to buy lean or low fat meats rather than to exclude meat from the diet.

Consumers have also expressed concern over possible residues of additives and hormones in meat and about its microbiological safety (Consumers Association, 1989).

(4) Meat is now more likely to be purchased in a supermarket than in a butcher's shop. The results in Figure 1.2 show that for all meats during the 1980s,

**Table 1.1** VARIATION IN THE FAT CONTENT OF PORK CHOPS BETWEEN RETAIL OUTLETS (EIGHT SAMPLES FROM EACH OF 11 OUTLETS)

| *Retail outlet no.* | | *Fat (%)*[a] |
|---|---|---|
| 1 | much leaner than average | 12 |
| 2–6 | leaner than average | 16 |
| 7 | average | 18 |
| 8–10 | fatter than average | 20 |
| 11 | much fatter than average | 30 |

Consumers Association (1989)
[a] Dissected fat as % chop weight

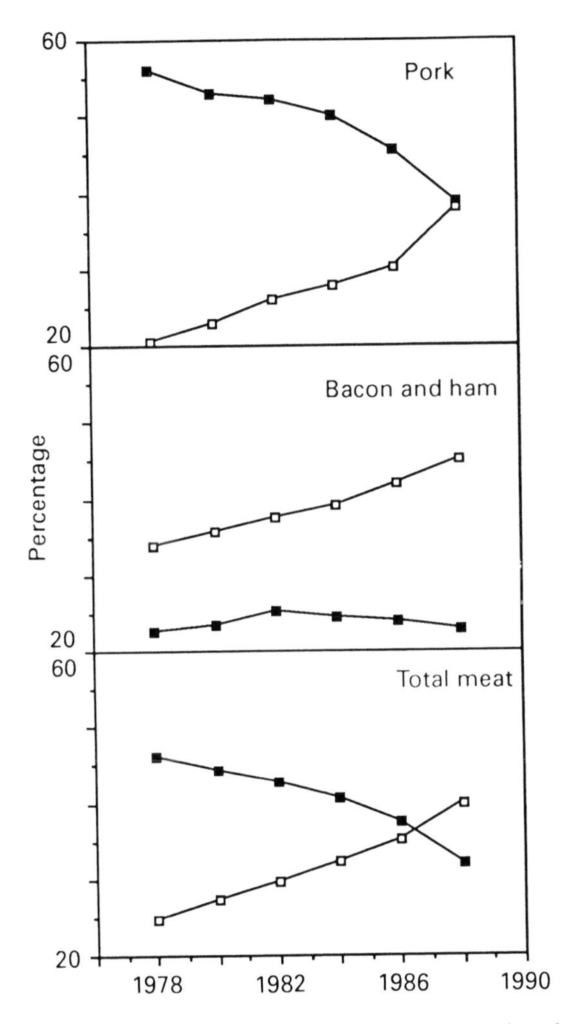


**Figure 1.2** Percentage of pork, bacon and ham and total meat purchased in butchers' shops (■) or supermarkets (□) in the UK (MLC Meat Demand Trends, 1989a)

supermarkets have been increasing their share of the meat market by about 1.5% per annum. The trend for pork has been even more marked, about 2.5% per annum. The share of the total meat market taken by the different types of outlet in 1988 is shown in Figure 1.3.

The 40% share of the meat market now held by the supermarkets puts them in a strong position to respond to changing consumer attitudes and indeed to influence them. Supermarkets have recognized the demand for easy-to-prepare low-fat cuts and have increasingly purchased primal cuts which are boneless, well trimmed and vacuum-packed from processors with modern equipment able to meet their demands precisely. Some of these same processors are

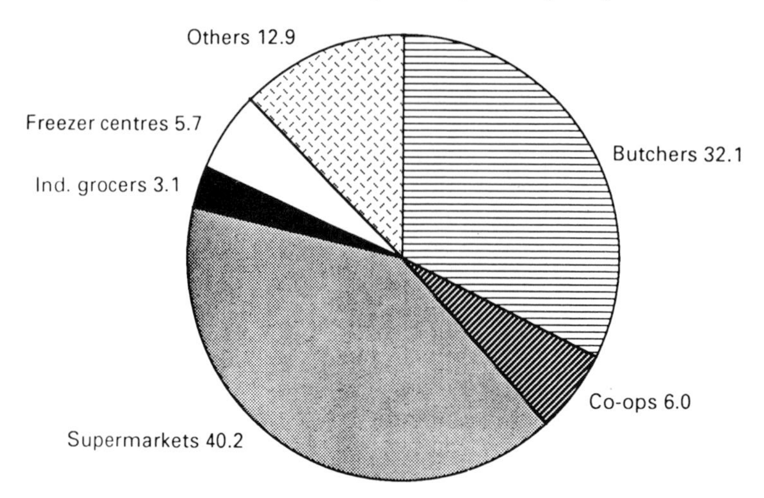


**Figure 1.3** Percentages of total meat purchased from different outlets in 1988 (MLC Meat Demand Trends, 1989a)

involved in the production of ready meals and other added-value products as well as more attractive fresh cuts. Supermarkets have placed the emphasis on 'quality' rather than 'quantity' and their increasing market share indicates the success of this formula.

It must be said, however, that many 'traditional' butchers have also followed the high quality route and have been equally successful.

(5) There is increasing public debate over the welfare of animals in intensive production systems. A greater attention to animal welfare coupled with a lower use of chemicals at all stages from birth to slaughter are major features of 'natural' or 'organic' meat production.

## Improving meat quality

It is accepted that the level of quality should be improved in order for meat to increase or retain its share of the food market in the future. The average consumer's definition of quality is broad although producers and processors should try to identify the main factors, especially those that can be controlled and are likely to be emphasized by supermarket buyers. For fresh pigmeat the main factors are

lean/fat ratio
appearance: the meat should not be pale, dark or wet
eating quality: the meat should be tender and juicy with good flavour
health and welfare aspects, sometimes referred to as 'the image' of meat

The factors controlling lean to fat ratio (genetics and nutrition) are broadly known but much less is known about eating quality and appearance. However, these are active fields of research at present, both in the UK and abroad, and some recent findings will now be discussed.

DIETARY EFFECTS ON QUALITY

Eating quality can be influenced by quantitative as well as qualitative dietary changes. Recent evidence collected by MLC in the first Stotfold trial showed that pigs fed *ad libitum* to 80 kg liveweight produced more tender meat than those grown slowly (Table 1.2). This was partly due to a higher concentration of intramuscular lipid (marbling fat) in pigs with higher P2 levels but other factors must also have been involved. The situation is similar to that in beef in which feedlot-fed beef is invariably more tender than that produced in more slowly-growing grazed animals.

**Table 1.2** EFFECTS OF SIRE TYPE AND FEEDING REGIMEN ON FATNESS AND EATING QUALITY OF PORK

| | *Sire type* | | *Feeding regimen* | |
|---|---|---|---|---|
| | *Meat* | *White* | *Ad lib* | *Restricted* |
| Fat thickness at P2 (mm) | 11.70 | 12.20* | 12.80 | 11.10* |
| Muscle lipid (%)[a] | 0.78 | 0.82 | 0.85 | 0.75* |
| Tenderness[b] | 4.86 | 5.07* | 5.20 | 4.73* |
| Juiciness[b] | 4.30 | 4.39 | 4.44 | 4.25* |
| Pork flavour intensity[b] | 4.56 | 4.53 | 4.52 | 4.57 |

MLC (1989c)
[a] Ether-extractable lipid in *m. longissimus* ('eye' muscle) of loin chop. Muscle lipid is more commonly referred to as 'marbling fat'
[b] Taste panel scores 1 to 8
* Indicates significance at 5% level or greater

The fatty acid composition of dietary fat has marked effects on the composition, consistency and possibly eating quality of meat. Increasing the proportion of unsaturated fatty acids in the diet reduces the firmness of fat tissue although increases in the P:S ratio in meat may have health benefits. Increasing the saturated fatty acid composition of the diet may improve eating charararacteristics.

In the USA claims have been made that a combination of dietary factors (whole-roasted soya beans, high fibre, low fat) reduces the cholesterol content of meat fat from about 70–30 mg/100 g (Marbery, 1989). This is a surprising finding which may have little medical relevance in view of the uncertain role of dietary cholesterol in cardiovascular disease. Nevertheless the approach recognizes that consumers are influenced by information suggesting a healthy product. The US pork industry has emphasized that in comparison with beef and lamb, pork is high in polyunsaturated fatty acids and can justify the label 'the other white meat'.

Feeding high levels of fibre to pigs increases the production of skatole in the intestine and the concentration of skatole in tissues. It is now believed that skatole rather than androstenone is the major factor contributing to 'boar taint' even though it is also found in gilts (Lundstrom *et al.*, 1988). Other ingredients, such as yeast solids, have also been shown to increase skatole concentrations above the threshold value of 0.25 ppm in backfat which is where most people find the meat unacceptably tainted.

## BREED EFFECTS ON QUALITY

A high level of pale soft exudative (PSE) muscle in pigs expressing the halothane gene is to be avoided since this results in tough meat because of low water binding characteristics. There was no evidence of a higher incidence of PSE muscle in the 'meat type' sires examined in the MLC Stotfold trial (Table 1.1). Durocs have some advantages in meat quality caused by a low level of genetic stress coupled with high levels of marbling fat. This has not consistently led to higher tenderness in comparative tests (e.g. Table 1.3 gives data from a recent UK study) although values for Duroc crosses are generally slightly higher than those for crosses of the white breeds when comparisons are made across several recent European studies (see Wood, 1989).

**Table 1.3** EATING QUALITY OF PORK IN CROSSBRED PIGS FROM DUROC OR LARGE WHITE SIRES

| | *Muscle lipid (%)*[a] | *Taste panel scores*[b] | | | |
|---|---|---|---|---|---|
| | | *Tenderness* | *Juiciness* | *Flavour* | *Overall acceptability* |
| Duroc | 1.4 | −0.3 | 1.3 | 2.1 | 1.0 |
| Large White | 1.0 | −0.0 | 1.3 | 2.0 | 1.1 |
| | | NS | NS | NS | NS |

Edwards *et al*. (1989)
[a] As Table 2.2
[b] Tenderness, flavour and overall acceptability scored −7 to +7; juiciness scored 0 to 4

## STUNNING AND SLAUGHTER EFFECTS ON QUALITY

It is generally accepted that high voltage stunning (i.e. above 200 V rather than 100 V) has advantages over other methods on welfare grounds because the animals are instantaneously stunned. However, there have been complaints of quality defects including PSE muscle, capillary rupture and bone fractures. Carbon dioxide stunning as used widely in Denmark and Sweden (compact stunner) is reported to have advantages on quality grounds (Table 1.4) but there are questions regarding animal welfare during the induction phase of anaesthesia.

**Table 1.4** MEAT QUALITY MEASURED UNDER COMMERCIAL CONDITIONS IN DENMARK IN PLANTS USING HIGH VOLTAGE OR $CO_2$ (COMPACT) STUNNING

| | *300 V manual* | *700V automatic* | *$CO_2$* |
|---|---|---|---|
| Bloody meat in shoulder (g) | 145 | 59 | 8 |
| Bone fracture (%) | 1.2 | 1.0 | 0 |
| PSE (%)[a] | 18.5 | 15.1 | 4.0 |
| DFD (%)[a] | 5.7 | 8.3 | 6.1 |

Larsen (1983)
[a] Based on measurements of light reflectance and pH

CARCASS CHILLING EFFECTS ON QUALITY

Rapid chilling of the carcass after slaughter can reduce tenderness in pork as well as beef and lamb. This is due to 'cold shortening' of muscles before they enter rigor. Results collected in Denmark showed important differences in the tenderness of loin chops between four factories (Table 1.5). This was thought to be due to different chilling procedures (e.g. air velocities) even though the temperature/time conditions were similar in factory 1 which had the toughest meat and factory 4 which had the most tender.

**Table 1.5** TENDERNESS OF GRILLED PORK CHOPS PRODUCED IN DIFFERENT DANISH FACTORIES

| | *Factory* | | | |
|---|---|---|---|---|
| | *1* | *2* | *3* | *4* |
| Time (and temperature) of chilling[d] | 47 min (–25.5) | 127 min (–10) | 60 min (–24) | 25 min (–10/+4) |
| Toughness[e] | 118[c] | 87[b] | 73[a] | 86[b] |
| Tenderness[f] | –0.51[c] | 1.86[a,b] | 2.27[a] | 1.59[b] |
| Chops with unacceptable eating quality (%) | 57 | 12 | 9 | 11 |

Barton-Gade *et al.* (1987)
[a,b,c] Means in a row with different superscripts are significantly different ($P < 0.05$)
[d] Time spent in chilling and average temperature of the chiller
[e] Shear force value, high figures indicate tougher meat
[f] Taste panel score –5 (v. tough) to +5 (v. tender)

As with beef carcasses, tenderness can be improved when chilling rates are high if the carcass is suspended from the pelvis rather than the achilles tendon (Moller *et al.*, 1987). This ensures that key muscles are under restraint during chilling and are not allowed to 'cold-shorten'.

INTEGRATED PRODUCTION TO ENSURE QUALITY. OUTDOOR REARING SYSTEMS

There have been several attempts to raise the quality of pork by ensuring that the several factors affecting quality are controlled. In practice this means specifying breed, diets and production system and employing special handling techniques at the abattoir. Many European countries have developed different systems emphasizing different aspects.

*Denmark*

It was recognized some time ago in Denmark that a higher level of eating quality was possible if certain aspects of production and processing were carefully controlled (Barton-Gade, 1984). The high quality demands of influential customers such as Marks and Spencer in the UK were important in developing the 'Antonius' system. In this integrated approach, breed, health status, feed, transportation and processing are specified and carcasses are assessed for PSE muscle, bruising, blemishes, etc. A premium, approximately 10% above the current average price, is paid for carcasses meeting these standards. Since 1984 the specifications have changed slightly with Duroc replacing Hampshire as the dominant sire breed and tighter controls over feed. Now no 'growth promoters' of any kind are allowed (Barton-Gade, personal communication).

It has been suggested that pigs reared under free range conditions will suffer less from quality defects associated with stress, e.g. PSE and dark firm dry (DFD) muscle. Recent Danish work showed that free-range pigs from the same production units as pigs reared indoors had slightly lower 24-h pH values and correspondingly a lower incidence of the DFD condition (Table 1.6). This was thought to be caused by less excitable behaviour in the lairage.

**Table 1.6** MEAT QUALITY IN PIGS REARED FREE-RANGE OR INDOORS[a]

| | *Free range* ($n$ = 29) | *Indoors* ($n$ = 28) | | *Free range* ($n$ = 29) | *Indoors* ($n$ = 28) |
|---|---|---|---|---|---|
| $pH_{ULT}$[b] | | | % DFD[c] | | |
| biceps femoris | 5.78 | 5.83 | biceps femoris | 3.4 | 10.7 |
| longissimus dorsi | 5.74 | 5.83* | longissimus dorsi | 3.4 | 21.4 |

Barton-Gade and Blaabjerg (1989)
[a] Results from 1 week of a 3-week trial
[b] Measured 24 h after slaughter
[c] Based on $pH_{ULT}$ measurements > 5.90
* $P < 0.05$

*West Germany*

The group Bauern Siegel emphasize PSE muscle, premiums being paid for carcasses having a muscle pH above 6.0 at 45 min after slaughter. The lowest price is paid for carcasses with a pH value 5.5 or below, these being the carcasses with pale watery muscle which are unsuitable for processing. Such premiums provide incentives for producers to make changes in breeding and to organize transportation and slaughter procedures in such a way as to minimize stress during the critical pre-slaughter period.

*Sweden*

The Swedish Farmers' Meat Marketing Association (SCAN) also pay strict attention to animal welfare at the preslaughter stage. Pigs are rested for at least 2 h before slaughter, rearing groups are transported together, electric goads are forbidden, etc. The predominant breed is Piggham, a cross between Swedish Landrace, Large White and Hampshire.

*UK*

At Dalehead Foods Ltd farmers are offered a 'special' contract for pigs which are produced in what can be described as a 'pig friendly' way. The criteria are fairly wide but can be summarized as follows:

(1) The pigs must be housed in straw-based systems with natural lighting and ventilation (outdoor systems would quality) and fed on cereal based feeds with no unnecessary additives.
(2) The pigs must have ample space, careful attention must be given to social grouping and there must be high standards of husbandry and management dedicated to providing a stress-free environment.
(3) There must be no unnecessary restraint of sows such as tethers or stalls, and no cages or totally slatted systems.
(4) There must be no castration or taildocking or any routine medication other than that prescribed by a veterinary surgeon for the treatment of specific health problems.
(5) Care and attention must be given to the handling of the pigs during loading and transportation.

At the abattoir every effort is made to minimize stress, the meat is chilled slowly and aged in vacuum packs. Great care is taken to preserve the integrity of the product by identifying it carefully, by carrying out routine analysis of the feed and the meat itself and by inspecting the farms at regular intervals.

No claims have been made regarding the eating quality of the meat, the scheme concentrates entirely on the animal welfare and residue-free aspects of the product.

There is a small but growing market prepared to pay a premium for the product. At the moment the scheme accounts for around 10% of the throughput of Dalehead Foods Ltd and we are able to pay farmers a premium of 5% over average market prices.

*France*

An increasing emphasis in several countries is now placed on outdoor rearing, both of breeding and growing pigs. A recent article in *Porc Magazine* (1989) describes an initiative in the French department of Sarthe involving producers, processors and retailers in which free range pigs (Porcs Fermiers de la Sarthe) qualify for 'labels rouge' when certain criteria are met:

outdoor rearing or access to outside at all stages from birth to slaughter at 75–100 kg liveweight. Age at slaughter at least 182 days
diet composed of 70% cereals with no 'growth additives'
careful handling at all stages
colour of lean pink; colour of fat white

*Netherlands*

The 'Scharrelvarkens' system of free-range production has rules which are rigidly monitored by inspectors employed by a consortium of producers, processors, retailers, government bodies, animal protection groups and consumers. It is envisaged that production will amount to about 4% of total Dutch production and command a price premium of 20% (G. Eikelenboom, personal communication).

## Implications for the future

The meat industry has had to face several problems over the last few years which have led to considerable and on-going rationalization. However, in the face of often adverse media attention it is responding to the changing market situation in the following positive ways:

(1) There is now much greater awareness of the problem of variable meat quality. Although it is a complex, multifactorial problem it is at least now being brought to the forefront of research and receiving much more attention.
(2) It has recognized that meat is going to become just another grocery item, attractively packaged and labelled, and competing on the supermarket shelf with a wide range of other alternative food products often at cheaper price. Poor and variable quality is, therefore, unacceptable.
(3) It understands that the consumer wants lean meat with good colour and a consistent, repeatable eating quality.
(4) It has realized that the 'image' of meat is also important and that in this context animal welfare considerations, dietary concerns and the safety of the product must be given higher priority.

A further complication facing the industry in the future is that labelling regulations are likely to become even more stringent, requiring ever increasing amounts of information about the product. Farmers, animal feed compounders and meat processors alike will all have to become fully accountable for and guarantee the 'quality' of their product (in all meanings of the word quality).

The most important implication arising from the trends that have been described is that the links between producer, processor and retailer will strengthen. Consumers identify a particular retailer with the level of quality they require, and retailers in turn expect this to be provided consistently. It follows that more emphasis will be placed on the 'blueprint' approach, i.e. combining the production, processing and marketing factors in an optimum way to produce the desired result. Also there will be an increasing demand for objective ways of measuring all aspects of meat quality. The degree of control over inputs at various stages from production to retail sale is likely to vary. Three different sorts of product requiring a different complexity of inputs can be envisaged:

(1) A lean product with acceptable quality and a low price.
(2) A lean product with guaranteed quality – medium price.
(3) A naturally produced product with high quality – high price.

In purchasing product 1 the abattoir will specify carcass weight and fat thickness. In moving to product 2 more emphasis will be given to breed, sex, feed, pre-slaughter handling and processing as they affect eating quality and this product may have a higher level of fat which can be trimmed. For product 3, guarantees will be sought regarding feed, ingredients, housing and containment systems. It will be extremely important for claims on 'additive-free' status, for example to be backed up by solid evidence based on routine sampling of feeds.

The meat industry is presently at an extremely critical stage of its development. Success will come to those who read the signs correctly and respond to them in time.

## References

Baron, P. (1988). *British Food Journal,* **1**, 10–14

Barton-Gade, P. A. (1984). Modern consumer trends regarding pork quality. Slagteriernes Forkningsinstitut Manuscript No. 666E

Barton-Gade, P. A., Bejerholm, C. and Borup, U. (1987). *Proceedings of the 33rd International Congress of Meat Science and Technology*, pp. 181–184

Barton-Gade, P. A. and Blaabjerg, L. O. (1989). *Proceedings of the 35th International Congress of Meat Science and Technology*, pp. 1002–1005

Consumers Association (1989). *Which*, September 1989, pp. 426–429

Edwards, S. A., Wood, J. D., Moncrieff, C. B. and Porter, S. J. (1989) *Animal Production* (in press)

Kempster, A. J. (1989). *Animal Production,* **48**, 483–496

Larsen, H. K. (1983). In *Stunning of Animals for Slaughter* (ed. G. Eikelenboom), pp. 73–81. Martinus Nijhoff, The Hague

Lundstrom, K., Malmfors, B., Malmfors, G., Stern, S., Petersson, H., Mortensen, A. B. and Sorensen, S. E. (1988). *Livestock Production Science,* **18**, 55–67

Marbery, S. (1989). *Hog Farm Management*, June, 1989, pp. 12–19

Meat and Livestock Commission (1988). Meat Demand Trends 88/3. Meat and Livestock Commission, Milton Keynes

Meat and Livestock Commission (1989a). Meat Demand Trends 89/1. Meat and Livestock Commission, Milton Keynes

Meat and Livestock Commission (1989b). Meat Demand Trends 89/2. Meat and Livestock Commission, Milton Keynes

Meat and Livestock Commission (1989c). Stotfold Pig Development Unit. First Trial. Meat and Livestock Commission, Milton Keynes

Moller, A. J., Kirkegaard, E. and Vestergaard, T. (1987). *Meat Science,* **21**, 275–286

*Porc Magazine* (1989). **215**, 66–74

Wood, J. D. (1989). In *The Voluntary Food Intake of Pigs* (eds. J. M. Forbes, M. A. Varley and T. L. J. Lawrence) Occasional Publication of the British Society for Animal Production No. 13, pp. 79–86

Woodward, J. (1988). Consumer attitudes to meat and meat products with regard to composition and labelling. Food Policy Research Unit, University of Bradford, Bradford

2

# NUTRITIONAL MANIPULATION OF CARCASS QUALITY IN PIGS

C.T. WHITTEMORE
*Institute of Ecology and Resource Management, Scotland, UK*

## Introduction

The business of producing meat from pigs requires that the producer has a clear view of his target end-product, has an adequate definition of that target, and has a means of manipulating the production system to achieve the stated goal. Usually, but not invariably, the pig buyer and meat packer will have some means of grading meat pigs received from producers; payment for the pigs will be related to the achievement of grading standards. This does not mean, however, that the grading target set by the producer to optimize his business should necessarily be the same as the grading standards set by the buyer. A market preferring, and paying premium rates for, a high standard of blockiness in the carcass (as would be attained with Pietrain or Belgian Landrace type pigs) may be best exploited, not by obtaining all pigs of premium grade from the use of pure bred animals, but by a lower percentage achieving the premium grade by the use of cross-bred Large White animals which would be more prolific, faster growing and less prone to stress. A market preferring and paying premium rates for pigs of less than 15 mm backfat depth at the P2 site (65 mm from the mid-line at the last rib), may often be best exploited by achieving less than 80 per cent top grade in circumstances where mixed groups of pigs require the castrated males to be grown so slowly that the naturally leaner females are unnecessarily held back.

Discrimination by a meat buyer against entire male pigs may be resolved by the producer castrating his animals, or by acceptance of the price penalty. Optimum tactics will depend upon cost benefit analysis and the ability of the producer to attain high feed intakes and rapid growth.

Pig producing businesses must therefore identify their target grading standards in the light of, but not necessarily identical with, grading and payment schemes set up by meat buyers and processors. The means by which producers can identify appropriate grading targets for pigs grown for meat are complex but becoming established (for example, the Edinburgh Model Pig (Whittemore, 1980 and 1983) and others). This chapter will discuss the ways in which producers can achieve those targets once the optimum grading standard for any particular production unit in any particular trading environment has been decided.

## Aspects of grading standards

Grading schemes usually contain a range of standards. In some the standard (such as level of fatness) may be the prime controller of value and payment received by the producer, whilst in others the standard (such as quality of fat) may be of considerable importance to carcass quality but not yet play a functioning part in the payment schedule.

In some markets conformation is a functional aspect of the grading standard. Shape is almost entirely dependent upon breed, and not much open to nutritional manipulation other than through the creation of over-fatness. Most aspects of meat quality, other than amount of fat, are also not open to nutritional control. For example, muscle quality (in particular PSE and DFD) is influenced primarily by breed and physical treatment around the time of slaughter. Soft fat in pigs is associated, in any breed or sex, with leanness itself, but there may also be a tendency for entire males to have slightly softer fat than females even at the same degree of fatness. Unsaturated fatty acids in the diet, especially linoleic (C18:2) may influence the type of fat deposited in the body, rendering it softer. Extreme leanness in pigs can also lead to the fat becoming lacey and splitting away from the lean. Splitting fat can be a problem where the packer wishes to prepare cuts with some fat left on, as for example with bacon, but is less of a problem where the joint is sold as lean alone (as would be the case for continental pig loins comprising solely longissimus dorsi (eye) muscle).

Important as the quality of the lean and fat is to meat consumers, grading standards take little account of them at present, although the imposition of minimum, rather than maximum, fatness levels would reduce problems associated with over-leanness. In the United Kingdom, good carcass shape may even be discriminated against. At equal percentage lean, the blocky breeds carry more backfat than conventional Large White and Landrace breeds. The consequence is that although some breeds of good conformation may contain more lean meat, having deeper hams and eye muscles, less bone (as well as an improved carcass yield of about 3 percentage units), and smaller heads, they are liable to be down-graded for overfatness at the P2 site. At equal P2 fat depth, total body fat may be 6 per cent less. As an approximation:

$$\text{Lean in carcass (\%)} = (B)\ \text{P2}^{-0.21}$$

where $B$ is the degree of blockiness and ranges from 90 for Large White and Landrace pigs to 100 for pure Pietrain/Belgian Landrace types. J.D. Wood (personal communication) estimates Pietrain pigs to have some 4 percentage units more lean meat than Large Whites (61 *vs.* 57 per cent), and to have a considerably more favourable lean:bone ratio (6:1 *vs.* 5:1).

Most grading standards and payment schedules relate to fatness grades within a given weight band. For example, one scheme may limit carcass weight to between 60 and 75 kg, and pay premium price for pigs of less than 15 mm P2 and impose a maximum price penalty for pigs of more than 20 mm. Equivalent schemes might pertain for pork carcasses of below 60 kg carcass weight and cutter and heavy pig carcasses of above 75 kg. There may be additional requirements, perhaps that the length of the carcass be greater than 775 mm, or that the fat measurements at shoulder, mid-back and loin each be contained within a certain limit. It follows from the natural growth pattern of the pig that lighter animals have a lower fat thickness, thus, as the potential value of the carcass increases with its weight, the likelihood of it being down-graded because of over-fatness is also increased. The consequence would be a diminished price paid for each kilogram of lean meat provided. This inconsistency would be ameliorated by a grading scheme which paid for the kilograms of lean meat yielded. This could be calculated from knowledge of the carcass weight and the P2 measurement (together with perhaps the breed, if of the blocky type). Some suggested predictions of percentage lean in the carcass side have been 63–0.51 P2 (Wood, J.D., personal communication); 61–0.52 P2 (Tullis, 1982); Rook, A.J. (unpublished data) examined a number of data sets for which the average multiplier for P2 was −0.58. Kempster and Evans (1979) presented three equations for 47, 72 and 93 kg carcasses. These were respectively; 60–0.73 P2, 60–0.63 P2 and 57–0.54 P2.

The phenomenon of increased fatness with increased weight would also indicate that the correct slaughter weight for the fatter castrated males should be towards the lighter end of the market, whilst that for entire males should be towards the heavier end. It is germane to production tactics that the natural tendency is for fatness to increase faster than body weight, thus fatness accelerates disproportionately rapidly. This has been well illustrated by Rook who gave the following prediction from one population of pigs:

$$\text{Subcutaneous fat (kg)} = 0.0002\ X^{2.54}$$

where X = carcass weight (kg)

Strict weight limits for any particular grading scheme are often not needed in effectively integrated producer/retailer organizations. Nevertheless, many buyers will take in pigs over a range of, say, 50–80 kg carcass weight, but do this through the medium of not one, but perhaps three, separate grading schedules; each penalizing producers for underweights and overweights. Achievement of maximum allowable weight consistent with an adequately low depth of backfat and an adequately low number of outgrades due to overweight, within any one scheme, is clearly a realistic producer target.

Standards of over-fatness (say a maximum of 15 mm P2) may come to be matched by standards of under-fatness. Minimum fat levels (about 10 mm P2) are required for the eating quality of the lean meat, as well as to help maintain the quality of the fat itself. Given modern techniques for measurement of carcass fatness in the live animal, pigs could, if required, be sent off to slaughter at a given target fatness of 10 mm P2. Targetting for sale within a given weight range can be antagonistic to targetting for sale within a given backfat depth range; the narrower these ranges become, the more difficult it is to meet both simultaneously.

Fatness is considered as being normally distributed through the population. However, as average fatness reduces, then so the possibility of normal distribution lessens and that for skewness increases. With a skewed distribution the likelihood of over-lean pigs rises disproportionately. Given the perceived biological distributions for fatness and for weight at any given age, it is difficult to see how producers could supply pigs to ever-decreasing ranges of weight, fatness and age. The benefits to the producer of sale for slaughter at pre-selected calendar dates may override the benefits of closely meeting arbitrary standards for weight and fatness.

## Nutrition and fatness

Whilst for many pigs there are few or no grading standards and for others the standards include measures such as conformation, pig grading standards at present in the UK can be considered as primarily dependent upon the thickness of backfat. Backfat depth is greatly influenced by breed and genetic merit; Large White pigs are leaner than Pietrain, and over the past 20 years of selection in the Large White breed some 10 mm of backfat has been removed from pigs slaughtered at 90 kg. The primary determinant of grade is the quality or the strain of the pig in use; genetically fat pigs will tend to be always fat within the feasible range of nutritional and environmental variation. However, whilst the long-range strategy for fatness reduction and meeting grading standards must be genetic, the tactics with animals of any given genetic composition must depend upon the knowledge that fatness is greatly influenced by both quality and quantity of food. The major mechanism open to producers to manipulate grade and achieve grading standards is therefore through the control of the nutrition of the growing pig.

### PROTEIN

Diets which do not adequately provide for the requirement of absolute amounts of ideal protein (ARC, 1981) fail to allow maximum lean tissue growth. Energy thus freed from protein synthesis is diverted to fatty tissue growth. Excess protein, on

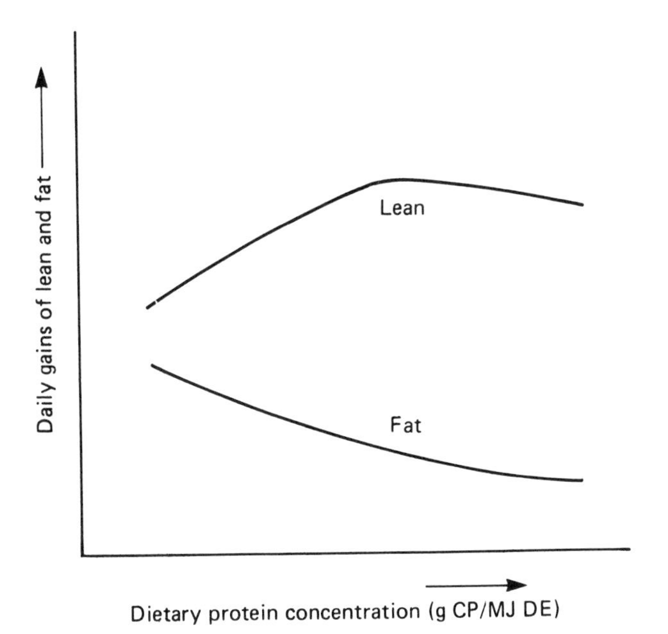


**Figure 2.1** The influence of increasing protein concentration (g CP/MJ DE) upon daily gains of lean and fat. The figure relates to an optimum level of feed intake; at lower levels the total growth response is less, while at higher levels of feed intake lean tissue growth will not increase above the potential limit of the pig but fatty tissue growth will increase markedly, and a lower ratio of lean to fat will be seen at all protein concentrations

the other hand, reduces energy status. The energy yielded from protein by deamination is about half of the assumed DE of protein. The effective energy value of a diet containing excess protein will therefore fall as deamination rate rises, with a resultant diminution of energy available for fat deposition. Excess supply of total protein over the requirements for protein maintenance and protein growth will have the consequence of enhancing leanness. Increasing diet protein therefore first reduces fat in the carcass by diverting energy into lean tissue growth and away from fatty tissue growth as a frank protein deficiency is reduced. Second, further increments of protein above the requirement will continue to increase leanness by effectively reducing the available energy yielded from the diet and thus pre-empting the conversion of energy to fat. This second method, of reducing carcass fatness by the supply of excess protein, while explaining linear responses of percentage lean to dietary crude protein, can be expensive to execute and also bring about a reduction in pig growth rate. This latter is consequent not only upon a reduction in the rate of fat gains; if the general level of feeding is not sufficiently generous, then lean growth itself may be curtailed through an inadequate energy supply to drive the metabolic motors of protein anabolism. The role of protein (CP) concentration is illustrated in *Figure 2.1*.

LEVEL OF FEED (ENERGY SUPPLIED)

As the amount of feed (balanced for protein, vitamins and minerals) consumed daily by the pig increases, then initially the daily gains of both lean and fat respond

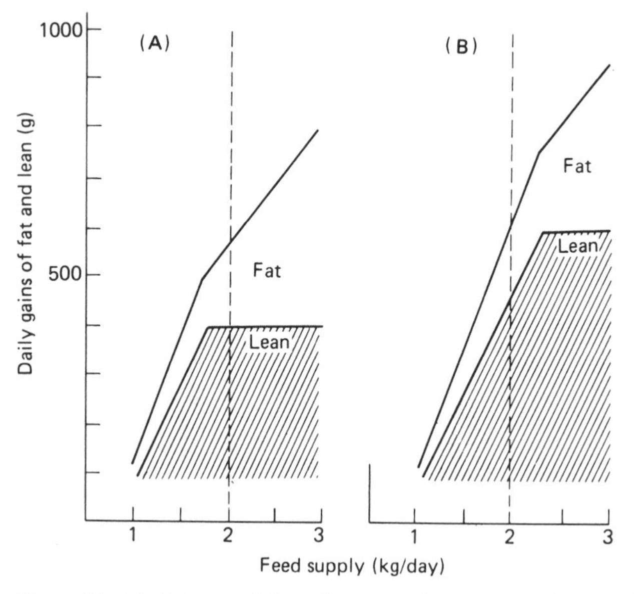


**Figure 2.2** Daily lean and fatty tissue growth responses of pigs to increases in daily feed supply. Increments of balanced feed induce linear growth of lean until the maximum potential daily lean growth is achieved. Up to that point fat growth proceeds at a lower rate determined by the minimum fat:lean ratio (*see also Table 7.1*). After maximum daily lean growth is achieved further increases in feed supply generate rapid growth of fatty tissue. Pigs differ genetically in their ability to grow lean, and this influences the level of feed intake at which fattening commences

linearly (*Figure 2.2*). Lean growth will cease to respond to increasing feed supply when the maximum lean tissue growth rate potential is reached; for pig A this is at 400 g, for pig B at 600 g. At this point the excess energy will be channelled to fat deposition with the consequence that, while total growth rate increases at a slower rate, the proportion of growth that is fat rapidly increases. This will mitigate against the achievement of a high percentage of pigs in the premium low-fat grades.

Even while feed supply is inadequate to maximize lean tissue growth there is nevertheless always some deposition of a minimum level of fat commensurate with normal positive daily gains. This minimum may be expressed in terms of a ratio to lean. This minimum fat:lean ratio gives the level of fatness in the pig which can only be undercut by creating abnormal conditions of fat catabolism such as would occur at very slow rates of growth. Should this minimum give backfat depths at slaughter weight that are in excess of the premium grade standard, then only considerable reduction in feed supply and growth rate could place carcasses in the top grade (such might be the case for castrated pigs of low genetic merit). Equally, should the minimum ratio give backfat depths at slaughter weight that are below a minimum fatness standard (as might be the case with entire males of high genetic merit), then adequate fatness can only be achieved by feed intake levels in excess of those that will maximize lean tissue growth rate. The higher the potential for lean tissue growth rate, the more difficult such a feed intake level would be to achieve.

Pigs of high genetic merit may have higher lean tissue growth rate potentials, lower minimum fat ratios, or both. Such animals will be thinner at low feed intakes and more difficult to fatten as feed level increases. Excessive fatness is feasible for all pigs, but only provided that voluntary feed intake is sufficient to put the maximum limit to potential lean tissue growth rate into range. The position as it relates to likely appetite and weight of pig is described in *Figure 2.3*, while some

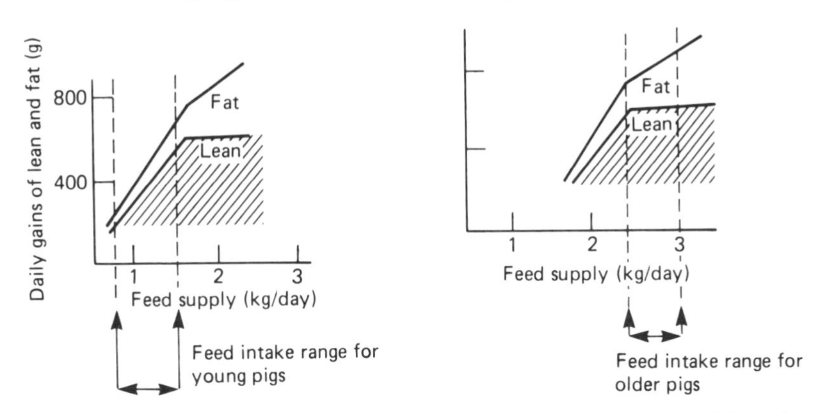


**Figure 2.3** Influence of age and appetite upon perceived rates of lean and fatty tissue growth

average values for maximum lean tissue growth rates and minimum fat:lean ratios are given in *Table 2.1*. It is evident that, in order to optimize production and achieve grading targets, the feed supply that will maximize lean tissue growth without generating excessive fat is crucial. It is likely that this point is out of appetite range for many young pigs, unless they are managed well and encouraged to eat; equally, for many older pigs a ration may need to be imposed unless the animal is either of high genetic merit or has a low appetite.

**Table 2.1** VALUES FOR MAXIMUM DAILY LEAN TISSUE GROWTH RATES (g)[a] AND FOR MINIMUM FAT:LEAN RATIOS. (THESE VALUES DETERMINE THE FORM OF THE RESPONSE TO FEED SUPPLY AS ILLUSTRATED IN *FIGURE 7.2*)

| *Genetic strain* | *Entire male* | | *Female* | | *Castrate male* | |
|---|---|---|---|---|---|---|
| | *Max. lean growth* | *Min. fat ratio* | *Max. lean growth* | *Min. fat ratio* | *Max. lean growth* | *Min. fat ratio* |
| Improved | 630 | 0.12 | 540 | 0.16 | 495 | 0.18 |
| Commercial | 585 | 0.16 | 520 | 0.20 | 475 | 0.22 |
| Utility | 540 | 0.20 | 475 | 0.24 | 405 | 0.26 |

[a]These values relate to the whole live pig, and not to the carcass alone; the latter values are usually about 60 per cent of the former.

ENERGY VALUE OF FEEDS

Achievement of target grading standards by alteration of the level of feeding implies knowledge of the energy density of the feed. Digestible energy values for compounded diets may be calculated from the known DE value of the ingredients, or where these are unknown from the chemical analysis of the mixed diet. The equations in *Table 2.2* have been proposed by Morgan and co-workers (1984).

**Table 2.2** EQUATIONS TO PREDICT DE (MJ/kg DM) FROM THE CHEMICAL COMPOSITION OF COMPOUNDED FEEDS (g/kg DM)

| | *r.s.d.*[a] |
|---|---|
| $DE^{b} = 3.8 - 0.019\ NDF^{c} + 0.76\ GE^{d}$ | 0.38 |
| $DE = 17.0 + 0.016\ EE^{e} - 0.018\ NDF$ | 0.44 |
| $DE = 17.0 + 0.011\ EE - 0.041\ CF^{f}$ | 0.66 |
| $DE = 18.0 + 0.016\ EE - 0.017\ NDF - 0.016\ Ash$ | 0.41 |
| $DE = 17.5 + 0.016\ EE - 0.008\ CP^{g} - 0.033\ Ash - 0.015\ NDF$ | 0.32 |

[a]r.s.d. = Residual standard deviation
[b]DE = Digestible energy
[c]NDF = Neutral detergent fibre
[d]GE = Gross energy
[e]EE = Oil by petroleum spirit extraction
[f]CF = Crude fibre
[g]CP = Crude protein

## Conclusion

The nutritional manipulation of carcass fatness is a major element in optimizing pig meat production tactics. First, the inherent genetic makeup of the animal must not predispose it to unacceptable levels of fatness even at low food intakes. This is becoming increasingly unlikely with the continued selection of improved strains of pigs and with the use of entire, rather than castrated, males. Second, the protein level in the diet must meet the appropriate energy:protein ratio. Third, the amount of food given should relate to the point at which daily lean tissue growth rate is maximized. Below that point pigs grow at less than optimum speed, but will not be fatter than the minimum fatness inherently pre-determined. Above the point of maximization of lean tissue growth rate, increments of food are diverted solely to the deposition of surplus fatty tissue with no benefit to lean growth. Invariably, limited voluntary feed intake levels dictate that maximum lean tissue growth rates are usually not achieved in commercial pigs of below 40 kg live weight. This situation can be eased by improving both the feed and the feeding method. Once voluntary feed intake no longer limits lean tissue growth rate, as may often be the case above 50 kg live weight, then the producer can restrict feed intake to maintain minimum fatness, or feed more to exceed it. Reduction of feed input will slow down growth rate, reduce fatness, and may improve grade and individual pig value. But the cost benefit of reduced feed inputs require careful assessment, particularly where throughput and overall feed efficiency is important to production optimization. Where problems of over-leanness pertain, there is a clear indication for enhancing feed intake to the highest possible level. Pigs of high genetic merit will also carry the potential for high daily lean tissue gains. Again, the consequence is that higher feed levels can be consumed with positive responses to growth rate without forgoing leanness.

Achieving grading standards or grading targets by nutritional manipulation is therefore relatively straightforward. However, to decide in the first place upon which of all the possible targets and standards is appropriate for an individual producer to optimize the production process is rather more complex.

## References

ARC (1981). *The Nutrient Requirements of Pigs.* Commonwealth Agricultural Bureaux; Farnham Royal, Slough, England

KEMPSTER, A.J. and EVANS, D. (1979). *Animal Production,* **28**, 87–96

MORGAN, C.A., WHITTEMORE, C.T., PHILLIPS, P. and CROOKS, P. (1984). *The Energy Value of Compound Foods for Pigs.* Edinburgh School of Agriculture; Scotland

TULLIS, J.B. (1982). PhD Thesis, University of Edinburgh

WHITTEMORE, C.T. (1980). *Pig News and Information,* **1**, 343–346

WHITTEMORE, C.T. (1983). *Agricultural Systems,* **11**, 159–186

3

# CONSEQUENCES OF CHANGES IN CARCASS COMPOSITION ON MEAT QUALITY

J.D. WOOD
*Department of Meat Animal Science, University of Bristol, UK*

## Introduction

Although the national consumption of pigmeat has remained high during the last ten years, particularly in relation to that of beef and lamb, some meat traders have said that the reduction in average fat thickness levels that has occurred in the same period has lowered meat quality. This chapter considers the role of fat in different aspects of meat quality, the changes in structure and composition of lean and fat tissues which occur as overall fatness is reduced and the consequences of these for changes in quality. Before proceeding however, it should be remembered that the recent success of pigmeat in consumption terms is closely linked with its relatively low price compared with other meats (Clark, 1984) which in turn is explained by high levels of feed efficiency, only possible in lean, rapidly-growing animals. Also, recent reports that the dietary intake of animal fat should be reduced for health reasons (e.g. DHSS, 1984) are likely to put even more emphasis on low-fat meats in the future.

## Definition of meat quality and the role of fat

There are three components of meat quality: visual, handling and eating quality. Visual quality is determined mainly by the relative proportions of lean and fat, all surveys showing that pigmeat with high ratios of lean to fat is most preferred and most likely to be bought (Baron and Carpenter, 1976).

Handling quality, which is possibly more important to butchers than consumers, refers to the firmness and cohesiveness of the tissues and meat when handled and cut. Fat tissue has a role in binding the muscles together and the fatty acid composition of lipid determines the firmness of fat tissue which is an important factor in overall firmness. The third and most important aspect of meat quality is eating quality. In this case, fat provides flavour components, prevents drying out during cooking and has a role in tenderness.

Most butchers agree that leaner meat is most likely to be bought and reserve their critical comments mainly for the handling and eating characteristics of lean pigmeat. Changes in these characteristics must be due to changes in the composition and structure of lean and fat tissues associated with differences in the overall fat content of the carcass.

COMPOSITION AND QUALITY OF LEAN TISSUE

The individual muscles of the pig carcass differ widely in lipid (intramuscular fat) content (Lawrie, Pomeroy and Cuthbertson, 1963; Davies and Pryor, 1977). *M. longissimus*, which is the heaviest muscle in the body (approximately 11 per cent of lean in the side) and is most frequently sampled, has a relatively low lipid content. For example, in the study of Davies and Pryor (1977) it varied from 2.1 (*m. adductor*) to 25.4 (*m. cutaneous omobrachialis*) per cent lipid with a mean value of 5.0. The lipid content of *m. longissimus* was 4.2 per cent. In general, lipid content was highest in the most superficial muscles.

As the fat content of the carcass increases so does the lipid content of individual muscles. The results in *Table 3.1* are taken from studies conducted at the Food

**Table 3.1** LIPID CONTENT OF *M. LONGISSIMUS* IN RELATION TO CARCASS COMPOSITION. THREE EXPERIMENTS INVOLVING RELATIVELY HIGH AND LOW LEVELS OF FEEDING

| | *Level of feeding* | | | |
|---|---|---|---|---|
| | *High* | *Low* | *S.E.D.* | *Significance of difference* |
| *Experiment 1[a] (47.0 kg carcass weight)* | | | | |
| C fat thickness (mm)[d] | 13.2 | 11.8 | 0.79 | NS |
| Lean (% of side) | 53.8 | 54.5 | 0.84 | NS |
| Lipid in *m. longissimus* (%) | 2.05 | 1.97 | 0.17 | NS |
| *Experiment 2[b] (50 kg carcass weight)* | | | | |
| C fat thickness (mm)[d] | 12.3 | 7.6 | 0.65 | *** |
| Lean (% of side) | 54.7 | 59.6 | 0.70 | *** |
| Lipid in *m. longissimus* (%) | 3.36 | 2.62 | 0.23 | *** |
| *Experiment 3[c] (66 kg carcass weight)* | | | | |
| C fat thickness (mm)[d] | 11.8 | 9.7 | 0.81 | ** |
| Lean (% of side) | 56.8 | 60.6 | 0.65 | *** |
| Lipid in *m. longissimus* (%) | 1.85 | 1.56 | 0.13 | * |

[a]Results of Wood, Dransfield and Rhodes (1979); 48 gilts (12 from each of four breeds)
[b]Unpublished results from 21 gilts, 13 boars and ten castrated males
[c]Results of Wood and Enser (1982); 64 pigs (32 boars and 32 gilts)
[d]Thickness of fat above deepest part of *m. longissimus* at last rib. Measured cold and excluding skin thickness

Research Institute Bristol (FRIB). When feeding at a low level reduced the thickness of backfat and increased the lean content of the carcass (experiments 2 and 3) it also reduced the concentration of lipid in *m. longissimus*. However, the relationship between muscle lipid and carcass fatness was quite variable, correlation coefficients of 0.61 and 0.40 between percentage lipid and C fat thickness being found in experiments 2 and 3, respectively.

Lipid is only obvious in *m. longissimus* as streaks of 'marbling fat' at high levels of carcass fatness. This does not mean that intramuscular fat is a 'late developing' fat depot, however, since studies in sheep by Broad and Davies (1981a and b) and in pigs by Davies and Pryor (1977 *Table 3.2*), showed that it is relatively slow growing compared with the other fat depots. The storage triglyceride component, approximately 80 per cent of the total in pigs of commercial weights, grows more rapidly than phospholipid, located in the cell membranes (Broad and Davies, 1981b).

**Table 3.2** GROWTH OF FAT DEPOTS IN THE SIDE (*Y*) RELATIVE TO TOTAL SIDE FAT (*X*) (SUBCUTANEOUS + INTERMUSCULAR + CAVITY) USING EQUATION $Y = aX^b$

| *Y* | *Growth coefficient*[b] | *S.E.* |
|---|---|---|
| Subcutaneous fat | 1.007 | 0.043 |
| Intermuscular fat | 0.972 | 0.038 |
| Cavity fat | 1.077 | 0.057 |
| Intramuscular fat | 0.910 | 0.045 |

(Davies and Pryor, 1977)

Genetic differences in the association between intramuscular fat and carcass fat would have important consequences for meat quality. Reports in the literature showing wide variations in muscle lipid content between individual animals (e.g., Lawrie, Pomeroy and Cuthbertson, 1963) and between groups of animals of the same fat thickness (*see Table 3.2*, experiments 1 and 2) suggest that such differences might exist. It has also been suggested that wild pigs have inherently less muscle lipid than domesticated pigs (Crawford, Hare and Whitehouse, 1984). However, there is no good evidence for clear genetic effects in pigs and it may well be that different lipid extraction and dissection procedures are the cause of much of the variation found between studies.

## INTRAMUSCULAR FAT AND EATING QUALITY

The importance of intramuscular (marbling) fat in the meat industry is epitomized in the USDA quality grading scheme for beef carcasses where carcasses with high levels of marbling are placed in the highest grade. In all species, including pigs (where concentrations of muscle lipid are lower than in beef), intramuscular fat affects the juiciness, flavour and tenderness of meat. The effect on juiciness is associated with the lubricating action of melted lipid during cooking; flavour is affected by the release of volatile compounds during cooking, some of which react with components from lean; and tenderness may be affected by the replacement of some fibrous protein by softer lipid. Early work at FRIB (e.g. Rhodes, 1970) showed that the measurable effect of fat thickness or intramuscular fat content on these aspects of eating quality, as determined in taste panel tests, was extremely

**Table 3.3** EATING QUALITY OF PORK LOINS IN 56 kg CARCASSES OF DIFFERENT FATNESS AS ASSESSED BY TASTE PANEL

| | *Fat thickness at 'C' (mm)* | | *S.E.D.* | *Significance of difference* |
|---|---|---|---|---|
| | 12 | 6 | | |
| Tenderness[a] | 1.8 | 0.9 | 0.69 | NS |
| Flavour[a] | 2.4 | 2.5 | 0.22 | NS |
| Juiciness[b] | 1.4 | 0.9 | 0.14 | * |
| Toughness (*J*)[c] | 0.14 | 0.16 | 0.02 | NS |

(Wood, Mottram and Brown, 1981)

[a]Scores −7 to +7 in steps of 2 where −7 is extremely tough or disliked extremely and +7 is extremely tender or liked extremely

[b]Scores 0 (dry) to 3 (extremely juicy)

[c]Measured by Instron materials testing instrument

small and led to the conclusion that 'selection programmes aimed at reducing fatness are in little danger of producing a less acceptable meat on the plate' (Rhodes, 1970). Similar conclusions were drawn in more recent studies (Wood, Dransfield and Rhodes, 1979; Wood, Mottram and Brown, 1981) although in the 1981 report a statistically significant difference in the taste panel score for juiciness was found between carcasses having C fat thickness measurements of 6 and 12 mm (10 and 16 mm P2 respectively) in carcasses of 56 kg (*Table 3.3*). These results for juiciness may be compared with those in the 1979 study in which four breeds had mean juiciness scores of 1.2 (these are the pigs described in *Table 3.1*, experiment 1; the comparison between studies is valid because the same panellists were involved in the two studies). There is therefore a hint that 6 mm C (10 mm P2) may represent a level of fat thickness below which some slight deterioration in eating quality is observed in carcasses of this weight. However, many more samples are required before this point can be established. A large-scale study currently being conducted by FRIB and the Meat and Livestock Commission should provide more information both on trained taste panel and consumer attitudes to lean pigmeat.

Danish studies suggest that around 2.0 per cent of lipid in *m. longissimus* is required for good eating quality (Buchter and Zeuthen, 1971; Jul and Zeuthen, 1980). A recent report (*Table 3.4*) showed that tenderness, flavour and juiciness

**Table 3.4** EFFECT OF LIPID (INTRAMUSCULAR FAT) CONTENT OF *M. LONGISSIMUS* ON EATING QUALITY[a] OF PORK CHOPS

| *Number of pigs* | *Intramuscular fat (%)* | *Flavour* | *Tenderness* | *Juiciness* | *Overall acceptability* |
|---|---|---|---|---|---|
| 24 | 1.47[b] | 2.5[b] | 1.3[b] | 1.7[b] | 0.6[b] |
| 43 | 2.89[c] | 2.9[c] | 3.1[d] | 3.2[d] | 2.0[c] |
| 51 | 4.34[d] | 2.8[c] | 2.4[c] | 2.5[c] | 2.0[c] |

(Bejerholm, 1984)
[a]Taste panel scores for each aspect on a scale of −5 to +5 in steps of 1 where −5 is poor and +5 ideal
[bcd]Means in a column with different superscripts are significantly different ($P<0.01$)

were scored higher in fried chops with 2.9 per cent lipid in *m. longissimus* than in those with 1.5 or 4.3 per cent. The pigs were Danish Landrace, Hampshire and Duroc but unfortunately no fat thickness measurements were given.

Few studies have attempted to define the exact role of lipid in eating quality. An exception is that of Mottram and Edwards (1983) which partitioned the characteristic aromas and flavours of cooked beef between the storage component triglyceride and the cellular component phospholipid. The conclusion was that phospholipids alone were required for the development of aroma and flavour. Translated into pig *longissimus* lipid values this suggests that less than 1 per cent of lipid is necessary for optimal flavour development.

## PALE SOFT EXUDATIVE (PSE) MUSCLE

Pigs which are homozygous for the gene which confers sensitivity to the anaesthetic halothane (*nn*) also produce PSE muscle which has poor visual and handling qualities (Webb, 1981). Their carcasses produce a lower yield of bacon after curing (Taylor, Dant and French, 1973) and are more likely to produce tough meat (e.g.

**Table 3.5** CARCASS COMPOSITION IN ENTIRE MALE PIETRAIN AND LARGE WHITE PIGS OF 90 kg LIVE WEIGHT

| | *Pietrain* | *Large White* |
|---|---|---|
| Lean[a] | 61.1 | 57.0 |
| Fat[a] | 22.6 | 25.6 |
| Bone[a] | 10.0 | 11.1 |
| Lean:bone ratio | 6.1 | 5.1 |
| Depth of *m. longissimus* (mm)[b] | 55.9 | 48.0 |
| Carcass length (mm) | 723 | 798 |

(Fortin, Wood and Whelehan, 1985)
[a]Percentage of side
[b]At last rib

Bejerholm, 1984). A high incidence of the halothane gene is found in pigs of the Pietrain breed which are lean, have 'blocky conformation' and a correspondingly high ratio of lean to bone (*Table 3.5*). Such results suggest a genetic link between PSE incidence and carcass lean and fat content although within breeds the genetic correlations are low (McGloughlin and McLoughlin, 1975; Kempster, Evans and Chadwick, 1984). It therefore cannot be said that the reported increase in PSE incidence in Britain during the last ten years (Chadwick and Kempster, 1983) is associated directly with the trend towards leaner pigs. Since the correlation with carcass conformation score is higher than that with carcass lean content, it may be that increased use of blocky breeding pigs partly accounts for the increased incidence of PSE muscle. The commercial breeding companies have suggested that *nn* sire lines (with above average leannesss and conformation) and *NN* dam lines (homozygous for the gene conferring halothane resistance) should be used to produce *Nn* slaughter progeny, which have some of the advantages in terms of carcass quality of the *nn* parent and none of the disadvantages in terms of stress sensitivity and meat quality (they do not react to halothane). However there are possible dangers in this approach. Apart from the possibility that some *Nn* progeny will be used for breeding rather than meat, thereby producing some *nn* progeny when mated to other *Nn* individuals, it has also been shown that *Nn* pigs are intermediate in muscle quality between the two extreme homozygotes (Frøystein *et al.*, 1981). A recent Danish report shows that *Nn* pigs are also intermediate in their reactions to pre-slaughter handling conditions (*Table 3.6*). Abattoir A had poor conditions with a high incidence of PSE muscle in *NN* and *Nn*. Abattoir C, on the other hand, had better (more considerate) conditions, resulting in a zero incidence in *NN*. However in all cases *Nn* pigs produced some PSE meat.

The results show that there is likely to be some penalty (in terms of meat quality) inherent in the promotion of the halothane gene in Britain. Other countries, e.g. Norway, have elected to reduce the incidence of the gene at the cost, in their case, of extremes in conformation and possibly leanness.

**Table 3.6** INCIDENCE OF PSE MUSCLE (% OF CARCASSES) IN 259 DANISH LANDRACE PIGS WITH KNOWN HALOTHANE GENOTYPE SLAUGHTERED AT THREE ABATTOIRS

| | *Abattoir* | | |
|---|---|---|---|
| | *A* | *B* | *C* |
| *nn* | 100 | 74 | 79 |
| *Nn* | 33 | 17 | 13 |
| *NN* | 33 | 8 | 0 |

(Barton-Gade, 1984)

## Composition and quality of fat tissue

### FACTORS CONTROLLING FAT QUALITY

The most important aspects of fat quality in pigmeat are firmness (i.e. hardness or softness), colour and cohesiveness (i.e. whether the elements of the tissue are bound together or show separation). Firm, white fat showing no separation is desired by butchers and consumers. This combination of characteristics is found in thick fat which is well developed but is more difficult to achieve in thin, underdeveloped fat because of the structural and chemical changes which occur as fattening proceeds (Wood, 1984).

During the development of fat tissue the concentration of lipid increases and the concentrations of water and connective tissue fall, although these changes are more gradual beyond about 30 days of age (10 kg live weight) than before (Wood, 1984). At the same time the lipid, more than 90 per cent of which is triglyceride, becomes more saturated. This is largely due to a decrease in the concentration of linoleic acid (C18:2), which is derived from the diet, and an increase in the concentration of stearic acid (C18:0) which is synthesized. Also the fat cells increase in size and become more turgid. All these factors contribute to the improvement in fat quality which occurs as the fat tissue develops but several studies show that fatty acid composition is the most critical (Enser, 1984; Wood *et al.*, 1984).

Similarly in pigs of a particular carcass weight the concentration of stearic acid provides the best prediction of instrumentally-measured firmness and lipid melting point (Enser, 1984) and directly affects colour through changing the opacity of lipid. Stearic acid and linoleic acid are important in explaining variation in cohesiveness (Wood *et al.*, 1984) although this is probably an indirect effect. Observation shows that separation occurs between the lobules of fat cells that are surrounded by sheaths of connective tissue (*Figure 3.1*). As yet very little work has

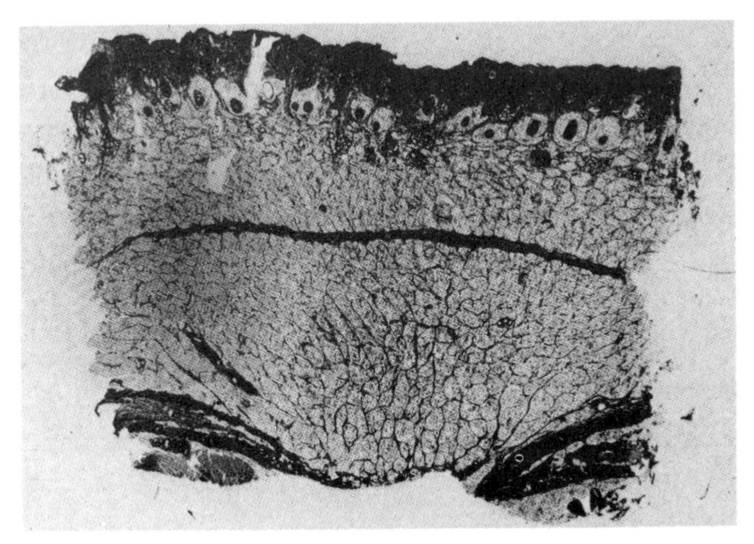

**Figure 3.1** Pig subcutaneous fat (backfat) is a highly organized tissue in which the fat cells are arranged in groups (lobules) surrounded by connective tissue. Connective tissue is concentrated in a band separating the outer and inner layers and in the skin dermis. (×10)

been done on factors affecting the strength of the connective tissue matrix in fat tissue.

### EFFECT OF FAT THICKNESS ON FAT QUALITY

The above suggests that the thickness of backfat is itself an important factor in fat quality, at least in subcutaneous fat. Recent results, collected in a joint study between FRIB and the Meat and Livestock Commission show that this is so (*Table 3.7*). Three hundred pork-weight pigs, half boars and half gilts, were sampled in ten abattoirs. The 30 pigs per abattoir were from five producers, each supplying three boars and three gilts falling into different fat thickness categories, respectively 8, 12 and 16 mm P2. There were significant effects of fat thickness on firmness assessed by experienced operators and by an instrumental method. Twenty-four per cent of

**Table 3.7** EFFECTS OF FAT THICKNESS ON FAT QUALITY IN 300 PORK-WEIGHT PIGS (58 kg CARCASS WEIGHT)

| | *Fat thickness at P2* (mm) | | | *S.E.D.* | *Significance of difference* |
|---|---|---|---|---|---|
| | *8* | *12* | *16* | | |
| Sensory firmness[a] | | | | | |
| Assessor 1 | 2.8 | 3.9 | 5.3 | 0.16 | *** |
| Assessor 2 | 2.8 | 3.8 | 5.3 | 0.13 | *** |
| Instrumental firmness (g)[b] | 432 | 637 | 913 | 46.7 | *** |
| Loin fat samples exhibiting separation (%) | 52 | 23 | 4 | | |

(Wood *et al.*, 1985)
[a]Using an 8-point scale to assess loin samples where 1 is very soft and 8 is very hard
[b]Loin fat samples assessed using materials testing instrument at 0 °C.

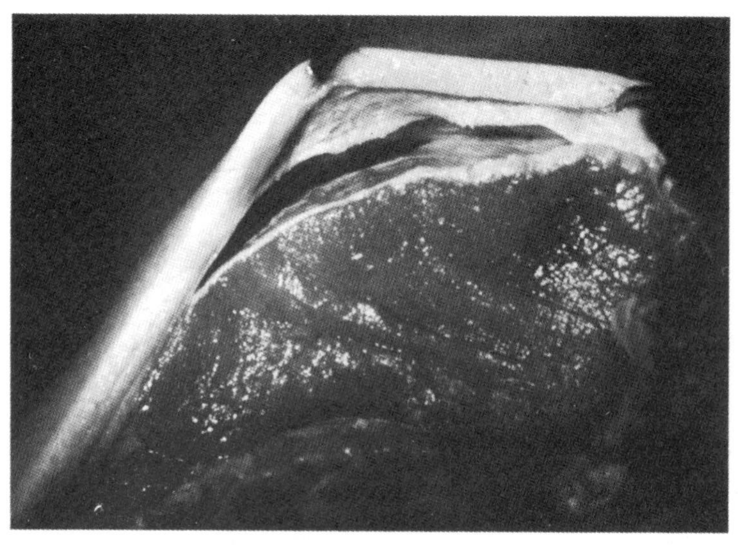

**Figure 3.2** An example of backfat separation from lean in the loin of a pork-weight carcass (58 kg) having 8 mm P2 fat thickness

loin samples were placed in categories 1 and 2 on the 8-point sensory scale and it may be that these samples are unsatisfactorily soft. More than half the samples in the 8 mm P2 group exhibited fat separation as determined by the physical separation of fat from lean when pulled upwards, away from *m. longissimus* (*Figure 3.2*).

The relevance of these findings to consumers in terms of overall satisfaction following eating cannot be assessed at present. Preliminary results suggest that consumers are less concerned with fat quality than butchers.

## EFFECTS OF SEX AND DIET ON FAT QUALITY

The study described in *Table 3.7* contained equal numbers of boars and gilts. Both groups had similar fat thickness (12 mm P2) and yet boars had significantly softer fat which tended to separate from lean more easily than that of gilts. However the differences were smaller than those between fatness categories. For example, the scores for sensory firmness of loin fat showed that boars with 12 mm P2 had fat as firm as all pigs (boars and gilts) with 11.5 mm P2. Gilts had fat as firm as all pigs with 12.5 mm P2. Previous work showed that boar backfat contains a higher percentage of water and a lower percentage of lipid than that of gilts or castrates (Wood and Enser, 1982) although no difference between the sexes in fat quality at the same fat thickness has been found before (also Wood *et al.*, 1984).

The fatty acid composition of the diet has long been recognized as an important factor in pig fat quality. For example, early American work showed that pigs which had eaten large quantities of soya beans produced soft oily fat because of incorporation of linoleic acid into body fat (Ellis and Isbell, 1926a and b). The more recent practice of incorporating high levels of vegetable oils in pig diets to increase energy density produces the same effect and has a particularly detrimental effect on fat quality if the diet is underfed or the pigs are lean, since thin backfat has a high ratio of linoleic to stearic acid anyway. Recent results show the consequences for fat quality of increasing the proportion of dietary energy derived from linoleic acid (*Table 3.8*). Boars and castrates were included in the experiment and although the results for both sexes were in the same direction, the effects of diet were more marked in the castrates. A number of FRIB studies summarized by Prescott

**Table 3.8** EFFECT OF CHANGING THE PROPORTION OF DIETARY ENERGY DERIVED FROM LINOLEIC ACID ON FAT QUALITY IN BACON-WEIGHT CASTRATE PIGS (90 kg LIVE WEIGHT)

| | *Percent of DE from linoleic acid* | | *S.E.D.* | *Significance of difference* |
|---|---|---|---|---|
| | *2* | *5* | | |
| Fat thickness at P2 (mm) | 15.4 | 11.1 | 0.93 | *** |
| Stearic acid (% of fatty acids)[a] | 14.6 | 12.5 | 0.45 | *** |
| Linoleic acid (% of fatty acids)[a] | 9.3 | 24.3 | 1.01 | *** |
| Firmness of backfat (N)[b] | 7.3 | 2.8 | 0.98 | *** |
| Cohesiveness of backfat ($Nm^{-2} \times 10^3$)[c] | 6.4 | 3.8 | 0.70 | ** |

(Wood *et al.*, 1984)
[a]Dorsal mid-line at last rib (inner layer)
[b]Instrumental method
[c]Instrumental method

(personal communication) suggests that the linoleic acid concentration in backfat is directly proportional to the ratio of linoleic acid energy to total energy consumed.

The effect of the concentration of a fatty acid in the diet on its concentration in backfat is more marked for linoleic acid than for the other fatty acids (Wood, 1984). Therefore, increasing the dietary concentration of stearic acid, apart from possibly reducing digestibility, does not have the same effect of 'hardening' fat as linoleic acid has in 'softening' it. Diets high in carbohydrate and protein and low in lipid which rely on synthesis rather than dietary incorporation for lipid deposition are a possible option for the retention of high fat quality in lean pigs.

## Conclusions

The recent reduction in fat thickness which has occurred in British pigs has produced a more attractive product in terms of appearance (visual quality) and puts pigmeat in a favourable position compared with other meats in the light of recent reports that consumers should reduce their intake of animal fat for health reasons. However as the fat tissue content of the carcass is reduced, the lipid content of lean and fat tissues also inevitably declines and becomes more unsaturated, thus changing handling and eating characteristics.

Lipid in muscle affects eating quality, but early British studies showed no clear association between the concentration of lipid in muscle and taste panel scores for tenderness, juiciness or flavour within the range of fat levels studied. More recent work with leaner pigs suggests that juiciness may be slightly reduced below 10 mm P2 fat thickness.

The incidence of PSE muscle has increased recently and this will have reduced tenderness in the 13 per cent or so of carcasses affected. There is no evidence that selection for leaner pigs has brought this about, rather, greater use of particular breeds which have good conformation and carry the halothane gene is implicated and further use of these types should be carefully controlled.

The firmness and cohesiveness of backfat decline with fat thickness, due mainly to a reduction in the saturation of fatty acids. Results of a recent study suggest that 8 mm P2 fat thickness in pork-weight pigs represents the point at which butchers' complaints of handling quality increase although there is doubt about the significance of this to consumers. Some of the problems can be overcome by manipulating the amount and type of dietary fat.

Taken together the information available at present shows no cause for concern over meat quality in lean pigs and there is wide variation at each fat thickness level. Rather than applying price penalties to pigs with thin backfat (perhaps less than 8 mm P2) the industry should seek solutions to particular problems and encourage the development of objective methods for evaluating each aspect of quality.

## References

BARON, P.J. and CARPENTER, E.M. (1976). *Report No. 23*. Department of Agricultural Marketing, University of Newcastle upon Tyne; Newcastle upon Tyne

BARTON-GADE, P. (1984). *Proceedings of 30th European Meeting Meat Research Workers*, pp. 8–9

BEJERHOLM, A.C. (1984). *Proceedings of 30th European Meeting Meat Research Workers*, pp. 196–197
BROAD, T.E. and DAVIES, A.S. (1981a). *Animal Production*, **31**, 63–71
BROAD, T.E. and DAVIES, A.S. (1981b). *Animal Production*, **32**, 234–243
BUCHTER, L. and ZEUTHEN, P. (1971). *Proceedings of 2nd International Symposium Condition and Meat Quality of Pigs*, pp. 247–254. Centre for Agricultural Publishing and Documentation; Wageningen, The Netherlands
CHADWICK, J.P. and KEMPSTER, A.J. (1983). *Meat Science*, **9**, 101–111
CLARK, A.G. (1984). In *Matching Production to the Markets for Meat*, pp. 35–38. Ed. A. Cuthbertson and R.G. Gunn. British Society of Animal Production Occ. Publ. No. 8; Edinburgh
CRAWFORD, M.A., HARE, W.R. and WHITEHOUSE, D.B. (1984). In *Fats in Animal Nutrition*, pp. 471–479. Ed. J. Wiseman. Butterworths; London
DAVIES, A.S. and PRYOR, W.J. (1977). *Journal of Agricultural Science*, **89**, 257–266
DEPARTMENT OF HEALTH AND SOCIAL SECURITY, (1984). *Diet and cardiovascular disease*. Report on Health and Social Subjects, No. 28. Her Majesty's Stationery Office; London
ELLIS, N.R. and ISBELL, H.S. (1926a). *Journal of Biological Chemistry*, **69**, 219–238
ELLIS, N.R. and ISBELL, H.S. (1926b). *Journal of Biological Chemistry*, **69**, 239–248
ENSER, M. (1984). In *Fat Quality in Lean Pigs*, pp. 53–57. Ed. J.D. Wood. Document No. EUR8901 EN, Commission of the European Communities; Brussels
FORTIN, A., WOOD, J.D. and WHELEHAN, O.P. (1985). *Animal Production* (in press)
FRØYSTEIN, T., NØSTVOLD, S.O., BRAEND, M., STORSETH, A. and SCHIE, K.A. (1981). In *Porcine Stress and Meat Quality*, pp. 161–176. Ed. T. Frøystein, E. Slinde and N. Standal. Agricultural Food Research Society; As, Norway
JUL, M. and ZEUTHEN, P. (1980). *Progress in Food and Nutrition Science*, **4**, 1–132
KEMPSTER, A.J., EVANS, D.G. and CHADWICK, J.P. (1984). *Animal Production*, **39**, 455–464
LAWRIE, R.A., POMEROY, R.W. and CUTHBERTSON, A. (1963). *Journal of Agricultural Science*, **60**, 195–209
MCGLOUGHLIN, P. and MCLOUGHLIN, J.V. (1975). *Livestock Production Science*, **2**, 271–280
MOTTRAM, D.S. and EDWARDS, R.A. (1983). *Journal of the Science of Food and Agriculture*, **34**, 517–522
RHODES, D.N. (1970). *Journal of Science of Food and Agriculture*, **21**, 572–575
TAYLOR, A.A., DANT, S.J. and FRENCH, J.W.L. (1973). *Journal of Food Technology*, **8**, 167–174
WEBB, A.J. (1981). In *Porcine Stress and Meat Quality*, pp. 105–124. Ed. T. Frøystein, E. Slinde and N. Standal. Agricultural Food Research Society, As, Norway
WOOD, J.D. (1984). In *Fats in Animal Nutrition*, pp. 407–435. Ed. J. Wiseman. Butterworths; London
WOOD, J.D., DRANSFIELD, E. and RHODES, D.N. (1979). *Journal of the Science of Food and Agriculture*, **30**, 493–498
WOOD, J.D. and ENSER, M. (1982). *Animal Production*, **35**, 65–74
WOOD, J.D., JONES, R.C.D., BAYNTUN, J.A. and DRANSFIELD, E. (1984). *Animal Production*, **40** (in press)
WOOD, J.D., JONES, R.C.D., DRANSFIELD, E. and FRANCOMBE, M.A. (1985). *Animal Production*, (in press)
WOOD, J.D., MOTTRAM, D.S. and BROWN, A.J. (1981). *Animal Production*, **32**, 117–120

4

# ENERGY–PROTEIN INTERACTIONS IN PIGS

A.C. EDWARDS and R.G. CAMPBELL
*Bunge Meat Industries, Corowa, NSW, Australia*

## Introduction

Knowledge of the factors influencing protein deposition capacity is crucial for the design of diets and feeding strategies for growing animals and for predicting the effects of change in feed or energy intake on growth performance and carcass composition. Protein deposition can be constrained by both dietary and intrinsic factors and in this chapter we have attempted to highlight the major factors affecting protein growth and how these affect requirements of growing pigs for dietary protein (amino acids) and the partition of energy between fat and protein.

The initial sections of the chapter concentrate on the interrelationship between nutrient intake and the various animal factors as they affect protein growth capacity and dietary requirements. The latter sections deal with dietary factors as they affect nutrient 'requirements' and attempt to integrate the animal and dietary factors to draw conclusions concerning the present state of knowledge and to identify areas requiring further work.

### PROTEIN AND ENERGY INTAKE EFFECTS OF PROTEIN DEPOSITION

The relationship between protein deposition and protein and energy intake consists of two phases: (i) an initial protein-dependent phase in which protein deposition is linearly related to protein intake and independent of energy intake or animal factors such as sex or genotype, and (ii) an energy-dependent phase in which additional protein is deposited only when energy intake is increased. These effects are illustrated in Figure 4.1.

When pigs of a given weight are fed increasing amounts of protein, of a constant quality, in conjunction with a set amount of energy (E1), protein deposition increases linearly until a maximum value (M1) is reached at a particular level of protein intake (A). Additional increments of protein will not produce any further rise in protein deposition. However, when more energy is supplied (E2) protein deposition increases up to new maximum value (M2) at a higher protein intake (B). Thus protein deposition is unaffected by energy intake when protein is limiting and, conversely, is driven by energy intake when dietary protein supply is equal to or above requirement.

The slope of the linear component of the response functions is determined by the digestibility and biological value of the dietary protein. The latter define

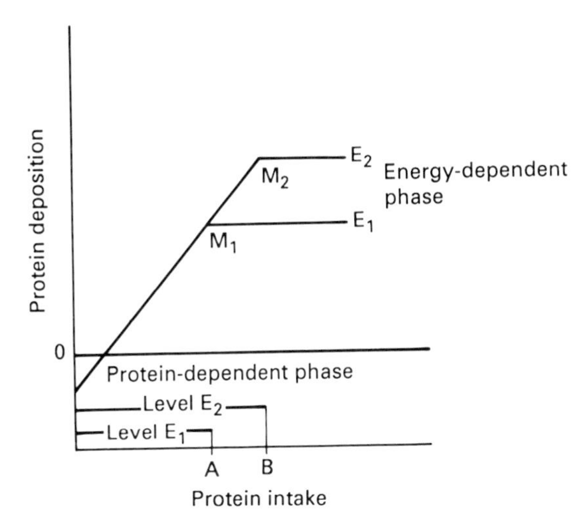


**Figure 4.1** Interrelationships between protein deposition and protein and energy intake

protein quality and this in turn determines the quantity of dietary protein required to support maximum protein deposition. In contrast M1 and M2 represent the requirement at a tissue level and are independent of protein quality, but are dependent on such animal factors as live weight, gender, breed or strain.

Consequently unless they are associated with improved dietary protein utilization these factors must alter the level of dietary protein required to support maximal rates of protein, or lean tissue, growth. For example the values M1 and M2 in Figure 4.1 could be used to depict the difference in protein accretion capacity comparing the intact male and female pigs provided with the same energy intake, and the values A and B the amounts of dietary protein required to support maximal protein deposition in the respective sexes.

The interrelationship between the pig's requirements for protein at the tissue and dietary levels can be described by the equation:

$$\text{DPR(g/d)} = \frac{\text{RPD} + \text{OPL}}{\text{Dig} \times \text{BV}}$$

where DPR = dietary protein requirement, RPD (g/d) = rate of protein deposition (tissue requirement for growth), OPL (g/d) = obligatory protein loss (tissue requirement for protein maintenance), Dig (%) = digestibility of dietary protein and BV (%) = biological value of dietary protein.

If all these factors were measured in experiments to assess the growing pigs' response to nutrient intake or any other factor which might influence growth performance it would be relatively easy to determine the extent to which these various factors might alter dietary protein requirements via their effects on either tissue requirements and/or alteration of dietary protein utilization. Unfortunately, such information is limited and these aspects should be seriously considered in the design of future experiments to elucidate the effects of animal or dietary factors on the growing pig's nutrient requirements.

It was mentioned previously that under conditions of dietary protein adequacy, protein deposition is a function of energy intake and it is the form of the relationship between energy intake and protein deposition which determines the partition of energy between protein and fat components. This relationship has consequent effects on energy intake (feeding level), growth performance and body composition. However, for reasons discussed previously it is essential when assessing the relationship between energy intake and protein deposition that the diet is not protein deficient. Otherwise any improvement in protein deposition resulting from increased feed intake will be in response to increased protein intake independent of energy intake, and the animal will not be able to express its inherent or metabolically enhanced capacity for protein growth.

In reviewing the available information for pigs the ARC (1981) mentioned the paucity of experiments of appropriate design to define the relationship and although favouring a linear relationship, which presumes there are no intrinsic limits to protein deposition, commented that there was some support for linear-plateau and curvilinear forms. These contrasting models imply markedly different rates of change in the fat:protein ratio and different expressions of the pig's requirement for dietary protein with change in energy intake.

It is now established that the relationship is essentially of the linear/plateau form (Campbell, Taverner and Curic, 1985a; Dunkin and Black, 1987) with the plateau value representing the animal's genetic or intrinsic limit for protein accretion. This relationship and the consequent effects on the partition of energy retained as protein and fat are shown in Figure 4.2.

Total energy retained increases linearly with energy intake (Figure 4.2a). Energy retained as protein also increases linearly to point Q beyond which it remains constant (maximal protein deposition). Energy deposited as fat is represented by the difference between total and protein energy deposition. At zero energy balance (maintenance energy requirement) protein gain is marginally positive but fat deposition is negative and does not commence until energy intake reaches some higher level (R as determined by factors such as gender and genotype). Figure 4.2b represents the corresponding change in the fat:protein ratio of weight gain in response to changing energy intake.

When protein deposition is linearly related to energy intake, which is the situation up to *ad libitum* energy intake for young pigs (<50 kg) and some superior genotypes, the fat:protein ratio and body fat content increase in a curvilinear fashion and approach a constant or steady-state value. However, if protein deposition reaches a plateau, further increases in energy intake result in a steep rise in the fat:protein ratio and body fat content. Growth rate increases linearly when protein deposition is similarly related to energy intake but the rate of increase declines once the plateau is reached.

Whether protein deposition responds linearly to increased energy intake to the limits of the animal's appetite or reaches a plateau at some intermediate level of feeding markedly affects the rate and composition of liveweight gain, feed conversion ratio and expression of dietary protein requirements. For example the concept of a constant protein (amino acid):energy ratio is only valid when protein deposition is a linear function of energy intake.

This is simply because energy intake is always the major determinant of maximal protein deposition and, providing the dietary protein:energy ratio is correct, changes in feed intake and thus in the pig's tissue demand for protein will always be satisfied by a concomitant change in dietary energy intake. Conversely, on the

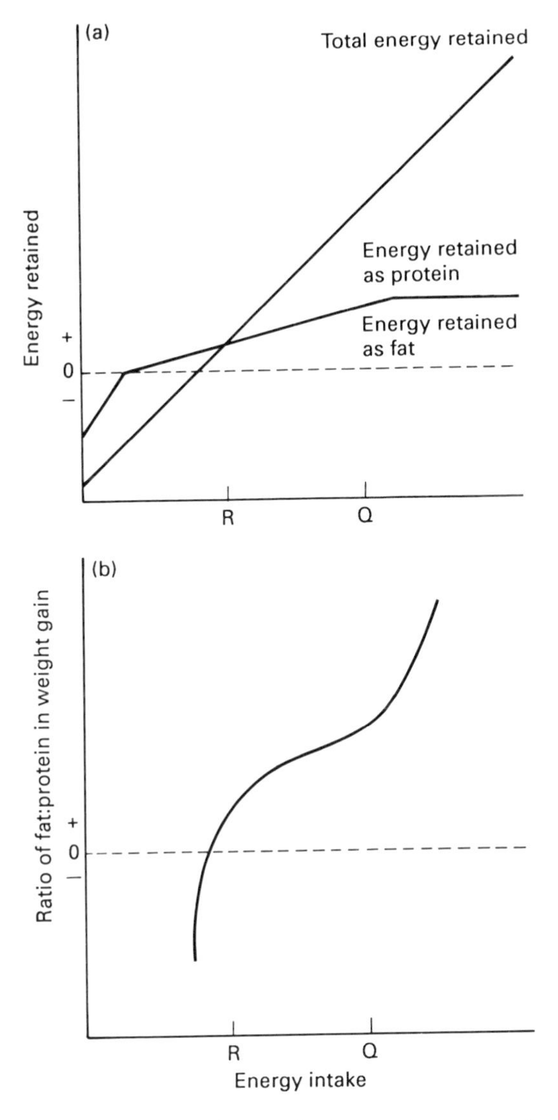


**Figure 4.2** Effect of energy intake on (a) retention of protein and fat where protein retention is of a linear/plateau form, and (b) the corresponding ratio of fat:protein in weight gain

plateau, the pig's tissue requirement is independent of energy intake and dietary requirements can only be expressed on a daily intake basis.

Nevertheless, the form of the relationship between energy intake and protein deposition is affected by factors such as sex and genotype, and by metabolism modifiers such as porcine somatotropin. Similarly the amount of dietary protein required to support maximum protein deposition even in pigs of a known age, sex or genotype is affected by amino acid availability and balance. Information is required on all these factors if the tissue and dietary components affecting performance and requirements are to be fully integrated. There is a need for nutritional programmes to be founded on a more biological basis than they are currently.

## Animal factors

### SEX

*Pigs between 20 and 50 kg live weight*

Females and castrated males have a lower capacity for muscle growth than entire males. Whilst this difference is reflected to some extent in the different degrees of energy restriction imposed on male and female finisher pigs, its practical implications have not been fully explored or exploited.

During the earlier stages of development (to 50 kg), rate of protein deposition increases linearly with increase in energy intake up to the limit of the animal's appetite (Campbell, Taverner and Curic, 1985a) and the differences between the sexes are generally only small but increase with live weight. Consequently, the most appropriate feeding strategy in the period 20–50 kg is that which promotes near maximum energy intake and thus most fully exploits this high potential for muscle growth. The implementation of such a strategy allows very rapid growth, but because rate of protein deposition is linearly related to energy intake, does not result in excessive fat deposition or deterioration in feed:gain ratio.

However, ensuring maximum energy intake is not merely a matter of offering pigs a 'grower' diet *ad libitum*. Between 20 and 50 kg, pigs eat to the limit of their ingestive capacity which lies between 1.8 and 2.0 kg/d. On the other hand, over the same liveweight range the pig's demand for energy, which is a reflection of its potential for protein and fat growth, lies between 30 and 32 MJ DE/d. Accordingly, unless offered diets with energy concentrations between 14 and 15 MJ/kg the animal is unable to satisfy its demand for energy or fully to express its potential for growth. The effect of dietary energy content on growth performance is shown in Table 4.1, which gives the results of an experiment in which entire male pigs were given five diets ranging in DE concentration from 11.8–15.1 MJ/kg between 22 and 50 kg live weight.

**Table 4.1** EFFECT OF DIETARY DE CONCENTRATION ON THE VOLUNTARY FEED INTAKE AND PERFORMANCE OF ENTIRE MALE PIGS GROWING FROM 22–50 kg

| | *Dietary energy content* (MJ/kg) | | | | |
|---|---|---|---|---|---|
| | *11.8* | *12.7* | *13.6* | *14.5* | *15.1* |
| Voluntary feed intake (kg/d) | 2.19 | 2.21 | 2.19 | 2.17 | 2.05 |
| Voluntary energy intake (MJ DE/d) | 25.7 | 27.7 | 29.7 | 31.3 | 30.9 |
| Daily gain (g) | 695 | 776 | 847 | 898 | 913 |
| Feed:gain | 3.16 | 2.89 | 2.61 | 2.39 | 2.25 |
| Carcass P2 (mm) | 14.4 | 15.3 | 15.6 | 16.0 | 16.4 |

Campbell and Taverner (1986)

*Effects of sex and level of energy intake on protein deposition in pigs between 50 and 90 kg*

Above 50 kg the genetic potential for protein deposition of entire male, female and castrated male pigs tends to lie within the limits of appetite. Table 4.2 gives the results of an experiment in which entire male and female pigs growing between 48 and 90 kg were given five levels of intake of a protein-adequate diet. The results showed that for both sexes, rate of protein deposition increased linearly with increase in DE intake up to 32 MJ/d but remained constant at 130 g/d for males and 102 g/d for females thereafter.

**Table 4.2** EFFECTS OF ENERGY INTAKE BETWEEN 48 AND 90 kg ON RATE OF PROTEIN DEPOSITION AND THE PERFORMANCE OF ENTIRE MALE (M) AND FEMALE (F) PIGS

| | | *Energy intake* (MJ DE/d) | | | | |
|---|---|---|---|---|---|---|
| | | *22.6* | *26.4* | *31.7* | *36.0* | *Ad lib*[a] |
| Protein deposition (g/d) | M | 69.5 | 94.8 | 129.5 | 130.0 | 132.0 |
| | F | 63.4 | 84.5 | 103.0 | 102.0 | 99.0 |
| Daily gain (g) | M | 418 | 576 | 793 | 842 | 884 |
| | F | 358 | 552 | 654 | 742 | 795 |
| Feed:gain | M | 3.9 | 3.4 | 2.9 | 3.1 | 3.5 |
| | F | 4.6 | 3.6 | 3.4 | 3.5 | 3.6 |
| Body fat (g/kg) | M | 203 | 249 | 257 | 315 | 332 |
| | F | 293 | 332 | 353 | 368 | 397 |

[a]39.8 MJ ME/d for M and 37.9 MJ ME/d for F ($P<0.05$).

The results (Table 4.2) show the adverse effects on carcass fatness and feed:gain of raising DE intake above the level at which the pig's potential for protein growth is achieved. Nevertheless, providing the relationship between energy intake and rate of protein deposition is known, these effects are predictable and can be taken into account when designing feeding strategies for heavier pigs. Because maximal rate of protein deposition was achieved at a DE intake of approximately 32 MJ/d, the level of dietary lysine required to support maximal muscle growth would decline with each increase in DE intake above 32 MJ/d. Thus, the possibility of using cheaper diets for pigs given high levels of energy intake would also have to be considered when deciding the most profitable nutritional strategy for heavier pigs. However, in the majority of cases the most profitable feeding level would be that which provided an average DE intake of between 32 and 34 MJ/d, and thus allowed the pig to express its potential for muscle growth but prevent the adverse effects of higher energy intakes on feed:gain and carcass fatness. The latter strategy may involve either the use of a relatively low energy diet (e.g. 12–12.5 MJ DE/d) offered *ad libitum* or the use of a restricted feeding programme.

The results in Table 4.2 also show that both maximum rate of protein deposition and the slope of the linear portion of the relationship between energy intake and rate of protein deposition was lower for females than for entire males. Thus, unless there was a marked difference in obligatory protein losses between the sexes the level of dietary protein required to support muscle growth in females would be expected to be lower than that for entire males.

**Table 4.3** EFFECT OF DIETARY LYSINE CONTENT ON THE FEED:GAIN OF ENTIRE MALE (M) AND FEMALE (F) PIGS GROWING FROM 20 TO 50 AND 50 TO 90 kg LIVE WEIGHT

| | | *Dietary lysine* (g/MJ DE) | | | | | | | |
|---|---|---|---|---|---|---|---|---|---|
| | | *0.4* | *0.5* | *0.6* | *0.67* | *0.76* | *0.83* | *0.94* | *1.02* |
| 20–50 kg | M | 3.3 | 2.9 | 2.6 | 2.4 | 2.2 | 2.2 | 2.3 | 2.3 |
| | F | 3.3 | 2.9 | 2.6 | 2.4 | 2.25 | 2.3 | 2.4 | 2.4 |
| 50–90 kg | M | 3.5 | 2.9 | 2.7 | 2.9 | 3.1 | 3.0 | 2.9 | 2.9 |
| | F | 3.5 | 2.9 | 2.9 | 3.2 | 3.2 | 3.3 | 3.5 | 3.3 |

Campbell, Tavener and Curic (1988)

The latter contention has been confirmed experimentally (Campbell, Taverner and Curic, 1988) and is illustrated in Table 4.3. These results show that between 20 and 50 kg live weight the level of dietary lysine required to support maximum growth performance was similar for both sexes. However, between 50 and 90 kg the level of dietary lysine required to support maximum growth performance in females was 15% below that required for entire males. Experimental results have also shown that between 50 and 90 kg female pigs are less tolerant of high protein intake than males, and that levels of dietary protein only marginally in excess of requirement tend to depress growth performance. This effect, however, is influenced by genotype.

In the short term there is considerable scope for improving the efficiency of pig production by using lower protein (amino acid) diets for female finisher pigs. The separate feeding of the sexes during the later stages of production would also enable high energy diets to be used for males and relatively low energy diets used for females which would further reduce the cost of production and prevent excess carcass fatness often observed in female pigs during the final stages of production. In the longer term the profitability of production would be most readily achieved by increasing the female's capacity for muscle growth. The latter might be achieved by genetic selection or by hormone manipulation, both of which are discussed below.

### EFFECT OF GENOTYPE OR STRAIN

Differences in growth performance and body composition have been reported between different strains and breeds of pigs. However, until recently there has been little information on the effect of genotype on energy and protein metabolism or on the extent of variation which might exist between commercial strains. The results of a study conducted at the ARI, Werribee provide an insight into the effects of genetic selection on muscle growth and the consequent effects on growth performance and energy utilization.

In the experiment conducted at Werribee, protein deposition was measured in two strains of entire male pig (Large White Landrace), given seven levels of intake of a protein-adequate diet between 45 and 90 kg live weight. One strain (strain A) was introduced to the experimental piggery by Caesarean section from sows obtained from a large commercial piggery (6000 sows) where all breeding stock have been selected on the basis of growth performance under *ad libitum* feeding for some 12 years. The other strain (strain B) was from the experimental herd of

**Table 4.4** EFFECTS OF ENERGY INTAKE BETWEEN 45 AND 90 kg LIVE WEIGHT ON PROTEIN DEPOSITION AND GROWTH PERFORMANCE IN FASTER (A) AND SLOWER (B) GROWING STRAINS OF ENTIRE MALE PIGS

| *Energy intake* (MJ DE/d) | *Strain* | *Protein deposition* (g) | *Dail gain* (g) | *Feed gain* | *Carcass fat* (%) |
|---|---|---|---|---|---|
| 22.2 | A | 92 | 567 | 2.60 | 18.8 |
| | B | 81 | 470 | 3.12 | 24.4 |
| 25.1 | A | 105 | 622 | 2.66 | 19.4 |
| | B | 87 | 595 | 2.80 | 26.6 |
| 27.6 | A | 119 | 764 | 2.39 | 21.0 |
| | B | 105 | 680 | 2.69 | 29.0 |
| 30.6 | A | 135 | 826 | 2.40 | 23.6 |
| | B | 115 | 734 | 2.77 | 28.9 |
| 33.5 | A | 148 | 944 | 2.36 | 25.4 |
| | B | 128 | 820 | 2.70 | 30.3 |
| 36.8 | A | 166 | 1110 | 2.23 | 25.8 |
| | B | 129 | 870 | 2.85 | 32.2 |
| *Ad libitum*[a] | A | 189 | 1202 | 2.26 | 26.0 |
| | B | 125 | 915 | 3.05 | 36.6 |

[a]*Ad libitum* energy intake was 40.6 and 40.7 MJ DE/d for strain A and B pigs respectively
Campbell and Taverner (1988)

some 50–60 sows and were representative of slower growing commercial genotypes.

The results for protein deposition (Table 4.4) showed that at all levels of feed intake strain A pigs deposited protein faster than strain B. The form of the relationship between energy intake and rate of protein deposition also differed between the strains. For strain B pigs, rate of protein deposition increased linearly with increase in DE intake up to a maximum rate of 129 g/d at approximately 33 MJ DE/d (80% *ad libitum*).

For strain A pigs there was no evidence of any genetic ceiling for rate of protein accretion which increased linearly with increases in energy intake up to 187 g/d on the *ad libitum* feeding treatment (40 MJ DE/d). It appears that selection under *ad libitum* feeding had raised the capacity for muscle growth of these pigs (strain A) beyond the upper limit of appetite. The slope of the relationship between DE intake and rate of protein deposition for strain A was also higher than that of the linear portion of the relationship for strain B. Given the marked difference in capacity for protein accretion between the strains, the levels of dietary lysine and other essential amino acids required to support muscle growth would also be expected to differ as would the responses of the two strains to change in level of feeding. The latter is obvious from Table 4.4, which shows the effects of energy intake on growth rate, feed:gain and P2 fat thickness of the two strains. For strain A pigs the most appropriate feeding strategy would be that which promoted maximum energy intake, and thus fully exploited the pig's high potential for muscle growth. This would probably involve the *ad libitum* feeding of diets with energy and lysine concentrations generally considered more appropriate for younger pigs. For strain B pigs however, the most appropriate and most profitable feeding strategy may involve a degree of energy restriction.

Field trials conducted in conjunction with the work at the ARI indicate the potential for muscle growth in commercial strains within Australia range from somewhat below that of the strain B pigs used in the experiment at Werribee up to 85–90% of that of strain A. Because of this wide variation there is a need, in the short term, for producers to measure the capacity for muscle growth of their stock and to use this information to design appropriate diets and feeding strategies. In the longer term, the profitability of individual enterprises and of pig production in general will be most readily improved by the identification and spread of genuinely superior stock. The results of the experiment conducted at Werribee show however, that unless the improvement in the pig's genetic capacity for muscle growth is matched by a concomitant improvement in its nutritional management, and in particular the level of dietary essential amino acids, much of the potential benefit offered by such animals will not be realized.

The results of the experiment presented in Table 4.4 suggest that in terms of growth performance, genetic improvement is associated with increases in both maximum muscle growth and in slope of the linear component of the relationship between energy intake and muscle growth. The relative rate of improvement in either of these characters can probably be influenced by the selection procedure employed, and in the future it may be more profitable to concentrate on increasing the slope of the relationship between energy intake and rate of protein deposition. This would enable more rapid and efficient growth to be obtained at the levels of energy intake achieved by pigs offered feed *ad libitum* under commercial conditions, which is often substantially lower than that achieved by pigs kept under experimental conditions or in performance testing situations.

## EXOGENOUS PORCINE SOMATOTROPIN (PST) ADMINISTRATION

The interrelationship between protein deposition capacity and dietary amino acid requirements is best illustrated by the interrelationship between exogenous PST administration and dietary lysine content on pig performance and protein accretion. Exogenous PST administration stimulates protein deposition and inhibits lipogenesis resulting in marked improvements in growth performance and reduction in carcass fat content (Campbell *et al.*, 1989a; Etherton *et al.*, 1987; Evock *et al.*, 1988). However, initiation and support of the higher rates of protein deposition able to be induced by PST technology require a concomitant increase in dietary lysine content. This is demonstrated in Table 4.5 which presents the results of an experiment in which the responses of control and PST treated boars were compared over six levels of dietary lysine between 60 and 90 kg live weight.

Exogenous PST administration increased maximal protein deposition (measured in the head off empty body) from 118 to 215 g/d (81%). However, the magnitude of the improvement induced by PST was directly related to dietary lysine content and on the two lowest lysine diets PST administration had no positive effect on protein deposition or growth performance. The level of dietary lysine required to support maximal protein accretion in control and PST treated pigs was approximately 0.75 and 1.2% respectively. The corresponding dietary lysine:DE values were 0.5 and 0.8 g/MJ respectively.

Apart from demonstrating the potential of PST technology to alter the efficiency of pig meat production, the results show that future improvements in the efficiency of growth will be dependent on identifying and removing the intrinsic factors constraining protein deposition capacity. This can be achieved by genetics and/or employment of new technologies such as exogenous PST administration. However, it is evident from the information discussed here that the advantages offered by

**Table 4.5** EFFECTS OF EXOGENOUS PORCINE SOMATOTROPIN ADMINISTRATION (PST) AND DIETARY LYSINE CONTENT BETWEEN 60 AND 90 kg ON THE GROWTH PERFORMANCE AND CARCASS P2 FAT THICKNESS OF ENTIRE MALE PIGS

| *PST* (mg/kg/d) | *Dietary lysine* (%) | *Protein deposition* (g/d) | *Daily gain* (g) | *Feed:gain* | *P2* (mm) |
|---|---|---|---|---|---|
| 0 | 0.45 | 67 | 628 | 3.71 | 20.5 |
| | 0.66 | 107 | 803 | 2.86 | 19.8 |
| | 0.88 | 118 | 862 | 2.65 | 18.6 |
| | 1.09 | 115 | 823 | 2.78 | 20.2 |
| | 1.31 | 119 | 887 | 2.63 | 17.2 |
| | 1.53 | 117 | 860 | 2.71 | 18.4 |
| 0.09 | 0.45 | 74 | 588 | 3.87 | 17.0 |
| | 0.66 | 104 | 760 | 3.02 | 15.0 |
| | 0.88 | 146 | 961 | 2.35 | 12.4 |
| | 1.09 | 175 | 1108 | 2.07 | 14.2 |
| | 1.31 | 216 | 1204 | 1.80 | 14.0 |
| | 1.53 | 213 | 1338 | 1.69 | 13.1 |

Campbell *et al.* (1989b)

enhancing the growing pig's capacity for protein deposition (muscle growth) will only be fully realized if dietary nutrient levels and feeding strategies are enhanced accordingly.

## Dietary factors

### AMINO ACID BALANCE AND THE IDEAL PROTEIN CONCEPT

The concept of ideal protein as promoted by the ARC (1981) has been successful in shifting the focus of commercial nutritionists away from individual amino acid requirements and toward a greater appreciation of the need for appropriate amino acid balance in diets.

In Australia, the industry has generally adopted the recommendations of the SCA (1988) which advocate an amino acid balance similar to the ARC (1981). More recently, some segments of the industry have adopted revised balances proposed by AEC (1987) and by Wang and Fuller (1987). The different amino acid patterns as proposed by these bodies are compared in Table 4.6.

The benefits of viewing the dietary protein requirements of growing pigs as a balanced pattern of amino acids, are obvious. However, some caution needs to be exercised with regard to the actual values chosen to represent the 'ideal' balance. The ARC (1981) and subsequently the SCA (1988) in proposing their estimate of the amino acid balance of 'ideal protein', though recognizing variations in digestibility between feedstuffs and between amino acids within feedstuffs, still defined the balance in terms of total amino acids. These estimates were based on experimental evidence and an appreciation of the amino acid profile of pig tissue. The SCA (1988) recommendations have partially addressed the complication of variable amino acid digestibility by couching the amino acid requirements of pigs in terms of available lysine:energy ratios in which lysine is used as proxy for all amino acids, linked by the pattern established for total amino acids. This

**Table 4.6** ESTIMATES OF AMINO ACID COMPOSITION 'IDEAL PROTEIN' RELATIVE TO LYSINE

| | *SCA (1988)*[a] | *AEC (1987)*[a] | *Wang and Fuller*[b] *(1987)* |
|---|---|---|---|
| Lysine | 100 | 100 | 100 |
| Methionine | 25 | 30 | – |
| Methinione + cystine | 50 | 55 | 63 |
| Threonine | 60 | 60 | 72 |
| Tryptophan | 14 | 18 | 18 |
| Isoleucine | 54 | 55 | 60 |
| Leucine | 100 | 100 | 110 |
| Histidine | 33 | 35 | – |
| Phenylalanine | 48 | – | – |
| Phenylalanine + tyrosine | 96 | 95 | 120 |
| Valine | 70 | 70 | 75 |

[a]Total amino acid
[b]Ileal digestible amino acid

unfortunately assumes that the 'availability' of each of the other essential amino acids is in parallel with the variations for lysine, and is invalid for many feedstuffs.

Fuller (1989) emphasized the point that the revised 'ideal protein' amino acid balance of Wang and Fuller (1987) was determined using materials which were digested virtually completely and as such 'describes the intrinsic needs of the pig for post absorptive metabolism'. Hence, before these recommendations can be adopted in diets employing conventional materials the amino acid content of such materials needs to be defined in terms which are unconfounded by the variations in digestibility.

In addition to revising the amino acid balance of ideal protein Fuller, McWilliam and Wang (1987) also established two further concepts of interest.

(1) That due to the very different amino acid requirements of maintenance and body protein accretion, the ideal protein will vary according to the level of protein growth and its changing relationship to maintenance. This is of particular interest in superior genotypes of high lean tissue deposition potential and animals whose protein deposition rates have been markedly elevated via treatments such as porcine somatotropin.
(2) That as well as a specific pattern of essential amino acids being necessary to achieve the 'ideal protein' there is also an optimal ratio of total essential to non-essential amino acids. It may well be in the future that certain combinations of non-essential amino acids may be identified which result in more efficient use of dietary nitrogen.

To the independent observer the debate on amino acid availability with all its complications of techniques employed (e.g. *in vitro vs in vivo*, faecal *vs* ileal recovery, apparent *vs* true *vs* real digestibilities, and digestibility *per se vs* utilization), appears to be as much an exercise in semantics as in science.

As a consequence many commercial nutritionists have opted to follow a conservative line formulating feeds on the basis of total amino acid proportions with some consideration of the 'availability' of lysine (often arrived at by very subjective and arbitrary means). Many would argue the ARC (1981) ideal protein amino acid pattern has served them well and in the absence of sufficient reliable amino acid availability values for all feedstuffs they see little point in changing their approach.

AMINO ACID AVAILABILITY

Although this area of animal nutrition has been a significant focus of research for many years there is still much to be resolved, not least of which should be agreement on an unambiguous definition of availability. In many instances the term availability is used interchangeably with digestibility yet should be limited to reference only to that proportion of the amino acid supply which can be fully utilized for protein synthesis at the cellular level. Where these two values are similar there is little concern but where the disparity is wide nutritionists are faced with a dilemma as to which definition and values they are going to adopt. An example of such a dilemma is the ongoing debate with regard to the 'availability' of lysine in sweet lupinseed (*L. angustifolius*). Taverner, Curic and Rayner (1983) reported a value for the apparent ileal digestibility of lysine in this material of 90% while Batterham, Murison and Anderson (1984) measured an availability of only 57% by the slope ratio, bioassay technique.

NET ENERGY—THE PREFERRED BASIS OF ENERGETIC EVALUATION OF FEEDS AND DIETARY REQUIREMENTS

Since energy and protein are so tightly interrelated in the nutrition of pigs it seems incongruous to pursue the 'availability' of amino acids such that feedstuffs can be described in terms of the levels of specific amino acids to which the animal responds directly, while the energy side functions with a relatively coarse descriptor (digestible energy). Just (1982) argues the case for a net energy system of feed evaluation and points out that net energy is the energetic parameter to which animals respond directly. Hence if tight control of animal responses and accurate descriptions of the nutritive value of feedstuffs are to be achieved then protein/energy requirements may need to be described in terms of units of available amino acids and net energy.

Just, Fernandez and Jorgenson (1983) showed that the ratio between the net energy and the metabolizable energy value of a diet decreased as the amount of hind gut digestion increased. Taverner and Curic (1983) applying the calculations of Just, Fernandez and Jorgenson (1983) demonstrated how digestible energy can be a poor estimate of the available energy in some feedstuffs (Table 4.7).

Hence, although lupins record higher DE and ME values than peas, their NE yield is considerably lower due to the large proportion of the energy being digested in the hind gut.

**Table 4.7** AVAILABLE ENERGY CONTENT OF FEED INGREDIENTS (MJ/kg DM)

| *Ingredient* | *Digestible energy* | *Metabolizable energy* | *Net energy* |
|---|---|---|---|
| Wheat | 15.6 | 15.1 | 10.5 |
| Peas | 14.1 | 13.6 | 9.7 |
| Lupins (*L. angustifolius*) | 15.1 | 13.9 | 7.2 |

After Taverner and Curic (1983)

Similar improvements in net energy yield from the digestible energy component of diets occurs as energy density is elevated via the replacement of carbohydrates with fats.

In situations where the energy density and proximate analysis (protein, fat, fibre) remain fairly static there is probably little advantage in adopting net energy as the basis of diet formulation since the relationship between digestible and net energy is fairly constant. However, in situations where the composition of diets is more diverse and the proportion of digestible energy disappearing in the hind gut varies, or in situations of elevated growth potential where appetite or heat stress are limiting performance, and can be overcome by an increase in energy density and a concomitant reduction in the heat increment, then net energy becomes the prime determinant of performance, and thus the preferred basis of feed formulation.

## Integration of the interactive factors—modelling

A number of computer simulation models have been developed in an attempt to integrate the many animal and dietary factors affecting the growth performance and nutrient requirements of growing pigs.

Undoubtedly the most comprehensive of these models to emerge to date is the AUSPIG model which was initially described by Black *et al.* (1986) and which has been in a constant phase of development ever since.

One of the real benefits of such models is that they identify the gaps in our knowledge which in the short term have to be covered by tentative assumptions to allow the continuity of the modelling process. However, in the longer term these have to be addressed by the appropriate research.

The model allows the integration of disciplines (nutrition/genetics/management) such that the objective mathematics of feed formulation using linear programming is more responsive to the biological feedback and the development of total system strategies to optimize production and/or profit, rather than the additive sequence of independent piecemeal decisions which currently characterize pig production management.

Validation exercises with the AUSPIG model where the predicted outputs have been compared to documented results of numerous growth studies, and production situations have proved most encouraging.

When the model was first constructed the main input areas for which valid and precise data were limited were as follows.

### GENOTYPIC DESCRIPTIONS

Most research projects with pigs generally target a specific growth phase and so although numerous trials may contribute information from the active growth phases, the model also requires a comprehensive description of the whole of life growth pattern/potential for accurate simulation. A series of extended growth studies have been undertaken with the major genotypes within Australia to generate the necessary information particularly at the mature end of the growth curves, to allow fine tuning of the algorithms which describe each genotype (G.T. Davies, CSIRO personal communication). Examples of such algorithms are shown in Figures 4.3a and 4.3b. The AUSPIG model in recognizing the dynamic nature of genotype development offers not only a series of base genotypes from which to

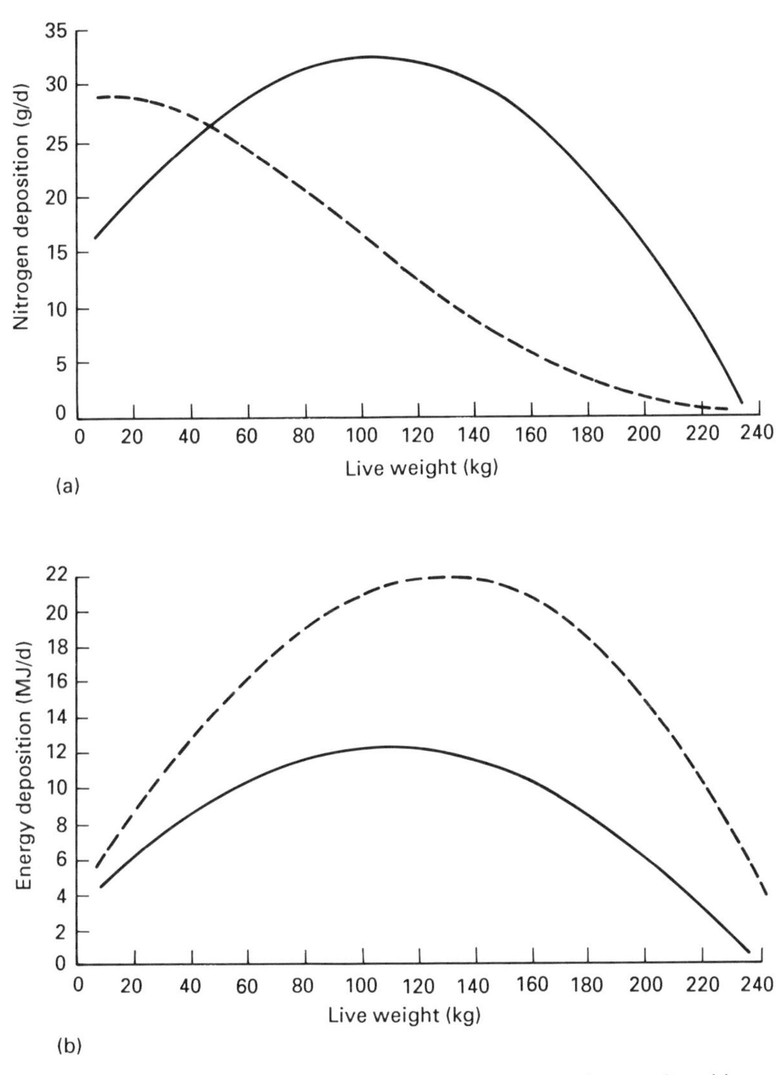


**Figure 4.3** Genotypic descriptions in terms of (a) potential nitrogen deposition and (b) potential energy deposition for (——) highly improved and (- - -) unimproved pigs with increasing live weight

choose to commence a simulation but also a facility to fine tune the genotype within the model.

## NUTRITIONAL DESCRIPTION OF FEEDSTUFFS

With the exception of genotype and other animal factors the main quantitative determinants of growth performance remain energy and available amino acids.

Hence the predictive precision of a model is directly dependent on the accuracy of the nutritional descriptors applied to the feedstuffs employed. This was a major area of concern in the original AUSPIG model and will remain an area requiring constant review due to the variable composition of feedstuffs.

Probably one of the major dilemmas currently faced by the commercial nutritionist is the choice of availability values for amino acids from various sources and how to apply them in day-to-day formulations. The nutrient requirement of the stock being fed and the nutrient content of the feedstuffs being employed need to be defined on the same basis. There is little point constructing a highly detailed and precise nutrient matrix for raw materials if the dietary specifications are drawn from textbook references with no qualifications as to how they were derived. A robust model should achieve a consistent accuracy of prediction across a range of radically different situations as long as the inputs are described in analogous terms.

One of the major areas of concern is the inefficient use of dietary amino acids. Despite attempts to match dietary amino acid balance with known tissue requirements, and allowing for differences in amino acid digestibility within individual feedstuffs, protein and amino acid utilization values of the order of only 50% are often recorded in both commercial and experimental diets (Campbell, Taverner and Curic, 1988).

This level of inefficiency may suggest that:

(1) The 'ideal' protein concept currently in vogue is far from perfect, and is not as simple or static as has been proposed to date.
(2) One or more of the essential amino acids other than lysine are less available than is generally presumed using conventional digestibility and availability values.
(3) Amino acids are being lost in the processes of intermediary metabolism for reasons not yet fully understood.
OR
(4) Additional factors such as health (particularly enteric disorders), stress, antinutritional factors (lectins, trypsin inhibitors, tannins, etc.) may be common sources of inefficient amino acid utilization in commercial situations. Although these factors are unlikely to affect amino acid utilization under experimental conditions they would nevertheless be difficult to incorporate into a model.

The AUSPIG model accommodates this apparent inefficiency of amino acid utilization by ascribing relatively low amino acid availability values to some raw materials in particular the grains. This could be considered as a temporary expedience awaiting further clarification of the factors involved.

Despite the apparent shortfalls and necessity for assumption in some aspects of modelling the quantitative precision of predictions of growth have been remarkably good especially in situations in which most of the major inputs have been accurately quantitated. Further to this, in the AUSPIG model, the incorporation of a profit maximizing facility allows not only the integration of animal performance with the most appropriate nutritional strategy but also through the incorporation of an adjustable statistical distribution of grading results and the simultaneous superimposition of various payment matrices, allows optimum profit projections detailing the required growing, feeding and marketing strategies. It even goes further to identify potential improvements in total enterprise profitability by defining the quantitative limits of capital, labour, specific facilities, etc.

## Conclusions

The growing pig's capacity for protein accretion is the major factor determining growth performance and dietary amino acid/energy requirements. In the short term, considerable improvement in the efficiency of production could be achieved by matching more closely the pig's nutritional management with its capacity for protein growth.

Future improvements in the efficiency and profitability of pig meat production will be dependent on identifying and removing intrinsic constraints to lean tissue growth capacity, particularly in heavier pigs. This can be achieved by conventional (e.g. genetics) or biotechnological (e.g. exogenous PST administration) techniques. However, in either case the potential advantages offered by pigs with increased lean tissue growth capacity will only be fully realized if dietary amino acid levels and energy intakes are adjusted accordingly.

The difficulties that confront the nutritionist attempting to meet these requirements include:

(1) An inadequate understanding of the protein deposition characteristics of the genotypes with which we are working. Included in this is an appreciation of the dynamic nature of the genotype (i.e. the rate of improvement by genetic or biotechnological means).
(2) An incomplete knowledge of the energy and amino acid values of feedstuffs as they translate into nutrients available for tissue growth at the cellular level.
(3) We do not address the problem of preparing diets and feeding strategies with the fully integrated (i.e. animal/environment/diet/market) approach required.

These points exemplify some of the main areas requiring further research.

## References

AEC (1987). *Recommendations for Animal Nutrition*, 5th edition, Rhone Poulenc Animal Nutrition, Commentry, France

ARC (1981). *The Nutrient Requirements of Pigs*. Commonwealth Agricultural Bureaux, Farnham Royal, UK

Batterham, E.S., Murison, R.D. and Anderson, L.M. (1984). Availability of lysine in vegetable protein concentrates as determined by the slope-ratio assay with growing pigs and rats and by chemical techniques. *British Journal of Nutrition*, **51**, 85–99

Black, J.L., Campbell, R.G., Williams, I.H., James, K.J. and Davies, G.T. (1986). Simulation of energy and amino acid utilisation in the pig. *Research and Development in Agriculture*, **3.3**, 121–145

Campbell, R.G. and Taverner, M.R. (1986). The effects of dietary source of fat and dietary energy content on the voluntary energy intake and performance of growing pigs. *Animal Production*, **42**, 327

Campbell, R.G. and Taverner, M.R. (1988). Genotype and sex effects on the relationship between energy intake and protein deposition in growing pigs. *Journal of Animal Science*, **66**, 676–686

Campbell, R.G., Taverner, M.R. and Curic, D.M. (1985a). Effects of sex and energy intake between 45 and 90 kg live weight on protein deposition in growing pigs. *Animal Production*, **40**, 497

Campbell, R.G., Taverner, M.R. and Curic, D.M. (1985b). The influence of feeding level on the protein requirements of pigs between 20–45 kg live weight. *Animal Production*, **40**, 489

Campbell, R.G., Taverner, M.R. and Curic, D.M. (1988). The effects of sex and live weight on the growing pigs response to dietary protein. *Animal Production*, **46**, 123

Campbell, R.G., Johnson, R.J., King, R.H. and Taverner, M.R. (1989a). Advances in the manipulation of pig growth. Interactions between porcine growth hormone administration and dietary potein. In: *Recent Advances in Animal Nutrition in Australia*. pp. 141–146. Ed. D.J. Farrell. University of New England

Campbell, R.G., Steele, N.C., Caperna, T.J., McMultry, J.P., Solomon, M.B. and Mitchell, A.D. (1989b). Interrelationships between sex and exogenous growth hormone administration of performance, body composition and protein and fat accretion of growing pigs. *Journal of Animal Science*, **67**, 177–186

Davies, G.T. CSIRO (personal communication)

Dunkin, A.C. and Black, J.L. (1987). The relationship between energy intake and nitrogen balance in the growing pig. In *Energy Metabolism of Farm Animals*. Eur. Ass. Anim. Prod. Publ. No. 32, 110–113

Etherton, T.D., Wiggins, J.P., Evock, C.M., Chung, C.S., Riebhun, J.F., Walton, P.E. and Steele, N.C. (1987). Stimulation of pig growth performance by porcine growth hormone: determination of the dose-response relationship. *Journal of Animal Science*, **64**, 433–442

Evock, C.M., Etherton, T.D., Chung, C.S. and Ivy, D.E. (1988). Pituitary porcine growth hormone (pGH) and a recombinant pGH analog stimulate pig growth performance in a similar manner. *Journal of Animal Science*, **66**, 1928

Fuller, M.F. (1989). *Ideal Protein: The Concept and its Application in Swine Diets*. Proc. of 10th Western Nut. Conf., Winnipeg, Canada

Fuller, M.F., McWilliam, R. and Wang, T.C. (1987). The amino acid requirement of pigs for maintenance and for growth. *Animal Production*, **44**, 476

Just, A. (1982). The net energy value of balanced diets for growing pigs. *Livestock Production Science*, **8**, 541–555

Just, A., Fernandez, J.A. and Jorgenson, H. (1983). The net energy value of diets for growth in pigs in relation to the fermentative processes in the digestive tract and the site of absorption of the nutrients. *Livestock Production Science*, **10**, 171–186

SCA (Standing Committee on Agriculture) (1988). *Feeding standards for Australian livestock. Pigs*. CSIRO

Taverner, M.R. and Curic, D.M. (1983). The influence of hind gut digestion on measures of nutrient availability in pig feeds. In *Feed Information and Animal Production*. pp. 295–298 Eds G.E. Robards and R.G. Packam. Commonwealth Agricultural Bureaux, Slough, UK

Taverner, M.R., Curic, D.M. and Rayner, C.J. (1983). A comparison of the extent and site of energy and protein digestion in wheat, lupin and meat and bone meal by pigs. *Journal of the Science of Food and Agriculture*, **34**, 122–128

Wang, T.C. and Fuller, M.F. (1987). An optimal dietary amino acid pattern for growing pigs. *Animal Production*, **44**, 486

5

# COMPARISON OF ARC AND NRC RECOMMENDED REQUIREMENTS FOR ENERGY AND PROTEIN IN GROWING PIGS

A.J. LEWIS
*University of Nebraska, USA*

## Introduction

The need for a set of national standards referring to the nutritional requirements of pigs has been recognized in both the UK and the USA for many years. Both countries have convened panels of experts to review scientific literature on pig nutrition and to publish tables listing nutritional requirements. In the UK, requirements are established by a working party of the Agricultural Research Council (ARC), while in the USA the comparable body is the swine nutrition subcommittee of the National Research Council (NRC). Before examining specific recommendations, a brief history of the two organizations and some comments on similarities and differences in approaches that they used will be presented.

## History

Scientific recommendations pertaining to the nutrition of pigs by a government agency in the UK date back to 1921 with the publication of the first edition of the government bulletin *Rations for Livestock*. It was not until 1967, however, that the first publication concerned specifically with pigs appeared. This was the first report of an ARC working party on nutrient requirements of pigs (ARC, 1967). The second edition, the current recommendations (ARC, 1981), was published 14 years later.

The NRC's *Nutrient Requirements of Swine* has a longer history. The first edition (NRC, 1944), which was only 11 pages long was published during the Second World War. Since then there have been seven revisions. The current edition (NRC, 1979) was published in 1979 and reprinted in 1983.

In this chapter the abbreviation ARC is used to refer to the current report of the ARC working party, and NRC is used to refer to the current report of the NRC subcommittee on swine nutrition.

## Similarities and differences

At first sight the ARC (1981) and NRC (1979) publications are quite different. This is somewhat misleading, however, as the two reports actually have much in common. Some of the important similarities are:

(1) Both were written by a panel of experts involved in the research or application of pig nutrition. The ARC panel was relatively large, containing 19 members; there were six members on the NRC panel.
(2) In both reports requirements were derived from world literature, not simply research from the home country or region.
(3) The recommendations made by both panels are requirements not allowances. That is, they do not contain a 'margin of safety'. The requirements of both groups are for total nutrients required in the diet, not for 'available' nutrients.
(4) Good to ideal conditions (in terms of environment, health of animals, etc.) were assumed in both reports.
(5) The recommendations for growing pigs are the same for different sexes, even though it is recognized that barrows, boars and gilts have different requirements for certain nutrients.
(6) Many values in both reports were obtained by calculation or extrapolation from other data, and not by direct experimentation.
(7) A bibliography is provided in both reports.

Some of the more significant differences between the two reports are:

(1) ARC (1981) contains an extensive literature review and full explanation of how it arrived at its requirements. The discussion by NRC (1979) is very brief with little or no direct explanation (other than a bibliography) of the derivation of its requirements.
(2) A 'factorial' method of estimating requirements was used many times by the ARC. In this method estimates are made for each of the major processes (e.g. maintenance, growth, lactation, etc.) that contribute to the requirements of pigs at a given stage of the life-cycle. The NRC seems to have relied more on direct experimental data, with extrapolation to different production situations where data were not available.
(3) The NRC used a set of consistent categories of the life-cycle (e.g. 20–35 kg, lactating sows and gilts, etc.) for all nutrient requirements. The categories within ARC are different for different nutrients.
(4) The NRC assumed a higher dietary energy density than the ARC. This presumably reflects differences in the energy density of the predominant cereal grains in the two countries.

## Energy

### CHOICE OF SYSTEM AND UNITS

The expert panels in both countries are in agreement that at the present time the digestible energy (DE) system represents the most desirable means of expressing the energy requirements of pigs and the energy value of diets. The metabolizable energy (ME) system was also discussed by both groups and there is good agreement about the relationship between DE and ME. The ARC estimated that the ME:DE ratio is 0.96 for diets based on cereals containing 16% crude protein (CP). The NRC used ME:DE ratios of 0.94 to 0.97 in converting its DE requirements to ME requirements. For some reason the NRC used the higher ratio for baby pig diets (containing high protein levels) and the lower ratio for finishing pig diets. This appears to be in error as there is known to be a negative relationship between crude protein level and ME:DE ratio. The negative relationship was recognized by the

NRC in its text where equation 1 (Asplund and Harris, 1969), is quoted, but it was not taken into account in the tables.

$$ME/DE = 0.96 - 0.000202\,CP\ (g/kg) \qquad (1)$$

The ARC listed two different equations that allow for the effect of protein level. They are those of May and Bell (1971) in equation 2 and Morgan, Cole and Lewis (1975) in equation 3.

$$ME/DE = 1.0121 - 0.00019\,CP\ (g/kg) \qquad (2)$$

$$ME/DE = 0.997 - 0.000189\,CP\ (g/kg) \qquad (3)$$

Net energy systems are not used in either country at present, largely because of inadequate data on the net energy content of feeds for pigs.

Energy requirements were listed in megajoules (MJ) by the ARC and kilocalories (kcal) by the NRC, but this presents no difficulty as both groups assume the conversion factor given in equation 4:

$$\text{joule/calorie} = 0.239 \qquad (4)$$

## ENERGY 'REQUIREMENTS' OF GROWING PIGS

Throughout most of North America growing pigs are given continuous access to feed from weaning until market weight. Although feed intake is often restricted during the later stages of the finishing period in European countries, growing pigs usually have unrestricted access to feed for the major part of their lives. Consequently the term 'requirement' is a misnomer when referring to energy for growing pigs. What is really meant, and what is needed in practice, are estimates of the voluntary energy intakes of growing pigs.

As a result of a compensatory mechanism, energy intake of growing pigs is generally considered to be relatively independent of dietary energy density (Cole, Hardy and Lewis, 1972). Consequently, if energy intakes can be predicted, then from a knowledge of energy contents of diets, feed intakes can be estimated. This information is fundamental to statements of requirements for all other nutrients, because for growing pigs given continuous access to feed the usual and most useful methods of expressing requirements are in relation to the energy content of the diet (e.g. g/MJ DE), or per unit of diet itself (e.g. %, g/kg, etc.).

In its chapter on energy requirements, ARC provided an excellent review of how the energy intake of young pigs is partitioned. Estimates of the energy requirements for maintenance and for accretion of a unit of protein and of fat were derived from available literature. The influence of effective environmental temperature is also discussed. From this 'factorial' approach estimates were made of the consequences of various energy intakes on growth and tissue deposition. In addition to its factorial approach the ARC also reviewed the literature pertaining to the effects of 'quantitative (empirical)' variations in energy intakes particularly on daily gain and backfat thickness.

Although the ARC found reasonable agreement between the two different approaches examined, the difficult question remaining is the choice of a criterion of requirement. The ARC suggested that the requirement should be 'the level of feeding during the appropriate growth phase at which the conversion of feed to lean tissue is at its most efficient'. However, as the ARC fully recognized, even this definition fails to take into account the complex economic factors (such as slaughter weight, value given to carcass quality, and costs of housing, labour, capital, etc.) that influence feeding levels in practical pig production. At the present time the net result of these factors dictates unrestricted feeding of pigs for most of their growth period.

The latter part of the ARC section on energy requirements of growing pigs reviewed the literature describing the voluntary intakes of pigs given continuous access to feed. It concluded that the DE intake of pigs can be adequately described by either of two equations. The first (equation 5) relates DE intake to the maintenance requirement

$$DE(MJ) = 4 \times \text{maintenance} \quad (5)$$

where maintenance (MJ) = $0.749 \times W^{0.63}$, and W = body weight in kilograms.

The second is an asymptotic equation involving body weight (equation 6)

$$DE(MJ) = 55(1 - e^{-0.0204W}) \quad (6)$$

In marked contrast to the ARC, the NRC provided essentially no information about how its estimates of energy requirements were derived. It seems that feed intakes for pigs of various weight ranges were first estimated from experimental data and practical experience. Then DE intakes were calculated using the DE contents of standard diets containing predominantly maize and soyabean meal.

A comparison of the estimates of the two groups is presented in *Figure 5.1*. Curves representing both of the ARC equations are given. For the NRC data the midpoint of each weight range was used to construct the line. It is clear that the agreement is surprisingly good. The only point at which the data differ substantially is the point for pigs weighing 47.5 kg where the NRC estimate is 17% lower than

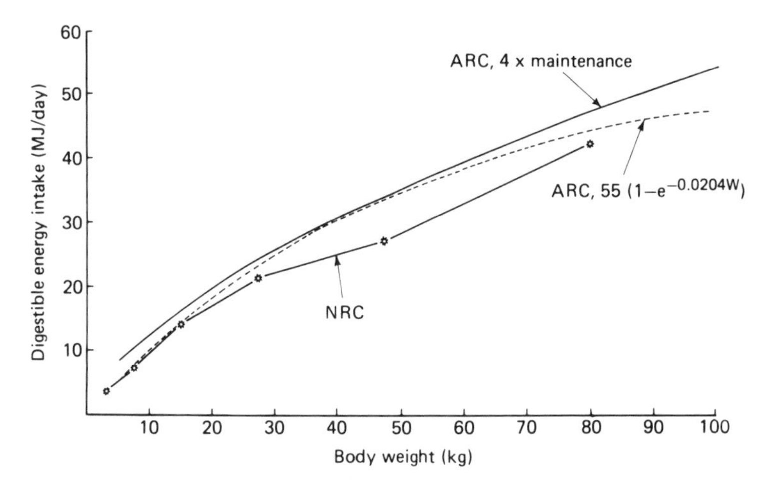


**Figure 5.1** Estimates of the digestible energy intakes of pigs of various weights when given continuous access to feed

that of the ARC. This point seems to be out of line with the remainder of the NRC values. Consequently, the two curves drawn from the ARC equations probably represent a better depiction of the normal DE intake of growing pigs.

## Protein

### CHOICE OF SYSTEM AND UNITS

In arriving at requirements for protein and amino acids quite different approaches were taken by the ARC and NRC.

The ARC, in a manner similar to its methods for determining energy requirements, utilized both a factorial and an empirical approach. An important concept in the development of the ARC requirements is that of 'ideal protein'.

An ideal protein (or ideally balanced protein) is one that contains an ideal balance both among essential amino acids and between essential and non-essential amino acids. The ideal balance is the one that leads to the highest utilization of the dietary protein measured by biological value (BV) or related indices. The ARC working party (Fuller and Chamberlain, 1983) have further defined an ideal protein 'as one which cannot be improved by any substitution of a quantity of one amino acid for the same quantity of another'. The application of this concept is very helpful in the elimination of variation in estimates of protein requirements caused by differences in protein quality.

One aspect of the ideal protein concept was also utilized by the NRC, although it did not specifically mention the term ideal protein. The NRC assumed that requirements for essential amino acids remain a constant percentage of the protein requirement regardless of age or weight of the growing pig. On the basis of this assumption, the majority of amino acid requirements in the NRC's table were extrapolated from the requirements for pigs weighing 20–35 kg.

Different weight (age) divisions and different units for listing requirements were utilized by the ARC and NRC panels. Thus, direct comparisons are difficult. To standardize things as much as possible, requirements for three weight categories are compared: starting (5 to 15/20 kg), growing (15/20 to 50/60 kg) and finishing (50/60 to 90/100 kg) pigs. In doing so, it is assumed that the ARC age range of three to eight weeks is approximately equivalent to 5 to 15 kg. Within each weight category requirements are presented both in terms of g/MJ DE (the primary method used by the ARC) and also percentage of the diet (the primary method used by the NRC).

### REQUIREMENTS OF STARTING PIGS

The requirements of pigs from 5 kg (three weeks) to 15/20 kg (eight weeks) are presented in *Table 5.1*. The dietary energy concentrations listed by the NRC for its two weight ranges were used to convert NRC requirements to g/MJ DE. To convert ARC requirements to percentage of the diet an energy concentration of 14.37 MJ/kg was chosen since that represents the mean of the two energy concentrations used by the NRC.

The protein recommendations of the two organizations are quite similar. Both groups agree that young pigs of three weeks of age and weighing 5 to 10 kg have a crude protein requirement of about 14 g/MJ DE or 20% of the diet. The NRC lists a

**Table 5.1** PROTEIN AND AMINO ACID REQUIREMENTS OF STARTING PIGS

| | *(Unit)* | (g/MJ DE) | | | *% of diet*[a] | | |
|---|---|---|---|---|---|---|---|
| | *(Source)* | *ARC (1981)* | *NRC (1979)* | | *ARC (1981)* | *NRC (1979)* | |
| *Nutrient* | *(Age or weight)* | *3–8 weeks* | *5–10 kg* | *10–20 kg* | *3–8 weeks* | *5–10 kg* | *10–20 kg* |
| Digestible energy (MJ/kg) | | — | — | — | 14.37[b] | 14.64 | 14.10 |
| Crude protein | | 14.00 | 13.66 | 12.77 | 20.12 | 20.00 | 18.00 |
| Arginine | | —[c] | 0.17 | 0.16 | —[c] | 0.25 | 0.23 |
| Histidine | | 0.32 | 0.16 | 0.14 | 0.46 | 0.23 | 0.20 |
| Isoleucine | | 0.53 | 0.43 | 0.40 | 0.76 | 0.63 | 0.56 |
| Leucine | | 0.98 | 0.51 | 0.48 | 1.41 | 0.75 | 0.68 |
| Lysine | | 0.98 | 0.65 | 0.56 | 1.41 | 0.95 | 0.79 |
| Methionine + cystine | | 0.49 | 0.38 | 0.36 | 0.70 | 0.56 | 0.51 |
| Phenylalanine + tyrosine | | 0.94 | 0.60 | 0.56 | 1.35 | 0.88 | 0.79 |
| Threonine | | 0.59 | 0.38 | 0.36 | 0.85 | 0.56 | 0.51 |
| Tryptophan | | 0.14 | 0.10 | 0.09 | 0.20 | 0.15 | 0.13 |
| Valine | | 0.69 | 0.43 | 0.40 | 0.99 | 0.63 | 0.56 |

[a]Concentrations and energy contents are based on amounts per unit of air-dry diet (i.e., 90% dry matter).
[b]Arbitrarily chosen, represents the mean of the concentrations for the two NRC weight ranges.
[c]Not listed.

protein requirement of 18% for pigs weighing 10 to 20 kg. A division of protein and amino acid requirements of starting pigs into two weight categories, as followed by the NRC, seems desirable because requirements are thought to change rapidly during the early stages of growth (AEC, 1978).

The requirements of the ARC and the NRC for amino acids differ greatly. In some cases (e.g. histidine, leucine) the ARC requirements for three to eight week old pigs are about twice as high as the comparable NRC requirements for 5 to 10 kg pigs. Even for the amino acid with the least difference (isoleucine) the ARC requirement is about 23% higher than that of the NRC. Differences are even greater when ARC requirements are compared to NRC requirements for 10 to 20 kg pigs.

The contrast between the two sets of requirements is so great that one wonders how two groups of experts examining essentially the same world literature on pig nutrition could settle on such different amino acid requirements. The answer seems to rest with the fact that in both cases recommended amino acid requirements of starting pigs have not been obtained directly from experiments designed for that purpose, but have been calculated from other data. Calculation and extrapolation are unavoidable; there are no experimental data with starting pigs for four of the essential amino acids. Furthermore, for three of the six other essential amino acids data are extremely limited.

A brief explanation of the methods used by the two groups to derive requirements is necessary to understand their different recommendations. Because the method used by the NRC is easier to describe, their method will be considered first.

With one exception, all of the NRC's requirements for 5 to 10 kg pigs and 10 to 20 kg pigs were extrapolated from amino acid requirements of 20 to 35 kg pigs. The exception is the lysine requirement of 5 to 10 kg pigs which is 8% higher than the extrapolated value. The requirements for 20 to 35 kg pigs (determined largely from direct experiments) were simply increased in proportion to the higher protein requirements of younger pigs. Thus to calculate requirements of 5 to 10 kg pigs a multiplication factor of 20/16 was used, and a factor of 18/16 was used for 10 to 20 kg pigs. The value for lysine for 5 to 10 kg pigs was increased from 0.88 to 0.95% because the committee felt that 'determined requirements have been somewhat higher than those predicted by extrapolation'. Several publications since the NRC report (Lewis *et al.*, 1980; Zimmerman, 1980; Aherne and Nielsen, 1983) have indicated that the requirement is even higher than 0.95%.

The accuracy of the NRC's amino acid requirements for starting pigs thus depends on three factors.

(1) The accuracy of its amino acid requirements for 20 to 35 kg pigs.
(2) The accuracy of its estimates of protein requirements of 5 to 10, 10 to 20 and 20 to 30 kg pigs.
(3) The validity of the assumption that amino acid requirements are a constant percentage of protein regardless of the stage of development of the growing pig.

In general, the accuracy of these three factors seems to be reasonable. Amino acid requirements of 20 to 35 kg pigs will be covered in the next section. The NRC's estimates of protein requirements agree reasonably well with other estimates (Homb, 1976; AEC, 1978) and the concept of constant proportions between amino acids is probably valid for most amino acids. The validity in the case of lysine has been questioned (Lewis *et al.*, 1977). If the lysine requirement is not a constant proportion of the protein requirement, this may explain why the extrapolated value for lysine did not fit well with determined values.

There were several steps involved in the calculation of protein and amino acid requirements by the ARC. The first step was to estimate the crude protein requirement. Relevant data were examined and the requirement was estimated to be 14 g crude protein/MJ DE (approximately equivalent to 20% of the diet); in good agreement with the NRC. The second step was to examine responses to supplemental crystalline lysine by pigs that were initially three to four weeks old. It was found that there were improvements in weight gain and feed efficiency up to about 7 g lysine/100 g crude protein. Because this represented the maximum response to lysine in relationship to protein, this was considered to be the lysine content of ideal protein. The quantities of other essential amino acids in ideal protein were set on the basis of their proportions to lysine. The proportions used were primarily the proportions in pig tissue and in sows' milk.

To calculate amino acid requirements in terms of g/MJ DE the ARC seems to have taken the following steps (although the first step was not explicitly stated in the text):

(1) The crude protein requirement (14 g crude protein/MJ DE) was multiplied by the lysine 'requirement' (7 g lysine/100 g crude protein) to obtain the lysine requirement in g/MJ DE.

$$14 \times (7/100) = 0.98 \text{ g/MJ DE} \qquad (7)$$

(2) Other amino acid requirements were set in relationship to the lysine requirement by using the proportions proposed as ideal protein.

The multiplication step appears to have been inappropriate. To calculate the lysine requirement in relationship to dietary energy concentration the appropriate multiplication would have been either to multiply the crude protein requirement by the lysine content (% of protein) of diets typically used to determine crude protein requirements, or alternatively, to multiply the lysine 'requirement' (g/100 g protein) by the ideal protein requirement. A maize–soyabean meal diet containing 20% crude protein has approximately 5.6 g lysine/100 g crude protein; a barley –fishmeal diet with 20% protein has approximately 5.8 g lysine/100 g crude protein. Use of a multiplication factor of 5.7 rather than 7 g lysine/100 g crude protein would have resulted in a lysine requirement of 0.80 g/MJ DE or 1.15% of the diet; a value supported by a good deal of recent experimental data. Requirements for all other essential amino acids would also have been 18% lower, and consequently closer to those of the NRC.

## REQUIREMENTS OF GROWING PIGS

There are also substantial differences in the requirements of the NRC and the ARC for growing pigs (i.e. pigs from 15/20 to 50/60 kg). A summary of the recommendations is presented in *Table 5.2*. For growing pigs (and also for finishing pigs) the ARC's protein requirements are listed in terms of ideal protein, whereas those of the NRC are listed as crude protein, thus direct comparison is not possible. One

**Table 5.2** PROTEIN AND AMINO ACID REQUIREMENTS OF GROWING PIGS

| | *(Unit)* | (g/MJ DE) | | | *% of diet*[a] | | |
|---|---|---|---|---|---|---|---|
| | *(Source)* | *ARC (1981)* | *NRC (1979)* | | *ARC (1981)* | *NRC (1979)* | |
| *Nutrient* | *(Weight)* | *15–50 kg* | *20–35 kg* | *35–60 kg* | *15–50 kg* | *20–35 kg* | *35–60 kg* |
| Digestible energy (MJ/kg) | | — | — | — | 14.16[b] | 14.14 | 14.18 |
| Protein | | 12.00[c] | 11.32[d] | 9.87[d] | 16.99[c] | 16.00[d] | 14.0[d] |
| Arginine | | —[e] | 0.14 | 0.13 | —[e] | 0.20 | 0.18 |
| Histidine | | 0.28 | 0.13 | 0.11 | 0.40 | 0.18 | 0.16 |
| Isoleucine | | 0.46 | 0.35 | 0.31 | 0.65 | 0.50 | 0.44 |
| Leucine | | 0.84 | 0.42 | 0.37 | 1.19 | 0.60 | 0.52 |
| Lysine | | 0.84 | 0.50 | 0.43 | 1.19 | 0.70 | 0.61 |
| Methionine + cystine | | 0.42 | 0.32 | 0.28 | 0.59 | 0.45 | 0.40 |
| Phenylalanine + tyrosine | | 0.80 | 0.50 | 0.43 | 1.13 | 0.70 | 0.61 |
| Threonine | | 0.50 | 0.32 | 0.28 | 0.71 | 0.45 | 0.39 |
| Tryptophan | | 0.12 | 0.08 | 0.08 | 0.17 | 0.12 | 0.11 |
| Valine | | 0.59 | 0.35 | 0.31 | 0.84 | 0.50 | 0.44 |

[a]Concentrations and energy contents are based on amounts per unit of air-dry diet (i.e. 90% dry matter).
[b]Arbitrarily chosen, represents the mean of the concentrations for the two NRC weight ranges.
[c]Ideal protein.
[d]Crude protein.
[e]Not listed.

would expect requirements in terms of ideal protein to be lower than in terms of crude protein, but this is not the case. The ARC's protein requirements for pigs of 15 to 50 kg are higher than the NRC's for both 20 to 35 kg and 35 to 60 kg pigs. The amino acid requirements of the ARC are also considerably higher than those of the NRC; again being twice as high in some cases.

As in the case of starting pigs, at first it is difficult to reconcile the substantial differences between the two sets of recommendations.

In the NRC table the values for 20 to 35 kg pigs are, in a sense, the basis for most of the rest of the table. These are the values from which almost all others were derived. Although there is no direct explanation of the methods used to obtain these values, they were, apparently, estimated from a direct examination of the available literature listed in the bibliography. Values for pigs of 35 to 60 kg were calculated by decreasing those of 20 to 35 kg by 14/16 (the ratio of the two protein requirements).

The ARC used both factorial and empirical approaches, and the concept of ideal protein. The primary method by which requirements were derived was as follows. The response (especially feed efficiency) of growing pigs to various concentrations of ideal protein was examined. The panel concluded that 'there was a continuously diminishing response, with maximum efficiency achieved, in most cases, when the diet supplied 12 g of ideal protein/MJ DE and a lysine concentration of 0.84 g/MJ DE'. Once the ideal protein and lysine requirements were established, then requirements for the other essential amino acids were again set on the basis of their proportions in ideal protein.

In the method used by the ARC, the selection of the ideal protein requirement is crucial; all of the amino acid requirements are based on it. Because the ARC identified the data that it considered, and also illustrated the response curve, readers are able to make their own decisions about what the data indicate. Although in some cases there was a response to a level of 12 g of ideal protein/MJ DE (and even higher), when the mean response at each protein level is examined, there is little increase above 10 g of ideal protein/MJ DE. If this level (equivalent to 14.2% of the diet) had been selected the lysine requirement would have been 0.7 g/MJ DE (0.99% of the diet) and requirements for all other essential amino acids would have been 17% lower.

## REQUIREMENTS OF FINISHING PIGS

A summary of the recommended requirements for finishing pigs (50/60 to 90/100 kg) is presented in *Table 5.3*. As for growing pigs, protein requirements were listed in terms of ideal protein by the ARC and crude protein by the NRC. Amino acid requirements of the two groups are more similar for this stage of the growth period than for other stages.

The NRC's amino acid requirements were extrapolated from values for 20 to 35 kg pigs using a multiplication factor of 13/16. The only exception to this was that the methionine + cystine requirement (0.30%) was set lower than the extrapolated value (0.37%) because of experimental data indicating that 0.30% was adequate.

For finishing pigs the ARC used the same approach as for growing pigs except that it did not provide an illustration of the data used. Based on the response of gain:feed energy ratio to concentration of ideal protein, it was concluded that the requirement was 8.6 g of ideal protein/MJ DE (12.22% of the diet) and 0.60 g lysine

**Table 5.3** PROTEIN AND AMINO ACID REQUIREMENTS OF FINISHING PIGS

| | *(Unit)* | *(g/MJ DE)* | | *% of diet*[a] | |
|---|---|---|---|---|---|
| | *(Source)* | *ARC (1981)* | *NRC (1979)* | *ARC (1981)* | *NRC (1979)* |
| *Nutrient* | *(Weight)* | *50–90 kg* | *60–100 kg* | *50–90 kg* | *60–100 kg* |
| Digestible energy (MJ/kg) | | — | — | 14.21[b] | 14.21 |
| Protein | | 8.60[c] | 9.15[d] | 12.22[c] | 13.00[d] |
| Arginine | | —[e] | 0.11 | —[e] | 0.16 |
| Histidine | | 0.20 | 0.11 | 0.28 | 0.15 |
| Isoleucine | | 0.33 | 0.29 | 0.47 | 0.41 |
| Leucine | | 0.60 | 0.34 | 0.85 | 0.48 |
| Lysine | | 0.60 | 0.40 | 0.85 | 0.57 |
| Methionine + cystine | | 0.30 | 0.21 | 0.43 | 0.30 |
| Phenylalanine + tyrosine | | 0.58 | 0.40 | 0.82 | 0.57 |
| Threonine | | 0.36 | 0.26 | 0.51 | 0.37 |
| Tryptophan | | 0.09 | 0.07 | 0.13 | 0.10 |
| Valine | | 0.42 | 0.29 | 0.60 | 0.41 |

[a]Concentrations and energy contents are based on amounts per unit of air-dry diet (i.e., 90% dry matter).
[b]Arbitrarily chosen, the same as the NRC.
[c]Ideal protein.
[d]Crude protein.
[e]Not listed.

(0.85% of the diet). Other essential amino acid requirements were set on the basis of their proportion in ideal protein. On average the ARC's amino acid requirements for finishing pigs were 48% higher than those of the NRC.

#### OTHER ESTIMATES

The ARC also summarized empirical data (mostly for pigs in the weight range of 10 to 50 kg) for four essential amino acids. It estimated requirements (% of the diet) from these data to be: methionine + cystine, 0.47 to 0.61; tryptophan, 0.14 to 0.18; threonine, 0.56 to 0.60 and isoleucine, 0.50. It is interesting to compare these values with ARC values for 15 to 60 kg pigs and NRC values for 20 to 35 kg pigs (*Table 5.2*). The range for methionine + cystine essentially covers both the ARC and the NRC values, the range for tryptophan is similar to the ARC value, the range for threonine is intermediate between the ARC and NRC, and the value for isoleucine is similar to the NRC.

## Conclusions

Estimates by the ARC and the NRC of energy intakes (DE) of growing pigs are in good agreement. The equations provided by the ARC are more versatile than the values of intakes for given weight ranges provided by the NRC. The ARC equations permit estimation of DE intakes of growing pigs of any weight. Estimates of both groups were based on data derived largely from research facilities. How closely intakes of pigs in commercial farm facilities correspond to these estimates is not well known.

Recommendations for amino acids differ markedly between the ARC and the NRC, the ARC values being considerably higher at all stages. Because of inadequate empirical data, both groups relied heavily on calculation and extrapolation from other values to derive recommended requirements. The differences in recommendations reflect, primarily, the different approaches taken and assumptions made.

The types of pigs (genotypes) in the UK and USA are quite similar, as are the environments in which they are raised. Although more emphasis is placed on carcass merit in the UK than in the USA, and the predominant feedstuffs are somewhat different, it is not reasonable to believe that there are actually such enormous differences in amino acid requirements as the recommendations imply. There is obviously need for more data. Two areas in particular seem crucial:

(1) further refinement and testing of the concept and application of ideal protein, and
(2) more well-designed, adequately replicated, experiments with pigs of various weights to determine the response to various levels of amino acids likely to be limiting in practical diets (i.e. lysine, tryptophan and threonine).

## References

AEC (1978). *Animal Feeding: Energy, amino acids, vitamins, minerals.* Document No. 4. AEC; Commentry, France

AGRICULTURAL RESEARCH COUNCIL (1967). *The Nutrient Requirements of Farm Livestock No. 3: Pigs*. Agricultural Research Council; London

AGRICULTURAL RESEARCH COUNCIL (1981). *The Nutrient Requirements of Pigs*. Commonwealth Agricultural Bureaux; Slough

AHERNE, F.X. and NIELSEN, H.E. (1983). *Can. J. Anim. Sci.*, **63**, 221–224

ASPLUND, J.M. and HARRIS, L.E. (1969). *Feedstuffs*, **41**(14), 38–40

COLE, D.J.A., HARDY, B. and LEWIS, D. (1972). In *Pig Production*. p. 243 Ed. D.J.A. Cole. Butterworths; London

FULLER, M.F. and CHAMBERLAIN, A.G. (1983). In *Recent Advances in Animal Nutrition—1982*. pp. 175–186. Ed. W. Haresign. Butterworths; London

HOMB, T. (1976). In *Protein Metabolism and Nutrition*. pp. 383–394. Ed. D.J.A. Cole *et al*. Butterworths; London

LEWIS, A.J., PEO, E.R. Jr., CUNNINGHAM, P.J. and MOSER, B.D. (1977). *J. Nutr.*, **107**, 1369–1376

LEWIS, A.J., PEO, E.R. Jr., MOSER, B.D. and CRENSHAW, T.D. (1980). *J. Anim. Sci.*, **51**, 361–366

MAY, R.W. and BELL, J.M. (1971). *Can. J. Anim. Sci.*, **51**, 271–278

MORGAN, D.J., COLE, D.J.A. and LEWIS, D. (1975). *J. Agric. Sci., Camb.*, **84**, 7–17

NATIONAL RESEARCH COUNCIL (1944). *Recommended Nutrient Allowances for Domestic Animals, No. II: Recommended Nutrient Allowances for Swine*. National Research Council; Washington, DC

NATIONAL RESEARCH COUNCIL (1979). *Nutrient Requirements of Domestic Animals, No. 2: Nutrient Requirements of Swine*. Eighth Revised Edn. National Academy of Sciences–National Research Council; Washington, DC

ZIMMERMAN, D.R. (1980). *J. Anim. Sci.*, **55**(Supplement 1), 97

6

# AMINO ACID NUTRITION OF PIGS AND POULTRY

D. H. BAKER
*Department of Animal Sciences, University of Illinois, Urbana, Illinois, USA*

## Introduction

Over the past decade a transition has been made by animal nutritionists to the formulation of diets on an amino acid rather than on a protein basis. Formulation for amino acids is nonetheless anything but simple or straightforward. This review attempts to discuss many of the factors complicating studies of amino acid requirements and their utilization.

In the simplest terms, amino acid needs of the pig (or any other species), can be depicted in a simple flow diagram shown in Figure 6.1.

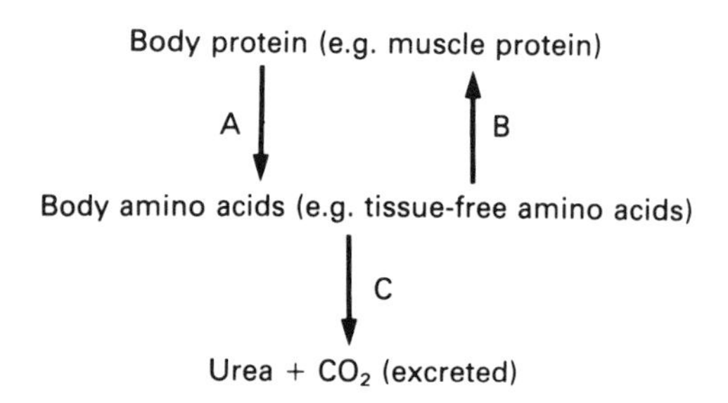


**Figure 6.1** Flow diagram of amino acid requirements where A = protein degradation, B = protein synthesis and C = amino acid oxidation

All of the body processes (A, B and C) in Figure 6.1 go on continuously. It is important to note that anywhere from 60% (young animal) to 80% (adult animal) of the amino acid needs for body protein synthesis come from endogenous protein degradation. Hence, the remaining 20–40% must be supplied in the diet. Because individual amino acids have different turnover rates (lysine slow, methionine high) and are therefore depleted via amino acid oxidation from tissue pools at different rates, one cannot assume that the amino acid composition of muscle tissue is predictive of the dietary amino acid requirement pattern. It is for this reason that requirements for individual amino acids must be determined by experiment.

**Table 6.1** RESPONSE OF YOUNG CHICKS TO DIETARY LYSINE[a]

| *Lysine*[b] *level* (%) | *Weight gain* (g) | *Feed intake* (g) | *Lysine intake* (mg) | *Feed conversion ratio*[d] | *Lysine conversion ratio*[e] |
|---|---|---|---|---|---|
| 0 | −2.43 | 39.8 | 0 | — | — |
| 0.1 | 0.13 | 45.7 | 46 | 352 | 354 |
| 0.2 | 4.56 | 50.0 | 100 | 10.96 | 21.9 |
| 0.3 | 16.50 | 69.6 | 209 | 4.22 | 12.7 |
| 0.4 | 26.60 | 82.4 | 330 | 3.10 | 12.4 |
| 0.5 | 47.20 | 111.1 | 543 | 2.35 | 11.5 |
| 0.6 | 68.60 | 132.6 | 797 | 1.93 | 11.6 |
| 0.7 | 97.60 | 164.6 | 1147 | 1.69 | 11.7 |
| 0.8 | 118.00 | 182.0 | 1456 | 1.54 | 12.3 |
| 0.9 | 118.90 | 174.8 | 1574 | 1.47 | 13.2 |
| 1.0 | 116.50 | 169.5 | 1702 | 1.45 | 14.6 |
| Pooled SEM | 4.92 | 7.2 | NA[c] | NA[c] | NA[c] |

[a] Data (Baker, 1978) represent means of triplicate groups of five male chicks per treatment during the period 8–16 days post-hatching; birds averaged 82 g at day 8 post-hatching
[b] Lysine provided as L-lysine.HCl (80%) lysine); the basal diet contained 3.8 Mcal ME (15.9 MJ) kg and 15% crude protein
[c] NA = not available due to heterogeneous variance
[d] feed/g liveweight gain
[e] mg lysine intake/g liveweight gain

Before addressing problems of assessing amino acid requirements, some comments should be made about response curves that result from incremental amino acid addition to a diet designed to be specifically deficient in a single amino acid.

## Response curves

Few studies have been carried out with pigs in which the entire growth curve has been covered in an amino acid supplementation study. Thus, the points to be made will be based upon an actual lysine experiment conducted with young growing chicks, and it is likely that pigs would respond similarly. Data in Table 6.1 show a lysine response curve. The purified crystalline amino acid diet (Baker, Robbins and Buck, 1979) to which lysine was added contained 3.8 Mcal ME/kg (15.9 MJ/kg) and 15% crude protein (N × 6.25).

A plot of the gain and feed intake data shows that a sigmoidal response occurs, incremental responses per unit of lysine intake being less in both the lower and higher portions of the curve than in the linear response area (between 0.5 and 0.7% lysine). The chicks used in this experiment were very homogeneous. In practice, animals for which requirements are sought are far more heterogeneous. The slope of the line below 0.5% lysine is lower than that between 0.5 and 0.7% lysine because in this region of the growth curve the lysine required for maintenance (zero gain requires about 45 mg lysine) is exerting a substantial effect on the overall efficiency of lysine utilization. At and above 0.7% lysine, some of the animals in the population are beginning to have their lysine requirement satisfied. The closer the lysine level gets to 0.8%, the greater will be the portion of the population that has been satiated insofar as lysine intake is concerned. Because of these phenomena, slope (gain per unit of lysine) should (and does) decrease at an increasing rate until the animal in the population requiring the most lysine has been satiated, at which point no further gain response will occur. This type of response is seen with all amino acids, as has been illustrated and discussed with histidine (Baker, 1986).

While growth of chicks is maximized at 0.80% dietary lysine, higher levels result in the same weight gain but at lower levels of dietary intake under *ad libitum* feeding conditions. Thus, a higher lysine level (0.90%) is required to maximize feed efficiency than is required to maximize weight gain. The same phenomenon occurs with pigs (Brown, Harmon and Jensen, 1973; Lin and Jensen, 1985), optimum feed conversion ratio or maximum carcass leanness requiring higher levels of a limiting amino acid (or crude protein) than that required for maximum weight gain. Because gilts and boars are leaner than barrows, the former have higher amino acid and crude protein requirements than the latter (Baker, 1986; Baker *et al.*, 1967). Also, exogenous protein anabolic substances such as Ractopamine (oral) or porcine growth hormone (injected) have a marked effect in increasing protein and amino acid requirements of pigs (Anderson *et al.*, 1987; Easter 1987).

## Factors complicating accurate assessment of amino acid requirements

Because amino acid requirements are not easy to determine with accuracy and precision and because a multitude of factors may influence the 'requirements', it has become popular to calculate amino acid requirements using mathematical models instead of actually determining the requirements by experiment. Factors underlying the modelling of amino acid requirements have been reviewed in a recent paper by Black *et al.* (1986). Certainly, calculated amino acid requirements are interesting, but the projected requirement estimates can only be as good, or as accurate, as the validity of the assumed factors going into construction of the model. Many of the assumptions themselves are in need of verification and further study. Some of the factors known to affect amino acid requirements of pigs and poultry are discussed below.

### CRITERIA OF RESPONSE

Requirements for amino acids are generally best defined in growing animals by growth data in *ad libitum* feeding studies. While pair or equal feeding circumvents several interpretive problems, it creates other problems (Baker, 1984). Thus, part of the growth response to an essential amino acid is due to its favourable effect on metabolism; but another part of its efficacy resides in its effect on food intake.

### STATISTICAL CONSIDERATIONS

With growth data, both weight gain over a defined period of time and efficiency of diet or nutrient utilization are the usual criteria of response. Requirement and bioavailability studies necessitate feeding graded levels of the nutrient in question. Statistical calculation is easier in these studies if increments are equally spaced. As a general rule there is only minimum utility in applying any kind of a range or paired-comparison test to these data (Peterson, 1977; Little, 1978). Indeed, such tests often lead to misinterpretation of the results. With bioavailability assays, the linear response area (i.e. constant utilization of the nutrient in question) is defined by a minimum of three levels of the independent variable (i.e. the nutrient)*. The significance of a difference between any two adjacent points is essentially meaningless. Instead the linearity of the response should be established together with the 'fit' ($r^2$ value) such that the determined slope of the response line has veracity. With requirement studies, a minimum of four levels is required (preferably six or more) such that the data can be fitted to a descriptive response curve (e.g. sigmoidal, asymptotic or broken line), thus facilitating objective assessment of the requirement (Robbins, Norton and Baker, 1979). It is improper procedure to use paired-comparison tests of adjacent points in requirement studies. Thus, some investigators select a requirement on the basis of a maximum level that no longer is statistically different from the level immediately below it. This is not a defensible procedure from a statistical standpoint. Moreover, all levels yielding growth performance data 'appearing' to reside on the plateau area of the growth curve must be evaluated carefully in order to arrive properly at what should be considered the true maximum response.

Aberrant results often occur in the upper curvilinear area of a growth curve where animal-to-animal variability is generally greatest. It is thus not uncommon to find a rat gaining 4 g/day at (ascending) level 3, 5 g/day at level 4, 6 g/day at level 5 and 5 g/day at level 6. Even though the gain at level 5 may be statistically better than the gain at level 4, this should not be taken as the requirement level of that nutrient. Because gain at level 6 was less than that at level 5, one must question whether level 4 might be a more proper selection as the minimum effective dose for maximum response. While curve fitting can be of great assistance in taking the subjectivity out of decisions such as these, a repeat of the experiment with additional, and possibly different, levels would seem called for.

Expressions of efficiency present more complicated problems because they involve both growth and consumption data. Unlike rodent studies where the animals are generally fed individually, studies with other species usually involve group feeding. While individual animal data are available for weight gain calculations, only group data are available for diet or nutrient intake calculations. This, of course, presents special problems when an animal dies or must be removed from a given pen for other reasons, since consumption by the animal removed cannot be determined accurately. Proper procedure for group feeding experiments dictates that both weight gain and diet efficiency data be based upon pen means. Hence, all statistical calculations should be made on a pen-means basis, with the total degrees-of-freedom equal to one less than the total number of pens.

## EXPRESSIONS OF NUTRIENT REQUIREMENTS

Requirements are generally defined for animals of a given age and for a specific physiological function (i.e. maintenance, growth, reproduction or production).

---

*Requirement studies can legitimately use growth as the dependent variable and concentration of the nutrient as the independent variable, but slope-ratio bioavailability studies should use growth rate regressed on absolute intake of the nutrient. The latter should be done in the constant slope region (i.e. generally between 30 and 70% of the requirement) of the growth curve.

Because amino acid requirements are most useful if defined for groups or populations, requirement expressions are generally based upon dietary concentration rather than absolute intake. Hence, a pen of 100 newly hatched broiler chicks may vary by 10–20% in body weight (e.g. from 100 to 120 g). The heavier chicks obviously require greater quantities of each essential amino acid than the lighter chicks, but the available evidence suggests that requirements expressed as a percentage of diet, percent of metabolizable energy (ME) or mg/kg do not vary greatly among heavy and light animals of the same age. The heavier animal merely meets its requirements by consuming more diet.

One opinion is that ‘concentration’ requirements are best expressed as g or mg/kJ ME. This is based upon the long-standing nutritional principle that animals will eat to meet their energy needs. While this principle is generally true in practice, there are many cases, some practical, where another factor overrides an animal’s tendency to eat to a given energy need. To use just one example of practical relevance, (Parsons, Edmonds and Baker, 1984; Edmonds, Parsons and Baker, 1985), broiler chicks increase rather than decrease their voluntary feed intake when dietary protein level in a corn-soyabean meal diet is reduced in decrements from 24% (normal level) to 16% by replacing soyabean meal (10.2 MJ ME/kg) with corn (14.1 MJ ME/kg). Thus, despite a considerable increase in energy density, the low protein diet is actually consumed in greater quantity than the high-protein diet. In effect, therefore, the birds appear to be trying to eat to meet their protein-amino acid needs rather than their energy needs. In so doing, they overeat energy in relation to ‘effective’ protein and deposit more body fat (Edmonds, Parsons and Baker, 1985). Low-protein or minimum soyabean meal diets have become a practical reality in the USA, but also in other countries whose growing seasons and climatic conditions are not amenable to production of high-lysine oilseeds such as soyabeans. The availability of economically priced methionine, lysine, threonine and tryptophan have served to encourage the use of amino acid fortified low-protein diets (Baker, 1985; Baker and Parsons, 1985). There are, therefore, instances in which expressing dietary requirements per unit of ME can result in over- or underformulation of a given diet.

## BODY COMPOSITIONAL FACTORS

Requirements for amino acids in growing animals, expressed in terms of dietary concentration, decrease as age and weight of the animal increase (NRC, 1979; Boomgaardt and Baker, 1973a). This occurs because body composition changes (e.g. more fat and less protein in the weight gain) as a growing animal matures. Whether protein percentage in the weight gain decreases in a straight line from weaning to market weight is questionable, although calculations of Whittemore (1987) assume a linear decrease does occur. For amino acids, weight gain generally correlates well with nitrogen retention in young, rapidly growing animals. After secondary sex characteristics have developed, however, body composition factors come into play such that a higher requirement is often predicted for maximal nitrogen retention than for maximum weight gain. Also, by the same reasoning, use of criteria such as maximum carcass leanness or optimum feed efficiency results in a higher requirement than that predicted for maximum weight gain (Baker, 1977; Baker *et al.*, 1967; Smith, Clawson and Barrick, 1967).

In species like the pig, the sex of the animal is a very important consideration. Thus, after secondary sex characteristics have developed, gilts, and boars, because they deposit more lean in relation to fat, exhibit higher protein and amino acid requirements than is the case for castrated males. By the same token, pigs bred for leanness require higher concentrations of amino acids in their diets than those not similarly possessing a high lean:fat ratio. Hence, should the use of materials such as recombinant porcine growth hormone become a practical reality in pig production (Boyd *et al.,* 1985; Etherton and Kensinger, 1984; Easter, 1987) then it is likely dietary requirements for amino acids will increase.

## FOOD COMPOSITION AND FOOD INTAKE FACTORS

Dietary protein level affects requirements for essential amino acids (Baker, 1977; Baker, Katz and Easter, 1975; Boomgaardt and Baker, 1973b). Thus, as protein level decreases, as would be the case with a minimum-soyabean meal, amino acid-fortified diet, food intake generally increases as the animal attempts to meet its protein or amino acid needs. Just as important, however, is that with low-protein diets fortified with a limiting amino acid in pure form, a greater portion of the total concentration of the limiting amino acid is bioavailable, since pure sources of amino acids are more bioavailable than intact-protein sources of the same amino acid.

The contribution of nutrients from protein- or energy-yielding ingredients relative to those from pure sources presents subtle and perplexing problems in diet formulation and nutrient requirement assessment. Most nutrients present in feed ingredients are bound in one form or another (e.g. amino acids in polymeric peptide linkage). Ideally, amino acids should be characterized in basal diets as to their bioavailability before a requirement study is undertaken.

## FREQUENCY OF FEEDING

The conditions of feeding studies can create bioavailability problems that contribute to confusion in amino acid requirement assessment. Dog, cat and adult pig feeding studies can be used as an example here. These animals are generally considered 'meal eating', as opposed to 'nibbling', species. As such, many requirement studies with them are conducted using a one-meal-per-day feeding regimen. Can it be assumed, however, that pure sources of the nutrient in question will be absorbed at the same time and rate as bound sources of the same nutrient? It probably cannot and pure lysine, for example, is absorbed much faster than intact-protein sources of lysine that require time in the gut for proteolysis to occur (Batterham, 1984; Baker, 1984; Baker and Izquierdo, 1985). As a result, pure lysine, generally assumed to be 100% bioavailable, may become only 50% bioavailable in a once-per-day feeding regimen (Batterham, 1984). Hence, much of the crystalline lysine may be wasted if it is absorbed from the gut 1–2 h before the rest of the intact-protein bound amino acids are similarly made available at the proper sites of tissue protein synthesis. Moreover, it cannot be assumed that the same magnitude of unavailability existing with lysine in a once-per-day feeding regimen will also occur with other essential amino acids. Thus, lysine is conserved in tissue pools during a deficit and its turnover rate is slow compared with that of, say, threonine and methionine. As a result, the magnitude of methionine or threonine wastage may exceed that of lysine wastage when pure sources of each are used to supplement intact protein diets in a once-per-day feeding regimen.

Selection of a proper feeding regimen for amino acid requirement studies is important. The researcher is presented with a dilemma, however, in that to be of greatest value, requirement studies should be carried out using feeding regimens typifying those used in practice. Yet, using these feeding regimens (e.g. one meal per day) may present problems in interpreting the results because, generally speaking, amino acid requirement assays, of necessity, require supplementing a deficient semi-purified diet (containing the deficient amino acid in bound form) with a free form of the amino acid in question. Yet, in practice, the nutrient in question may be provided in a grain-soyabean meal diet totally in the bound form. If such is likely to be the case, a perspicacious research scientist will probably decide that a two- or three-times-per-day feeding regimen is preferable to a once-per-day feeding for the requirement bioassay.

## PRECURSOR MATERIALS

With several amino acids, precursor materials may be present in the diet used to assess a requirement. Classic examples are (precursor/product) methionine/cysteine, phenylalanine/tyrosine, cysteine/taurine, glutathione/cysteine, and carnosine/histidine (Czarnecki, Halpin and Baker, 1983).

Errors are frequently made in assessing the maximum portion of the total sulphur amino acid requirement that can be furnished by cysteine. Knowledge of this exact value is important in least-cost diet formulation for several species, but particularly for poultry. A series of three experiments is necessary to quantify accurately the percentage of the sulphur amino acid requirement met by cysteine (Table 6.2). Unfortunately, many investigators have merely done assays 1 and 2 without doing assay 3. They have concluded, therefore, that 55% (w/w) of the sulphur amino acid requirement can be furnished by cysteine instead of the actual value of 50%. Moreover, some have even suggested, based upon a two-assay approach, that the sulphur amino acid requirement is 0.60% of the diet (methionine alone) rather than the correct requirement of 0.54% (methionine + cysteine, 1:1). Thus, because the molecular weight of cysteine is only about 80% that of methionine, the total sulphur amino acid requirement (i.e. methionine + cysteine) is lower with what would be considered a proper combination of methionine + cysteine than with

**Table 6.2** EXPERIMENTS REQUIRED TO QUANTIFY THE MAXIMUM PERCENTAGE OF THE CHICK'S SULPHUR AMINO ACID REQUIREMENT MET BY CYSTEINE AND THE EFFICIENCY OF METHIONINE AS A CYSTEINE PRECURSOR – AN EXAMPLE

| *Assay number and order* | *Requirement study* | *Dietary condition* | *Resulting requirement (% of diet)* |
|---|---|---|---|
| 1 | Methionine | 0 cysteine | 0.60 |
| 2 | Methionine | Excess cysteine | 0.27 |
| 3 | Cysteine | 0.27% methionine | 0.27 |

methionine alone furnishing the sulphur amino acids (Graber and Baker, 1971; Halpin and Baker, 1984). Failure to consider carefully precursor:product interrelationships and their implications can also result in improper assessment of the bioefficacy of analogue materials that may be metabolized into one or more useful products (e.g. methionine hydroxy analogue serving as a precursor of methionine, cysteine, or both (Baker, 1976; Boebel and Baker, 1982).

### MATHEMATICAL AND JUDGEMENTAL CONSIDERATIONS

Requirement bioassays should be designed in such a way as to allow objective estimation of desired maxima, e.g. maximum weight gain (Robbins *et al.*, 1977; 1979) or minima, e.g. minimum plasma urea concentration (Lewis *et al.*, 1977). Some investigators have assumed that an objective requirement can be obtained by calculating a linear regression line through the selected points or levels appearing to reside in the linear response surface, then determining the intercept of this line with a horizontal (zero slope) line calculated from points appearing to reside in the plateau region of the response curve. This procedure is unacceptably marred by subjectivity in that it involves some form of selection of the points or levels to use in the linear and plateau regression lines. A continuous broken line calculated by least squares, is the preferred method because it objectively selects the breakpoint in the response line (Robbins, Baker and Norton, 1977; Robbins, 1986).

Broken lines, even if properly calculated, may not adequately describe the response. In previous work from our laboratory with histidine, for example, involving 10 increments of L-histidine ranging from 0 to 0.57% of the diet, the response was best described by a sigmoidal fit (Robbins, Norton and Baker, 1979). The problem with fitted curvilinear response lines (e.g. sigmoid or asymptotes) is that breakpoints (i.e. an abrupt change in slope, representing a requirement) are not a component. Only if an arbitrary point is selected, such as 95% of the upper asymptotic value, can one allow the computer rather than the investigator to select an objective requirement value.

Bias can and should be taken out of the requirement-selection process. It is thus not defensible to construct two independent straight lines by selecting specific points for the ascending linear line and the zero-slope horizontal line. The subjective selection process can and generally does have a material effect on where the two lines intersect (i.e. the assumed requirement) which can lead to unintentional, or intentional, bias.

## Extrapolating literature amino acid requirements to practice

Assuming a proper objective method is used to assess levels of an amino acid that result in optimum measures of animal performance (e.g. growth, feed efficiency, carcass yield), and assuming the animals used in the experiments are representative of the defined population as a whole (e.g. UK meat-type pigs between 20 and 50 kg body weight), how can one translate literature requirement estimates to the real world (i.e. on-farm conditions)? The author has found that including the following as factors often leads to an accurate translation of a requirement determined experimentally to one that would apply to on-farm conditions: (1) dietary metabolizable energy level, (2) dietary protein level and (3) bioavailability.

It is known that both dietary protein level and energy density of the diet exert effects on amino acid requirements expressed in terms of dietary concentration, i.e. % of diet (Mitchell *et al.*, 1965; Boomgaardt and Baker, 1973b, c; Baker, Katz and Easter, 1975). Research in this laboratory indicates, however, that the relationships are not directly proportional, i.e. doubling energy level does not double the requirement, nor does a doubling of the crude protein level double the requirement for an essential amino acid. Thus, while the directional nature is apparent, the quantitative relationships have not been clearly delineated nor have associative effects, either positive or negative, between energy level (and source) and protein level (and source) been defined. Sometimes animals appear to eat to meet energy needs, but sometimes they appear to eat to meet amino acid needs.

Some have suggested that amino acid requirements are affected by environmental temperature extremes. Although there is some logic in this reasoning, there also is a basic fallacy in the relationship. Pigs or chickens do eat less feed in a warm environment (e.g. 35°C) than in a cold environment (e.g. −10°C). In so doing they eat less of the limiting amino acid. Does adding more of the first-limiting amino acid (e.g. methionine for poultry, lysine for pigs) result in increased response? The author's experience would suggest that an increase in gain will occur only if addition of the first-limiting amino acid stimulates voluntary feed intake. Thus, given the almost perfect relationship between feed intake and weight gain (Allen *et al.*, 1972), how can more of the limiting amino acid bring about more weight gain unless more of the remaining growth-requiring raw materials (especially energy) are not present as well? In reality, a 20% reduction in voluntary feed intake due to heat stress may require only a 10% increase in the dietary concentration of the limiting amino acid in order to obtain an approximate 5% increase in weight gain of the heat-stressed animals. Certainly, the 20% reduction in feed intake will not necessitate a 20% increase in the need for the limiting amino acid. The best that can be hoped for is that increasing the limiting amino acid by 10% may increase feed intake by 10% and weight gain by 5%.

Most purified diets used to establish amino acid requirements are lower in crude protein but higher in metabolizable energy than conventional practical diets. Thus, an adjustment upwards needs to be made to translate the lower crude protein purified diet work to the higher crude protein practical diet basis. An adjustment downwards needs to be made, on the other hand, to correct for the higher metabolizable energy level present in the purified diet. In the author's experience, whatever errors are present in the linear relationship between crude protein level and the amino acid requirement, and between metabolizable energy level and the same requirement, these tend to cancel out when both factors are included in the adjustment equation together.

An additional factor must be considered before a proposed correction equation can be established. Purified diets frequently contain amino acids that are 100% available while practical diets do not. Thus, an average amino acid bioavailability factor of 85% will be used for the correction equation that follows:

$$R_p = \frac{R_e}{BV_p} \times \frac{CP_p}{CP_e} \times \frac{ME_p}{ME_e} \qquad (6.1)$$

where: $R_p$ = Amino acid requirement in practical diet (%).
$R_e$ = Requirement determined with experimental diet (%).
$BV_p$ = Bioavailability in practical diet relative to that in purified experimental diet (%).

$CP_p$ = Crude protein level in practical diet (%).
$CP_e$ = Crude protein level in experimental diet (%).
$ME_p$ = Metabolizable energy level in practical diet (MJ/kg).
$ME_e$ = Metabolizable energy level in experimental diet (MJ/kg).

This equation almost exactly translates broiler-chick requirements established for sulphur amino acids and lysine using the Illinois amino acid diet (15% CP, 15.9 MJ ME/kg) to the practical setting (i.e. NRC = 23% CP, 13.4 MJ ME/kg). Thus, for sulphur amino acids where the purified diet requirement is 0.60%:

$$\frac{0.60}{0.85} \times \frac{23}{15} \times \frac{13.4}{15.9} = 0.91\% = R_p \qquad (6.2)$$

The estimate of 0.91% sulphur amino acids is very close to the NRC estimate of 0.93% sulphur amino acids for broiler chicks fed a 23% crude protein corn-soyabean meal diet containing 13.4 MJ ME/kg.

For lysine, the purified diet requirement estimate is 0.80% (cf. Table 6.1). Using the same translation formula, a practical diet requirement of 1.21% lysine is predicted, very close to the NRC estimate of 1.20%. Work in our laboratory with arginine suggests that this amino acid, too, fits well into the practical-diet prediction equation.

For young chicks and poults, sulphur amino acids, lysine and arginine are the first-, second- and third-limiting amino acids in lower protein corn–soyabean meal diets. Thus, the aforementioned prediction equation may provide a useful means of interpreting purified diet requirements for these amino acids in terms of relating them to the practical setting.

In lieu of resorting to some form of prediction equation, one could attempt to convince all pig nutritionists doing requirement work that amino acid requirement experiments should consist of the following features (Izquierdo, Wedekind and Baker, 1988).

(1) A defined age and weight range of pigs should be covered in the experiments.
(2) The pigs should be identified as to sex and meat characteristics.
(3) The experimental diets employed should contain both protein and metabolizable energy levels consistent with those used in on-farm commercial pig production.
(4) The experimental diet should be frankly and singly deficient in the amino acid under investigation so that a substantial growth response can be demonstrated.
(5) The diet when adequately fortified with the amino acid under study should promote near optimum (maximum?) growth in *ad libitum*-fed pigs.
(6) Both the linear and plateau portions of the growth curve should be covered in the amino acid levels employed in the growth trials.
(7) Both *total* and *bioavailable* levels of the limiting amino acids in the basal experimental diet must be known.

It would indeed be helpful if pig experiments involved many amino acid levels (i.e. six or more) such that an objective curve-fitting procedure could be used to identify the dose that yields optimum (most profitable?) response. Because the purified and semipurified diets necessarily required for pig amino acid requirement studies are often very costly, most investigators compromise on the 'ideal' insofar as dosage levels are concerned.

AN EXAMPLE

An attempt has recently been made at Illinois to establish a meaningful histidine requirement for pigs in the weight range of 10–20 kg (Izquierdo, Wedekind and Baker, 1988). A somewhat bizarre and far-from-practical basal diet (based upon corn, dried whey and feather meal) had to be used in order to achieve a diet that was deficient in histidine. After proving experimentally that 10-kg pigs would grow well on this diet when it contained adequate histidine, analytical and bioassay studies were carried out to establish both the total (0.22% of diet) and bioavailable (0.19% of diet) histidine concentration in the basal experimental diet. The next step involved experiments where graded levels of histidine (covering both deficiency and adequacy) were fed. Because the basal diet was costly (it contained crystalline methionine, tryptophan and lysine in addition to corn starch and the aforementioned protein-supplying ingredients), only four carefully chosen doses were employed in the first experiment (Table 6.3). Moreover, based upon the results of experiment 1, it was reasoned that only three doses would be sufficient in experiment 2. Thus, experiment 2 was really needed to clear up confusion associated with experiment 1. A brief summary of experiments 1 and 2 is presented in Table 6.3.

If experiment 1 had been the only requirement study conducted, we would have been forced to conclude that the histidine requirement was greater than 0.25% but not greater than 0.31%. Thus, since the magnitude of the gain and gain:feed response between 0.19% and 0.25% bioavailable histidine was substantially greater than that between 0.25% and 0.31, one could suggest that maximal growth might

**Table 6.3** GROWTH PERFORMANCE OF PIGS FED GRADED LEVELS OF HISTIDINE[a]

| *Diet*[b] | *Bioavailable histidine* (%) | *Daily weight gain* (g) | *Food conversion ratio*[e] |
|---|---|---|---|
| *Exp. 1* | | | |
| 1. Basal diet | 0.19 | 188 | 2.46 |
| 2. As 1 + 0.06% L-histidine | 0.25 | 383 | 1.80 |
| 3. As 1 + 0.12% L-histidine | 0.31 | 467 | 1.59 |
| 4. As 1 + 0.18% L-histidine | 0.37 | 438 | 1.68 |
| Pooled SEM | | 22[c] | 0.08[c] |
| *Exp. 2* | | | |
| 2. Basal + 0.06% L-histidine | 0.25 | 254 | 2.21 |
| 2. As 1 + 0.03% L-histidine | 0.28 | 322 | 2.02 |
| 3. As 1 + 0.06% L-histidine | 0.31 | 458 | 1.69 |
| Pooled SEM | | 29[d] | 0.08[d] |

[a] Data (Izquierdo, Wedekind and Baker, 1988) represent means of six individually fed pigs during the period 5–8 weeks of age; average initial weight was 10 kg
[b] The basal histidine-deficient diet contained 20% feather meal, 20% corn and 10% dried whey; it was also fortified with methionine, lysine and tryptophan. Supplemental L-histidine was provided as L-histidine·HCl·$H_2O$ (74% histidine)
[c] Histidine linear and quadratic effect ($P<0.01$)
[d] Histidine linear effect ($P<0.01$)
[e] kg feed/kg liveweight gain

have occurred somewhere between 0.25% and 0.31%. Experiment 2 was conducted, therefore, to establish whether this was the case. In this trial involving narrower increments, the gain and feed efficiency responses between 0.28% and 0.31% bioavailable histidine were as great or greater than those occurring between 0.25% and 0.28%. Hence, when the two histidine requirement trials are viewed as a whole, 0.31% bioavailable histidine emerges as a defendable requirement estimate.

Because pig amino acid requirements are based upon those levels needed for optimum rate and efficiency of weight gain of pigs consuming grain-soyabean meal type diets, extrapolation of the bioavailable histidine requirement to a grain-soyabean meal basis is necessary. If ileal histidine digestibility data are available for dietary ingredients used in the grain-soyabean meal diet, these should be used to adjust the requirement estimate. If unavailable, an educated bioavailability estimate should be made, e.g. 85%. Using 85%, one then can correct the bioavailable histidine requirement estimate to a more meaningful value of 0.36% (0.31% ÷ 0.85 = 0.36%).

## Factors complicating assessment of amino acid limitations

There are pitfalls in chemical scoring as well as biological evaluative methods of determining limiting amino acids in feedstuffs or diets. Quite obviously, chemical scoring methods such as 'chemical score' (whole egg standard) or amino acid indexes do not take account of either palatability or bioavailability. As such, methods like these often predict wrongly the order of limiting amino acids in a feedstuff or diet (Harper, 1959; Boomgaardt and Baker, 1972). Less obvious are the pitfalls inherent in using animal tests to assess amino acid limitations. Because of the increase in commercial availability of synthetic amino acids, determination of limiting indispensable amino acids in low-protein grain-soyabean meal diets is becoming very important to the poultry and pig industries. The two most common methods used to assess limiting amino acids in diets are amino acid addition and amino acid deletion assays. Amino acid addition studies involve adding amino acids individually and in combination to a low-protein or amino acid-deficient diet. In amino acid deletion studies, individual amino acids are deleted sequentially from a diet containing a full complement of supplemented amino acids.

Amino acid addition trials, though less efficient in experimental design, are probably more relevant to practical pig and poultry production (Edmonds, Parsons and Baker, 1985). Indeed, sequential supplementation studies have led to the conclusion that the three most limiting amino acids in reduced protein (i.e. lower soyabean meal) corn-soya diets, in order of limitation, are (1) lysine, (2) tryptophan and (3) threonine for growing pigs (Russell, Cromwell and Stahly, 1983; Russell *et al.*, 1986) but are (1) methionine, and (2) lysine and arginine (equally second-limiting) for young chicks (Edmonds, Parsons and Baker, 1985). Using barley/wheat, or sorghum, in place of most of the corn in grain-soya diets for growing pigs results in lysine being first-limiting, threonine being second-limiting and tryptophan or methionine being third-limiting.

Amino acid deletion experiments conducted with chicks have yielded conclusions different from amino acid addition experiments as to the order of limiting aamino acids in low-protein corn-soya diets (Edmonds, Parsons and Baker, 1985). These authors added eight essential amino acids plus glutamic acid to a 16% protein corn-soya diet fed to 1-week-old chicks (23% CP corn-soya diets are normally fed to chicks of this age). Deletion of methionine from the amino acid supplement produced the same growth depression as that obtained from deletion of lysine. Thus, that these two amino acid deletions produced the greatest growth depressions (which were equal) led to the conclusion that these two amino acids were equally first-limiting. Three other amino acids (arginine, valine and threonine), upon individual deletion from the complete amino acid supplement, also caused growth depressions that were approximately equal but of lesser magnitude than those resulting from either methionine or lysine deletion. Hence, the conclusion was that arginine, valine and threonine were equally third-limiting – after methionine and lysine.

It is clear when examining the order of amino acid limitation resulting from 'addition' vs 'deletion' experiments in the work of Edmonds *et al.* (1985) that different conclusions resulted from the two methods. Deletion experiments have a far more efficient experimental design. The key question asks if this approach accurately predicts the situation as it would occur in commercial production? The answer is, that it probably does not. Deletion experiments potentially impose amino acid imbalances on amino acid deficiencies. While the amino acid excesses over and above each deficiency (i.e. each individual amino acid deletion) are not likely to cause metabolic inefficiency, they may, particularly with certain amino acid deficiencies, reduce voluntary food intake and therefore growth rate. Thus, addition experiments, because they more nearly simulate the practical setting, should be considered the more appropriate method of determining the order of limiting amino acids in a feedstuff or diet.

## Amino acid excesses

The scientific literature contains many articles on effects of excesses of individual amino acids, or amino acid mixtures, on performance of rats and chicks. Most of the studies, however, have used semi-purified or purified diets. Only recently has information been generated using pigs and poultry where individual amino acid excesses have been evaluated in animals fed practical grain-soya diets (Southern and Baker, 1982; Hagemeier, Libal and Wahlstrom, 1983; Anderson *et al.*, 1984a,b; Rosell and Zimmerman, 1984a,b; Edmonds and Baker, 1987a,b,c; Edmonds, Gonyou and Baker, 1987).

A brief summary of findings in experiments where excess amino acids were added to practical diets for pigs and poultry is:

(1) A 1% excess addition of any single essential amino acid depresses neither growth nor feed efficiency in pigs or chicks. Methionine is the only essential amino acid that depresses weight gain in pigs at a supplemental level of 2%.
(2) At a supplemental level of 4%, leucine, isoleucine and valine are not growth depressing.
(3) At supplemental levels of 4%, methionine is the most growth depressing essential amino acid. In choice studies, however, diets containing 4% excess tryptophan are rejected to a greater extent than diets containing 4% excess methionine.

(4) A 4% supplementary level of lysine is tolerated better by pigs than by chicks; the reverse is true with arginine, which is far more growth depressing in pigs than in poultry.
(5) A 4% supplementary level of threonine is growth depressing in chicks but not in pigs.
(6) Excess essential amino acids [i.e. arginine (pigs), leucine, phenylalanine-tyrosine] found in practical grain-soya diets for pigs and poultry are not deleterious to animal performance.
(7) Excess amino acids were not found to increase heat production in pigs by Fuller *et al.* (1987), although Kerr (1988) did observe an increase in heat production in pigs fed excess amino acids during summer months but not during winter months.

## References


ALLEN, N.K., BAKER, D.H., SCOTT, H.M. and NORTON, H.W. (1972). *Journal of Nutrition,* **102**, 171–180

ANDERSON, D.B., VEENHUIZEN, E.L., WAITT, W.P., PAXTON, R.E. and YOUNG, S.S. (1987). *Federation Proceedings,* **46**, 1021 (abstract)

ANDERSON, L.C., LEWIS, A.J., PEO, E.R. and CRENSHAW, J.D. (1984a). *Journal of Animal Science,* **58**, 362–368

ANDERSON, L.C., LEWIS, A.J., PEO, E.R. and CRENSHAW, J.D. (1984b). *Journal of Animal Science,* **58**, 369–377

ANONYMOUS (1985). *Nutrition Reviews,* **43**, 88–90

BAKER, D.H. (1976). *Journal of Nutrition,* **106**, 1376–1377

BAKER, D.H. (1977). In *Advances in Nutrition Research*, Vol. I, pp. 229–335. Ed. Draper, H.H. Plenum Publishing Company, New York, NY

BAKER, D.H. (1978). *Proceedings of the Georgia Nutrition Conference*, pp. 1–12

BAKER, D.H. (1984). *Nutrition Reviews,* **42**, 269–273

BAKER, D.H. (1986). *Journal of Nutrition,* **116**, 2339–2349

BAKER, D.H. and IZQUIERDO, O.A. (1985). *Nutrition Research,* **5**, 1103–1112

BAKER, D.H. and PARSONS, C.M. (1985). *Recent Advances in Amino Acid Nutrition*, 48 pp. Ajinomoto Publishing Company, Tokyo, Japan

BAKER, D.H., JORDAN, C.E., WAITT, W.P. and GOUWENS, D.W. (1967). *Journal of Animal Science,* **26**, 1059–1066

BAKER, D.H., KATZ, R.S. and EASTER, R.A. (1975). *Journal of Animal Science,* **40**, 851–856

BAKER, D.H., ROBBINS, K.R. and BUCK, J.S. (1979). *Poultry Science,* **58**, 749–750

BATTERHAM, E.S. (1984). *Pig News and Information,* **5**, 85–88

BLACK, J.L., CAMPBELL, R.G., WILLIAMS, I.H., JAMES, K.J. and DAVIES, G.T. (1986). *Research and Development in Agriculture,* **3**, 121–145

BOEBEL, K.P. and BAKER, D.H. (1982). *Poultry Science,* **61**, 1167–1175

BOOMGAARDT, J. and BAKER, D.H. (1972). *Poultry Science,* **51**, 1650–1655

BOOMGAARDT, J. and BAKER, D.H. (1973a). *Poultry Science,* **52**, 592–599

BOOMGAARDT, J. and BAKER, D.H. (1973b). *Poultry Science,* **52**, 586–592

BOOMGAARDT, J. and BAKER, D.H. (1973c). *Journal of Animal Science,* **36**, 307–311

BOYD, R.D., HARKENS, M., BAUMAN, D.E. and BUTLER, W.R. (1985). *Proceedings of the Cornell Nutrition Conference*, pp. 10–19

BROWN, H.D., HARMON, B.G. and JENSEN, A.H. (1973). *Journal of Animal Science,* **37**, 708–712

CZARNECKI, G.L., HALPIN, K.M. and BAKER, D.H. (1983). *Poultry Science,* **62**, 371–375

EASTER, R.A. (1987). In *Repartitioning Resolution: Impact of Somatotropin and Beta Adrenergic Agonists on Future Pork Production*, pp. 193–199. University of Illinois Pork Industry Conference

EDMONDS, M.S. and BAKER, D.H. (1987a). *Journal of Animal Science,* **64**, 1664–1671

EDMONDS, M.S. and BAKER, D.H. (1987b). *Journal of Animal Science,* **65**, 699–705

EDMONDS, M.S. and BAKER, D.H. (1987c). *Journal of Nutrition,* **117**, 1396–1401

EDMONDS, M.S., GONYOU, H.W. and BAKER, D.H. (1987). *Journal of Animal Science,* **65**, 179–185

EDMONDS, M.S., PARSONS, C.M. and BAKER, D.H. (1985). *Poultry Science,* **64**, 1519–1526

ETHERTON, T.D. and KENSINGER, R.S. (1984). *Journal of Animal Science,* **59**, 511–528

FULLER, M.F., CADENHEAD, A., MOLLISON, G. and SEVE, B. (1987). *British Journal of Nutrition,* **58**, 277–285

GRABER, G. and BAKER, D.H. (1971). *Journal of Animal Science,* **33**, 1005–1011

HAGEMEIER, D.L., LIBAL, G.W. and WAHLSTROM, R.C. (1983). *Journal of Animal Science,* **57**, 99–105

HALPIN, K.M. and BAKER, D.H. (1984). *Journal of Nutrition,* **114**, 606–612

HARPER, A.E. (1959). *Journal of Nutrition,* **67**, 109–122

IZQUIERDO, O.A., WEDEKIND, K.J. and BAKER, D.H. (1988). *Journal of Animal Science,* **66**, 2886–2892

KERR, B.J. (1988). PhD Thesis. University of Illinois, Urbana, Illinois, USA

LEWIS, A.J., PEO, E.R., CUNNINGHAM, P.J. and MOSER, B.D. (1977). *Journal of Nutrition,* **107**, 1369–1376

LIN, C.C. and JENSEN, A.H. (1985). *Journal of Animal Science,* **61** (Supplement 1), 298–299 (abstract)

LITTLE, T.M. (1978). *Hortscience,* **13**, 504–506

MITCHELL, J.R., BECKER, D.E., JENSEN, A.H., NORTON, H.W. and HARMON, B.G. (1965). *Journal of Animal Science,* **24**, 977–980

NRC (1979). *Nutrient Requirements of Swine.* National Academy of Science, Washington DC

PARSONS, C.M., EDMONDS, M.S. and BAKER, D.H. (1984). *Poultry Science,* **63**, 2438–2443

PETERSON, R.G. (1977). *Agronomy Journal,* **69**, 205–208

ROBBINS, K.R. (1986). *University of Tennessee Agriculture Experiment Station Research Report, 86-09*, pp. 1–8

ROBBINS, K.R., BAKER, D.H. and NORTON, H.W. (1977). *Journal of Nutrition,* **107**, 2055–2061

ROBBINS, K.R., NORTON, H.W. and BAKER, D.H. (1979). *Journal of Nutrition,* **109**, 1710–1714

ROSELL, V.L. and ZIMMERMAN, D.R. (1984a). *Journal of Animal Science,* **59**, 135–140

ROSELL, V.L. and ZIMMERMAN, D.R. (1984b). *Nutrition Reports International,* **39**, 1345–1352

RUSSELL, L.E., CROMSELL, G.L. and STAHLY, T.S. (1983). *Journal of Animal Science,* **56**, 1115–1123

RUSSELL, L.E., EASTER, R.A., GOMEZ-ROJAS, V., CROMWELL, G.L. and STAHLY, T.S. (1986). *Animal Production,* **42**, 291–295

SMITH, J., CLAWSON, A.J. and BARRICK, E.R. (1967). *Journal of Animal Science,* **26**, 752–758

SOUTHERN L.L. and BAKER, D.H. (1982). *Journal of Animal Science,* **55**, 857–866

WHITTEMORE, C.T. (1987). *Elements of Pig Science*, p. 43. Longman Group UK Limited, Essex, England

7

# METHODS OF DETERMINING THE AMINO ACID REQUIREMENTS OF PIGS

H.S. BAYLEY
*Department of Nutritional Sciences, University of Guelph, Canada*

Much has been written on this subject describing the many experiments which have attempted to measure the response of the growing pig to variations in the intake of a single amino acid. Indeed, the subject would be of little further interest to nutritionists were it not for two problems: different conclusions have been drawn from the available data by scientists in the UK and those in the USA (ARC, 1981; NRC, 1979), and the requirements for baby pigs have been determined either by extrapolation from the requirements of older pigs, or from analyses of amino acid concentrations in piglet tissue.

The second of these two problems has been addressed by colleagues at the University of Guelph following the observation that the catabolism of an essential amino acid such as methionine depends on the balance of the other amino acids in the diet (Newport *et al.*, 1976). This provided a basis for measuring an acute response to dietary manipulation. A similar approach had been used by Brooks, Owens and Garrigus (1972) to study lysine.

Very early weaning of the piglet has been a goal for researchers, because the high level of mortality in the first few days of life makes this the major area for improvement in production efficiency. While morbidity of the piglets and reproductive problems of the sows have often prevented attainment of this goal (McCallum *et al.*, 1977), healthy pigs have been produced with weaning a few hours after birth onto diets largely based upon cows' milk (Braude *et al.*, 1970). However, attempts to replace the milk with an alternative protein source greatly reduced performance (Newport and Keal, 1983).

## Free amino acid diets for young piglets

The dependence upon cows' milk as a basis for successful diets for young pigs has been a major obstacle in research designed to measure their amino acid requirements: the milk based diet contains such an abundance of the amino acids that no response to amino acid supplementation can be observed. Experiments to measure the amino acid requirements of the rat and the chicken used diets containing mixtures of free amino acids, but such diets are not consumed by young pigs. In addressing this problem Ball *et al.* (1984) found that 60% of the dietary nitrogen could be replaced by a mixture of free amino acids with only a marginal reduction in growth response. While inclusion of these levels of free amino acids reduced the pH of the liquid diet from 6.5 for the all skim milk diet to 4.5, giving the liquid diet a bitter taste to the human palate, increasing the pH to 6.5 with sodium hydroxide reduced the bitterness of the diet. In the preparation of diets with free amino acids it is important to ensure that the sulphur containing amino acids used are free of foul smelling sulphur containing compounds. In addition, the use of cystine rather than cysteine improved the taste of the diet. Details of the diet are shown in *Table 7.1*. The amounts of the essential amino acids added were equal to those provided

**Table 7.1** COMPOSITION (g/kg) OF EXPERIMENTAL DIETS

| *Diet* | *Skim milk* | *Amino acid* | |
|---|---|---|---|
| *Ingredients* | | | |
| Skim milk (330 g crude protein/kg) | 740 | 296 | |
| Glucose | 13 | 301 | |
| Maize oil | 200 | 200 | |
| Minerals and vitamins | 47 | 47 | |
| Free amino acids | — | 156 | |
| *Protein and amino acids* | *From skim milk* | *From skim milk* | *From free amino acids* |
| Protein (N × 6.25) | 240 | 99 | 141.0 |
| Arginine | 8.6 | 3.4 | 5.1 |
| Histidine | 6.4 | 2.6 | 3.8 |
| Isoleucine | 16.5 | 6.6 | 9.9 |
| Leucine | 23.6 | 9.4 | 14.2 |
| Lysine | 20.1 | 8.0 | 12.1 |
| Methionine | 5.8 | 2.3 | 3.5 |
| Cystine | 3.6 | 1.4 | 2.2 |
| Phenylalanine | 10.7 | 4.3 | 6.4 |
| Tyrosine | 9.3 | 3.7 | 5.6 |
| Threonine | 10.0 | 4.0 | 6.0 |
| Tryptophan | 2.9 | 1.2 | 1.7 |
| Valine | 15.8 | 6.3 | 9.5 |
| Non-essential amino acids[a] | — | — | 76.1 |

[a] As g/kg diet: alanine 8.3, aspartic acid 5.3, asparagine 5.3, glutamic acid 16.3, glutamine 16.3, glycine 16.4, proline 4.1, serine 4.1

by the skim milk diet, but the non-essential amino acids were mixed in the ratios used by Robbins and Baker (1978) except that equal mixtures of asparagine and aspartic acid, and of glutamine and glutamic acid were used. The diets were homogenized with water (1:4 w/v) and stored at 4°C overnight. The diets were given to piglets which had been weaned at three days of age and transferred to individual cages. The growth curves (*Figure 7.1*) for the piglets receiving the two diets between 3 and 18 days of age show that the piglets gained 282 and 236 g/day for the skim milk and amino acid diets respectively, corresponding to body weights of 6.0 and 5.3 kg at 18 days of age. The mean 21 day weight of the university herd is 6 kg. The diet with 60% of the nitrogen as free amino acid was used in all the subsequent studies of the response of the young pig to changes in amino acid intake.

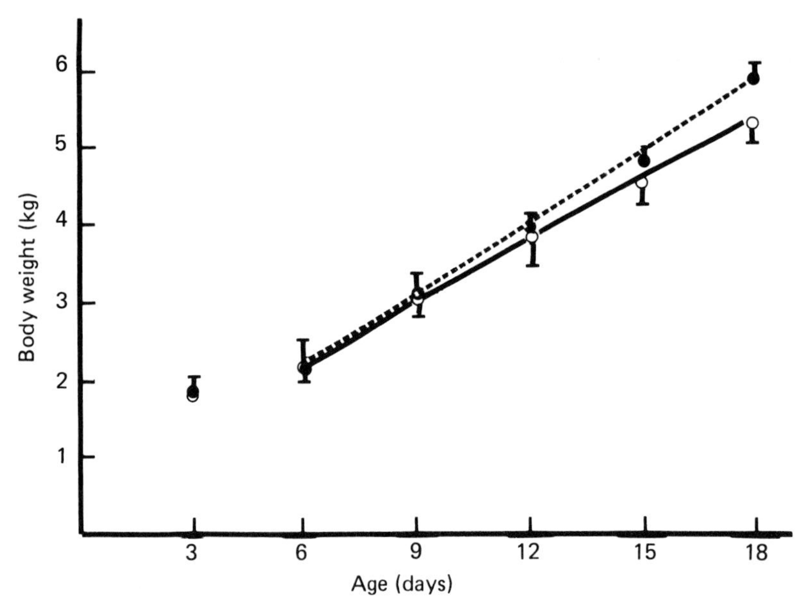


**Figure 7.1** Growth of pigs receiving diets in which the protein was supplied as either cows' skim milk (- - - • - - -) or as a 40:60 mixture of cows' skim milk and a mixture of free amino acids (—— ○ ——)

## Amino acid oxidation and dietary protein adequacy

Chavez and Bayley (1976) interpreted the response of pigs to graded levels of lysine in the diet upon the oxidation of labelled lysine to indicate a dietary lysine requirement. Kang-Lee and Harper (1977) had done the same with the rat for histidine. Attempts to use this approach to measure a methionine requirement of the rat by Aguilar, Harper and Benevenga (1972) were unsuccessful, attempts by the author with the pig were equally unsuccessful. These failures prompted a careful evaluation of the influence of dietary amino acid intake on protein synthesis and amino acid catabolism. Evaluations of the metabolic fate of an amino acid using a tracer dose of the labelled amino acid require definition of the size of the metabolic pools of the amino acid at the various sites where it is oxidized: predominantly the liver for most amino acids. Clearly, using a tracer dose of the amino acid being investigated introduces a systematic change in the dilution of this label. The use of a tracer dose of a different amino acid would reduce this problem because of the constant intake of all the amino acids except the one being investigated. The concept of studying the metabolic fate of one amino acid to indicate the effect of graded levels of another amino acid in the diet (Kim, Elliott and Bayley, 1983) offered the possibility of systematic study of the responses to dietary levels of a number of amino acids.

The theoretical basis of this system is that a deficiency of one essential amino acid limits the amount of all the other amino acids which can be incorporated into protein while an increment in the level of the limiting amino acid allows a greater proportion of the other amino acids to be used for protein synthesis, with a corresponding decline in the amount oxidized. However, when the dietary level of the amino acid being studied is no longer limiting, i.e. at, or above, the dietary requirement, further increases will not influence the partition of the other amino acids between protein synthesis and oxidation. For this system to be sensitive to the change in dietary intake of the amino acid being studied, the intake of the indicator amino acid must be above (but not in excess of) the requirement for this amino acid. The amino acid chosen as the indicator was phenylalanine because it is predominantly oxidized in the liver, and is available with the $^{14}C$ label on the carboxyl group.

## Histidine requirement of the young piglet

The use of the indicator amino acid to measure a dietary requirement is best illustrated by the determination of dietary histidine requirement. Diets containing graded levels of histidine: 2.6, 3.0, 3.5, 4.0, 5.0 and 6.0 g histidine/kg were prepared by supplementing the diet with L-histidine. The concentration of phenylalanine in the diets was 8 g/kg. The oxidation of 20 μCi L-[1-$^{14}C$] phenylalanine was studied by collecting the $^{14}CO_2$ expired in a 1 h period. The plot of the $^{14}CO_2$ recovered against dietary histidine level (*Figure* 7.2) shows that 125 000 disintegrations/minute (dpm) were recovered as $^{14}CO_2$ from the pigs receiving the diet with 2.6 g histidine/kg and this was reduced by supplementing to a dietary level of 4 g histidine/kg. Further increase in dietary histidine level to 6 g/kg had no effect on $^{14}CO_2$ production. Analysis of the data using a two-phase, linear regression indicated that the break point occurred at a dietary histidine level of 3.95 g/kg with 95% confidence limits of 3.6 to 4.2 g/kg for the experiment with 28 piglets.

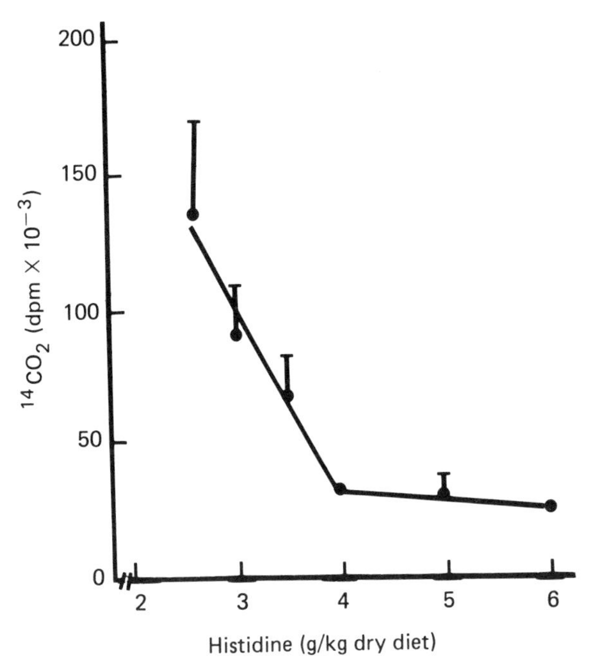


**Figure 7.2** Influence of dietary histidine level on $^{14}C$-carbon dioxide release from $^{14}C$-phenylalanine

Histidine was chosen for this demonstration because it can be purchased with the $^{14}C$ label in the carboxyl group, allowing a comparative determination of the effects of dietary histidine level on the oxidation of histidine to be carried out. The results of this experiment are shown in *Figure 7.3*. Less than 25 000 dpm were recovered as carbon dioxide from the pigs receiving the diet with 2.6 g histidine/kg; supplementing the diet with histidine resulted in a slight increase in radioactivity recovery in the carbon dioxide, but the rate of increase rose dramatically for the diets containing 4 g histidine/kg or more. Statistical analysis of these data showed that the two lines crossed at the dietary histidine level of 4.3 g/kg: 95% confidence limits were 4.0 to 4.5 g/kg for the experiment with 20 piglets. This was not statistically different from the value obtained using $^{14}C$-phenylalanine as an indicator.

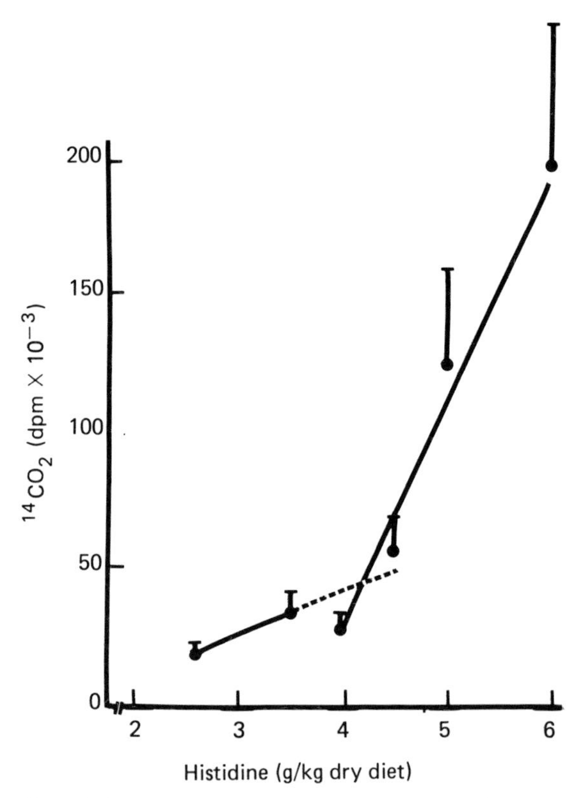


**Figure 7.3** Influence of dietary histidine level on $^{14}C$-carbon dioxide release from $^{14}C$-histidine

These two determinations show that the dietary histidine requirement of young pigs was 4 g/kg diet, much below the level of 6.4 g/kg provided in a milk-based diet containing 240 g protein/kg and explaining the lack of any response to supplementation of such a diet with histidine.

Further experiments were carried out to determine requirements for sulphur-containing amino acids (Kim and Bayley, 1983), lysine and threonine (Kim, McMillan and Bayley, 1983), tryptophan (Ball and Bayley, 1984) and total protein in the diet (Ball and Bayley, 1986), so that a comparison can be made between the results of these acute determinations of protein and amino acid requirements with the estimates of the ARC and the NRC (*Table 7.2*).

**Table 7.2** COMPARISON OF DIFFERENT PROTEIN AND AMINO ACID REQUIREMENTS

| | *ARC*[a] | *NRC* | *Guelph* |
|---|---|---|---|
| Protein | 248 | 270 | 240 |
| Arginine | — | 3.3 | — |
| Histidine | 5.7 | 3.1 | 3.8 |
| Isoleucine | 9.4 | 8.5 | — |
| Leucine | 17.4 | 10.1 | — |
| Lysine | 17.4 | 12.8 | 12.0 |
| Methionine and cystine | 8.7 | 7.6 | 4.4 |
| Phenylalanine and tyrosine | 16.6 | 11.8 | 9.0 |
| Threonine | 10.4 | 7.6 | 6.1 |
| Tryptophan | 2.5 | 2.0 | 2.0 |
| Valine | 12.1 | 8.5 | — |

[a] Calculated for a diet with an energy level of 15.5 MJ digestible energy/kg

## Comparison of protein and amino acid requirement estimates

The NRC (1979) cautions that a 'substantial level of milk products should be included in the diet'. Calculation of the level of skim milk protein necessary to provide the estimated requirements for each of the amino acids (*Table 7.3*) shows that the ARC (1981) estimate of a requirement for 248 g milk protein/kg diet is necessary to fulfil the threonine requirement. However, the estimates of the NRC (1979) and the results of the Guelph experiments fail to show why the piglets responded to dietary protein levels of over 200 g/kg.

**Table 7.3** LEVEL OF DIETARY PROTEIN FROM DRIED SKIM MILK REQUIRED TO MEET CURRENT RECOMMENDED AMINO ACID REQUIREMENTS (g/kg) FOR A PIG OF 1–5 kg

| *Source* | *ARC*[a] | *NRC* | *Guelph* |
|---|---|---|---|
| Recommended protein level | 248 | 270 | 240 |
| Arginine | — | 92 | — |
| Histidine | 206 | 116 | 149 |
| Isoleucine | 132 | 123 | — |
| Leucine | 171 | 103 | — |
| Lysine | 198 | 151 | 142 |
| Methionine + cystine | 217 | 196 | 124 |
| Phenylalanine + tyrosine | 193 | 142 | 157 |
| Threonine | 242 | 183 | 144 |
| Tryptophan | 198 | 165 | 161 |
| Valine | 179 | 129 | — |

To examine this question, a diet containing 200 g/kg protein, of which 60% was provided as free amino acids, was supplemented with a variety of amino acids to raise their levels to those in the 240 g protein/kg diet, and the catabolism of $^{14}C$ labelled phenylalanine measured for pigs receiving either the unsupplemented or

**Table 7.4** $^{14}C$-PHENYLALANINE OXIDATION BY PIGLETS RECEIVING DIETS CONTAINING 200 g/kg PROTEIN EITHER UNSUPPLEMENTED, OR SUPPLEMENTED WITH VARIOUS AMINO ACIDS TO THE LEVELS IN A SKIM MILK DIET CONTAINING 240 g/kg PROTEIN

| | *Supplement* | | | | | | |
|---|---|---|---|---|---|---|---|
| | *None* | *Threonine* | *Methionine* | *Lysine* | *Non-essential amino acids*[a] | *Proline* | *Arginine* |
| Level of addition (g/kg) | — | 1.67 | 1.01 | 3.35 | 40.0 | 1.30 | 1.43 |
| $^{14}C$ carbon dioxide released (dpm/kg) | 1000 | 1006 | 978 | 1032 | 828 | 681[b] | 744[b] |

[a] Alanine 4.6, aspartic acid 2.96, asparagine 2.96, glutamic acid 9.04, glutamine 9.04, glycine 9.12, serine 2.28 g
[b] Significantly ($P<0.05$) less than the unsupplemented control

supplemented diets; a decrease in labelled carbon dioxide release would indicate a positive response to the supplement. *Table 7.4* shows the results of this experiment: the only supplements which gave a significant reduction were proline and arginine. Thus, the Guelph estimate of a protein requirement of 240 g/kg diet was based upon the need of the piglets for proline and arginine in the diet containing skim milk and a mixture of free amino acids. Proline has not been considered as essential for the pig; however, proline has been shown to be essential for maximum growth in the chick (Baker, 1977). There are several other reports of a response to arginine in the pig (Southern and Baker, 1983). The lack of response to supplementation with either threonine, methionine or lysine confirms that these were not limiting the capacity of the piglets to retain protein, as shown by the oxidation of the indicator amino acid.

## Proline and arginine requirements of the young piglet

Having established that the piglets required proline, their response to graded levels of proline was studied using diets with either inadequate (200 g/kg) or excess (260 g/kg) protein to ensure that the response was to the proline in particular, and not to its amino nitrogen. Using both diets, the pigs responded to increased dietary proline levels up to 14 g/kg diet. The response to the addition of arginine to the diet containing 240 g protein/kg showed an arginine requirement of 5.3 g arginine/kg diet.

## Tryptophan requirement of growing boars

One of the criticisms of the application of the oxidation of an indicator amino acid to the determination of the requirement of another amino acid for the baby pig is the short-term nature of the assay. It is difficult to carry out long-term experiments with piglets up to 5 kg. Thus, to validate this approach, the response of bigger pigs to dietary amino acid concentration has been carried out using $^{14}C$-phenylalanine oxidation and other response parameters, including growth. Lin, Smith and Bayley (1986) showed that a dietary tryptophan level of 1.3 g/kg minimized phenylalanine oxidation in growing (40 kg) boars. In a subsequent (unpublished) experiment it was found that supplementation of a diet containing 0.5 g tryptophan/kg to a tryptophan level of 1.3 g/kg resulted in significant increases in the growth of boars between 20 and 50 kg, but supplementation to higher levels resulted in no further growth response. These two experiments provide some validation of the acute, indicator amino acid oxidation procedure.

## Conclusions

The significance of this work is that it has provided some directly determined values of the amino acid needs of the young pig. However, the digestive system requires the provision of a clot-forming diet, necessitating the use of 'substantial' amounts of milk products in the diet. Inclusion of 160 g milk protein/kg diet, equivalent to an inclusion in the diet of approximately 500 g/kg skim milk which contained 335 g protein/kg, would supply all the amino acids at the requirements determined by the indicator oxidation procedure. Development of low-milk, or milk-free diets needs evaluation of the clot-forming properties of alternative protein sources.

## References

AGUILAR, T.S., HARPER, A.E. and BENEVENGA, N.J. (1972). *Journal of Nutrition*, **102**, 1199–1208

ARC (1981). *The Nutrient Requirements of Pigs*. Commonwealth Agricultural Bureaux, Slough, England

BAKER, D.H. (1977). In *Advances in Nutrition Research*, pp. 299–335. Ed. Draper, H.H., Plenum Press, New York

BALL, R.O. and BAYLEY, H.S. (1984). *Journal of Nutrition*, **114**, 1741–1746

BALL, R.O. and BAYLEY, H.S. (1986). *British Journal of Nutrition*, **55**, 651–658

BALL, R.O., KIM, K.I. and BAYLEY, H.S. (1984). *Canadian Journal of Animal Sciences*, **64**, 1019–1022

BRAUDE, R., MITCHELL, K.F., NEWPORT, M.J. and PORTER, J.W.G. (1970). *British Journal of Nutrition*, **24**, 501–516

BROOKS, I.M., OWENS, F.N. and GARRIGUS, U.S. (1972). *Journal of Nutrition*, **102**, 27–36

CHAVEZ, E.R. and BAYLEY, H.S. (1976). *British Journal of Nutrition*, **36**, 369–380

KANG-LEE, Y.A. and HARPER, A.E. (1977). *Journal of Nutrition*, **107**, 1427–1443

KIM, K.I. and BAYLEY, H.S. (1983). *British Journal of Nutrition*, **50**, 383–390

KIM, K.I., ELLIOT, J.I. and BAYLEY, H.S. (1983). *British Journal of Nutrition*, **50**, 391–399

KIM, K.I., MCMILLAN, I. and BAYLEY, H.S. (1983). *British Journal of Nutrition*, **50**, 369–382

LIN, F.D., SMITH, T.K. and BAYLEY, H.S. (1986). *Journal of Animal Sciences*, **62**, 660–664

MCCALLUM, I.M., ELLIOT, J.I. and OWEN, B.D. (1977). *Canadian Journal of Animal Science*, **57**, 151–158

NRC (1979). *Nutrient Requirements of Swine*, 8th Edition. National Academy of Sciences — National Research Council, Washington DC

NEWPORT, M.J., CHAVEZ, E.R., HORNEY, F.D. and BAYLEY, H.S. (1976). *British Journal of Nutrition*, **36**, 87–99

NEWPORT, M.J. and KEAL, H.D. (1983). *British Journal of Nutrition*, **49**, 43–50

ROBBINS, K.R. and BAKER, D.H. (1978). *Canadian Journal of Animal Sciences*, 533–535

SOUTHERN, L.L. and BAKER, D.H. (1983). *Journal of Animal Science*, **57**, 402–412

8

# ILEAL DIGESTIBILITIES OF AMINO ACIDS IN PIG FEEDS AND THEIR USE IN FORMULATING DIETS

T.D. TANKSLEY, Jr. and D.A. KNABE
*Texas A & M University, USA*

## Introduction

One of the major goals of the nutritionist when formulating diets for non-ruminants is to provide the essential amino acids needed to support maximum growth adequately and economically. To do this effectively, one must be able to evaluate feeds as a source of biologically available amino acids. Unfortunately, most of the values in the literature are not suitable for this purpose because they represent the total amino acid composition of feeds, determined by chemical methods. Nutritionists have known for many years that all the nutrients in feedstuffs, as determined by chemical analyses, are not biologically available to the pig. Amino acids may be 'unavailable' because of incomplete protein hydrolysis resulting from inaccessibility of the protein to proteolytic enzymes, inhibition of enzymes (such as the trypsin inhibitor in raw soyabeans) or inhibition of amino acid absorption. In high-protein feedstuffs, the most common reasons for this 'unavailable' component are excessive heat treatment during processing, indigestible cell walls which prevent the proteases from entering the cell, and the presence of protein inhibitors in many feeds of plant origin.

Knowledge of amino acid digestibilities of feedstuffs is particularly important in formulating pig diets since the minimum quantity of high-protein feedstuffs is included to meet the requirement for the first limiting amino acid. However, if the digestibilities of the first two or three most limiting amino acids are different, formulations may not provide the levels needed for optimum growth. This is the reason that much work has been directed toward determining amino acid digestibilities in the past two decades. Excellent reviews discussing the merits of the different *in vivo* and *in vitro* methods have been prepared by Ebersdobler (1976) and Zebrowska (1978).

Before discussing the merits of the various methods of determining amino acid digestibility the meaning and use of two words needs to be agreed. Most people use the words 'digestibility' and 'availability' synonymously when referring to amino acids. However, some scientists contend that in a strict interpretation they do not mean the same. Digestible means that a nutrient has been absorbed (disappeared from the gastrointestinal tract). On the other hand, availability means that a nutrient has not only been absorbed but is available to the cell for synthesis when it is needed. Since protein synthesis is an 'all or nothing' type of synthesis, it is mandatory that all amino acids be available in the proper amounts for maximum synthesis by the animal. Since the technology needed to routinely measure the amount of amino acids that are actually available at the cellular level has not been developed, the word digestibility appears to better describe the amino acid values that are being used today.

The most commonly used procedure for determining amino acid digestibility for pigs has been the faecal index method that was first suggested by Kuiken and Lyman (1948). Using this procedure, digestible amino acid values represent the amount of amino acids in the feedstuff that disappear over the total digestive tract. Digestibilities of amino acids in a wide range of feedstuffs have been determined using this procedure (Dammers, 1964; Poppe, Meier and Wiesemuller, 1970; Eggum, 1973; Poppe, 1976). Their values have been used extensively, particularly in European and Scandinavian countries, in formulating diets.

However, recent experiments in which protein and free amino acids were infused into the pig's large intestine (Zebrowska, 1973a, 1975; Just, Jorgensen and Fernandez, 1981) have shown that most of the nitrogen disappearing from the large intestine is not retained, indicating that the compounds absorbed are not used for protein synthesis by the pig. The bacterial flora hydrolyse the nitrogenous compound and most of the nitrogen is absorbed as ammonia, amines or amides. Since the amino acids that are hydrolysed in the large intestine are not used by the pig for protein synthesis, digestibilities determined at the end of the small intestine should be a more accurate indicator of the amino acids that are available for use by the pig.

Because of this, several European and Canadian workers, as well as our laboratory, have determined amino acid digestibilities of feedstuffs at the end of the small intestine. Examples of this work include Zebrowska (1973a, 1973b); Holmes *et al.* (1974); Ivan and Farrell (1976); Sauer, Strothers and Parker (1977); Sauer, Strothers and Phillips (1977); Zebrowska and Buraczewski (1977); Low (1979); Just (1980); and Taverner and Farrell (1981).

Essentially two methods have been used to determine amino acid digestibilities at the terminal ileum in pigs. In the first, pigs are equipped with a re-entrant cannula at the terminal ileum (some go from ileum to caecum). The total amount of digesta passing the proximal cannula is collected, sampled, and the remaining digesta are returned through the distal cannula to the animal.

In the second method, pigs are fitted with a single T-cannula at the terminal ileum (about 15 cm from the ileocaecal valve). Digesta are collected, sampled, and the remaining digesta are returned through the cannula. Digestibilities are calculated based on the concentration of indigestible marker (usually chromic acid) in the feed and digesta.

A more sophisticated fistulation technique called post-valvular ileocolic fistulation, that preserves the role of the ileo-caeco-colic sphincter, has been developed by Darcy, Laplace and Villers (1980). Initial results indicate that the time for food to reach the large intestine is increased slightly (7 to 10%) compared with re-entrant cannulation. Digestibilities determined using this technique show little variation from those obtained using the single T-cannula (Darcy and Rerat, 1983).

## Nitrogen and amino acid digestibilities in selected cereal grains

Apparent digestibilities of nitrogen (N) and essential amino acids (EAA) in wheat, yellow maize, low and high-tannin sorghums, barley, oat groats and some wheat

**Table 8.1** APPARENT DIGESTIBILITY OF NITROGEN AND ESSENTIAL AMINO ACIDS IN SELECTED CEREAL GRAINS AND MILLING BY-PRODUCTS MEASURED AT THE END OF THE SMALL INTESTINE IN GROWING PIGS

| *Reference* | *Wheat* | | | | *Maize yellow* | | *Sorghum* | | | *Barley* | | | *Oat groats* | *Wheat* | | | |
|---|---|---|---|---|---|---|---|---|---|---|---|---|---|---|---|---|---|
| | *Hard* | *Hard* | *Hard* | *Soft* | | | *Low-tannin* | | *High tannin* | | | | | *Whole* | *Flour* | *Offal* | *Midds* |
| | *1* | *2* | *5* | *1* | *2* | *5* | *5* | *6(a)* | *6(a)* | *2* | *4* | *5* | *5* | *3* | *3* | *3* | *5* |
| Nitrogen | 85.6 | 82.9 | 84.8 | 80.4 | 82.4 | 79.6 | 81.5 | 76.6 | 68.7 | 74.9 | 76.0 | 80.0 | 83.6 | 85.3 | 90.5 | 69.8 | 70.1 |
| Essential amino acids | | | | | | | | | | | | | | | | | |
| Arginine | 90.0 | 85.8 | 90.4 | 79.0 | 87.4 | 88.6 | 84.6 | 81.4 | 72.8 | 81.5 | 82.0 | 85.7 | 89.8 | 87.1 | 90.7 | 94.6 | 83.9 |
| Histidine | 92.0 | 89.1 | 88.0 | 86.0 | 88.3 | 85.1 | 80.9 | 81.1 | 70.0 | 80.4 | 75.0 | 83.1 | 86.0 | 88.4 | 90.9 | 78.5 | 79.1 |
| Isoleucine | 85.0 | 85.3 | 88.0 | 81.0 | 87.5 | 84.3 | 86.9 | 77.9 | 71.0 | 79.1 | 77.0 | 83.2 | 85.6 | 89.1 | 94.0 | 72.9 | 75.6 |
| Leucine | 85.0 | 86.9 | 85.5 | 81.0 | 92.5 | 89.7 | 91.0 | 81.0 | 73.2 | 81.5 | 78.0 | 82.7 | 84.8 | 89.9 | 94.7 | 74.4 | 72.4 |
| Lysine | 78.0 | 75.7 | 84.2 | 64.0 | 82.0 | 77.5 | 73.8 | 82.7 | 78.6 | 73.3 | 70.0 | 79.4 | 82.2 | 79.5 | 84.2 | 66.4 | 75.5 |
| Methionine | 83.0 | 86.6 | 89.9 | 80.0 | 91.9 | 87.7 | 88.0 | 78.5 | 80.0 | 80.4 | — | 86.9 | 89.0 | 92.4 | 93.7 | 77.8 | 78.7 |
| Phenylalanine | 92.0 | 88.8 | 92.4 | 88.0 | 90.5 | 90.6 | 92.1 | 84.1 | 77.0 | 82.2 | 80.0 | 88.7 | 90.0 | 91.5 | 95.5 | 76.0 | 81.5 |
| Threonine | 78.0 | 76.5 | 77.5 | 74.0 | 78.9 | 72.4 | 74.7 | 71.8 | 65.4 | 71.2 | 67.0 | 73.5 | 77.9 | 78.4 | 85.4 | 54.0 | 63.4 |
| Tryptophan | — | — | 81.4 | — | — | 69.6 | 79.7 | 71.5 | 59.9 | — | — | 73.0 | 80.6 | — | — | — | 77.7 |
| Valine | 85.0 | 82.8 | 86.0 | 82.0 | 84.9 | 83.7 | 85.0 | 79.1 | 71.4 | 78.0 | 75.0 | 82.1 | 85.7 | 86.7 | 92.7 | 71.5 | 73.4 |
| Average essential amino acid digestibility | 85.3 | 84.2 | 86.3 | 79.4 | 87.1 | 82.9 | 83.7 | 78.9 | 71.9 | 78.6 | 75.5 | 81.8 | 85.2 | 87.0 | 91.3 | 74.0 | 76.1 |

References:
1 Ivan and Farrell (1976) re-entrant cannula, ileum-ileum.
2 Sauer, Strothers and Phillips (1977) re-entrant cannula, ileum-caecum.
3 Sauer, Strothers and Parker (1977) re-entrant cannula, ileum-caecum.
4 Zebrowska (1973b) re-entrant cannula, ileum-ileum.
5 Lin (1983) single T-cannula.
6 Cousins *et al.* (1981) single T-cannula, diets included 6% casein. (a) The low-tannin sorghum had 0.83 catechin equivalents compared to 3.40 for the high-tannin sorghum (Ga615) determined by the modified-HCl method (Maxon and Rooney, 1972).

**Table 8.2** APPARENT DIGESTIBILITY OF NITROGEN AND ESSENTIAL AMINO ACIDS IN SELECTED CEREAL GRAINS AND MILLING BY-PRODUCTS MEASURED OVER THE TOTAL DIGESTIVE TRACT OF GROWING PIGS

| | *Wheat* | | | | *Maize yellow* | | *Sorghum* | | | *Barley* | | *Oat groats* | *Wheat* | | | |
|---|---|---|---|---|---|---|---|---|---|---|---|---|---|---|---|---|
| | *Hard* | *Hard* | *Hard* | *Soft* | | | *Low-tannin* | | *High tannin* | | | | *Whole* | *Flour* | *Offal* | *Midds* |
| *Reference* | *1* | *2* | *5* | *1* | *2* | *5* | *5* | *6(a)* | *6(a)* | *2* | *5* | *5* | *3* | *3* | *3* | *5* |
| Nitrogen | 89.2 | 91.2 | 88.9 | 87.1 | 89.4 | 86.9 | 86.4 | 85.4 | 72.0 | 85.9 | 83.9 | 90.6 | 93.3 | 95.6 | 80.9 | 76.0 |
| Essential amino acids | | | | | | | | | | | | | | | | |
| Arginine | 93.0 | 92.7 | 92.5 | 91.0 | 92.2 | 91.7 | 86.3 | 85.7 | 73.7 | 89.4 | 89.2 | 93.2 | 94.6 | 95.6 | 90.0 | 87.8 |
| Histidine | 94.0 | 94.9 | 92.7 | 92.0 | 93.6 | 91.8 | 85.6 | 87.1 | 70.9 | 91.9 | 88.3 | 91.5 | 94.0 | 95.6 | 88.9 | 86.4 |
| Isoleucine | 90.0 | 89.4 | 87.8 | 87.0 | 88.1 | 84.7 | 85.6 | 83.2 | 73.3 | 83.1 | 85.2 | 89.4 | 91.6 | 94.7 | 74.6 | 71.4 |
| Leucine | 90.0 | 91.5 | 88.9 | 88.0 | 93.8 | 92.2 | 92.2 | 87.1 | 77.1 | 86.6 | 86.4 | 91.1 | 93.2 | 95.6 | 78.7 | 75.6 |
| Lysine | 80.0 | 80.7 | 80.0 | 71.0 | 83.0 | 77.1 | 69.7 | 83.6 | 77.6 | 77.5 | 76.6 | 86.9 | 86.1 | 86.0 | 75.5 | 70.3 |
| Methionine | 88.0 | 88.9 | 86.7 | 86.0 | 89.5 | 86.5 | 83.3 | 84.2 | 78.6 | 79.9 | 79.3 | 88.8 | 93.4 | 93.4 | 81.8 | 70.5 |
| Phenylalanine | 94.0 | 92.5 | 91.7 | 92.0 | 91.3 | 89.6 | 89.1 | 87.0 | 77.8 | 87.9 | 88.4 | 92.0 | 94.3 | 96.4 | 79.5 | 79.5 |
| Threonine | 86.0 | 86.7 | 84.9 | 85.0 | 86.3 | 82.5 | 80.0 | 81.9 | 70.8 | 81.4 | 82.8 | 88.5 | 89.2 | 92.3 | 71.3 | 70.6 |
| Tryptophan | — | — | 89.2 | — | — | 79.6 | 82.2 | 82.3 | 68.8 | — | 84.2 | 91.4 | — | — | — | 82.4 |
| Valine | 91.0 | 88.9 | 88.3 | 90.0 | 88.2 | 86.7 | 84.8 | 84.9 | 74.4 | 84.3 | 86.4 | 90.8 | 91.8 | 94.3 | 76.1 | 75.4 |
| Average essential amino acid digestibility | 89.6 | 89.6 | 88.3 | 86.9 | 89.6 | 86.3 | 83.9 | 84.7 | 74.3 | 84.7 | 84.7 | 90.4 | 92.0 | 93.8 | 79.6 | 77.0 |

References:
1 Ivan and Farrell (1976) re-entrant cannula, ileum-ileum.
2 Sauer, Strothers and Phillips (1977) re-entrant cannula, ileum-caecum.
3 Sauer, Strothers and Parker (1977) re-entrant cannula, ileum-caecum.
4 Zebrowska (1973b) re-entrant cannula, ileum-ileum.
5 Lin (1983) single T-cannula.
6 Cousins *et al*. (1981) single T-cannula, diets included 6% casein. (a) The low-tannin sorghum had 0.83 catechin equivalents compared to 3.40 for the high-tannin sorghum (Ga615) determined by the modified-HCl method (Maxon and Rooney, 1972).

milling by-products, measured at the end of the small intestine and over the total tract of growing-finishing pigs are given in *Tables 8.1* and *8.2*. Several observations are warranted:

(1) Ileal N and most amino acid digestibilities tended to be lower than total digestibilities, indicating further hydrolysis of nutrients in the large intestine. The amount of disappearance was quite variable, both within and among grain samples. Average ileal digestibilities of the EAA were 4.5, 3.0, 4.5, 5.2 and 3.0% lower than faecal values for wheat, maize, barley, oat groats and low-tannin sorghum. Threonine and tryptophan were consistently degraded the most in the large intestine. Methionine and lysine showed an apparent increase in the large intestine for about one-half of the grains suggesting a synthesis of these by the microflora of the large intestine.
(2) Ileal digestibilities of all amino acids tended to be higher for hard than for soft wheat. Significantly more lysine, arginine, methionine, and isoleucine were absorbed from hard than from soft wheat. Averages for the EAAs were 85.3, 84.2 and 86.3% in the hard wheats compared with 79.4% in the soft wheat. Differences were much smaller when measured over the total tract.
(3) Nitrogen and most amino acids in low-tannin sorghums were more digestible than those in the high-tannin sorghum. This is consistent with pig feeding experiments which indicate that high-tannin sorghum has only about 90% of the value of low-tannin sorghum in terms of feed required per unit gain. Fortunately, most (> 95%) of the sorghum grown in the USA is the low-tannin varieties.
(4) Ileal amino acid digestibilities in barley were generally lower than those in wheat, maize and sorghum. This is attributed primarily to its higher fibre content. The barley from Canada (Sauer, Strothers and Phillips, 1977) and Poland (Zebrowska, 1973b) tended to have lower ileal digestibilities than the sample from the USA (Lin, 1983).
(5) Oat groats appear to be well digested by the pig based on this one sample. Ileal N and EAA digestibilities were similar to those for wheat and slightly higher than those for maize and sorghum. This sample of oat groats was low in crude fibre (2.08%) and high in ether extract (5.70%).
(6) Comparison of ileal amino acid digestibilities in whole wheat, wheat flour, and wheat offal showed a decrease from wheat flour to whole wheat to wheat offal. Improved digestibilities from fine grinding have also been demonstrated in sorghum (Owsley, Knabe and Tanksley, 1981). Ileal digestibilities of wheat offal and wheat midds were severely depressed compared with wheat flour and whole wheat. Again, this is attributed to the higher fibre content of these feeds compared with wheat (9 to 10% vs 3.0%).
(7) Without exception, threonine, lysine and tryptophan (when tryptophan was measured) were the three least digestible of the EAA in all cereal grains. This has profound nutritional importance since these are generally the first three limiting amino acids in grain–soyabean meal diets for growing pigs.

In summary, ileal digestibilities appear to be highest for wheat, maize and oat groats, followed by low-tannin sorghum and barley. The absolute differences in digestibilities among these grains were relatively small. This fact plus the relatively low lysine, tryptophan and threonine content of the grains suggests that use of ileal amino acid digestibilities will not significantly improve the precision of formulating practical pig diets when these grains are used. On the other hand, ileal amino acid digestibilities of the high-tannin sorghum and the wheat milling by-products were considerably lower than for the other grains. The wheat by-products are also relatively rich in lysine, tryptophan and threonine compared with the cereal grains; therefore the use of ileal amino acid digestibility values for these products may improve the precision of formulating practical pig diets.

These conclusions must be tentative due to the small number of samples involved and the large number of factors (e.g. variety of grain, fertilizer application, and differences in procedures between laboratories) that can affect digestibility values.

## Nitrogen and amino acid digestibilities in high-protein feedstuffs

In the past six years, we have determined digestibilities of amino acids in 17 commercially processed high-protein feedstuffs at both the end of the small intestine and over the total digestive tract of growing-finishing pigs. We have included tabular data for feedstuffs for which we have evaluated at least two samples. We have placed major emphasis on high-protein feedstuffs because they provide more than 70% of the first limiting amino acid (lysine) in a maize:soyabean meal, growing-pig diet (16% CP) and because there is more variation in digestibilities among high-protein feedstuffs than predominantly energy-yielding feedstuffs.

A review of the procedures used appears warranted before discussing the results. A single T-cannula placed about 15 cm anterior to the ileo-caecal valve was used in all experiments. This single cannula is preferred to a re-entrant cannula because of maintaining a more normal physiological condition (transecting the small intestine disrupts the migrating myoelectric complex). The surgical technique is easy and straightforward and the pigs eat and drink similarly to non-cannulated pigs that are confined to metabolism crates. All the data were collected from replicated 3 × 3 or simple 4 × 4 Latin square-designed experiments that yielded either four or six observations for each feedstuff. The test feedstuff was the only source of protein in maize starch-based diets. Within each trial, diets were formulated to contain the same amounts of lysine and nitrogen. Chromic oxide was added as an indigestible marker for calculating nutrient digestibilities. During each experimental period, all pigs were fed the same amount of diet. During each period in each experiment, pigs were fed the amount that the pig eating the least would consume. Daily intake amounted to 3 to 4% of the pigs' body weight. Feeding occurred at 12-h intervals. After a four to five day adjustment phase, faeces were collected for a minimum of three days, followed by three days of ileal digesta collection. Digesta were collected for 12 h each day, beginning at the morning feeding and ending at the night feeding.

Nitrogen and EAA digestibilities for six soyabean meals (SBM), four cottonseed meals (CSM) and two meat and bone meals (M&BM) are shown in *Table 8.3* (end of small intestine) and *8.4* (over the total digestive tract). Apparent digestibilities of N and most amino acids at the end of the small intestine for 44 and 48.5% CP SBM were similar and consistently higher than digestibilities found for CSM and M&BM. Differences in digestibilities for specific amino acids among the SBMs were small, generally ranging from 2 to 5 percentage units. Compared with the average digestion coefficients for all SBMs, the mean EAA digestibilities of CSM were 1 (arginine) to 24 (lysine) percentage units lower and M&BM 14 (arginine) to 28 (threonine) percentage units lower. Among the essential amino acids, threonine was the least digestible in the SBMs; lysine was the least digestible in CSM and tryptophan least digestible in M&BM. Threonine also had the second lowest digestibility in CSM and M&BM. Ironically, these are the first three limiting amino acids in grain–SBM diets.

The low apparent digestibility of threonine is consistent with the results of other workers (Cho and Bayley, 1972; Buraczewska, Buraczewski and Zebrowska, 1975; Zebrowska *et al.*, 1977) and is possibly due to the high concentration of threonine in the endogenous secretions (Buraczewska, Buraczewski and Zebrowska, 1975; Sauer, Strothers and Parker, 1977; Taverner, Hume and Farrell, 1981).

Lysine digestibility at the end of the small intestine was considerably higher for the SBMs (87 and 84%) than for CSM (63%) or M&BM (64%). This has special interest since the amount of high-protein feedstuffs added to cereal-based diets is normally determined by the level of dietary lysine. Lysine digestibilities in the SBMs are in general agreement with the ileal values of 90.7% reported by Holmes *et al.* (1974) and 80.9 and 82.2% reported by Zebrowska, Buraczewska and Buraczewski (1977). Batterham, Murison and Lewis (1979) also reported an 87% availability for lysine in SBM based on a slope-ratio growth assay with pigs. Ileal digestibility of lysine in CSM by pigs, has not been reported except by this laboratory; however, the low lysine availabilities (49%) determined from chick growth assays by Ousterhout, Grau and Lundholm (1959) and from a pig growth assay (39%) by Batterham, Murison and Lewis (1979) are in general agreement with our values. Zebrowska and Buraczewski (1977) reported an ileal lysine digestibility of 58% for M&BM while Batterham, Murison and Lewis (1979) reported a value of 50% based on a growth study.

Tryptophan was the least digestible (53%) EAA at the end of the small intestine for M&BM. Zebrowska and Buraczewski (1977) reported a similar value (51.5%) and also found it to be the least digestible among the EAAs. The low digestibility of tryptophan in M&BM has added significance in that the total tryptophan in M&BM is only about half of that present in 44% SBM and 41% CSM.

Average digestibilities of EAAs were more uniform for the SBMs than CSM or M&BM (*Table 8.3*). Digestibilities for the SBMs ranged from 73 to 91%, while CSMs ranged from 62 to 89% and M&BMs ranged from 53 to 76%. A comparison of digestibilities across the four CSMs and the four 44% SBMs indicates less variation among the SBMs than the CSMs.

*Table 8.4* shows the apparent digestibilities of N and amino acids measured over the total digestive tract. In general, digestibilities followed the same pattern found at the end of the small intestine, but values tended to be higher when measured over the total tract. Digestibilities for the EAAs in the SBMs were still higher than for CSM and M&BM, but the magnitude of difference changed due to the extent of digestion of the meals in the pig's hindgut. The 44 and 48.5% SBM values became closer while differences between SBM and CSM increased, and differences between SBM and M&BM remained essentially the same or decreased slightly. The N and amino acids in CSM were much less digested in the large intestine than the protein from SBM and M&BM. This differential effect of digestion in the large intestine again indicates that amino acid digestibilities determined at the end of the small intestine more accurately reflect those available to the pig.

*Table 8.5* shows the differences between small intestine and total tract digestibilities that were calculated by subtracting small intestine values from total tract digestibilities. A positive value in *Table 8.5* indicates the amount of disappearance or extent of digestion in the large intestine (in percentage units) while a negative value indicates a synthesis of that amino acid in the large intestine. In most

**Table 8.3** APPARENT ILEAL DIGESTIBILITIES OF NITROGEN AND AMINO ACIDS IN SOYABEAN, COTTONSEED, AND MEAT AND BONE MEALS

| | *44% SBM* | | | | *48.5% SBM* | | *CSM*[a] | | | | *M&BM* | | *Average values* | | | |
|---|---|---|---|---|---|---|---|---|---|---|---|---|---|---|---|---|
| | *1* | 2 | 3 | *4* | *1* | 2 | *1* | 2 | 3 | *4* | *1* | 2 | *44% SBM* | *48.5% SBM* | *CSM* | *M&BM* |
| Nitrogen | 79 | 89 | 82 | 85 | 78 | 79 | 69 | 72 | 79 | 75 | 59 | 66 | 82 | 79 | 74 | 63 |
| Essential amino acids | | | | | | | | | | | | | | | | |
| Arginine | 89 | 89 | 92 | 92 | 90 | 88 | 85 | 87 | 92 | 90 | 74 | 78 | 91 | 89 | 89 | 76 |
| Histidine | 87 | 87 | 89 | 90 | 85 | 85 | 75 | 79 | 85 | 81 | 65 | 68 | 88 | 85 | 80 | 67 |
| Isoleucine | 82 | 82 | 83 | 85 | 82 | 79 | 61 | 66 | 76 | 70 | 59 | 65 | 83 | 81 | 68 | 62 |
| Leucine | 81 | 81 | 83 | 85 | 79 | 78 | 64 | 69 | 77 | 73 | 62 | 69 | 83 | 79 | 71 | 66 |
| Lysine | 86 | 85 | 89 | 89 | 83 | 84 | 53 | 62 | 70 | 64 | 61 | 67 | 87 | 84 | 62 | 64 |
| Methionine | 87 | 90 | 77 | 88 | 89 | 89 | 66 | 65 | 82 | 66 | 72 | 74 | 86 | 89 | 70 | 73 |
| Phenylalanine | 83 | 86 | 88 | 88 | 84 | 83 | 78 | 77 | 87 | 72 | 66 | 71 | 86 | 84 | 81 | 69 |
| Threonine | 73 | 75 | 77 | 79 | 73 | 73 | 55 | 62 | 69 | 65 | 50 | 61 | 76 | 73 | 63 | 56 |
| Tryptophan | 81 | 78 | 82 | 84 | 76 | 79 | — | 69 | 78 | 68 | 55 | 50 | 81 | 78 | 72 | 53 |
| Valine | 81 | 80 | 81 | 79 | 79 | 77 | 66 | 68 | 79 | 71 | 61 | 68 | 81 | 78 | 71 | 65 |
| Average essential amino acid digestibility | 83.0 | 83.3 | 84.1 | 85.9 | 82.0 | 81.5 | 67.0 | 70.4 | 79.5 | 72.0 | 62.5 | 67.1 | 84.2 | 82.0 | 72.7 | 65.1 |

[a]Meals 1–3 were direct solvent processed; meal 4 was a screwpress product.

**Table 8.4** APPARENT FAECAL DIGESTIBILITIES OF NITROGEN AND AMINO ACIDS IN SOYABEAN, COTTONSEED, AND MEAT AND BONE MEALS

| | *44% SBM* | | | | *48.5% SBM* | | *CSM*[a] | | | | *M&BM* | | *Average values* | | | |
|---|---|---|---|---|---|---|---|---|---|---|---|---|---|---|---|---|
| | *1* | *2* | *3* | *4* | *1* | *2* | *1* | *2* | *3* | *4* | *1* | *2* | *44% SBM* | *48.5% SBM* | *CSM* | *M&BM* |
| Nitrogen | 87 | 90 | 89 | 92 | 86 | 90 | 71 | 77 | 82 | 78 | 72 | 77 | 90 | 88 | 77 | 75 |
| Essential amino acids | | | | | | | | | | | | | | | | |
| Arginine | 92 | 95 | 94 | 96 | 93 | 95 | 86 | 87 | 93 | 90 | 76 | 82 | 94 | 94 | 89 | 79 |
| Histidine | 92 | 94 | 93 | 96 | 92 | 95 | 87 | 80 | 88 | 83 | 72 | 78 | 94 | 94 | 82 | 75 |
| Isoleucine | 85 | 88 | 87 | 90 | 85 | 89 | 58 | 65 | 76 | 69 | 66 | 71 | 88 | 87 | 67 | 69 |
| Leucine | 86 | 88 | 88 | 91 | 86 | 89 | 64 | 68 | 78 | 72 | 75 | 88 | 88 | 88 | 71 | 72 |
| Lysine | 88 | 90 | 89 | 93 | 99 | 91 | 49 | 58 | 69 | 61 | 68 | 76 | 90 | 90 | 59 | 72 |
| Methionine | 78 | 87 | 81 | 88 | 83 | 87 | 63 | 65 | 80 | 67 | 67 | 79 | 84 | 85 | 69 | 73 |
| Phenylalanine | 88 | 90 | 89 | 93 | 88 | 90 | 77 | 78 | 86 | 81 | 70 | 77 | 90 | 89 | 81 | 74 |
| Threonine | 85 | 87 | 85 | 90 | 84 | 87 | 61 | 64 | 76 | 67 | 66 | 73 | 87 | 86 | 67 | 70 |
| Tryptophan | 89 | 91 | 91 | 94 | 88 | 92 | — | 71 | 84 | 74 | 67 | 55 | 91 | 90 | 76 | 61 |
| Valine | 86 | 88 | 86 | 90 | 85 | 88 | 67 | 68 | 80 | 72 | 70 | 76 | 88 | 87 | 72 | 73 |
| Average essential amino acid digestibility | 86.9 | 89.8 | 88.3 | 92.1 | 88.3 | 90.3 | 68.0 | 70.4 | 81.0 | 73.6 | 69.7 | 75.5 | 89.4 | 89 | 73.3 | 71.8 |

[a]Meals 1–3 were direct solvent processed; meal 4 was a screwpress product.

**Table 8.5** DIFFERENCES BETWEEN N AND AMINO ACID DIGESTIBILITIES AT THE TERMINAL ILEUM AND OVER THE TOTAL TRACT (TOTAL TRACT VALUE—ILEAL VALUE)

| | *44% SBM* | | | | *48.5% SBM* | | *CSM*[a] | | | | *M&BM* | | *Average values* | | | |
|---|---|---|---|---|---|---|---|---|---|---|---|---|---|---|---|---|
| | *1* | *2* | *3* | *4* | *1* | *2* | *1* | *2* | *3* | *4* | *1* | *2* | *44% SBM* | *48.5% SBM* | *CSM* | *M&BM* |
| Nitrogen | 8 | 10 | 7 | 7 | 8 | 11 | 2 | 5 | 3 | 3 | 13 | 11 | 8 | 9 | 3 | 12 |
| Essential amino acids | | | | | | | | | | | | | | | | |
| Arginine | 3 | 6 | 2 | 4 | 3 | 7 | 1 | 0 | 1 | 0 | 2 | 4 | 3 | 5 | 0 | 3 |
| Histidine | 5 | 7 | 4 | 6 | 7 | 10 | 2 | 1 | 3 | 2 | 7 | 10 | 6 | 9 | 2 | 8 |
| Isoleucine | 3 | 6 | 4 | 5 | 3 | 10 | −3 | −1 | 0 | −1 | 7 | 6 | 5 | 6 | −1 | 7 |
| Leucine | 5 | 7 | 5 | 6 | 7 | 11 | 0 | −1 | 1 | −1 | 7 | 6 | 5 | 9 | 0 | 6 |
| Lysine | 2 | 5 | 0 | 4 | 5 | 7 | −4 | −4 | −1 | −3 | 7 | 9 | 3 | 6 | −3 | 8 |
| Methionine | −9 | −3 | 4 | 0 | −6 | −2 | −3 | 0 | −2 | 1 | −5 | 5 | 8 | 6 | −1 | 0 |
| Phenylalanine | 5 | 4 | 1 | 5 | 4 | 7 | −1 | 1 | −1 | −1 | 4 | 6 | 4 | 5 | 0 | 5 |
| Threonine | 12 | 12 | 8 | 11 | 11 | 14 | 6 | 2 | 7 | 2 | 16 | 12 | 11 | 13 | 4 | 14 |
| Tryptophan | 8 | 13 | 9 | 10 | 12 | 13 | – | 2 | 6 | 6 | 12 | 5 | 10 | 12 | 4 | 8 |
| Valine | 5 | 8 | 5 | 6 | 6 | 11 | 1 | 0 | 1 | 1 | 9 | 8 | 7 | 9 | 1 | 8 |
| Average essential amino acid digestibility | 3.9 | 6.5 | 4.2 | 5.7 | 5.2 | 8.8 | −0.1 | 0 | 1.5 | 0.6 | 6.6 | 7.1 | 6.2 | 8.0 | −0.6 | 6.7 |

[a]Meals 1–3 were direct solvent processed; meal 4 was a screwpress product.

**Table 8.6** TOTAL AND DIGESTIBLE PROTEIN AND SELECTED AMINO ACID CONTENT OF SOME HIGH-PROTEIN FEEDSTUFFS MEASURED AT THE END OF THE SMALL INTESTINE[a]

| | *Total*[b] | | | | *Digestible*[b] | | | |
|---|---|---|---|---|---|---|---|---|
| | *44%* SBM | *48.5%* SBM | CSM[c] | *M&BM* | *44% SBM* | *48.5% SBM* | *CSM*[c] | *M&BM* |
| Crude protein | 44.0 | 48.5 | 41.0 | 50.0 | 36.08 | 38.32 | 30.34 | 31.50 |
| Essential amino acids | | | | | | | | |
| Arginine | 3.04 | 3.48 | 4.14 | 2.99 | 2.77 | 3.10 | 3.68 | 2.27 |
| Histidine | 1.07 | 1.17 | 0.97 | 1.03 | 0.94 | 0.99 | 0.78 | 0.69 |
| Isoleucine | 1.79 | 2.03 | 1.14 | 1.22 | 1.49 | 1.64 | 0.78 | 0.76 |
| Leucine | 3.01 | 3.53 | 2.05 | 3.34 | 2.50 | 2.79 | 1.46 | 2.20 |
| Lysine | 2.54 | 2.84 | 1.54 | 2.71 | 2.21 | 2.39 | 0.95 | 1.73 |
| Methionine | 0.68 | 0.67 | 0.60 | 0.75 | 0.58 | 0.67 | 0.42 | 0.55 |
| Cystine | 0.63 | 0.66 | — | — | 0.49 | 0.54 | — | — |
| Phenylalanine | 2.03 | 2.29 | 1.96 | 1.91 | 1.75 | 1.92 | 1.59 | 1.32 |
| Tyrosine | 1.52 | 1.73 | 1.11 | 1.19 | 1.25 | 1.40 | 0.81 | 0.68 |
| Threonine | 1.52 | 1.68 | 1.12 | 1.67 | 1.16 | 1.23 | 0.71 | 0.94 |
| Tryptophan | 0.51 | 0.59 | 0.46 | 0.24 | 0.41 | 0.46 | 0.33 | 0.13 |
| Valine | 3.01 | 3.43 | 2.58 | 3.71 | 2.44 | 2.68 | 1.83 | 2.41 |

[a]Condensed molecular weight of amino acids used in calculating values.
[b]Based on average small intestine digestibilities given in Table 8.3.
[c]Average for three direct-solvent processed and one screwpress meal.

instances, N and amino acids disappeared from the large intestine which is consistent with previously reported data. The greatest disappearance in the large intestine occurred, in general, for amino acids with the lowest digestibilities at the end of the small intestine. In all feedstuffs, the threonine that passed into the large intestine was highly digested in the hindgut. Interestingly, the lysine in CSM and tryptophan in M&BM (the EAAs that had the lowest digestibility in the small intestine) did not follow the usual digestion pattern in the large intestine. Tryptophan in M&BM was less degraded than other amino acids with low digestibilities, and lysine in CSM showed a net synthesis in the large intestine instead of a disappearance. This suggests that the amino acids in CSM which were resistant to enzymatic digestion in the small intestine were also resistant to bacterial degradation in the large intestine.

A net synthesis of methionine in the large intestine was noted for most SBMs and CSMs. Synthesis of methionine and other amino acids in the large intestine has been reported by other workers (Holmes *et al.*, 1974; Zebrowska, Buraczewska and Buraczewski, 1977).

Based on the average ileal digestibilities reported in *Table 8.3*, the digestible protein and amino acid content of 44 and 48.5% SBMs, CSM and M&BM are given in *Table 8.6*. The total amino acid contents are analysed values for the high-protein feedstuffs used in these studies, corrected to the protein contents shown.

## Use of ileal digestibilities in formulating diets

The important question is: Can ileal digestibility values help in formulating diets more precisely? Based on present research, the following guidelines appear sound:

(1) If approximately the same ingredients are used in formulating diets as were used in determining the amino acid requirements, digestible values offer little help unless one decides to reduce the amount of the high-quality protein feedstuff to a minimum and supplement with specific essential amino acids. In the USA, many of the NRC requirements have been determined using maize-SBM diets. Therefore, if only maize and SBM are used in formulating diets, digestible values cannot help because the present requirements compensate for the digestibilities of amino acids in maize and SBM.

(2) If other high protein feedstuffs (M&BM, CSM, peanut or sunflower meal) or grain products with a high fibre content (wheat offal, maize bran, rice bran, etc.) are used in the diet, the use of digestible amino acid values offer great potential in increasing the precision of diet formulation.

In the past, attempts to substitute other high-protein feedstuffs such as CSM, M&BM, peanut or sunflower meals for all or part of the SBM in pig diets have resulted in less-than-expected performance. Some producers become dissatisfied if they find that a feed company has included some of the lower quality meals in their feed—perhaps rightly so, if substitutions have not been made to ensure the same protein quality. Many producers have a negative attitude because of the overall poor pig performance experienced when attempts to use lower-quality protein meals have been made.

No one advocates using lower-quality meals except when they can provide as good or better amino acid profile than that provided by SBM at a lower price. Moreover, as the world demand for soyabean products increases, the supply and price of SBM (at least in the short term) may force the pig industry to use more of the lower-quality meals for survival.

The wide variation in amino acid digestibilities among SBM and the other meals suggests that digestibility values should be used when the lower-quality meals replace a portion of the SBM. This is true but there are other considerations, such as the high level of calcium and phosphorus in M&BM, the presence of free gossypol in CSM, and the wide variations in processing methods which affect availability of nutrients. However, diets can be formulated to overcome these limitations. More feeding trials must be conducted before the ultimate value of formulating practical diets on a digestible amino acid basis can be assessed and a decision reached on how far one can go in substituting lower-quality meals for SBM without sacrificing pig performance.

In this chapter, the results of three trials with growing pig diets are presented. In these CSM or M&BM protein replaced about one-half of the SBM protein in simple maize–SBM or sorghum–SBM diets. Before initiating the first two trials, adequate quantities of 44% CP SBM, 50% CP M&BM, and 41% CP CSM (direct solvent processed) and yellow maize were obtained for digestibility and growing-finishing trials. The high-protein meals and maize were characterized by proximate and amino acid analyses.

Apparent digestibilities of amino acids in the three high-protein meals were determined at the end of the small intestine of growing pigs (Jones, 1981). Digestibilities of selected essential amino acids for the three meals plus the values for maize determined in a previous trial (Cousins *et al.*, 1981) are shown in *Table 8.7*. The total content and ileal values shown in *Table 8.8* were used in formulating diets.

**Table 8.7** ILEAL DIGESTIBILITIES OF SELECTED AMINO ACIDS IN FEED INGREDIENTS[a]

| *Amino acids* | *Maize* | *SBM* | *M&BM* | *CSM* |
|---|---|---|---|---|
| Lysine | 84.6 | 86.1 | 67.1 | 69.8 |
| Tryptophan | 71.7 | 80.9 | 49.7 | 77.6 |
| Threonine | 75.0 | 72.9 | 60.8 | 69.3 |
| Isoleucine | 81.5 | 81.5 | 64.5 | 76.2 |

[a]Means for six observations.

**Table 8.8** TOTAL AND DIGESTIBLE VALUES FOR SELECTED AMINO ACIDS IN MAIZE AND HIGH-PROTEIN MEALS USED[a]

| *Amino acids* | *Maize* | *SBM* | *M&BM* | *CSM* |
|---|---|---|---|---|
| Lysine | 0.26 (0.22) | 2.59 (2.23) | 2.43 (1.63) | 1.54 (1.07) |
| Tryptophan | 0.046 (0.033) | 0.44 (0.36) | 0.23 (0.11) | 0.42 (0.33) |
| Threonine | 0.28 (0.21) | 1.55 (1.13) | 1.50 (0.91) | 1.10 (0.76) |
| Isoleucine | 0.27 (0.22) | 1.75 (1.43) | 1.10 (0.71) | 1.11 (0.85) |

[a]As-fed basis.

FIRST COTTONSEED MEAL TRIAL

The basis used for formulating diets in the CSM trial are outlined below:

| | | |
|---|---|---|
| Diet 1. Control | all SBM | 14.0% CP maize–SBM (0.63% lysine). |
| Diet 2. CSM | CP basis | 10% CSM included in diet formulated to contain same CP as diet 1. |
| Diet 3. CSM | digestible lysine basis | As diet 2 plus enough of the first limiting amino acid (lysine) was added to give same digestible lysine as diet 1. |
| Diet 4. CSM | digestible lysine basis + 10% | As diet 3 plus enough synthetic lysine to give 110% of the digestible lysine content as diet 1. |
| Diet 5. Maximum CSM | | Contained same amount of SBM as diets 2 to 4, but enough CSM was included to provide the same total lysine as diet 1. |

The all SBM control diet (diet 1) was marginally deficient in total lysine compared to the 1979 NRC requirement (0.63 vs 0.70%). Such a control diet was used as the base for comparing the CSM-containing diets so that small changes in amino acid content would be reflected in pig performance.

Pig performance for the control and four CSM-containing diets are shown in *Table 8.9*. When 10% CSM was included and the diet formulated on a CP basis (diet 2) average daily gains (ADG) and feed:gain ratios were less desirable ($P < 0.01$) than for diet 1 (0.62 vs 0.74 kg/day and 2.92 vs 2.50 kg feed/kg liveweight gain. The digestible lysine content of diet 2 was much less than 1 (0.44 vs 0.54%) which was probably the primary factor limiting growth and feed efficiency. When supplementary lysine was added (diet 3) to give the same digestible lysine as diet 1, both liveweight gain and feed:gain ratio were markedly better than diet 2 although the advantages for diet 1 over diet 3 were significant. This reflects the higher crude fibre content of CSM.

Diet 4, which was formulated to provide 10% more lysine than diet 3, gave improved gain and feed:gain ratio over diet 3 (0.75 vs 0.71 kg/day and 2.56 vs 2.69) which suggests that lysine was still the first limiting amino acid. Diet 5 contained the same percentage of SBM as diets 2–4, but additional CSM was included (16.4% of the total diet) to provide as much of the lysine requirement from CSM as possible. As expected, the daily gain and feed:gain ratio of pigs fed diet 5 most closely paralleled those that were fed the diet with a similar digestible lysine content (diet 3). Feed intake, daily gain and feed:gain ratio of diets 3 and 5 were, respectively, 1.91 vs 1.94 kg/day, 0.71 vs 0.71 kg/day and 2.69 vs 2.74.

These results suggest that CSM (direct-solvent processed) can replace up to one-half of the SBM protein when synthetic lysine is added to give the same digestible lysine content as a 14.0% CP maize SBM diet and to obtain similar gains. However, feed efficiencies of the CSM-containing diets will be less desirable because of CSM's lower metabolizable energy content.

**Table 8.9** EFFECT OF SUBSTITUTING CSM FOR SBM ON DIGESTIBLE LYSINE BASIS ON GROWING PIG PERFORMANCE[a] (FIRST COTTONSEED MEAL TRIAL)

| *Supplementary protein:* | *SBM* | *10% CSM + SBM* | | | *16.4% CSM + CSM* | *Comparisons*[b] | | | |
|---|---|---|---|---|---|---|---|---|---|
| *Basis for diet formulation:* | *Control* | *Same CP as 1* | *Same dig. lysine as 1* | *As 4 + 10% lysine* | *Add CSM to same total lysine as 1* | *1 vs* | *1 vs* | *3 vs* | *1 vs* |
| *Diet no:* | *1* | *2* | *3* | *4* | *5* | *2* | *3* | *4* | *5* |
| Daily gain (kg) | 0.74 | 0.62 | 0.71 | 0.75 | 0.71 | ** | ** | NS | NS |
| Feed intake (kg/day) | 1.86 | 1.81 | 1.91 | 1.93 | 1.94 | NS | NS | NS | NS |
| Feed: gain ratio (kg feed/kg liveweight gain) | 2.50 | 2.92 | 2.69 | 2.56 | 2.74 | ** | ** | ** | ** |
| Calculated analyses (%) | | | | | | | | | |
| Crude protein (N × 6.25) | 14.01 | 14.01 | 14.13 | 14.19 | 16.12 | | | | |
| Total lysine | 0.63 | 0.54 | 0.64 | 0.70 | 0.63 | | | | |
| Digestible lysine | 0.54 | 0.44 | 0.54 | 0.59 | 0.50 | | | | |

[a]12 pigs per treatment. (six pens, two pigs per pen), 21.0 kg initial wt, 35 day trial.
[b]$*P < 0.05$, $**P < 0.01$.

MEAT AND BONE MEAL TRIAL

The basis used for formulating diets in the M&BM trial are outlined below:

| | | |
|---|---|---|
| Diet 1. Control | (all SBM) | 15.3% CP maize-SBM (0.70% lysine and 0.115% tryptophan) |
| Diet 2. M&BM | CP basis | M&BM replaced one-half of the SBM, diet formulated to contain same CP as diet 2. |
| Diet 3. M&BM | digestible tryptophan | As diet 2 plus enough of the first limiting amino acid (tryptophan) was added to give same digestible tryptophan as diet 1. |
| Diet 4. M&BM | digestible tryptophan + digestible lysine basis | As diet 3 plus enough of the second limiting acid (lysine) to give same digestible lysine as diet 1. |
| Diet 5. M&BM | digestible tryptophan + digestible lysine + digestible isoleucine + digestible threonine basis | As diet 4 plus enough synthetic isoleucine and threonine added to give same digestible isoleucine and digestible threonine as diet 1. |

Pig performance for the four M&BM-containing diets and the control diet are shown in *Table 8.10*. When M&BM replaced one-half of the SBM protein and the diet formulated to contain the same CP (diet 2) as diet 1, daily gain and feed:gain

**Table 8.10** EFFECT OF SUBSTITUTING 50% M&BM FOR SBM PROTEIN ON A DIGESTIBLE AMINO ACID BASIS ON GROWING PIG PERFORMANCE[a]

| *Supplementary protein:* | *100% SBM* | *M&BM for 50% SBM* | | | |
|---|---|---|---|---|---|
| *Basis for diet formulation:* | *Control* | *CP basis* | *Digestible tryptophan* | *Digestible tryptophan + lysine* | *Digestible tryptophan + lysine + isoleucine + threonine* |
| *Diet no:* | *1* | *2* | *3* | *4* | *5* |
| Daily gain (kg) | 0.69 | 0.59 | 0.61 | 0.71 | 0.69 |
| Feed intake (kg/day) | 1.67 | 1.53 | 1.60 | 1.78 | 1.72 |
| Feed: gain ratio (kg feed/kg liveweight gain) | 2.42 | 2.63 | 2.63 | 2.49 | 2.49 |
| Calculated analyses (%) | | | | | |
| Crude protein (N × 6.25) | 15.2 | 15.2 | 15.2 | 15.3 | 15.4 |
| Total tryptophan | 0.115 | 0.091 | 0.118 | 0.118 | 0.118 |
| Digestible tryptophan | 0.091 | 0.064 | 0.091 | 0.091 | 0.091 |
| Total lysine | 0.700 | 0.662 | 0.662 | 0.738 | 0.738 |
| Digestible lysine | 0.603 | 0.527 | 0.527 | 0.603 | 0.603 |
| Total isoleucine | 0.537 | 0.464 | 0.464 | 0.464 | 0.541 |
| Digestible isoleucine | 0.438 | 0.361 | 0.361 | 0.361 | 0.438 |
| Total threonine | 0.505 | 0.490 | 0.490 | 0.490 | 0.516 |
| Digestible threonine | 0.371 | 0.345 | 0.345 | 0.345 | 0.371 |

[a]Ten individually-fed pigs per treatment, 19.6 kg initial wt, 35 day trial.

ratio were less desirable than those fed the control diet (0.59 vs 0.69 kg/day and 2.63 vs 2.42). When tryptophan was added to provide the same digestible L-tryptophan (diet 3) as diet 1, daily gain was only slightly higher than those fed diet 2 (0.61 vs 0.59 kg/day). This suggests that tryptophan is first limiting but other amino acid(s) are also limiting since pig performance on diet 3 was much less desirable than those fed the control diet. When both tryptophan and lysine were added to give the same digestible tryptophan and digestible lysine (diet 4) as in the control diet, ADG increased sharply and was slightly higher than for pigs fed the control diet (0.71 vs 0.69 kg/day) and feed:gain ratio approached that of the control diet (2.49 vs 2.42). Addition of isoleucine and threonine in addition to tryptophan and lysine to give the same digestible amino acid levels as the control diet (diet 5) failed to improve pig performance over that realized when only tryptophan and lysine were added. A small response to isoleucine was expected since diet 4 contained only 0.46% total isoleucine which is slightly below the suggested NRC requirement (0.50%) for the 20 to 40 kg weight pig.

These results suggest that M&BM can replace up to one-half of the SBM protein in growing-finishing diets when synthetic tryptophan and lysine are added to give the same digestible tryptophan and lysine content as a 15.2% CP maize SBM diet and obtain similar gains and feed efficiency.

## SECOND COTTONSEED MEAL TRIAL

The third trial also substituted 50% CSM protein for SBM protein in a growing pig trial but the procedure varied from the first trial in three distinct ways. First, a commercial CSM (direct, solvent-processed) was purchased, analysed and the average ileal amino acid digestibility values determined in previous trials were used to estimate digestible amino acid levels in the meal (*Table 8.11*). Secondly, diets were made isocaloric by addition of animal fat, and thirdly, sorghum was used in place of maize.

**Table 8.11** PERCENTAGES OF CRUDE PROTEIN AND SELECTED AMINO ACIDS OF FEEDSTUFFS (ILEAL DIGESTIBILITIES)[a]

| | *SBM* | *CSM* | *Sorghum* |
|---|---|---|---|
| Crude protein (N × 6.25) | 39.2 | 39.2 | 9.0 |
| Lysine | 2.32 (86) | 1.56 (62) | 0.22 (73) |
| Tryptophan | 0.45 (80) | 0.44 (72) | 0.07 (80) |
| Threonine | 1.36 (75) | 1.09 (63) | 0.30 (77) |
| Isoleucine | 1.59 (82) | 1.09 (68) | 0.38 (87) |

[a]N and amino acid values were determined. Digestibility values are means of previously determined values.

A marginally lysine-deficient diet for the growing pig (0.60%) was used as the control (diet 1, *Table 8.12*). In diets 2–4, CSM protein replaced one-half of the SBM protein. Synthetic lysine was added to diet 2 to give it the same level of digestible lysine (0.495%) as diet 1. The same amount of SBM was used in diet 3 as in diets 1 and 2 but CSM was increased from 9.35 to 14.69% in an effort to obtain the same level of total lysine (0.60%) as diet 1. Diet 4 was the same as diet 3 except synthetic lysine was added to give the same digestible lysine content as diets 1 and 2 (0.495%).

**Table 8.12** COMPOSITION OF DIETS USED IN SECOND COTTONSEED MEAL TRIAL

| *Supplemental protein:* | *100% SBM* | *50:50 CSM:SBM protein* | | |
|---|---|---|---|---|
| *Basis of substitution:* | *Control* | *Digestible lysine* | *Total lysine* | *Total lysine + digestible lysine* |
| *Diet no:* | *1* | *2* | *3* | *4* |
| Sorghum | 75.54 | 75.54 | 69.85 | 69.85 |
| Soyabean meal (39.2%) | 18.70 | 9.35 | 9.35 | 9.35 |
| Cottonseed meal (39.2%) | — | 9.35 | 14.69 | 14.69 |
| Fat | — | 1.81 | 3.00 | 3.00 |
| Starch | 2.85 | 0.828 | 0.07 | 0.071 |
| $FeSO_4.H_2O$ | — | 0.09 | 0.13 | |
| Lysine monohydrochloride (98%) | — | 0.122 | — | 0.069 |
| [a]Other | 2.91 | 2.91 | 2.91 | 2.91 |
| Total | 100.00 | 100.000 | 100.00 | 100.000 |
| Analysis (%) | | | | |
| Crude protein (N × 6.25) | 14.13 | 14.24 | 15.72 | 15.79 |
| Total lysine | 0.600 | 0.625 | 0.600 | 0.654 |
| Digestible lysine | 0.495 | 0.495 | 0.441 | 0.495 |
| Metabolizable energy (kcal/kg) | 3076 | 3076 | 3076 | 3073 |
| Crude fibre | 3.20 | 3.76 | 4.32 | 4.32 |

[a]All diets contained 0.64% ground limestone, 1.52% defluorinated phosphate, 0.35% salt, 0.15% trace mineral mix, and 0.25% vitamin premix.

**Table 8.13** EFFECT OF SUBSTITUTING 50% CSM FOR SBM PROTEIN ON GROWING PIG PERFORMANCE

| *Protein supplement:* | *100% SBM* | *50:50 CSM:SBM protein* | | | |
|---|---|---|---|---|---|
| *Basis of substitution:* | *Control* | *Digestible lysine* | *Total lysine* | *Total lysine + digestible lysine* | *NRC requirement* |
| Average daily gain (kg) | 0.70 | 0.64 | 0.62 | 0.67 | 20–35 |
| Average feed intake (kg/day) | 1.75 | 1.68 | 1.69 | 1.72 | kg |
| Feed:gain ratio (kg feed/ kg liveweight gain) | 2.51 | 2.61 | 2.72 | 2.56 | pig |
| Calculated analyses, % | | | | | |
| Crude protein (N × 6.25) | 14.13 | 14.24 | 15.72 | 15.79 | |
| Total lysine | 0.600 | 0.625 | 0.600 | 0.654 | 0.70 |
| Digestible lysine | 0.495 | 0.495 | 0.441 | 0.495 | |
| Total isoleucine | 0.584 | 0.538 | 0.574 | 0.574 | 0.50 |
| Digestible isoleucine | 0.492 | 0.440 | 0.461 | 0.461 | |
| Total threonine | 0.481 | 0.456 | 0.497 | 0.497 | 0.45 |
| Digestible threonine | 0.365 | 0.334 | 0.357 | 0.357 | |
| Crude fibre | 3.20 | 3.76 | 4.32 | 4.32 | |

[a]Ten individually-fed pigs per treatment; average initial weight was 24.2 kg; 42 day trial.

Pig performance for the three CSM-containing diets and the control diet are shown in *Table 8.13*. When CSM replaced one-half of the SBM protein and synthetic lysine was added to give the same digestible lysine content (diet 2) as diet 1, pigs fed the CSM-diet tended to consume less feed (1.68 vs 1.75 kg/day), gain less (0.64 vs 0.70 kg/day) and exhibited slightly higher feed:gain ratios (2.61 vs 2.51) than those fed diet 1. However, pig performance was slightly less desirable for diet

3 (additional CSM added to give equal total lysine to diet 1) than diet 2 (0.62 vs 0.64 kg/day daily gain and 2.72 vs 2.61 feed:gain ratio). However, when synthetic lysine was added (diet 4) to give the same digestible lysine as diet 1, daily gain was more similar to diet 1 (0.67 vs 0.70 kg%day) as was feed:gain ratio (2.56 vs 2.51). Although the total isoleucine and threonine content of diet 2 should have been adequate based on the NRC requirements (0.538% vs 0.50% for isoleucine and 0.456% vs 0.45% for threonine), diets 1 and 4 which supported better performance contained higher amounts of isoleucine and threonine which were calculated to be the second and third limiting amino acids in the diets.

Although pig performance for the CSM-containing diets failed to reach the performance of the SBM control diet, performance was much better than if diets had been formulated on a crude protein or total lysine basis.

## Future activities

Efforts will continue to determine the practicality and economic value of substituting lower-quality meals for a portion of the SBM on a digestible amino acid basis in practical pig diets. The lower ileal lysine digestibility for wheat offal than wheat and wheat flour (66.4 vs 78 and 82.2%; Sauer, Strothers and Parker, 1977) suggests that using ileal digestibility values may give improved performance when such grain by-products containing a high percentage of fibre are used. Digestibility values at the end of the small intestine need to be determined for the grain by-products that are commonly used in pig feeding. Growing-finishing trials then need to be conducted to determine pig performance when various levels of these by-products are included in diets on a digestible amino acid basis.

Reliable ileal digestibility values may also help in formulating diets when the level of SBM is reduced to a minimum and synthetic amino acids in addition to lysine are added. Application of this possibility is a moot point at present because lysine and methionine are the only synthetic amino acids available on a feed-grade basis. However, if tryptophan and threonine become available (as they ultimately will), ileal digestibility values provide the most precise basis for formulating diets that require the least amount of synthetic amino acid additions to optimize pig performance and costs.

### AIMS

What needs to be accomplished to enable ileal digestibility data to be widely used in diet formulation? Experiments to obtain digestibility values are costly and require much time and effort, therefore it appears that our future goals should be to:

(1) predict amino acid digestibilities in a feedstuff from N digestibility,
(2) predict N digestibility in a feedstuff by *in vitro* tests.

Danish scientists use the N digestibility of a feedstuff as the digestibility value for each amino acid in formulating diets. However, a careful study of *Tables 8.3* and *8.4* shows rather wide variations in digestibilities between N and individual amino acids among the meals; but the variations of individual amino acids from N digestibility appear to be fairly consistent for a specific feedstuff. With enough data on specific meals, it will be possible to develop relationships between individual amino acids and N digestibility to make this an effective procedure.

If this could be accomplished, the next goal would be to develop one or more *in vitro* tests that would accurately predict N digestibility of a high-protein meal. Much work will be required before this is possible, but it appears to be a realistic goal, that hopefully can be accomplished in the next few years.

## Acknowledgements

The authors recognize the contributions of several people in obtaining the data presented. Foremost is Dr Teresa Zebrowska of the Polish Academy of Science, who spent a year (1978) in their Texas A&M University laboratory, and the large number of graduate students who have contributed greatly to the total effort. These include Bob Easter, Ken Purser, Bart Cousins, Bill Vandergrift, Jim Corley, John Hamstreet, Bryan Rudolph, Frank Owsley, Lynne Boggs, Keith Haydon, Frank Lin, Bob Jones and Carl Dobler.

## References

BATTERHAM, E.S., MURISON, R.D. and LEWIS, C.E. (1979). *J. Nutr.*, **41**, 383–391

BURACZEWSKA, L., BURACZEWSKI, S. and ZEBROSKA, T. (1975). *Roczniki Nauk Rolniczych*, **B-97**, 103–115

CHO, C.Y. and BAYLEY, H.S. (1972). *Can. J. Physiol. Pharmacol.*, **50**, 513–522

COUSINS, B.W., TANKSLEY, T.D. Jr., KNABE, D.A. and ZEBROWSKA, T. (1981), *J. Anim. Sci.*, **53**, 1524–1537

DAMMERS, J. (1964). *Leuven. Inst. Veevoedingsonderzoek. Hoorn.* 152 pp.

DARCY, B., LAPLACE, J.P. and VILLERS, P.A. (1980). *Ann. Zootech.*, **29**, 147–177

DARCY, B. and RERAT, A. (1983). *Proc. 4th Int. Symp. Protein Metab. Nutr.* pp. 233–244. Clermont-Ferrand; France

EBERSDOBLER, H. (1976). In *Protein Metabolism and Nutrition*, pp. 139–158. Eds D.J.A. Cole, K.N. Boorman, P.J. Buttery. D. Lewis, R.J. Neale and H. Swan. Butterworths; London

EGGUM, B.O. (1973). *Beretning fra Forsogslaboratoriet. Landokonomisk Forsogslaboratorium, 1958.* Copenhagen V.; Denmark

HOLMES, J.H.G., BAYLEY, H.S., LEADBEATER, P.A. and HORNEY, F.D. (1974). *Br. J. Nutr.*, **32**, 479–489

IVAN, M. and FARRELL, D.J. (1976). *Anim. Prod.*, **23**, 111–119

JONES, R.W. (1981). Undergraduate Honors Thesis. Texas A&M University, College Station; Texas

JUST, A. (1980). In *Current Concepts of Digestion and Absorption in Pigs*. Eds Low, A.G. and Partridge, I.G. Technical Bulletin 3. NIRD; Reading, England

JUST, A., JORGENSEN, M. and FERNANDEZ, J.A. (1981). *Br. J. Nutr.*, **46**, 209–219

KUIKEN, K.A. and LYMAN, C.M. (1948). *J. Nutr.*, **36**, 359–368

LIN, F.D-T. (1983). PhD Dissertation. Texas A&M University, College Station; Texas

LOW, A.G. (1976). *Proc. Nutr. Soc.*, **35**, 57–62

LOW, A.G. (1979). *Br. J. Nutr.*, **41**, 147–156

MAXON, E.D. and ROONEY, L.W. (1972). *Crop. Sci.*, **12**, 253

NATIONAL RESEARCH COUNCIL (1979). *Nutrient Requirements of Swine,* Eighth revised edition. National Academy of Sciences; Washington

OUSTERHOUT, L.E., GRAU, C.R. and LUNDHOLM, B.D. (1959). *J. Nutr.*, **69**, 65–73

OWSLEY, W.F., KNABE, D.A. and TANKSLEY, T.D. Jr. (1981). *J. Anim. Sci.*, **52**, 557–566

POPPE, S., MEIER, M.H. and WIESEMULLER, W. (1970). *Archiv fur Tierernahrung*, **21**, 572

POPPE, S. (1976). *Protein Metabolism and Nutrition*, p. 369. EAAP Publication 16

SAUER, W.C., STROTHERS, S.C. and PHILLIPS, G.D. (1977). *Can. J. Anim. Sci.*, **57**, 585–597

SAUER, W.C., STROTHERS, S.C. and PARKER, R.J. (1977). *Can. J. Anim. Sci.*, **57**, 775–784

TAVERNER, M.R. and FARRELL, D.J. (1981). *Br. J. Nutr.*, **46**, 159–171

TAVERNER, M.R., HUME, I.D. and FARRELL, D.J. (1981). *Br. J. Nutr.*, **46**, 149–158

ZEBROWSKA, T. (1973a). *Roczniki Nauk Rolnizych*, **B-95**, 85–90

ZEBROWSKA, T. (1973b). *Roczniki Nauk Rolnizych*, **B.-95**, 115–134

ZEBROWSKA, T. (1973c). *Roczniki Nauk Rolnizych*, **B-95**, 135–156

ZEBROWSKA, T. (1975). *Roczniki Nauk Rolnizych*, **B-97**, 117–123

ZEBROWSKA, T. and BURACZEWSKI, S. (1977). *Vth International Symposium on Amino Acids*, Budapest

ZEBROWSKA, T., BURACZEWSKA, L. and BURACZEWSKI, S. (1977). *Vth International Symposium on Amino Acids*, Budapest

ZEBROWSKA, T., BURACZEWSKA, L., PASTUSZEWSKA, B., CHAMBERLAIN, A.G. and BURACZEWSKI, S.(1977).*Vth International Symposium on Amino Acids*,Budapest

ZEBROWSKA, T. (1978). *Feedstuffs*, **50**(53), 15–17, 43–44

9

# USE OF SYNTHETIC AMINO ACIDS IN PIG AND POULTRY DIETS

K. E. BACH KNUDSEN and H. JØRGENSEN
*National Institute of Animal Science, Copenhagen, Denmark*

## Introduction

The pig and poultry industries are the two most intensive livestock industries in the world. Intensive livestock production requires, primarily, two main inputs, namely net energy and digestible amino acids, together with vitamins and trace minerals. The energy of the feed is most commonly derived from cereals and to a lesser extent from cereal by-products and other industrial by-products. In addition to supplying energy, the cereals also supply about 50% of the digestible amino acids in typical pig and poultry diets. The remainder is supplied by protein concentrates of which fish meal, soyabean meal, skim milk powder, cottonseed and rape seed are the most important.

Recent advances have been made in developing industrial processes for economical production of synthetic amino acids. Lysine and methionine, have been available economically for several years.

A typical problem in feed formulation is the achievement of an optimum balance of nutrients to meet the animal's requirement from a particular range of raw materials. As the ratios between the individual amino acids in protein concentrates vary substantially, there may be occasions when it is not possible, within the range of raw materials available, to meet the animal's requirement for all amino acids. In these situations supplementation with free synthetic amino acids would be very useful. However, the *in vivo* utilization of added free synthetic amino acids is still open to considerable discussion. Studies by Batterham (1974) and Batterham and O'Neill (1978) have indicated a significantly lower utilization of free synthetic amino acids than of protein bound amino acids.

Several recent papers have reviewed the processes involved in the digestion, absorption and metabolism of amino acids and their nutritional implications (Low, 1981; Rérat, 1981; Just, 1983; Riis, 1983). This chapter deals with the use of free synthetic amino acids. The topic is discussed in relation to digestion and absorption, protein and energy metabolism and the uses of free synthetic amino acids in pig and poultry diets.

## The physiological basis for utilization of synthetic amino acids

### DIGESTION AND ABSORPTION

The supply of amino acids to the sites of protein synthesis is influenced by several factors. Among these the separation of the individual dietary constituents in the stomach and their rate of passage from the stomach may play an important role. Feeding a diet comprising barley, wheat and fish meal to pigs, Low (1979) found that during the first hour after the meal, 60% of the nitrogen passing the pylorus was in the form of free amino acids or short peptides, 16% was soluble protein and long peptides, and 24% was insoluble proteins. These figures were significantly different from the mean values found 24 h after feeding, when free amino acids or short peptides contributed 43%, soluble protein and long peptides 11% and insoluble protein 46%. Thus, easily hydrolysable nitrogen passes through the stomach very shortly after feeding. Free synthetic amino acids will pass through the stomach together with the soluble fraction of the protein. This was demonstrated in the studies of Buraczewska, Zebrowska and Buraczewski (1978). Pigs were fitted with re-entrant cannulae placed 20 to 110 cm distal to the pylorus and fed twice daily (08.00 and 20.00) with diets containing different levels of lysine. A basal diet, containing 84.2% barley, 13.3% groundnut oilmeal and 2% minerals and vitamins but deficient in lysine (3.5 g/16 g N), was supplemented with synthetic lysine to obtain diets containing 4.7 and 5.6 g/16 g N. The crude protein content of the diets was kept constant at 150 g/kg. The amount of nitrogen leaving the stomach declined with time after feeding and averaged 2.7 g N/30 min for all three diets in digesta collected between 16.00 and 20.00. The lysine content of digesta of pigs fed the basal diet (diet A) changed little with time after feeding and was in the range 3.5–3.6 g/16 g N (*Figure 9.1*). In contrast pigs fed lysine supplemented diets (diets B and C) showed a gradual decrease in lysine concentration in digesta as time after feeding increased. The initial lysine concentration of digesta collected within the first half hour after feeding was positively correlated to dietary lysine concentration. Furthermore, the decrease in lysine concentration of digesta was

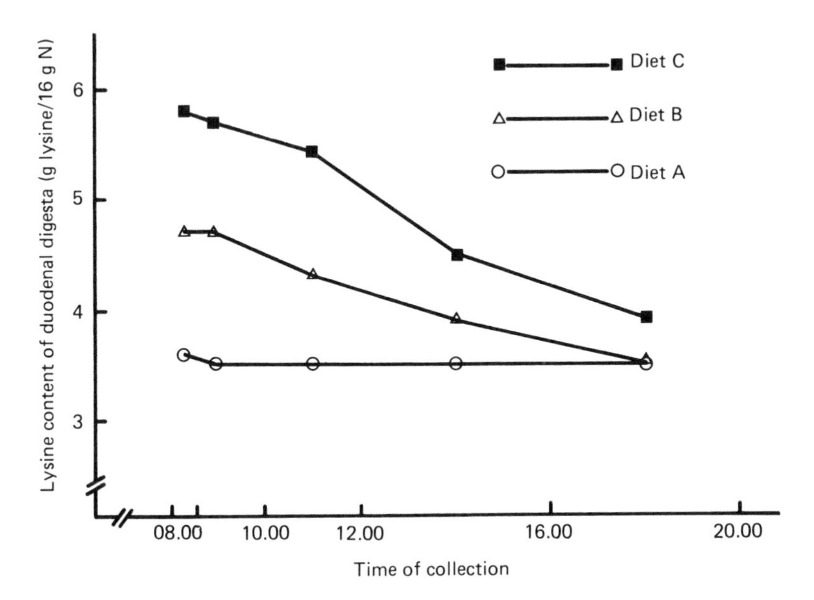


**Figure 9.1** The content of lysine (g/16 g N) in five consecutive samples of duodenal digesta of pigs fed twice daily (08.00 and 20.00) and fed diets without (diet A) or with supplemented synthetic lysine (diets B and C). (Adapted from Buraczewska, Zebrowska and Buraczewski, 1978)

**Table 9.1** THE INFLUENCE OF TIME ON THE APPARENT ABSORPTION OF DIETARY METHIONINE (PERCENTAGE OF INTAKE)

| *Time after feeding* (min) | *Protein-bound methionine* | *Free methionine* | *Ratio free methionine: bound methionine* |
|---|---|---|---|
| 15 | 29.1 | 36.5 | 1.25 |
| 30 | 33.2 | 43.6 | 1.39 |
| 60 | 40.9 | 54.0 | 1.32 |
| 120 | 62.2 | 77.3 | 1.24 |
| 240 | 95.7 | 96.1 | 1.00 |

(Adapted from Canolty and Nasset, 1975)

inversely correlated to both initial and dietary lysine concentrations. Thus, the rate of passage of the synthetic amino acids from the stomach to the small intestine was faster than that of protein-bound amino acids.

Investigations which have compared the rate of absorption of free synthetic amino acids and protein-bound amino acids have generally given the same results. *Table 9.1* gives the result of an experiment with rats offered methionine in the free form or as a protein-bound methionine derived from dried egg white (Canolty and Nasset, 1975). Early in the postprandial phase (15–60 min), 25–40% more of the free than the protein-bound methionine had been absorbed. The differences were negligible after 2 h. In an experiment with pigs Rérat, Corring and Laplace (1976) compared free methionine with methionine from fish meal. They found that the differences in absorption rate were restricted to the first 2–3 h after feeding. The absorption coefficient (amount absorbed/amount ingested × 100) 1 h after feeding was 16.3% for free methionine compared with 9.3% for protein-bound methionine. However, after 3 h the absorption was the same whatever the origin of methionine. Similarly Walz (1976a,b) reported that free synthetic lysine added to oat-based diets was absorbed much more rapidly than the protein-bound amino acids.

Frequency of feeding seems to influence the absorption rate of amino acids from the gastrointestinal tract. Buraczewska, Lachowicz and Buracrewski (1980) studied one and four time daily feeding of a barley, sesame oil meal diet supplemented with synthetic lysine. When feeding once a day the rate of passage of nitrogen through the intestine (4 m distal to the pylorus) varied from 0.4 to 3.2 g N/h, while the lysine content of digesta was 3.2 to 4.5 g/16 g N. When feeding four times a day the rate of passage was less variable, ranging from 0.7 to 1.8 g N/h and the lysine content of digesta was in the range 3.4 to 4.1 g/16 g N.

Thus, current knowledge suggests that added free synthetic amino acids pass the stomach together with the easily hydrolysable nitrogen fraction very shortly after feeding and are absorbed from the gut faster than the bulk of protein-bound amino acids.

## PROTEIN METABOLISM

It is difficult to assess the nutritional importance of the absorption rate of free amino acids compared with protein-bound amino acids, because utilization depends on the buffering capacity of the body's metabolic pool of amino acids. From the gut they are absorbed to the portal vein and passed to the liver. The direct role of the liver in protein metabolism is as a short-term pool of amino acids buffering the variation in supply from the gastrointestinal tract. Pool size may vary 20% between meals (Garlick, Millward and James, 1973). Studies of protein synthesis and degradation show a high fractional rate of synthesis in the gastrointestinal tract and in the liver (Garlick, Burk and Swick, 1976; Simon, Muchmeyer and Bergner, 1978). Although the fractional rate of protein synthesis is significantly lower in muscle, the skeletal muscles are the most important tissues in whole body protein metabolism (Riis, 1983). The rate of whole body protein synthesis and of amino acid oxidation in pigs weighing 35–90 kg, has been measured by Reeds *et al.* (1980). They estimated that in the 35 kg pigs, which apparently absorbed 39.8 g N/day, 82 g protein expressed as nitrogen was synthesized daily, i.e. absorbed protein was 49% of the amount used in synthesis. In terms of nitrogen, 11 g protein was oxidized daily. This nitrogen was almost entirely accounted for by urea excreted in the urine, showing that 68% of the apparently absorbed nitrogen in the 35 kg pigs was retained. At 95 kg the relative proportion of absorbed nitrogen to synthesized N was approximately the same as at 35 kg, while retained nitrogen declined to 28% (Reeds *et al.*, 1980). These data demonstrate the relative increase in catabolism over synthesis as the pigs get older. However, it should be stressed that in spite of growth being zero or negative, a substantial amount of protein is synthesized and broken down daily.

**Table 9.2** THE INFLUENCE OF GRADUAL SUBSTITUTION OF SOYABEAN MEAL (DIET 1) IN SYNTHETIC AMINO ACIDS (DIET 2, 3 AND 4) ON NITROGEN METABOLISM DETERMINED IN BALANCE EXPERIMENTS WITH PIGLETS

| *Diet* | *1* | *2* | *3* | *4* |
|---|---|---|---|---|
| Nitrogen intake (IN) (g) | 23.6$^{a}$ | 21.4$^{b}$ | 18.1$^{c}$ | 12.4$^{d}$ |
| Apparent digestible nitrogen (ADN) (g) | 20.5$^{a}$ | 18.4$^{b}$ | 15.5$^{c}$ | 10.4$^{d}$ |
| Nitrogen retained (RN) (g) | 14.4$^{a}$ | 13.8$^{ab}$ | 12.8$^{b}$ | 8.0$^{c}$ |
| RN/IN (%) | 59.5$^{a}$ | 64.0$^{b}$ | 69.6$^{c}$ | 64.6$^{b}$ |
| RN/ADN (%) | 70.2$^{a}$ | 75.0$^{b}$ | 82.6$^{c}$ | 76.9$^{b}$ |

(Adapted from Eggum *et al.*, 1985b)
Groups in the same row with the same superscript do not differ significantly ($P<0.05$)

Although the data of Reeds *et al.* (1980; 1981) showed that nitrogen absorbed from the gastrointestinal tract accounts for less than 50% of the amino acids used in protein synthesis, several studies indicate that free synthetic amino acids added to diets based on protein-bound amino acids are used less efficiently (Batterham, 1974; Batterham and O'Neill, 1978; Eggum *et al.*, 1985a,b). The results of energy and nitrogen balance experiments with fast growing piglets, initially weighing 9 kg and with final body weight of 16–18 kg are given in *Table 9.2* (Eggum *et al.*, 1985b). The basal diet (1) consisted of soyabean meal, wheat, maize, barley, skim milk powder, fish meal and lard. In diets 2–4, soyabean meal was gradually substituted by a mixture of wheat, maize and free synthetic amino acids to provide a constant level (g/kg DM) of lysine (14), methionine + cystine (8) and threonine (10). As a result of these dietary changes the protein content decreased from 254 g/kg in diet 1 to 158 g/kg in diet 4. Although the utilization of apparently absorbed nitrogen (RN/ADN) was significantly higher in the amino acid supplemented diets (75.0 to 82.6% in diets 2–4 compared with 70.2% in diet 1) the nitrogen retention in absolute terms was significantly lower in diet 3 and especially in diet 4 than in the control diet. The same conclusion could be drawn from the

**Table 9.3** INTAKE OF NET ENERGY ($FU_p$), DIGESTIBLE PROTEIN AND AMINO ACIDS AND DEPOSITION OF PROTEIN PER DAY

| *Diet* | | *Barley +* | | | | |
|---|---|---|---|---|---|---|
| | *Barley* | *Soyabean meal* | *Lysine* | *Threonine Lysine* | *Methionine Threonine Lysine* | *Tryptophan Methionine Threonine Lysine* |
| Net energy intake ($FU_p$/day) | 1.76 | 1.79 | 1.83 | 1.82 | 1.91 | 1.89 |
| *Digestible intakes (g/day)* | | | | | | |
| Crude protein | 121 | 237 | 134 | 136 | 144 | 140 |
| Lysine | 3.4 | 11.8 | 10.9 | 10.7 | 11.2 | 11.4 |
| Threonine | 2.5 | 8.3 | 3.8 | 7.8 | 8.2 | 8.4 |
| Methionine | 1.9 | 3.4 | 1.9 | 1.9 | 4.7 | 4.8 |
| Tryptophan | 1.6 | 3.1 | 1.7 | 1.6 | 1.7 | 3.1 |
| *Deposited protein* | | | | | | |
| (g/day) | 37 | 128 | 70 | 81 | 90 | 93 |
| (% of digested protein) | 31 | 54 | 52 | 59 | 62 | 67 |

(Jørgensen, 1985 unpublished data)

experiment in *Table 9.3* (Jørgensen, unpublished results). Barley, with and without supplementation with soyabean meal, was used as the positive and negative controls respectively. Four diets were prepared by addition to the barley diet of free synthetic amino acids: (i) lysine; (ii) lysine and threonine; (iii) lysine, threonine and methionine; (iv) lysine, threonine, methionine and tryptophan. The chemical composition and the digestibility of the barley and soyabean meal used were estimated before the balance experiment and the free synthetic amino acids were added to obtain the same daily intake as in the positive control. The utilization of absorbed protein and protein deposition were significantly higher with the supplemented diets than with the negative control. However, the daily intake of digestible lysine, threonine, methionine and tryptophan in the supplemented diets was as high as in the positive control; protein deposition in absolute terms was 35 g/day lower than with the positive control. Thus, data from balance experiments with pigs indicated that the utilization of free synthetic amino acids was lower than that of protein-bound amino acids when added in the quantities in *Tables 9.2* and *9.3* (up to 80% of daily intake). This lower utilization seems to be the case irrespective of body weight or the relative rate of protein synthesis to catabolism.

Several factors might be responsible for the lower protein deposition when feeding diets supplemented with high levels of synthetic amino acids. As discussed by Eggum *et al.* (1985b), it is most likely that protein level, and consequently amino acids other than those kept constant due to supplementation, was limiting for maximum protein deposition when feeding diets 3 and 4 (*Table 9.2*). Due to feed refusals, significantly lower intakes of protein and energy were obtained in pigs fed diet 4. This might indicate imbalance in the amino acid composition of absorbed protein. Compared with an earlier investigation (Eggum *et al.*, 1985a), protein deposition when feeding diets 3 and 4 was improved from 11.3 and 6.7 g N/day to 12.8 and 8.0 g N/day by adding threonine to a diet already supplemented with lysine and methionine. However, as seen from the study of Jørgensen (*Table 9.3*), addition of tryptophan to a lysine, threonine and methionine supplemented diet did not improve protein deposition further. Batterham (1974) and Batterham and

O'Neill (1978) stressed the importance of frequent feeding to achieve an optimum utilization of synthetic amino acids. Batterham (1974) reported that the utilization of free synthetic lysine by pigs fed once a day was only 0.43 that of pigs given the same ration in six portions at 3-h interval. Similarly Batterham and O'Neill (1978) found that the response of growth to free synthetic lysine with once daily feeding was 0.67 of that obtained when feeding six times a day. In those trials no improvement in performance was seen when feeding the control diet six times compared with once daily. In contrast Walz (1981; 1983) could not verify any improvement in pig performance by frequent administration of free synthetic lysine. A basal diet deficient in lysine (5 g/kg) was offered in two meals at 07.00 and 17.00.

The lysine supplementation of 4 g/kg was adminstered:

(1) with the basal meal;
(2) 2 g/kg with the basal meal + 2 g/kg 2 h after meal consumption;
(3) 1 g/kg with the basal meal + 1 g/kg 1 h after + 2 g/kg 2 h after the meal consumption.

As expected, supplementation with synthetic amino acids significantly improved N-utilization of all groups. However, it was only for pigs weighing 80–100 kg that the delayed lysine administration improved nitrogen retention compared with the group where lysine was administered twice daily with the basal meal. A similar conclusion was drawn by Partridge, Low and Keal (1985) in an investigation where a meal supplemented with lysine was offered one, two and four times a day. With once daily feeding a significantly poorer nitrogen utilization was seen than with two and four times feeding. Two and four times did not differ significantly from each other. The interpretation of these data is that to achieve an optimum utilization of free synthetic amino acids the meal must be offered at least twice daily. Furthermore, the results of Buraczewska and Buraczewski (1980) demonstrated significantly higher values for nitrogen retention in pigs fed four times compared with once daily. This was the case both for control animals as well as animals given synthetic lysine.

## ENERGY METABOLISM

Any surplus of absorbed amino acids and the obligatory losses of protein have to be oxidized; the carbon skeletons of the amino acids are added to the energy yielding substances, while nitrogen, after transamination and deamination, is excreted from the body in the form of urea. The formation of 1 mol urea requires 4 mol ATP. The synthesis of each mol of ATP needs 85 kJ metabolizable energy. This value corresponds to 10.2% of the energy in protein used for oxidation (19.0 kJ/g). However, the experimental estimates of Martin and Blaxter (1965) were appreciably higher. They estimated energy cost of urea synthesis and excretion to be 11 and 4% of metabolizable energy respectively. Even higher values were obtained by Schiemann *et al.* (1971), and Just (1982). In the experiment of Just (1982) pigs were fed up to twice the daily protein requirement (*Table 9.4*). The heat loss in this experiment corresponded to 35% of the energy in protein. The reason for the big differences between theoretical estimates and the results in these two exeriments is likely to result from a substantial recycling of urea between the

**Table 9.4** THE INFLUENCE OF CRUDE PROTEIN ON THE UTILIZATION OF METABOLIZABLE ENERGY (ME)

| *Diet* | *1* | *2* |
|---|---|---|
| Crude protein (% DM) | 13.2 | 29.4 |
| ME (% DE) | 97 | 94 |
| Deposited energy (% ME) | 36 | 30 |
| Net energy (% ME) | 64 | 57 |

(Adapted from Just, 1982)

liver and the gut (Rérat, 1978; Thacker, Sauer and Jørgensen, 1984). In experiments with pigs fed barley-based diets supplemented with soyabean meal or sunflower meal, at normal protein levels for growing pigs, it was found that up to 30% of the urea produced entered the gut and was degraded to ammonia and carbon dioxide (Thacker, Sauer and Jørgensen, 1984). Some of the ammonia produced from this breakdown enters the portal vein and is transported to the liver where it adds to the nitrogen pool available for the synthesis of non-essential amino acids or reconversion into urea. The net result of recycling from the gut to the liver is a higher requirement of ATP/mol urea excreted than that calculated from the biochemical reactions. Martin and Blaxter (1965) estimated the requirement to be $4.8 \pm 0.7$ mol ATP/mol urea excreted.

The high energy cost of urea synthesis and excretion implies several nutritional advantages of balancing the amino acid composition of absorbed protein close to the requirement. The results of Just (1982) clearly demonstrate that protein is a poor energy source as the net energy of crude protein was only 50% of the gross energy (23.9 kJ/g). The low net energy value is due on the one hand to the energy content of urea and on the other hand to an increased heat loss due to more urea being synthesized and excreted (*Table 9.4*). The net effect in the work of Just (1982) was leaner pigs. However, it should be stressed that overfeeding the pigs with protein is a very expensive way of producing leaner carcasses. In contrast to these results, energy metabolism studies of diets supplemented with amino acids and given to piglets (*Table 9.2*) could not verify a reduced heat loss when the pigs were fed amino acids at requirement levels (*Table 9.5*). Retained energy was the same on all four diets. However, retained protein energy was significantly lower in piglets given diets 3 and 4 compared with control pigs, while the opposite was the case with energy retained as fat. Although it might be believed that the reduced

**Table 9.5** THE INFLUENCE OF GRADUAL SUBSTITUTION OF SOYABEAN MEAL (DIET 1) WITH SYNTHETIC AMINO ACIDS (DIETS 2, 3 AND 4) ON ENERGY METABOLISM DETERMINED IN BALANCE EXPERIMENTS WITH PIGLETS

| *Diet* | *1* | *2* | *3* | *4* |
|---|---|---|---|---|
| Gross energy (MJ/day) | $11.23^{a}$ | $11.43^{a}$ | $11.51^{a}$ | $10.06^{b}$ |
| Metabolizable energy (MJ/day) | $9.47^{a}$ | $9.62^{a}$ | $9.79^{a}$ | $8.63^{b}$ |
| Heat production (MJ/day) | $4.84^{a}$ | $4.76^{ab}$ | $5.02^{a}$ | $4.19^{b}$ |
| Retained energy (RE) (MJ/day) | $4.52^{a}$ | $4.80^{a}$ | $4.70^{a}$ | $4.59^{a}$ |
| Retained protein energy (RPE) (MJ) | $2.14^{a}$ | $2.06^{ab}$ | $1.90^{b}$ | $1.20^{c}$ |
| Retained fat energy (RFE) (MJ) | $2.39^{a}$ | $2.75^{ab}$ | $2.81^{b}$ | $3.35^{c}$ |
| RFE/RE (%) | $50.8^{a}$ | $57.0^{b}$ | $58.4^{b}$ | $70.4^{c}$ |

(Adapted from Eggum *et al.*, 1985b)
Groups in the same rows with the same letters do not differ significantly ($P<0.05$).

urea synthesis and excretion, when feeding diets 3 and 4, would result in a lower heat loss, it seems more likely that the reduction in retained protein energy was due to a reduction in protein synthesis as a consequence of a lower protein and hence amino acid intake. As shown by Reeds *et al.* (1980) daily protein synthesis is highly correlated to digestible protein (g/day), which in turn was correlated to heat production.

## Practical use of synthetic amino acids in pig and poultry diets

The effect of supplementation of practical pig diets with free synthetic amino acids has formed the basis for numerous investigations over the years. However, the results of many of these investigations have frequently appeared to be inconsistent. This might be due, in part, to limitations of other amino acids than those supplementing the diets, unreliable amino acid analyses, or because estimates of amino acid composition of diets are based on 'table values' rather than analyses of the actual diets used.

It appears that the utilization of synthetic amino acids was lower than that of protein-bound amino acids when given in quantities up to 80% of daily requirement (*Tables 9.2* and *9.3*). However, it is possible to obtain approximately the same performance when the limiting amino acids are provided in the synthetic form as when they originate from intact protein, as shown by the experiments of Madsen and Mortensen (1977) with growing pigs (*Table 9.6*). In this experiment the control diet comprised barley and soyabean meal (diet A). In diets B and C part of the soyabean meal was substituted with free synthetic amino acids; in diet B with lysine, threonine and methionine to give the same concentration as in diet A. Diet C was further supplemented with histidine, leucine and isoleucine to obtain the same total concentration as in diet A. It appears from the results that it is possible to obtain the same performance when feeding diet C as when feeding diet A. In contrast, when feeding diet B, daily gain was significantly lower than for the control. However, from a summary of several trials with free synthetic amino acids fed to growing pigs (Madsen and Mortensen, 1977) it appears that performance is slightly lower when diets based on free synthetic amino acids are fed compared with diets solely based on protein-bound amino acids.

The physiological basis of utilization of free synthetic amino acids is generally more favourable for poultry than for pigs. Firstly, poultry are usually fed a more concentrated ration *ad libitum*. Although there may be some variation in the feed

**Table 9.6** PERFORMANCE OF PIGS FED DIETS SUPPLEMENTED WITH SYNTHETIC AMINO ACIDS (DIETS B AND C)

| *Diet* | *A* | *B* | *C* |
|---|---|---|---|
| Number of pigs | 12 | 12 | 12 |
| *20–90 kg* | | | |
| Feed intake (kg/day) | 2.04[a] | 2.00[b] | 2.04[a] |
| Daily gain (g) | 704[a] | 674[b] | 696[ab] |
| $FU_p$/kg gain | 2.90[a] | 2.98[a] | 2.94[a] |
| Eye muscle area ($cm^2$) | 29.7[a] | 29.3[a] | 28.9[a] |
| Lean content (%) | 53.7[a] | 52.7[a] | 53.4[a] |

(Adapted from Madsen and Mortensen, 1977)
Groups in the same row with the same letters do not differ significantly ($P<0.05$).

intake from morning to afternoon, it is generally believed that the diurnal variation of nutrients in plasma is lower in poultry than in pigs (Riis, 1983). This forms the basis of a better utilization of synthetic amino acids in protein synthesis. Secondly, poultry are more sensitive to amino acid imbalance, which has been shown to have an adverse effect on the feed intake of the animal (Woodham and Deans, 1977). Experiments with chickens have supported the hypothesis that diets formulated to minimize excess of amino acids over the chicks' known requirements would improve the efficiency of protein and energy utilization (Waldroup *et al.*, 1976). Moreover, in some instances the relative proportions of essential amino acids may be of greater significance than the absolute amounts because of the complex relationship between amino acids (Brewer, Halworson and Clark, 1978). Therefore, the advances in balancing amino acid composition of diet and the physiological basis for a high utilization of added free synthetic amino acids are generally more favourable in poultry than in pigs. This interpretation was confirmed in a recent study with chickens by Sibbald and Wolynetz (1985). The utilization of synthetic lysine was 0.92, which was significantly ($P<0.05$) higher than that of protein-bound lysine (0.88). However, a synthetic lysine utilization of 0.92 is significantly lower than 1.00 which is often used when estimating amino acid requirements of poultry.

## Conclusions

The absorption of free synthetic amino acids is more rapid than that of protein-bound amino acids due to the fact that the added free amino acids pass the stomach together with the easily hydrolysable nitrogen fraction very shortly after feeding.

Balance experiments with pigs have demonstrated reduced nitrogen retention when diets have been supplemented with high amounts of synthetic amino acids. The explanation is probably partly due to an imbalance of other amino acids or a limitation in the capacity of the body to buffer the short-term variation in absorption of free synthetic amino acids compared with the bulk of protein-bound amino acids.

In growth trials with pigs where part of the protein supplement was substituted by synthetic amino acids, approximately the same performance in terms of daily gain and feed utilization was obtained when the diets were adjusted for lysine, threonine, methionine, histidine, leucine and isoleucine.

The utilization of synthetic lysine in poultry is higher than that of protein-bound lysine. However, the utilization is lower than 1 which is usually assumed when calculating requirements for poultry

It is concluded that synthetic amino acids can be used to balance the amino acid composition of pig and poultry diets provided they are given in amounts not exceeding the buffering capacity of the body.

## References

BATTERHAM, E.S. (1974). *British Journal of Nutrition,* **31**, 237–242
BATTERHAM, E.S. and O'NEILL, G.H. (1978). *British Journal of Nutrition,* **39**, 265–270

BREWER, M.F., HALVORSON, J.D. and CLARK, H.E. (1978). *American Journal of Clinical Nutrition,* **31**, 786–792

BURACZEWSKA, L., ZEBROWSKA, T. and BURACREWSKI, S. (1978). *Roczniki Nauk rolniczych, Seria B,* **99**, 107–113

BURACZEWSKA, L., LACHOWICZ, J. and BURACZEWSKI, S. (1980). *Archiv für Tierernährung,* **30**, 751–758

BURACZEWSKA, L. and BURACZEWSKI, S. (1980). *Proceedings of the 3rd Symposium on Protein Metabolism and Nutrition*, pp. 307–312. Ed. Oslage, H.J. and Rohr, K. European Association for Animal Production, Publication, No 27

CANOLTY, N.L. and NASSET, E.S. (1975). *Journal of Nutrition,* **105**, 867–877

EGGUM, B.O., CHWALIBOG, A., NIELSEN, H.E. and DANIELSEN, V. (1985a). *Zeitschrift für Tierphysiologie, Tierernährung und Futtermittelkunde,* **53**, 113–123

EGGUM, B.O., CHWALIBOG, A., NIELSEN, H.E. and DANIELSEN, V. (1985b). *Zeitschrift für Tierphysiologie, Tierernährung und Futtermittelkunde,* **53**, 124–134

GARLICK, P.J., BURK, T.L. and SWICK, R.W. (1976). *American Journal of Physiology,* **230**, 1108–1112

GARLICK, P.J., MILLWARD, O.J. and JAMES, W.P.T. (1973). *Biochemistry Journal,* **136**, 935–946

JUST, A. (1982). *Livestock Production Science,* **9**, 349–360

JUST, A. (1983). *Proceedings of the 4th International Symposium on Protein Metabolism and Nutrition*, Les Colloques de l'INRA, 16, pp. 289–309. Ed. Arnal, M., Pion, R. and Bonin, D.

LOW, A.G. (1979). *British Journal of Nutrition,* **41**, 137–156

LOW, A.G. (1981). In *Recent Advances in Animal Nutrition—1981*, pp. 141–156. Ed. Haresign, W. Butterworths, London

MADSEN, A. and MORTENSEN, H.P. (1977). *US Feed Grains Council.* Hamburg, 20 pp

MARTIN, A.K. and BLAXTER, K.L. (1965). In *Energy Cost of Urea Synthesis in Sheep*, pp. 83–90. Ed. Blaxter, K.L. Academic Press, London

PARTRIDGE, I.G., LOW, A.G. and KEAL, H.P. (1985). *Animal Production,* **40**, 375–377

REEDS, P.J., CADENHEAD, A., FULLER, M.F., LOBLEY, G.E. and McDONALD, J.D. (1980). *British Journal of Nutrition,* **43**, 445–455

REEDS, P.J., FULLER, M.F., CADENHEAD, A., LOBLEY, G.E. and McDONALD, J.D. (1981). *British Journal of Nutrition,* **45**, 539–546

RÉRAT, A. (1978). *Journal of Animal Science,* **46**, 1808–1837

RÉRAT, A. (1981). *World Review of Nutrition and Dietetics,* **37**, 229–287

RÉRAT, A., CORRING, T. and LAPLACE, J.P. (1976). In *Protein Metabolism and Nutrition*, pp. 97–138. Ed. Cole, D.J.A., Boorman, K.N., Buttery, P.J., Lewis, D., Neal, R.J. and Swan, H. Butterworths, London

RIIS, P.M. (1983). In *Dynamic Biochemistry of Animal Production*, pp. 151–171. Ed. Riis, P.M. Elsevier, Amsterdam

SCHIEMANN, R., NEHRING, K., HOFFMANN, L., JENTSCH, W. and CHUDY, A. (1971). *Energetische Futterbewertung und Energienormen.* VEB Deutscher Landwirtschaftsverlag, Berlin, 344 pp

SIBBALD, I.R. and WALYNETZ, M.S. (1985). *Poultry Science,* **64**, 1972–1975

SIMON, O., MÜNCHMEYER, R. and BERGNER, H. (1978). *British Journal of Nutrition,* **40**, 243–252

THACKER, P.A., SAUER, W.C. and JØRGENSEN, H. (1984). *Journal of Animal Science,* **59**, 409–415

WALDROUP, P.W., MITCHELL, R.J., DAYNE, J.R. and HAZEN, K.R. (1976). *Poultry Science,* **55**, 243–253

WALZ, O.P. (1976a). *Zeitschrift für Tierphysiologie, Tierernährung und Futtermittelkunde,* **36**, 119–138

WALZ, O.P. (1976b). *Zeitschrift für Tierphysiologie, Tierernährung und Futtermittelkunde,* **36**, 139–150

WALZ, O.P. (1980). *Zeitschrift für Tierphysiologie, Tierernährung und Futtermittelkunde,* **46**, 113–124

WALZ, O.P. (1983). *Zeitschrift für Tierphysiologie, Tierernährung und Futtermittelkunde,* **49**, 181–194

WOODHAM, A.A. and DEANO, P.S. (1977). *British Journal of Nutrition,* **37**, 289–308

10

# TOWARDS AN IMPROVED UTILIZATION OF DIETARY AMINO ACIDS BY THE GROWING PIG

P. J. MOUGHAN
*Department of Animal Science, Massey University, Palmerston North, New Zealand*

## Introduction

A primary aim of modern pig production is to fully realize the animal's potential for daily lean gain. It is also important that the maximum rate of body protein deposition be achieved with as little wastage of the ingested amino acids as possible. This is important for two reasons. Firstly, because an avoidable loss of amino acids is inherently wasteful biologically and secondly because the excretion of nitrogen contributes to environmental pollution. The optimization of dietary amino acid utilization during pig growth, however, takes place within a commercial framework and the process is thus subject to certain economic constraints.

In spite of there being compelling reasons for maximizing the utilization of dietary protein, values for the efficiency of utilization determined for a range of pig grower diets used in modern commercial practice (Table 10.1) indicate that

**Table 10.1** EFFICIENCY OF UTILIZATION[a] OF DIETARY CRUDE PROTEIN (CP) AND LYSINE IN SIX COMMERCIAL PIG GROWER DIETS, GIVEN AT TWO FEEDING LEVELS TO 50 kg LIVEWEIGHT GILTS

| | *Diet* | | | | | |
|---|---|---|---|---|---|---|
| | *1* | 2 | 3 | 4 | 5 | 6 |
| *Feeding level = 1710 g meal/d:* | | | | | | |
| Digestible CP intake (g/d) | 175 | 281 | 235 | 232 | 182 | 215 |
| Protein deposited (g/d) | 48.9 | 110.0 | 73.5 | 106.9 | 74.3 | 115.0 |
| Pe (%)[b] | 20.4 | 30.0 | 23.1 | 33.8 | 32.3 | 42.1 |
| Le (%)[c] | 37.2 | 38.3 | 38.5 | 43.5 | 54.0 | 59.0 |
| *Feeding level = 2270 g meal/d:* | | | | | | |
| Digestible CP intake (g/d) | 232 | 374 | 312 | 309 | 242 | 285 |
| Protein deposited (g/d) | 71.4 | 115.0 | 104.1 | 115.0 | 105.2 | 115.0 |
| Pe (%)[b] | 22.4 | 23.7 | 24.6 | 27.4 | 34.4 | 31.7 |
| Le (%)[c] | 40.9 | 30.2 | 41.1 | 35.3 | 59.0 | 45.0 |

[a]Predicted values (Moughan, 1984) from a pig growth simulation model. Assumes healthy animals growing in a thermoneutral environment

$$^{b}\text{Pe} = \frac{\text{Body protein deposited}}{\text{Diet crude protein intake}} \times \frac{100}{1}$$

$$^{c}\text{Le} = \frac{\text{Body lysine deposited}}{\text{Diet total lysine intake}} \times \frac{100}{1}$$

utilization efficiency remains low. The efficiency of utilization of dietary crude protein intake (Pe) ranged from 20 to 42% at a low level of meal intake and from 22 to 34% at a higher level. On average the ingested dietary protein was utilized with an efficiency close to 30%. The equivalent of around 70% of the ingested nitrogen was excreted from the pig's body. Part of this inefficiency can be explained by dietary amino acid imbalance, which may be purposeful and economically justifiable.

Lysine was the first limiting amino acid in each of the six diets, so it is pertinent to examine (Table 10.1) the efficiency of utilization of ingested lysine (Le), whereby the effect of amino acid imbalance is removed. As expected, the values for Le are higher than the comparable values for Pe. On average the ingested lysine was utilized with an efficiency close to 44%, but still over half the dietary lysine was not used for the net deposition of lean tissue. Data such as those presented in Table 10.1 highlight the importance of understanding the physiological processes which lead to losses of amino acids from the body, and thus explain inefficiency of utilization of the dietary first-limiting amino acid for growth.

This chapter outlines these processes and consideration is given to ways of increasing the efficiency of utilization of dietary amino acids.

## Physiological processes affecting amino acid utilization

The absorption and metabolism of amino acids in mammals is complex and highly integrated, with continuous flux within and between body cells. It is useful, however, to view the metabolism in terms of several discrete physiological processes. The various processes which impinge upon the pig's ability to utilize the dietary first-limiting amino acid are listed in Table 10.2. These processes are discussed in the following sections.

**Table 10.2** PROCESSES CONTRIBUTING TO INCOMPLETE UTILIZATION[a] OF THE DIETARY FIRST-LIMITING AMINO ACID

| *Physiological process* |
|---|
| (1) Excretion of unabsorbed dietary amino acids. |
| (2) Absorption of chemically-unavailable compounds derived from amino acids. |
| (3) Use of amino acids for body protein maintenance. |
| (4) Use of amino acids for synthesis of non-protein nitrogenous compounds. |
| (5) Preferential catabolism of amino acids for energy supply. |
| (6) Inevitable amino acid catabolism. |
| (7) Catabolism of amino acids supplied above the amount required for the maximum rate of protein retention. |

[a]Efficiency of utilization for net whole-body protein deposition

### EXCRETION OF UNABSORBED DIETARY AMINO ACIDS

Not all of the amino acids in feed proteins are released during digestion and absorbed from the gut. The measurement of amino acid digestibility is discussed later in this chapter, where it is concluded that true ileal amino acid digestibility coefficients offer the best practical means of predicting the amounts of amino acids absorbed from the digestive tract of the pig.

Recent studies in which the true ileal amino acid digestibility of various feed ingredients has been determined (Furuya, Nagano and Kaji, 1986; Green *et al.*, 1987; Green and Kiener, 1989) indicate that between zero and 35% of the intake of an amino acid may remain unabsorbed and pass from the small intestine. For more poorly digestible feed ingredients, such as meat and bone meal for example, the flow of unabsorbed nitrogen from the terminal ileum can be as high as 45% of dietary intake (Moughan *et al.*, 1989). For a normal mixed diet, however, around 15% of dietary amino acid intake leaves the small intestine unabsorbed.

## ABSORPTION OF CHEMICALLY-UNAVAILABLE COMPOUNDS DERIVED FROM AMINO ACIDS

Ingredients used to manufacture compounded pig diets have often been subjected to physical, biological or chemical treatments while being processed. During processing, particularly that involving heat, and during subsequent storage of the material, protein-bound amino acids can react with reducing sugars, fats and their oxidation products, polyphenols or chemical additives such as alkali. The amino acids lysine, methionine, cysteine and tryptophan are particularly reactive. Lysine, which is commonly the first-limiting amino acid in pig diets, is most susceptible to damage because of the ready involvement of its epsilon-$NH_2$ group in intra- and intermolecular crosslinking and in reaction with other compounds.

The reaction between lysine and reducing sugars (the Maillard reaction) which occurs under mild processing conditions and during storage is well-documented.

The chemically-modified amino acids may be released during digestion and absorbed across the small intestinal mucosa. The uptake of these compounds, however, which are excreted in the urine appears to be quantitatively low (Austic, 1983). Nevertheless, the presence of chemically-modified amino acids in feeds does contribute significantly to a lowered efficiency of utilization of the determined levels of dietary amino acids. It does so in two ways. Firstly, when the feed is subjected to acid hydrolysis during conventional amino acid analysis, a variable proportion of the modified amino acid may revert back to the amino acid and thus its dietary level is overestimated. Secondly, the chemical modifications may result in a lowered digestion and absorption of amino acids generally. By way of example, after early Maillard reaction a variable proportion of the deoxyketosyl derivative (Amadori compound) formed reverts back to lysine in the presence of strong acid. Part of the determined lysine will represent deoxyketosyl lysine, which although being partly absorbed from the gut is of no nutritional value to the pig (Hurrell and Carpenter, 1981). In the more advanced stages of the Maillard reaction (brown pigment formation) the original lysine molecule will not be recovered following acid hydrolysis during amino acid analysis. There may, however, be a generally lowered *in vivo* digestibility of all amino acids due to the formation of enzyme-resistant cross-linkages and a possible direct effect of the advanced Maillard compounds on the digestive enzymes (Hurrell and Finot, 1985; Oste *et al.*, 1986).

In a recent study in which processing conditions giving rise to early Maillard reaction were simulated (M. Gall, M.N. Wilson, P.J. Moughan, unpublished), it was found (see Table 10.8) that total lysine determined by conventional analytical methods overestimated the unaltered or reactive lysine present in the diet by 30%. Furthermore, only 70% of the actual lysine was absorbed from the small intestine of the pig, presumably due to the effect of cross-linkages having been formed within the protein molecule. Values such as these demonstrate that the occurrence of chemically-modified amino acids in feed proteins can have a large influence on the determined efficiency of utilization of the dietary first-limiting amino acid.

## AMINO ACIDS REQUIRED FOR BODY PROTEIN MAINTENANCE

The body protein of the growing pig is continually being broken down and resynthesized. This turnover is quantitatively very significant and is also rather efficient (Simon, 1989). The process is not completely efficient, however, there being a continuous loss of amino acid nitrogen of body origin via the urine. Ultimately this loss must be met by absorbed dietary amino acids. Tissue amino acids are also lost via the digestive tract mucus, digestive enzymes, shed mucosal cells, albumin) and from the shedding of skin and hair. Absorbed amino acids used to meet these costs for maintenance (and it is usually assumed that maintenance processes have a priority for nutrients over growth processes) will not be available for the synthesis of new body protein.

An appreciation of the quantitative significance of the daily losses of body amino acids can be obtained by simulating the physiological processes involved (Moughan, 1989). In the latter model, amino acid loss from the integument was considered as a constant proportion of metabolic body weight. Losses from the digestive tract were assumed to be linearly related to food dry-matter intake and to be influenced by the presence of dietary protein in the gut. The amount of body amino acid nitrogen lost in the urine was estimated as a constant proportion of the rate of whole-body protein synthesis, which was assumed to be related to the rate of dietary protein retention. Correction was made for preferential retention within the cell of some amino acids during body protein turnover. The predicted (simulation model) amounts of lysine being lost daily from the body of a 50 kg liveweight pig, growing at three different rates of body protein retention, are given in Table 10.3. The amount of lysine required to offset the total losses from the body, expressed as a proportion of the amount of absorbed lysine required for growth plus maintenance, is higher for slower growing pigs. For a rapidly growing animal with a body protein retention of 145 g/d, around 10% of absorbed lysine is needed for maintenance purposes. Losses of amino acids from the integument are relatively minor but the inefficiency in body protein turnover and losses of amino

**Table 10.3** PREDICTED[a] DAILY LOSSES OF LYSINE (MAINTENANCE) FROM THE BODY OF THE 50 kg LIVEWEIGHT PIG RETAINING DIFFERENT LEVELS OF BODY PROTEIN, RELATED TO THE UNDERLYING PHYSIOLOGICAL PROCESSES

| *Protein retention* (g/d)[b] | *Lysine lost from the body* | | | | | | | |
|---|---|---|---|---|---|---|---|---|
| | *Loss due to turnover* | | *Gut loss* | | *Integument loss* | | *Total loss* | |
| | (g/d) | (% absorbed)[c] | (g/d) | (% absorbed) | (g/d) | (% absorbed) | (g/d) | (% absorbed) |
| 60 | 0.60 | 11.1 | 0.71 | 13.1 | 0.08 | 1.47 | 1.39 | 25.6 |
| 90 | 0.67 | 8.2 | 0.93 | 11.4 | 0.08 | 0.98 | 1.68 | 20.6 |
| 145 | 0.80 | 4.4 | 1.00 | 5.5 | 0.08 | 0.44 | 1.88 | 10.3 |

[a]Based on a model simulating lysine partitioning in growing pigs (Moughan, 1989) for a 12.50 MJ DE/kg, 90% dry-matter diet

[b]Assuming body lipid deposition (Ld) = 1.6 body protein deposition (Pd) except for Pd = 145 g/d where Ld was limited by *ad libitum* food intake to 1.3 Pd

[c]Respective lysine loss (g/d) expressed as a percentage of the estimated total absorbable dietary lysine requirement of the pig

acids from the gut each constitute a significant drain on the supply of amino acids for new protein synthesis. It should also be noted that the losses of lysine from the gut, shown in Table 10.3, correspond to the use of a high-quality pig grower diet. Where pigs are fed either high-fibre diets or diets where ingredients contain anti-nutritional factors, the gut losses may be substantially higher.

### AMINO ACIDS AS PRECURSORS FOR THE SYNTHESIS OF COMPOUNDS OTHER THAN PROTEINS

Amino acids are used in the synthesis of a wide array of non-protein compounds, which are essential for the normal functioning of animals. Such amino acids, which ultimately must be supplied from the diet, are irreversibly lost from the body amino acid pool and are thus unavailable for body protein synthesis. It is usually considered (Reeds, 1988) that the use of amino acids for these purposes is quantitatively very low. However, for certain specific amino acids, such as methionine or glycine for example, this pathway of amino acid loss may not be negligible. It has been argued (Millward and Rivers, 1988; Reeds, 1988) that the use of amino acids for the synthesis of non-protein compounds, during the post-absorptive state, may lead to amino acid imbalance within the cell and thus at least partly explain inefficiency in the process of whole-body protein turnover.

### THE PREFERENTIAL CATABOLISM OF AMINO ACIDS FOR ENERGY SUPPLY

If the growing pig receives a diet in which the ratio of non-protein energy to balanced amino acid energy is low then amino acids can be catabolized preferentially (as opposed to inevitably) to supply energy. Such catabolism not only saps amino acid supply but is also inherently wasteful energetically.

Whittemore (1983) proposed a biological basis for the interaction between dietary protein and non-protein energy, in terms of a physiological requirement for a certain minimal level of body lipid.

If following the 'classic' approach to energy partitioning, residual energy (residual ME = ME − maintenance energy cost − total energy cost of depositing protein) is insufficient to meet the minimum lipid deposition requirement, then deamination of amino acids is triggered and amino acids are degraded to supply energy. Such deamination must also be directly related to the maintenance of blood glucose levels and the role of amino acids in gluconeogenesis.

With well-balanced diets it should be possible to avoid 'preferential' catabolism of amino acids. However, this potential source of loss of absorbed diet amino acids may become of greater quantitative significance, as the potential lean growth rate of pigs is increased by genetic means and producers adopt diets formulated to minimize the deposition of body fat during growth.

### INEVITABLE AMINO ACID CATABOLISM

A proportion of the absorbed level of the first-limiting amino acid will be degraded, with the rate of amino acid catabolism being modulated by glucose and fatty acid levels (Fuller, 1980). This degradation occurs, even though non-protein energy is supplied well in excess of the requirement for essential metabolic processes (maintenance, protein synthesis, essential lipid synthesis), and would seem to be an 'inevitable' consequence of the presence in cells of an active catabolic machinery.

Indirect evidence for such amino acid metabolism comes from the common observation that the determined biological value of high-quality proteins (e.g. milk, egg) can be lower than unity at moderate levels of dietary inclusion and with a concomitant high ratio of dietary non-protein to protein energy (Millward and Rivers, 1988; Pellett, 1990). Further indirect evidence has been discussed by Millward (1989). The accepted average protein and amino acid requirements of humans of all ages are considerably higher than can be explained by the needs for growth and replacement of endogenous losses.

There is little comprehensive information on the magnitude of 'inevitable' amino acid catabolism in the pig. Average estimates which have been published range from the negligible to 40% of the absorbed level of the first-limiting amino acid. There is even less quantitative information available about the pattern of change of inevitable catabolism with change in the absorbed amounts of amino acids, from low levels to those corresponding to the maximal rate of body protein retention. This is particularly true for the response to amino acid supply about the maximum rate of protein retention.

Inevitable catabolism could be constant with increasing amino acid supply, and thus be a declining proportion of the amount of absorbed amino acid (Young *et al.*, 1985) or it could increase in a proportional manner (Brookes, Owens and Garrigus, 1972; Bayley, 1987). Alternatively, the degree of inevitable catabolism may increase in a disproportionate manner, with curvilinear response being marked around the maximum level of protein retention (Heger and Frydrych, 1985).

The work of Heger and Frydrych (1985) with the growing rat, is of particular interest because inevitable catabolism was determined over a wide range of dietary amino acid levels and there was control of a number of important experimental variables. Male rats were fed semi-synthetic diets containing crystalline amino acids, and relationships between nitrogen balance and amino acid intakes were determined. The amino acid of interest was included (0–120% of NRC requirement) in iso-nitrogenous diets which supplied adequate levels of all other amino acids and a high level of energy in the form of non-amino compounds. From information on nitrogen retention and the amino acid composition of body protein, the net retentions of body amino acids were determined. Third degree polynomials were fitted to the data and the intake of each amino acid corresponding to the maximum response was considered to be the optimum requirement, while the maintenance requirement was estimated from the point of intersection of the regression curve with the zero balance point. The data of Heger and Frydrych (1985) allow calculation of the amounts of first-limiting amino acid (AA) inevitably catabolized, using the following equation:

[inevitable catabolism = AA absorbed − AA retained in body − AA used for maintenance − AA used for synthesis of non-protein compounds]

The loss of most amino acids for synthesis of non-protein compounds can be regarded as negligible for most amino acids. Further, it can be assumed that the crystalline amino acids were completely absorbed and chemically available, and that there was minimal catabolism of amino acids for the express purpose of energy supply. Estimates of the inevitable catabolism of lysine are given in Table 10.4. The data for lysine were fairly representative of those found for the other essential amino acids except methionine, and demonstrate that the inevitable catabolism of

**Table 10.4** AMOUNTS OF LYSINE INEVITABLY CATABOLIZED BY THE GROWING RAT[a]

| | *Lysine absorbed (mg/d)* | | | | | | | |
|---|---|---|---|---|---|---|---|---|
| | *10* | *20* | *30* | *40* | *50* | *60* | *70* | *80*[b] |
| Lysine catabolysed (mg/d) | 0.85 | 1.82 | 3.06 | 4.92 | 7.80 | 12.06 | 18.20 | 26.48 |
| % absorbed | 8.5 | 9.1 | 10.2 | 12.3 | 15.6 | 20.1 | 26.0 | 33.1 |

[a]Calculated from the data presented by Heger and Frydrych (1985)
[b]Level of absorbed lysine corresponding to the maximum rate of body protein retention

amino acids may be considerable, particularly at levels of amino acid absorption supporting near maximal body protein deposition.

Of course, the conclusions that are drawn depend upon the type of function fitted to the data, and more work is required to define the pattern of response. However, the data of Heger and Frydrych (1985) do serve to illustrate that at least 10% of the absorbed limiting-amino-acid flow is subject to inevitable degradation in mammals, with the rate of loss possibly being much greater at higher levels of amino acid intake. The study of Batterham *et al.* (1990) in which lysine retention in growing pigs was determined, gives support to the findings of Heger and Frydrych (1985).

In a recent review on the process of inevitable catabolism, Heger and Frydrych (1989) concluded that even in the case of dietary deficient amino acids and well-balanced proteins, a portion of amino acids is subject to oxidation. The loss increases with increasing intake of proteins from levels allowing only maintenance and minimum growth to levels supporting growth. These authors viewed this catabolism as 'an inevitable consequence of the operation of mechanisms controlling the degradation of amino acids in the body' and an 'unavoidable tax which the animal must pay for the ability to respond quickly to quantitative and qualitative changes in protein supply'. Millward and Rivers (1988) and Millward (1989), on the other hand, have introduced the concept of an 'anabolic drive' to explain the process of inevitable catabolism of amino acids which is seen as beneficial to the organism. They suggest that there are advantages to the animal in consuming levels of essential amino acids in excess of the level required to match identifiable needs, since before their oxidation, these amino acids exert a transitory regulatory influence (the anabolic drive) on growth and maintenance.

### CATABOLISM OF AMINO ACIDS SUPPLIED ABOVE THE MAXIMAL RATE OF PROTEIN RETENTION

Cells have a finite capacity for protein synthesis (resynthesis and new synthesis) and are not able to store amino acids as such, for later use. If, after a meal, the uptake of balanced amino acids exceeds the animal's capacity for protein synthesis, the excess supply will be deaminated and the carbon skeletons eventually degraded.

The existence of an intrinsic upper-limit to whole-body protein retention in growing pigs, influenced by genotype and sex, has recently been demonstrated empirically by Campbell, Taverner and Curic (1984) and Dunkin, Black and James (1984). In pigs growing between 20 and 90 kg live weight, a plateau to daily body protein deposition was found, when the intake of a protein-adequate diet was increased towards *ad libitum*. If in these trials it is assumed that body protein deposition at plateau was not constrained by nutrient intake, environmental factors or carryover effects from early nutrition, which is a reasonable assumption, then protein deposition must have been set by an intrinsic ceiling.

In practice, the catabolism of amino acids due to over-supply, may be considerable, partly due to animal genotype not always being adequately considered during dietary formulation, but also due to the current lack of hard descriptive data on the upper-limits to protein retention of different types of growing boars and gilts.

## Means of improving the efficiency of utilization of dietary amino acids

Having outlined various ways by which ingested amino acids may not become available for new body protein synthesis in the growing pig, it is important to examine how improvement in utilization efficiency can most readily be obtained.

The excretions of amino acids of body origin (maintenance) via the urine and digestive tract would appear to be inevitable costs associated with body-protein turnover, digestion and gut maintenance. The controls of these processes are not well-defined, however, and it may be possible, with more knowledge, to elicit decreases in these routes of amino acid loss. There is a pressing need to identify and describe the mode of action of dietary factors (e.g. fibre, anti-nutritional factors) which may heighten gut amino acid losses. Ingredients containing such compounds can be avoided or their inclusion rate limited. Alternatively, the undesirable compounds may be able to be removed or deactivated.

With respect to amino acids being 'preferentially' catabolized for energy supply or being catabolized because of oversupply in relation to the upper limit to body protein retention, there is a role for computerized models simulating pig growth (Moughan and Verstegen, 1988), to facilitate diet formulation. Models may be used to 'tailor-make' diets for pigs of different genotype and sex and thus avoid unnecessary catabolism of amino acids. Also, and concerning the 'preferential' catabolism of amino acids for energy supply, it is important when selecting animals for higher lean growth rate, to avoid a concomitant decrease in voluntary food intake (appetite). Ultimately, the implications of genetic selection programmes on the intrinsic minimum body lipid to protein ratios will need to be examined.

The process of 'inevitable' catabolism needs further study with emphasis being given to description of the response between catabolism and the level of absorbed first-limiting amino acid. If loss of the first-limiting amino acid via this process is as quantitatively significant as some recent reports suggest, then considerable gains in efficiency of amino acid utilization may ensue from elucidating and manipulating the control of amino acid catabolism.

Of all the factors listed in Table 10.2, however, it is the first two (digestibility and chemical unavailability) that are possibly of most immediate concern to the feed compounder. The digestibility of protein in feeds can be enhanced by various processing techniques, but these must be carefully controlled, to avoid the production of chemically modified amino acids. The addition of enzyme preparations to dry feeds or enzyme pre-digestion of the feed or feed ingredients (Puia-Negulescu, Stavri and Burlacu, 1989) also holds some promise. The storage of feedstuffs under suitable conditions for as short a time as is practically possible will minimize the formation of chemically-unavailable amino acids. Finally, Low (1989) has suggested that it may eventually be possible to increase the absorption of amino acids in the small intestine of the pig by reducing or modifying bacterial activity by dietary means rather than by using antibiotics.

In spite of it being possible to take some steps when purchasing, handling, processing or storing feedstuffs, to avoid products having or developing high levels of undigestible chemically-unavailable amino acids, it remains that batches of ingredients used in preparing diets are of variable quality. It thus becomes critical that the digestibility and chemical availability of the amino acids in the feeds be known, so that these factors can at least be accounted for in diet formulation. Assays are required for routinely determining amino acid digestibility and chemical availability. It is also necessary to have such methods, to allow proper evaluation of feed processing technologies. Considerable effort has been expended on the development of suitable methods over the past years and much progress has been made, in spite of a number of difficulties.

The development of amino acid digestibility and availability assays is considered in the following sections.

## The development of amino acid digestibility assays

### BACKGROUND

The traditional and simplest way of determining amino acid digestibility is based on measurement of faecal amino acid excretion. This approach, however, is now known to lead to considerable error and it is preferable to measure loss at the terminal ileum.

After ingestion by the animal, dietary protein becomes mixed with endogenous proteins within the lumen of the gut, and the total mixture is subjected to digestive breakdown in the upper alimentary tract with amino acids being absorbed in the small intestine as either free amino acids or small peptides. These peptides are hydrolysed within the enterocyte, though some may pass intact into the portal blood (Alpers, 1986). Material remaining unabsorbed at the end of the small intestine (terminal ileum) passes into the large intestine whereby it is subject to the action of a significant population of microbes. It is as a result of this microbial metabolism, that estimates of undigested dietary amino acid flow based on faecal measurement are ambiguous.

During digestion, some protein, peptides and free amino acids pass through the hindgut unaltered and are excreted in the faeces, but a considerable proportion of the nitrogenous material entering the hindgut will be metabolized by the microflora. Also, non-protein nitrogen (mainly urea) may enter the gut from the animal's blood, and be used for the synthesis of microbial protein and amino acids. This is likely to occur to a significant degree when the nitrogen flow into the large intestine from the ileum is low but the ratio of fermentable carbohydrate to protein of the material entering the hindgut is high.

The hindgut microbes are capable of intense proteolytic activity, but it appears that at least for the growing pig (20–90 kg live weight) amino acids are not absorbed through the large intestinal mucosa to any significant extent (Buraczewski, 1980; Wrong, Edmonds and Chadwick, 1981; Krawielitski, Schadereit and Zebrowska, 1984; Low and Zebrowska, 1989).

The metabolism of nitrogenous compounds by the microflora of the pig has been the subject of several comprehensive reviews (Rerat, 1981; Just, 1983; Mason, 1984; McNeil, 1988; Low and Zebrowska, 1989). For most amino acids in most feeds there is a net loss of amino acid between the terminal ileum and the rectum. The amino nitrogen is absorbed from the hindgut as ammonia, which under normal circumstances is of no nutritional value to the animal. Especially for methionine, and sometimes lysine, there may be a net gain due to microbial action. The danger in basing conclusions about dietary amino acid excretion on faecal measurements, is highlighted by the observation that up to 80% of faecal nitrogen is found in microbial bodies (Mason, Just and Bech-Andersen, 1976; Stephen and Cummings, 1980; Low and Zebrowska, 1989). This means that only a very small proportion of the faecal amino acid flow directly relates to the flow of undigested dietary amino acids entering the hindgut.

Measurement of amino acid flow and digestibility at the end of the ileum (Payne *et al.*, 1968) is now generally recognized as a more acceptable approach (Rerat, 1981; Tanksley and Knabe, 1984, 1985; Sauer and Ozimek, 1986; Sauer *et al.*, 1989, van Weerden, 1989) and a literature on the ileal digestibility of amino acids in foods for pigs is accumulating.

A potential criticism of the ileal measure is that there may be interference from a population of microorganisms present in the upper digestive tract (Horvath *et al.*, 1958; Williams Smith, 1965; Wiesemuller, 1983; Bergner *et al.*, 1986). Amino acids may be catabolized or synthesized or incorporated into microbial protein. The *in vitro* and *in vivo* studies of Dierick *et al.* (1986a,b) indicate that there may be a small but measurable catabolism of amino acids by the flora in the upper digestive tract of the pig. Nevertheless, ileal digestibility coefficients have been shown to be sensitive in detecting small differences in protein digestibility due to the processing of foods (van Weerden *et al.*, 1985; Sauer and Ozimek, 1986) and several studies (Tanksley and Knabe, 1980; Low *et al.*, 1982; Just, Jorgensen and Fernandez, 1985; Moughan and Smith, 1985; Dierick *et al.*, 1988) have demonstrated that ileal values are reasonably accurate in describing the extent of uptake of amino acids from the gut, at least for some feedstuffs.

## THE COLLECTION OF ILEAL DIGESTA

In most studies with the pig where ileal amino acid digestibility has been determined, ileal digesta have been collected following the surgical implantation of cannulas into the experimental animals. There are several methods of cannulation, some allowing total collection of digesta while with others digesta are sampled, thus necessitating inclusion of an indigestible marker in the diet. The different approaches to cannulation have been reviewed recently (Sauer *et al.*, 1989; Low, 1990). It was concluded that further studies are required before drawing firm conclusions about which is the superior procedure. However, some general comments can be made.

Ileo-ileo and ileo-caecal re-entrant cannulation involve total transection of the ileum, which is obviously undesirable. The ileo-colic (post-valve) re-entrant cannulation, post-valvular T-caecum cannulation and simple T-ileum cannulation all share the distinct advantage that the function of the ileo-caecal valve is preserved and the ileum is not transected. Simple T-cannulation of the ileum is less surgically invasive compared with the other two approaches, but the post-valve T-caecum method has the advantage that during collection most of the digesta pass through the cannula because the ileo-caecal valve protrudes directly into the cannula. The ileo-colic (post-valve) re-entrant procedure would seem to offer no advantages over the latter apparently more straight-forward technique. Although, the post-valve T-caecum technique (van Leeuwen *et al.*, 1988) would appear to be the current method-of-choice, its superiority over the simple T-cannulation of the ileum remains to be demonstrated. If care is exercised, representative digesta samples can be obtained with the latter method, and indeed with continuous collection of digesta high recoveries (70–80%) of dietary chromic oxide have been obtained (C. Butts, P.J. Moughan, W.C. Smith, unpublished).

The potential impact of any form of cannulation on the normal physiological functioning of the animal, however, should not be overlooked. Livingstone and McWilliam (1985) reported that pigs with simple T cannulas placed in the terminal ileum had similar voluntary feed intakes to their non-cannulated sisters but exhibited lower growth rates and less efficient feed utilization. Further, Wenham and Wyburn (1980) who conducted radiological studies with sheep found that several forms of cannulation, including simple T-piece cannulation, caused some disruption to the normal digesta flow.

The slaughter method, whereby samples of ileal digesta are collected under anaesthesia before sacrifice of the animal, involves minimal disruption of normal digestive function and samples of digesta for any type of diet may be obtained from several parts of the digestive tract. The main criticism of this method concerns the potential difficulty of obtaining representative samples of digesta. However, in the author's experience, digestibility data obtained using this technique (coupled with a frequent feeding regime) are not necessarily any more variable than those obtained from cannulated animals. Moreover, Moughan and Smith (1987) found no differences in the ileal digestibility of amino acids in ground barley as determined using simple T cannulated pigs or by the slaughter method.

## ROUTINE METHODS OF ILEAL DIGESTA COLLECTION

All of the above described methods of digesta collection are laborious and expensive and there is need for more rapid and routine methods. It was for this reason that the ileo-rectal anastomosis technique was developed and has been widely adopted in practice. There are several variations of this method, which generally involve transection of the distal ileum anterior to the ileo-caecal valve. The ileostomy operation as used in humans, has also been applied to pigs (A. Rowan, P.J. Moughan, M.N. Wilson and C. Tasman-Jones, unpublished), and is similar in principle to the ileo-rectal anastomosis. The large intestine is left intact but the ileum is exteriorized to the side of the animal and attached to the skin by preparation of a stoma.

According to Sauer *et al.* (1989), pigs prepared with ileo-rectal anastomosis require less time and effort to maintain than pigs fitted with, at least, re-entrant cannulas and the collection of digesta is easy. The animals would seem to have normal appetites and there is no restriction on the type of diet that can be investigated. Moreover, the preparations last for several months. Ileo-rectal anastomosis, however, may be criticized on three grounds. Firstly, complete transection of the ileum anterior to the ileo-caecal valve (as with classic re-entrant cannulation) may affect gut motility and digesta flow rate, though Green (1988) found no differences in dietary amino acid digestibility between pre- and post-valve ileo-rectal anastomized pigs. Secondly, there may be a proliferation of bacteria in the small intestine (Sauer *et al.*, 1989) of the anastomized animal and a degree of physical and functional adaptation of the small intestine. Such effects are well documented for small animals and human ileostomates. The third criticism stems from the observation that a normally-functioning large intestine exerts considerable control over digestive function (Sakata, 1987; Engelhardt *et al.*, 1989) and affects the animal's physiological status generally, especially with respect to mineral and water uptake.

Despite these potential difficulties, Hennig *et al.* (1986), who conducted a detailed study on ileo-rectal anastomized pigs, found few significant anatomical, histological or functional effects due to the operation and concluded that anastomized pigs may be used with confidence in digestibility studies. Van der Walt, Meyer and van Rensburg (1990) also concluded that the ileo-rectal anastomized sheep is a satisfactory model for protein digestibility studies in this species. Moreover, Leterme *et al.* (1990) found only minor differences in the digestibility of the amino acids in peas between anastomized and T-cannulated pigs. Contrary to this, however, is the finding of Darcy-Vrillon and Laplace (1990) who reported lower apparent digestibility of total nitrogen and amino acids in a semi-synthetic diet containing beet pulp, with ileo-rectal anastomized pigs compared with pigs prepared with an ileo-colic post-valve cannulation which was considered as a reference method. Further, very recent results indicating physiological disturbances in anastomized pigs (Kohler *et al.*, 1990; M.W.A. Verstegen, personal communication) give cause for concern.

An alternative approach to the routine measurement of ileal amino acid digestibility is the development of assays using small animals such as the laboratory rat. Ileal digesta samples can be obtained quickly and easily from the rat after slaughter and this species lends itself to relatively inexpensive, well-controlled experimentation with large numbers of animals being able to be studied at any one time.

**Table 10.5** APPARENT ILEAL DIGESTIBILITY OF ESSENTIAL AMINO ACIDS IN GROUND BARLEY FOR THE GROWING RAT AND PIG[a]

| *Amino acid* | *Apparent digestibility* | | |
|---|---|---|---|
| | *Rat* (*n*=12) | *Pig* (*n*=11) | *Statistical significance* |
| Threonine | 0.68 | 0.72 | NS |
| Valine | 0.79 | 0.80 | NS |
| Methionine | 0.75 | 0.84 | *** |
| Isoleucine | 0.80 | 0.78 | NS |
| Leucine | 0.78 | 0.80 | NS |
| Tyrosine | 0.77 | 0.79 | NS |
| Phenylalanine | 0.76 | 0.82 | *[b] |
| Lysine | 0.75 | 0.78 | NS |

[a]Adapted from Moughan *et al.* (1987) with permission of the New Zealand Department of Scientific and Industrial Research

[b]In a subsequent study with barley, no significant difference was found for phenylalanine

There is general agreement between the rat and pig for the apparent ileal digestibility of protein in several feed ingredients (Moughan, Smith and James, 1984; Picard *et al.*, 1984). Results from a further study (Moughan *et al.*, 1987), in which ileal contents from rats and pigs fed a barley diet were obtained using the slaughter method, are given in Table 10.5. There was close agreement between the species for all amino acids except methionine. No difference in the digestibility of methionine in barley was observed by French workers (M. Picard, personal communication). Based on these and other results from our Centre it is concluded that the laboratory rat offers much promise for the development of a routine ileal digestibility assay for a range of feed ingredients, possibly with the exception of some legumes and plant foods containing high levels of anti-nutritional factors. More work is in progress to develop and evaluate this approach.

A further simple and rapid method which could be used for allowing determination of ileal amino acid digestibility in pig feeds, is the nylon-bag technique (Sauer *et al.*, 1989). This method warrants further investigation and may have value as a rapid screening test.

## CORRECTION FOR ILEAL ENDOGENOUS EXCRETION

Significant quantities of endogenous amino acids are present in the digesta at the terminal ileum of the pig, and to obtain a 'true' estimate of the unabsorbed dietary amino acid flow, correction needs to be made for the endogenous component. True digestibility is a fundamental property of a feed ingredient, regardless of the dietary conditions under which that ingredient was fed to the animal. This is not so for apparent digestibility. With removal of the effect of the confounding variable of endogenous excretion, true digestibility values should be more precise in detecting differences in digestibility arising from the processing of a material. Also, there is an increasing trend with the use in practice of pig growth simulation models, towards expressing daily requirements in units of grams of absorbed amino acid. The requirement value is likely then to include maintenance amino acid cost, and in this case it is more appropriate to use true amino acid digestibility coefficients in diet formulation.

It has been clearly demonstrated, using a variety of different experimental techniques (de Lange, Souffrant and Sauer, 1990; Darragh, Moughan and Smith, 1990; Moughan and Rutherfurd, 1990) that the traditional protein-free method for determining endogenous loss in simple-stomached animals leads to serious underestimation of ileal endogenous amino acid excretion. Further, and given that the regression technique generates similar values to those obtained after feeding pigs a protein-free diet (Leibholz and Mollah, 1988), use of this method may also lead to error. An alternative approach which allows measurement of endogenous excretion under conditions of peptide alimentation has been proposed (Moughan *et al.*, 1990) and evaluated. The latter method would appear to generate estimates of ileal endogenous amino acid loss, applicable to the correction of ileal flows for ingredients not containing fibre and anti-nutritional factors, such as meat meals, fish meals, milk powder, etc. Ileal amino acid flows for the growing pig obtained with this new technique are shown compared with protein-free values in Table 10.6.

Research into methods for determining ileal endogenous amino acid excretion is required, but true ileal amino acid digestibility coefficients should ultimately provide more meaningful data on amino acid absorption in the pig.

**Table 10.6** MEAN ($n$ = 5) ENDOGENOUS AMINO ACID FLOWS[a] AT THE TERMINAL ILEUM OF THE GROWING PIG, DETERMINED USING TWO METHODS

| *Amino acid* | *Endogenous flow* (g/kg dry-matter intake) | |
|---|---|---|
| | *Technique* | |
| | *Enzyme-hydrolysed casein*[b] | *Protein-free* |
| Lysine | 0.46 | 0.31 |
| Histidine | 0.32 | 0.23 |
| Threonine | 0.91 | 0.57 |
| Isoleucine | 0.50 | 0.23 |

[a]The flows were significantly ($P$ <0.05) different
[b]Amino acids in the retentate (>10 000 MW) of digesta from pigs fed an enzyme hydrolysed casein based diet (P.J. Moughan and G. Schuttert, unpublished data)

IN VITRO METHODS OF AMINO ACID DIGESTIBILITY

The *in vitro* approach can provide protein or amino acid digestibility data rapidly, cheaply and with precision. It is very difficult, however, to simulate adequately the complex processes occurring in the mammalian gut, which calls into question the accuracy of such *in vitro* assays. In the past, numerous methods have been proposed and sometimes high correlations with *in vivo* data have been found, but only too often a strong *in vitro*, *in vivo* association has not been confirmed in subsequent studies. It is also important that comparison be made with the appropriate *in vivo* baseline, which given that there is no endogenous component in the *in vitro* system is true ileal digestibility. Three *in vitro* techniques (pronase assay, pH drop assay and a multi-enzyme assay) have been tested against *in vivo* true ileal digestibility data for meat and bone meals (Moughan *et al.*, 1989), but low correlations were found in all cases. Recently, a number of new *in vitro* methods have been described in the literature (Pedersen and Eggum, 1983; Porter, Swaisgood and Catignani, 1984; Liang and Zhang, 1988; Graham, Lowgren and Aman, 1989), which are deserving of thorough independent evaluation.

## The development of assays to determine digestible levels of chemically-available amino acids

Numerous methods have been developed (Whitacre and Tanner, 1989) to allow measurement of the amounts of amino acids in a feedstuff which are bioavailable (i.e. able to be released from the protein and absorbed in a form which can potentially be utilized for protein synthesis). *In vivo* measures such as the slope-ratio assay would be expected to be the most exact indicators of amino acid bioavailability, but such methods appear quite unreliable in practice (Austic, 1983; Whitacre and Tanner, 1989). The potential difficulties of the slope-ratio assay are well illustrated by recent work in Australia (Batterham, 1987). In a collaborative study involving two research centres, very different estimates of lysine availability

**Table 10.7** MEAN ($n$ = 6) AMOUNTS OF TOTAL LYSINE[a], FDNB REACTIVE LYSINE AND TRUE ILEAL DIGESTIBLE LYSINE IN 12 SAMPLES OF MEAT AND BONE MEAL[b]

| | *Sample number*[c] | | | | | | | | | | | |
|---|---|---|---|---|---|---|---|---|---|---|---|---|
| | *1* | *2* | *3* | *4* | *5* | *6* | *7* | *8* | *9* | *10* | *11* | *12* |
| Total lysine (g/100 g) | 2.65 | 2.59 | 2.82 | 2.73 | 3.89 | 2.68 | 3.40 | 4.39 | 2.72 | 2.65 | 2.93 | 4.78 |
| FDNB available lysine (g/100 g) | 2.07 | 1.91 | 2.53 | 2.32 | 2.57 | 2.11 | 2.39 | 3.31 | 2.18 | 1.99 | 2.10 | 3.19 |
| True ileal digestible lysine (g/100 g) | 1.72 | 1.75 | 1.93 | 1.97 | 2.88 | 2.03 | 2.69 | 3.47 | 2.23 | 2.21 | 2.47 | 4.15 |

[a]After conventional acid hydrolysis
[b]Adapted from Moughan *et al.* (1989) with permission of the Society of Chemical Industry
[c]Sample numbering is based on the nitrogen digestibility ranking, Meals 1–6 had been subjected to higher temperatures for longer periods of time than meals 7–12

(0.27 *vs* 0.69) were obtained using the assay, on the same sample of cottonseed meal. Further, Leibholz (1989) has discussed the general imprecision of the assay (high standard error about the mean) and how results may vary markedly depending upon the criterion of response adopted. At least for lysine, an alternative approach is to determine the amount of chemically-reactive lysine present in the feed. The most important and practically the only source of utilizable lysine in feedstuffs is the lysine residue with its epsilon-amino group free to react with various chemical reagents (Finot and Hurrell, 1985). The direct Carpenter method, involving reaction between lysine and 1-fluoro-2,4-dinitrobenzene (FDNB), remains one of the most useful methods in practice (Hurrell and Carpenter, 1981; Erbersdobler and Anderson, 1983; Finot and Hurrell, 1985).

The ileal digestibility of lysine has been assumed to be a satisfactory measure of lysine availability in heat-treated feeds. It would seem, however, that the digestibility coefficient would only be an accurate indicator of the absorption of available (un-reacted) lysine molecules, if deoxyketosyl lysine was absorbed to the same extent as the available lysine or was completely unabsorbed, and if the regeneration of deoxyketosyl lysine during hydrolysis was proportional for the diet and ileal digesta samples. This would appear to be unlikely. Total FDNB and true ileal digestible levels of lysine in meat and bone meals, which are heated during processing, are compared in Table 10.7. Total lysine determined by ion-exchange chromatography considerably overestimated the amounts of chemically available or reactive lysine (FDNB method). Also, and as expected based on theoretical considerations the estimates of true ileal digestible lysine overestimated the amounts of reactive lysine, at least for samples 7–12. Samples 1–6 which represented meals subjected to higher temperatures for longer times, however, showed low digestibilities for lysine and the FDNB available lysine was higher than the digestible lysine contents. Although, as discussed earlier, the ileal digestibility data could be aberrant this consistent effect does suggest the possibility that chemically-reactive (FDNB) lysine may overestimate the actual lysine bioavailability. It is now becoming increasingly clear that this is so, even with mildly heat-treated proteins. Not all of the reactive lysine may be absorbed from the small intestine (Hurrell and Carpenter, 1981) and indeed all of the estimates of FDNB available lysine given in Table 10.7 are probably overestimates of the actual bioavailability. The values for ileal digestible lysine of meals 1–6 (Table 10.7) indicate a generally low digestibility of protein, but it should not be assumed that they provide realistic estimates of bioavailability. The absorption (measured at terminal ileum) of reactive lysine has been determined in a recent study (M. Gall, M.N. Wilson and P.J. Moughan, unpublished) with the growing pig (Table 10.8). A casein–glucose mixture was heated to produce early Maillard compounds, and the amount of epsilon-*n*-deoxy-fructosyl-lysine (blocked lysine) and lysine regenerated after acid hydrolysis in the resulting material was calculated from the determined level of furosine. The amount of unaltered or reactive lysine was found by difference between the total lysine (acid hydrolysis) and regenerated lysine. The FDNB method allowed accurate assessment of the amount of chemically-reactive

**Table 10.8** AMOUNTS OF ACID-HYDROLYSED LYSINE[a], FDNB LYSINE, REACTIVE LYSINE[b] AND ABSORBED REACTIVE LYSINE[c] IN HEATED CASEIN-GLUCOSE MIXTURES

| *Study* | *Lysine* (g/100 g) | | | |
|---|---|---|---|---|
| | *Acid-hydrolysed* | *FDNB* | *Reactive* | *Absorbed reactive* |
| 1[d] | 2.60 | 1.91 | 1.98 | 1.40 |
| 2[e] | – | – | 3.64 | 3.35 |

[a]After conventional amino acid analysis
[b]Lysine units remaining chemically reactive after heating, determined from furosine levels in study 1 and from homoarginine levels after guanidination in study 2
[c]Reactive lysine absorbed by the end of the small intestine
[d]M. Gall, M.N. Wilson and P.J. Moughan, unpublished data
[e]Calculated from the data of Schmitz (1988) assuming that the heated casein/glucose mixture contained 5.6 g lysine/100 g

lysine, which was grossly over-estimated by conventional amino acid analysis (acid-hydrolysed lysine), but the reactive lysine was not completely absorbed. An incomplete absorption of reactive (guanidinated) lysine by growing pigs was also reported by Schmitz (1988) (see Table 10.8) and by Desrosiers *et al.* (1989) using an *in vitro* technique.

There is need for an assay with particular application to processed feeds, which would allow measurement of the levels of chemically-available amino acids which are able to be released from the protein and absorbed.

## Conclusion

It is concluded that, in current commercial practice, the efficiency of utilization of dietary amino acids by the growing pig is rather low, and there is considerable scope for improvement. Improvement is likely to come about through a better and more mechanistic understanding of the physiological processes involved in amino acid digestion and metabolism and through developing chemical and animal assays which allow clearer definition of the bioavailable levels of dietary amino acids.

## References

Alpers, D.H. (1986). *Federation Proceedings*, **45**, 2261–2267

Austic, R.E. (1983). In: *Feed Information and Animal Production*, pp. 175–189. Ed. Robards, G.E. and Packham, R.G. Commonwealth Agricultural Bureaux, Farnham Royal, Slough

Batterham, E.S. (1987). In *Manipulating Pig Production*, pp. 129–133. Ed. Australasian Pig Science Association Committee. Australasian Pig Science Association, Werribee, Australia

Batterham, E.S., Andersen, L.M., Bargent, D.R. and White, E. (1990). *British Journal of Nutrition*, **64**, 81–94

Bayley, H.S. (1987). In *Recent Advances in Animal Nutrition, 1987.* pp. 117–125. Ed. Haresign, W. and Cole, D.J.A. Butterworths, London

Bergner, H., Simon, O., Zebrowska, T. and Munchmeyer, R. (1986). *Archiv fur Tierernährung*, **36**, 479–490

Brookes, I.M., Owens, F.N. and Garrigus, U.S. (1972). *Journal of Nutrition*, **102**, 27–36

Buraczewski, S. (1980). In *Protein Metabolism and Nutrition, Proceedings of the Third EAAP Symposium*, pp. 179–195. Ed. Oslage, H.J. and Rohr, K. EAAP, Germany

Campbell, R.G., Taverner, M.R. and Curic, D.M. (1984). *Animal Production*, **38**, 233–240

Darcy-Vrillon, B. and Laplace, J.P. (1990). *Animal Feed Science and Technology*, **27**, 307–316

Darragh, A.J., Moughan, P.J. and Smith, W.C. (1990). *Journal of the Science of Food and Agriculture*, **51**, 47–56

de Lange, C.F.M., Souffrant, W.B. and Sauer, W.C. (1990). *Journal of Animal Science*, **68**, 409–418

Desrosiers, T., Savoire, L., Bergeron, G. and Parent, G. (1989). *Journal of Agricultural and Food Chemistry*, **37**, 1385–1391

Dierick, N.A., Vervaeke, I.J., Decuypere, J.A. and Henderickx, H.K. (1986a). *Livestock Production Science*, **14**, 161–176

Dierick, N.A., Vervaeke, I.J., Decuypere, J.A. and Henderickx, H.K. (1986b). *Livestock Production Science*, **14**, 177–193

Dierick, N.A., Vervaeke, I., Decuypere, J., van der Heyde, H. and Henderickx, H. (1988). In *European Association for Animal Production Publication No. 35*, pp. 50–51, EAAP, Rostock

Dunkin, A.C., Black, J.L. and James, K.J. (1984). *Animal Production in Australia*, **15**, 672

Engelhardt, W. von., Ronnau, K., Rechkemmer, G. and Sakata, T. (1989). *Animal Feed Science and Technology*, **23**, 43–53

Erbersdobler, H.F. and Anderson, T.R. (1983). In *The Maillard Reaction in Foods and Nutrition*, pp. 419–427. Ed. Waller, G.R. and Feather, M.S. American Chemical Society

Finot, P.A. and Hurrell, R.F. (1985). In *Digestibility and Amino Acid Availability in Cereals and Oilseeds*, pp. 247–258. Ed. Finely, J.W. and Hopkins, D.T. American Association of Cereal Chemists Inc., Minnesota

Fuller, M.F. (1980). *Proceedings of the Nutrition Society*, **39**, 193–203

Furuya, S., Nagano, R. and Kaji, Y. (1986). *Japanese Journal of Zootechnical Science*, **57**, 859–870

Graham, H., Lowgren, W. and Aman, P. (1989). *British Journal of Nutrition*, **61**, 689–698
Green, S. (1988). *Animal Production*, **47**, 317–320
Green, S. and Kiener, T. (1989). *Animal Production*, **48**, 157–179
Green, S., Bertrand, S.L., Duron, M.J.C. and Maillard, R.A. (1987). *Journal of the Science of Food and Agriculture*, **41**, 29–43
Heger, J. and Frydrych, Z. (1985). *British Journal of Nutrition*, **54**, 499–508
Heger, J. and Frydrych, Z. (1989). In *Absorption and Utilization of Amino Acids*, Volume I, pp. 31–56, Ed. Friedman, M. CRC Press, Florida
Hennig, U., Noel, R., Herrmann, U., Wunsche, J. and Mehnert, E. (1986). *Archives of Animal Nutrition*, **36**, 585–596
Horvath, D.J., Seeley, H.W., Warner, R.G. and Loosli, J.K. (1958). *Journal of Animal Science*, **17**, 714–722
Hurrell, R.F. and Carpenter, K.J. (1981). In *Maillard Reactions in Foods*, pp. 159–176. Ed. C. Eriksson, Pergamon Press, Oxford
Hurrell, R.F. and Finot, P.A. (1985). In *Digestibility and Amino Acid Availability in Cereals and Oilseeds*, pp. 233–246. Ed. Finely, J.W. and Hopkins, D.T. American Association of Cereal Chemists Inc., Minnesota
Just, A. (1983). In *Metabolisme et Nutrition Azotes*, pp. 289–309. Ed. Arnal, M., Pion, R. and Bonin, D. INRA, Paris
Just, A., Jorgensen, H. and Fernandez, J.A. (1985). *Livestock Production Science*, **12**, 145–159
Kohler, T., Mosenthin, R., Verstegen, M.W.A., den Hartog, L.A., Huisman, J. and Ahrens, F. (1990). In *Tagung der Gesellschraft fur Ernahrungsphysiologie der Haustiere in Gottingen*
Krawielitski, K., Schadereit, R. and Zebrowska, T. (1984). *Archiv fur Tierernahrung*, **34**, 1–18
Leibholz, J. (1989). In *Absorption and Utilization of Amino Acids*, Volume III, pp. 175–186. Ed. Friedman, M. CRC Press Inc, Florida
Leibholz, J. and Mollah, Y. (1988). *Australian Journal of Agricultural Research*, **39**, 713–719
Leterme, P., Thesis, A., Beckers, Y. and Baudart, E. (1990). *Journal of the Science of Food and Agriculture*, **52**, 485–497
Liang, G.Y. and Zhang, Z.Y. (1988). *Chinese Journal of Animal Science*, **4**, 13–17
Livingstone, R.M. and McWilliam, R. (1985). *British Veterinary Journal*, **141**, 186–191
Low, A.G. (1989). In *Nutrition and Digestive Physiology in Monogastric Farm Animals*, pp. 1–15. Ed. van Weerden, E.J. and Huisman, J. Pudoc, Wageningen
Low, A.G. (1990). In: *Feedstuff Evaluation*, pp. 91–114. Ed. Wiseman, J. and Cole, D.J.A. Butterworths, London
Low, A.G., Partridge, I.G., Keal, H.D. and Jones, A.R. (1982). *Animal Production*, **34**, 403
Low, A.G. and Zebrowska, T. (1989). In *Protein Metabolism in Farm Animals*, pp. 53–121. Ed. Bock, H.D., Eggum, B.O., Low, A.G., Simon, O. and Zebrowska, T. Oxford University Press, Berlin
Mason, V.C. (1984). *Proceedings of the Nutrition Society*, **43**, 45–53
Mason, V.C., Just, A. and Bech-Andersen, S. (1976). *Zeitschrift fur Tierphysiologie, Tierernahrung und Futtermittelkunde*, **36**, 311–324
McNeil, N.I. (1988). *World Review of Nutrition and Dietetics*, **56**, 1–42
Millward, D.J. (1989). In *Milk Proteins*, pp. 49–61. Ed. Barth, C.A. and Schlimme, E. Steinkopff, Darmstadt

Millward, D.J. and Rivers, J.P.W. (1988). *European Journal of Clinical Nutrition*, **42**, 367–393
Moughan, P.J. (1984). Aspects of dietary protein quality for the growing pig, PhD Thesis, Massey University, New Zealand
Moughan, P.J. (1989). *Research and Development in Agriculture*, **6**, 7–14
Moughan, P.J., Darragh, A.J., Smith, W.C. and Butts, C.A. (1990). *Journal of the Science of Food and Agriculture*, **52**, 13–21
Moughan, P.J. and Rutherfurd, S.M. (1990). *Journal of the Science of Food and Agriculture*, **52**, 179–192
Moughan, P.J., Schrama, J., Skilton, G.A. and Smith, W.C. (1989). *Journal of the Science of Food and Agriculture*, **47**, 281–292
Moughan, P.J. and Smith, W.C. (1985). *New Zealand Journal of Agricultural Research*, **28**, 365–370
Moughan, P.J. and Smith, W.C. (1987). *Animal Production*, **44**, 319–321
Moughan, P.J., Smith, W.C. and James, K.A.C. (1984). *New Zealand Journal of Agricultural Research*, **27**, 509–512
Moughan, P.J., Smith, W.C., Kies, A.K. and James, K.A.C. (1987). *New Zealand Journal of Agricultural Research*, **30**, 59–66
Moughan, P.J. and Verstegen, M.W.A. (1988). *Netherlands Journal of Agricultural Science*, **36**, 145–166
Oste, R.E., Dahlqvist, A., Sjostrom, H., Noren, O. and Miller, R. (1986). *Journal of Agricultural and Food Chemistry*, **34**, 355–358
Payne, W.L., Combs, G.F., Kifer, R.R. and Snyder, D.G. (1968). *Federation Proceedings*, **27**, 1199–1203
Pedersen, B. and Eggum, B.O. (1983). *Zeitschrift fur Tierphysiologie, Tierernahrung und Futtermittelkunde*, **49**, 265–277
Pellett, P.L. (1990). *American Journal of Clinical Nutrition*, **51**, 723–737
Picard, M., Bertrand, S., Duron, M., Maillard, R. (1984). In *Proceedings of the IVth European Symposium on Poultry Nutrition*, p. 165. Tours, France
Porter, D.H., Swaisgood, H.E. and Catignani, G.L. (1984). *Journal of Agricultural and Food Chemistry*, **32**, 334–339
Puia-Negulescu, G., Stavri, J. and Burlacu, G. (1989). *Archiva Zootechnica*, **1**, 81–87
Reeds, P.J. (1988). In *Comparative Nutrition*, pp. 55–72. Ed. Blaxter, K. and Macdonald, I. John Libbey, London
Rerat, A. (1984). *World Review of Nutrition and Dietetics*, **37**, 229–287
Sakata, T. (1987). *British Journal of Nutrition*, **58**, 95–103
Sauer, W.C., den Hartog, L.A., Huisman, J. van Leeuwen, P. and de Lange, C.F.M. (1989). *Journal of Animal Science*, **67**, 432–440
Sauer, W.C., Dugan, M., de Lange, C.F.M., Imbeah, M. and Mosenthin, R. (1989). In *Absorption and Utilization of Amino Acids*, pp. 217–230. Ed. Friedman, M. CRC Press, Florida
Sauer, W.C. and Ozimek, L. (1986). *Livestock Production Science*, **15**, 367–388
Schmitz, M. (1988). Moglichkeiten und grenzen der homoargininmarkierungsmethode zur messung der proteinverdaulichkeit beim schwein. PhD Thesis, Christian-Albrechts-Universitat, Kiel
Simon, O. (1989). In *Protein Metabolism in Farm Animals*, pp. 273–366. Ed. Bock, H.D., Eggum, B.O., Low, A.G., Simon, O. and Zebrowska, T. Oxford University Press, Berlin
Stephen, A.M. and Cummings, J.H. (1980). *Journal of Medical Microbiology*, **13**, 45–56

Tanksley, T.D. and Knabe, D.A. (1980). *Proceedings of the 1980 Georgia Nutrition Conference*, pp. 157–168
Tanksley, T.D. and Knabe, D.A. (1984). In *Recent Advances in Animal Nutrition, 1984*, pp. 75–95. Ed. Haresign, W. and Cole, D.J.A. Butterworths, London
Tanksley, T.D. and Knabe, D.A. (1985). In *Digestibility and Amino Acid Availability in Cereals and Oilseeds*, pp. 259–273. Ed. Finely, J.W. and Hopkins, D.T. American Association of Cereal Chemists Inc., Minnesota
Van der Walt, J.G., Meyer, J.H.F. and van Rensburg, I.B.J. (1990). *Animal Production*, **50**, 277–290
Van Leeuwen, P., Huisman, J., Verstegen, M.W.A., Baak, M.J., van Kleef, D.J., van Weerden, E.J. and den Hartog, L.A. (1988). In *Digestive Physiology in the Pig*, pp. 289–296. Ed. Buraczewska, L., Buraczewski, S., Pastuzewska, B. and Zebrowska, T. Polish Academy of Sciences, Jablonna
Van Weerden, E.J. (1989). In *Nutrition and Digestive Physiology in Monogastric Farm Animals*, pp. 89–101. Ed. van Weerden, E.J. and Huisman, J. Pudoc, Wageningen
Van Weerden, E.J., Huisman, J., van Leeuwen, P. and Slump, P. (1985) In *Digestive Physiology in the Pig*, pp. 392–395. Ed. Just, A., Jorgensen, H. and Fernandez, J.A. National Institute of Animal Science, Copenhagen
Wenham, G. and Wyburn, R.S. (1980). *Journal of Agricultural Science*, **95**, 539–546
Whitacre, M.E. and Tanner, H. (1989). In *Absorption and Utilization of Amino Acids*, Volume III, pp. 129–141. Ed. Friedman, M. CRC Press Inc., Florida
Whittemore, C.T. (1983). *Agricultural Systems*, **11**, 159–186
Wiesemuller, W. (1983). In *Metabolisme et Nutrition Azotes*, pp. 405–431. Ed. Arnal, M., Pion, R. and Bonin, D. INRA, Paris
Williams Smith, H. (1965). *Journal of Pathology and Bacteriology*, **86**, 387–412
Wrong, O.M., Edmonds, C.J. and Chadwick, V.S. (1981). *The Large Intestine: Its Role in Mammalian Nutrition and Homeostasis*. MTP Press Ltd, Lancaster
Young, V.R., Meredith, C., Hoerr, R., Bier, D.M. and Matthews, D.E. (1985). In *Substrate and Energy Metabolism in Man*, pp. 119–134. Ed. Garrow, J.S. and Halliday, D. John Libbey, London

11

# ROLE OF DIETARY FIBRE IN PIG DIETS

A.G. LOW*
*56 Reigate Rd, Brighton, East Sussex BN1 5AH*

## Introduction

Contemporary pig production in western countries is heavily dependent upon diets based on grain and high-quality protein supplements, which can also be used directly to provide a nutritious diet for man. At a time when the world population is increasing, while more than half of the human race is inadequately fed, it is clear that the future of feeding pigs on high quality feedstuffs will be increasingly questioned. Attention is therefore being given to the ability of pigs to consume and use feedstuffs which are unacceptable to man. Many of these have a high content of plant cell walls and are rich in dietary fibre. The aim of this review is to consider our present knowledge about the effects that dietary fibre may have upon intake, digestion, absorption, metabolism and growth of pigs.

## Definitions of dietary fibre

Many attempts have been made to define dietary fibre, but each definition is lacking in some way or other because of the variety and complexity of the chemical components of plant cell walls, their physical composition, and their metabolic effects. A widely accepted definition is 'the sum of lignin and the polysaccharides that are not digested by the endogenous secretions of the digestive tract' (Trowell *et al.*, 1976). This broad conceptual definition combines both chemical and physiological aspects of dietary fibre and does not apply directly to an entity that can be easily measured. Thus, it is important to have a practical definition which describes attributes of dietary fibre which can be analysed by existing methods. For this purpose dietary fibre may be defined as 'non-starch polysaccharides and lignin'. The importance of describing dietary fibre in as much chemical and physical detail as possible needs to be emphasized. The lack of such detailed information makes comparison of most published studies on the effects of dietary fibre very difficult. Much of the problem arises because different components of dietary fibre

*Previous address: The Animal and Grassland Research Institute, Shinfield, Reading, Berks RG2 9AQ, UK.

are measured by different analytical methods. A further difficulty is that the same method used by different laboratories on samples of the same ingredient can lead to a surprisingly wide scatter of estimates.

## Some properties of dietary fibre

The primary cell wall of plants contains (mg/g fresh weight) water 600, hemicellulose 50–150, cellulose 100–150, pectic substances 20–80, lipid 5–30, protein 10–20. The cellulose is a highly ordered fibrillar component while the rest is less ordered and may contain lignin. As the cell matures the cellulose and lignin content increase while the other components decrease; the botanical composition of cell walls in different parts of the plant varies greatly. The outer coat or husk of mature grain is made up in part of compressed cellulose and cuticle. The husk (about 130 g/kg of the grain) also includes the pericarp (about 40 g/kg of the grain); the outer epidermis and cell layers (containing about 320 g/kg cellulose and 350 g/kg hemicellulose) and the aleurone layer (60–70 g/kg of the grain) and containing about 560 g/kg hemicellulose, 290 g/kg cellulose and lignin.

Dietary fibre swells to a variable extent in water: for example, isolated pectin swells greatly, but when contained within a mesh of less hydrophilic substances it swells much less. The water-holding capacity is determined by the physicochemical structure of the molecule, and also by the pH and electrolyte concentration of the surrounding fluid; thus during passage through the gut, dietary fibre may swell to a very variable extent. There are several different methods of measuring the water-holding capacity of dietary fibre (centrifugation, dialysis bags, filtration) each leading to different results. Furthermore, the particle size and method of preparation of samples are important determinants of water-holding capacity, as measured in the laboratory.

The acidic sugars of polysaccharides confer ion exchange properties; several cations are known to bind to dietary fibre, but so far no anion exchange properties have been found. Adsorption of bile acids to dietary fibre, especially in the colon, is well established as a pH-dependent process, the degree of adsorption varying between types of dietary fibre.

## Analytical methods

### CRUDE FIBRE

This method was developed over 150 years ago as a means of measuring the indigestible fraction of feedstuffs. The sample is treated sequentially with petroleum ether, hot sulphuric acid, boiling water and alkali. It is now clear that the resultant insoluble residue contains mainly cellulose and lignin but the recovery is not always complete (Van Soest and McQueen, 1973). In spite of these shortcomings this is still an official method of measuring dietary fibre in animal feedstuffs in many countries.

### NEUTRAL DETERGENT FIBRE

During this method developed by Van Soest (1963a), the sample is digested by boiling in a neutral detergent solution, filtered, dried and weighed. Although lignin and cellulose are completely recovered, there may be some loss of hemicelluloses while water soluble carbohydrates are normally completely lost during the procedure.

ACID DETERGENT FIBRE

This method, also developed by Van Soest (1963b) involves digestion by boiling in an acid detergent solution, followed by filtration, drying and weighing. This method is usually considered to provide a reasonably reliable estimate of the sum of cellulose and lignin; thus almost all other components of fibre are excluded.

NON-STARCH POLYSACCHARIDES

Following starch removal (by enzymic hydrolysis) the residue in the sample is separated into cellulose, non-cellulosic polysaccharides and lignin, followed by acid hydrolysis and colorimetric measurement of the component sugars (Southgate, 1969). A modified version of this procedure in which alditol acetate derivatives of the sugars are measured by gas-liquid chromatography has recently been developed by Englyst, Wiggins and Cummings (1982).

ENZYMIC ASSAY OF INSOLUBLE AND SOLUBLE DIETARY FIBRE

A rapid gravimetric method of enzymic hydrolysis of the sample has been described by Asp *et al.* (1983). Initial gelatinization by boiling is followed by incubation with pepsin and then pancreatin. Insoluble dietary fibre is separated by filtration and the components are then analysed in as much detail as required.

**Table 11.1** TYPICAL CRUDE FIBRE AND POLYSACCHARIDE CONTENTS OF FEEDSTUFFS (AS % OF DRY MATTER)

| | | *Crude fibre* | | | |
|---|---|---|---|---|---|
| Alfalfa | 28 | Cowpea | 6 | White rice | 0.4 |
| Barley hay | 26 | Fishmeal | 1 | Rye straw | 48 |
| Barley grain | 5 | Swill (restaurant) | 3 | Rye grain | 2 |
| Barley screenings | 9 | Millet | 9 | Rye grass | 22 |
| Kidney bean | 5 | Oat hulls | 29 | Sorghum grain | 2 |
| Sugar beet | 20 | Oat grain | 12 | Sorghum fodder | 27 |
| Bermuda grass | 30 | Oat meal | 4 | Soyabean meal | 7 |
| Buckwheat | 20 | Orchard grass hay | 34 | Sunflower meal | 14 |
| Milk (bovine) | <0.1 | Pea | 10 | Timothy grass | 33 |
| Citrus pulp | 15 | Peanut meal | 14 | Wheat straw | 42 |
| Clover | 30 | Peanut shells | 65 | Wheat bran | 11 |
| Maize | 2 | Potato (cooked) | 3 | Wheat grain | 3 |
| Cottonseed | 18 | Rapeseed | 15 | Yeast | 3 |
| | | Rice bran | 12 | | |

| | *Polysaccharides content of some feedstuffs*[a] | | | | |
|---|---|---|---|---|---|
| | *Starch* | *Cellulose* | *Non-cellulosic polysaccharides* | | *Total non-starch polysaccharides* |
| | | | *Soluble* | *Insoluble* | |
| Barley grain | 72.1 | 1.44 | 3.89 | 6.50 | 11.83 |
| Oat meal | 64.0 | 0.40 | 3.93 | 2.96 | 7.29 |
| Rye grain | 66.7 | 1.52 | 4.47 | 7.24 | 13.23 |
| Wheat bran | 16.4 | 8.17 | 4.25 | 28.60 | 41.06 |
| Wheat grain | 64.6 | 1.52 | 2.57 | 5.48 | 9.58 |

[a]Data from Englyst, Anderson and Cummings (1983).

A critical review of the many methods of dietary fibre analysis now available has been edited by James and Theander (1981).

Some examples of the crude fibre and non-starch polysaccharide content of feedstuffs are shown in *Table 11.1*. The large differences that can be seen serve to emphasize the need for an agreed standard method of chemical analysis. Reliable methods of characterizing the physical properties of dietary fibre are not well developed and are urgently needed. There are signs that the present lack of agreement on standard analytical methods may be resolved before long in clinical circles, and it is to be hoped that agricultural science will follow suit.

## Fibre and the digestive processes

### STOMACH

The possibility that dietary fibre may prevent the development of oesophagastric ulcers was examined by Henry (1970) who found a decreased incidence of lesions when wood cellulose was added to the diet of growing-finishing pigs. The effect was greater with coarsely-ground than finely-ground cellulose. However, the addition of unground or ground bran to barley based diets did not significantly reduce the incidence of oesophagogastric ulceration in growing pigs (Potkins, Lawrence and Thomlinson, 1984). It is probable that several factors contribute to the development of oesophagogastric ulcerations and no firm conclusions about the role of dietary fibre in this process can be drawn at present.

### DIGESTIVE SECRETIONS

The effects of dietary fibre on gastric, biliary and pancreatic secretion appear to be considerable. For example, Zebrowska *et al.* (1983) and Sambrook (1981) found significantly higher outputs of all three secretions in pigs fed a barley-based diet (A) containing a wide variety of types of dietary fibre, than in the same pigs when they received a semi-purified diet containing cellulose as the only dietary fibre source (B). Although crude fibre intakes were similar in both cases, neutral detergent fibre intakes were 180 (diet A) and 50 g per day (diet B), which emphasizes the large contribution of non-cellulosic components of dietary fibre in diet A. Some of the main results of these studies are shown in *Table 11.2*.

**Table 11.2** EFFECT OF DIETARY FIBRE ON GASTRIC AND PANCREATIC FUNCTION IN 40 kg PIGS DURING 24 h PERIODS

| | *Diet A*<br>*High fibre*<br>*(barley–soya)* | *Diet B*<br>*Low fibre*<br>*(starch–casein–cellulose)* |
|---|---|---|
| Gastric juice (l) | 8.0 | 4.0 |
| Pepsin (units $\times 10^{-6}$) | 1.47 | 0.76 |
| Pancreatic juice (l) | 2.18 | 1.20 |
| Ash (g) | 17.3 | 9.5 |
| Trypsin (units $\times 10^{-3}$) | 114 | 138 |
| Chymotrypsin (units $\times 10^{-3}$) | 84 | 84 |
| Amylase (units $\times 10^{-3}$) | 981 | 1061 |
| Bile (l)[a] | 1.72 | 1.17 |

Data from Zebrowska, Low and Żebrowska (1983) and [a]Sambrook (1981).

Certain types of soluble dietary fibre such as guar gum, pectin and sodium carboxymethyl cellulose increase the viscosity of solutions in which they are dissolved. Our recent studies on the effect of these gums on gastric emptying in growing pigs showed that while they did not affect the rate of emptying of dry matter (Rainbird, Low and Sambrook, 1983; Rainbird and Low, 1983) they did delay the rate of digesta emptying. In other words these materials reduced the rate of water passage into the duodenum, partly because of their large water holding capacity.

Further secretory responses to dietary fibre have been found by Low and Rainbird (1984) in isolated loops of jejunum in conscious growing pigs. Addition of guar gum to glucose solutions perfused through such loops increased nitrogen secretion from 35 to 67 mg/m/h (equivalent to a calculated increase in the whole small intestine from 15 to 27 g per day). The nature of the secreted nitrogen has not yet been fully elucidated but it is included in proteins and DNA. It has been estimated that 20–25 per cent of total body protein synthesis in growing pigs occurs in the gut (for example Reeds *et al.*, 1980; Simon *et al.*, 1978), and much of this is secreted into the gut lumen. These endogenous secretions are a major component of those amino acids which are not absorbed by the end of the small intestine (after which they have no further nutritional value to the pig), and have been estimated to amount to half of the amino acids in the terminal ileal digesta (Zebrowska *et al.*, 1982). Studies in other species of animals have shown that dietary fibre can increase the rate of synthesis and migration of epithelial cells along intestinal villi. Sauer, Stothers and Parker (1977) and Taverner, Hume and Farrell (1981) observed increases in the ileal digesta content of nitrogen and amino acids when graded levels of cellulose were added to protein-free diets for pigs indicating that insoluble dietary fibre can also markedly influence gut secretion. Similar studies by Behm (1954) showed that graded cellulose addition increased faecal N output. Whiting and Bezeau (1957) again found a similar effect of cellulose on faecal N; oat hulls had less effect and methylcelluloses only led to small increases. It is thus possible that dietary fibre in both soluble and insoluble forms may be an important determinant of apparent protein digestibility, and thus may also influence the efficiency of conversion of dietary protein into carcass protein. However, the mechanisms by which such effects of dietary fibre may be mediated are not yet understood.

Several types of dietary fibre increase the water content of both the small and large intestines and of faeces: Cooper and Tyler (1959) found this to be so for bran and fibrous cellulose but not for finely powdered cellulose. Partridge (1978) similarly found much larger volumes of digesta in the ileum of pigs given additional wood cellulose. Whether this effect is one of increased water secretion into the gut, reduced water absorption or a combination of both effects, in some way mediated by dietary fibre, is not understood.

## EFFECT OF DIETARY FIBRE ON ABSORPTION FROM THE SMALL INTESTINE

Recent studies by Rainbird, Low and Zebrowska (1984) have shown that the addition of guar gum to a glucose solution perfused through isolated loops of pig jejunum halved the rate of glucose absorption: the mechanism by which this occurs is not fully understood but it is thought to be associated with reduced diffusion from the intestinal lumen to the epithelial cells or inhibition of the absorption process.

The appearance of glucose and α-$NH_2$ nitrogen in blood plasma is delayed and peak concentrations are lower, following meals containing guar gum (Sambrook, Rainbird and Low, 1982) indicating that this source of dietary fibre has an effect on both carbohydrate and protein digestion and absorption. The more rapid rate of absorption of glucose and amino acids following wheat meals than barley meals, found by Rerat, Vaissade and Vaugelade (1979), may have been due in part to the higher dietary fibre content (especially soluble components; see *Table 11.1*) of barley than of wheat. Guar gum is an interesting 'model' of soluble dietary fibre because it can be obtained in a purified form and because its physiological effects are probably very similar to the various soluble types of dietary fibre found in substantial amounts in most feedstuffs for pigs, and especially cereal grains.

### EFFECT OF DIETARY FIBRE ON NUTRIENT ABSORPTION MEASURED AT THE END OF THE SMALL INTESTINE

The effects of cellulose (100 g/kg), the gel-forming methyl cellulose (60 g/kg) and pectin (60 g/kg) on the apparent digestibility of nitrogen in the terminal ileum of 50–100 kg pigs given a barley–soya–starch diet were measured by Murray, Fuller and Pirie (1977). Cellulose had no effect, but methyl cellulose reduced the value from 76 (control) to 48 per cent, and increased the rate of passage. Pectin had an intermediate effect. The authors suggested that impaired protein digestion and absorption were due to inhibition of protein hydrolysis when the gel-forming polysaccharides were given, because they did not affect the absorption of free lysine given in the diet.

The apparent digestibility of dry matter, organic matter, nitrogen, gross energy, crude fibre and amino acids was measured in the terminal ileum of growing pigs

**Table 11.3** EFFECT OF SOYA BEAN MEAL, MALT CULMS, DARK GRAINS AND WEATINGS ON DRY MATTER (DM), ORGANIC MATTER (OM), NITROGEN (N), GROSS ENERGY (GE), CRUDE FIBRE (CF) AND APPARENT DIGESTIBILITY (%) IN THE TERMINAL ILEUM AND OVERALL OF GROWING PIGS

| | *Diet* | | | |
|---|---|---|---|---|
| | *Barley + soya* | *Barley + malt culms* | *Barley + dark grains* | *Barley + weatings* |
| DM | | | | |
| ileum | 72 | 68 | 64 | 59 |
| overall | 80 | 76 | 70 | 66 |
| OM | | | | |
| ileum | 74 | 70 | 66 | 62 |
| overall | 82 | 78 | 73 | 68 |
| N | | | | |
| ileum | 75 | 66 | 63 | 67 |
| overall | 76 | 68 | 66 | 62 |
| GE | | | | |
| ileum | 74 | 68 | 65 | 61 |
| overall | 79 | 74 | 70 | 65 |
| CF | | | | |
| ileum | 17 | 38 | 38 | 26 |
| overall | 30 | 19 | 9 | −2 |
| (CF content of diet, g/kg) | 3.88 | 5.04 | 5.65 | 5.85) |

(Data from Zoiopoulos, Topps and English, 1983b)

**Table 11.4** APPARENT DIGESTIBILITY (%) OF NUTRIENTS IN THE TERMINAL ILEUM AND OVERALL OF GROWING PIGS GIVEN DIETS CONTAINING 33-161 g CRUDE FIBRE/kg

| | *Diet* | | | | | |
|---|---|---|---|---|---|---|
| | *1* | *2* | *3* | *4* | *5* | *6* |
| Nitrogen diet (g/kg) | 35.7 | 35.8 | 36.0 | 35.0 | 35.4 | 35.7 |
| ileum | 78 | 84 | 82 | 78 | 76 | 75 |
| faeces | 93 | 91 | 87 | 84 | 81 | 77 |
| Stoldt fat diet (g/kg) | 72 | 72 | 72 | 72 | 77 | 78 |
| ileum | 75 | 81 | 81 | 79 | 81 | 79 |
| faeces | 85 | 83 | 80 | 76 | 80 | 78 |
| Crude fibre diet (g/kg) | 33 | 57 | 84 | 109 | 137 | 161 |
| ileum | 0 | 0 | 0 | 0 | 0 | 0 |
| faeces | 55 | 60 | 68 | 63 | 51 | 59 |
| Gross energy diet (MJ/kg) | 19.36 | 19.19 | 19.25 | 19.09 | 19.27 | 19.2 |
| ileum | 75 | 72 | 69 | 65 | 56 | 52 |
| faeces | 92 | 90 | 87 | 84 | 79 | 78 |
| Lysine diet | | | (not shown in paper) | | | |
| ileum | 0.92 | 0.93 | 0.91 | 0.90 | 0.87 | 0.86 |
| faeces | 0.95 | 0.93 | 0.90 | 0.87 | 0.84 | 0.80 |
| % of that digested disappearing in large intestine | | | | | | |
| Nitrogen | 16 | 8 | 6 | 7 | 6 | 2 |
| Stoldt fat | 12 | 2 | –1 | –4 | –1 | –1 |
| Crude fibre | 100 | 100 | 100 | 100 | 100 | 100 |
| Gross energy | 18 | 20 | 21 | 23 | 29 | 33 |
| Lysine | 3 | 0 | –1 | –3 | –4 | 17 |

(data from Just, Fernandez and Jørgensen, 1983)

fitted with a 'T' cannula and given a barley-based diet supplemented with soya bean meal, malt culms, dark grains or weatings by Zoiopoulos, Topps and English (1983). The results are summarized in *Table 11.3*. In general the higher the crude fibre content of the diet, the lower was the apparent digestibility of each component. A similar conclusion was reached by Just, Fernandez and Jorgenson (1983) who gave pigs a series of six diets containing 33 to 161 g crude fibre/kg (96–285 g neutral detergent fibre), and based on combinations of casein, soya bean meal, meat and bone meal, oats, barley, maize starch, potato starch cellulose, soya oil, sugar beet molasses, minerals and vitamins. The apparent digestibility of a variety of dietary components is shown in *Table 11.4*. Kass *et al.* (1980) investigated the digestion of diets containing 0, 200, 400 or 600 g/kg alfalfa meal in pigs after slaughter at 48 or 89 kg. Increasing depression of digestibility of dry matter, nitrogen and cell wall components was found in the small intestine (measured in a single sample from the whole organ) and in the caecum, colon and faeces, was found as the level of alfalfa in the diet rose.

The disappearance of dietary fibre components from diets containing 300 g/kg of cell wall material from alfalfa, grain sorghum, Texas Kleingrass and Coastal Bermuda Grass was examined in four 82–90 kg pigs with 'T' cannulas in the terminal ileum by Keys and Debarthe (1974). Faeces were also collected. The results are summarized in *Table 11.5*.

The effect of supplementary cellulose in a semi-purified diet on mineral absorption of pigs measured in the terminal ileum and overall was studied by Partridge (1978) and is summarized in *Table 11.6*. The supplementary cellulose significantly increased the amounts of digesta, faeces and organic matter.

**Table 11.5** APPARENT DIGESTIBILITY (%) IN THE ILEUM AND OVERALL OF DRY MATTER, CELL WALLS, CELLULOSE AND HEMICELLULOSE IN PIGS GIVEN DIETS CONTAINING ALFALFA, GRAIN SORGHUM, TEXAS KLEINGRASS AND COASTAL BERMUDA GRASS

| | *Diet* | | | |
|---|---|---|---|---|
| | *Alfalfa* | *Grain sorghum* | *Texas Kleingrass* | *Coastal Bermuda grass* |
| Dry matter | | | | |
| intake (g/kg) | 912.5 | 928.6 | 926.6 | 928.9 |
| ileum | 39 | 41 | 38 | 47 |
| faeces | 68 | 70 | 66 | 75 |
| Cell walls | | | | |
| intake (g/kg) | 275.9 | 281.3 | 305.6 | 328.7 |
| ileum | -5 | -3 | -6 | 39 |
| faeces | 32 | 31 | 22 | 50 |
| Cellulose | | | | |
| intake (g/kg) | 174.6 | 143.5 | 148.9 | 144.9 |
| ileum | -9 | -8 | -7 | 33 |
| faeces | 38 | 33 | 21 | 48 |
| Hemicellulose | | | | |
| intake (g/kg) | 64.5 | 119.9 | 127.9 | 161.9 |
| ileum | 10 | 4 | 5 | 47 |
| faeces | 33 | 32 | 25 | 54 |

(Data from Keys and Debarthe, 1974)

**Table 11.6** EFFECT OF CELLULOSE ON APPARENT DIGESTIBILITY (%) OF ORGANIC MATTER, WATER AND MINERALS IN THE TERMINAL ILEUM AND OVERALL IN GROWING PIGS

| | *Cellulose* (g/kg) | | | |
|---|---|---|---|---|
| | *30* | | *90* | |
| | *Ileum* | *Faeces* | *Ileum* | *Faeces* |
| Digesta/faeces | 84 | 99 | 76 | 97 |
| Organic matter | 94 | 99 | 90 | 95 |
| Water | 81 | 99 | 72 | 98 |
| Sodium | 46 | 99 | 15 | 98 |
| Potassium | 90 | 97 | 88 | 86 |
| Calcium | 43 | 74 | 44 | 63 |
| Phosphorus | 64 | 81 | 69 | 74 |
| Magnesium | -1 | 73 | 3 | 62 |
| Zinc | 10 | 60 | 24 | 37 |

(Data from Partridge, 1978)

The effects of supplementary cellulose, pectin and dried sugar beet pulp in semi-purified diets on dry matter, nitrogen and amino acid digestibility in the ileum and faeces of growing pigs and on nitrogen retention were studied by Dierick *et al.* (1983). As the level of fibre in the diet increased, so the nutrient digestibility in the ileum fell, to a much greater extent with pectin and dried sugar beet pulp than with cellulose. However, although a similar pattern of effects of dietary fibre could be seen in the faeces, the differences between treatments were much less than in ileal digesta. The principal results are shown in *Table 11.7*.

**Table 11.7** THE EFFECTS OF CELLULOSE, PECTIN AND SUGAR BEET PULP ON THE APPARENT DIGESTIBILITY (%) OF DRY MATTER (DM), NITROGEN (N), ESSENTIAL (EAA) AND NON-ESSENTIAL (NEAA) AMINO ACIDS IN GROWING PIGS, MEASURED IN THE ILEUM AND OVERALL

| *Diet* | *Ileal digesta* | | | | *Faeces* | | | |
|---|---|---|---|---|---|---|---|---|
| | *DM* | *N* | *EAA* | *NEAA* | *DM* | *N* | *EAA* | *NEAA* |
| Experiment 1 | | | | | | | | |
| Fibre-free | 91 | 88 | 90 | 89 | 95 | 94 | 93 | 93 |
| Cellulose (75 g/kg) | 84 | 86 | 90 | 88 | 84 | 90 | 91 | 92 |
| Pectin (50 g/kg) | 80 | 76 | 81 | 75 | 93 | 92 | 91 | 92 |
| Experiment 2 | | | | | | | | |
| Fibre-free | 92 | 89 | 94 | 93 | 95 | 95 | 96 | 96 |
| Cellulose (50 g/kg) | 85 | 89 | 93 | 92 | 93 | 95 | 96 | 96 |
| Cellulose (100 g/kg) | 79 | 88 | 93 | 92 | 89 | 93 | 95 | 96 |
| Cellulose (150 g/kg) | 74 | 84 | 94 | 93 | 84 | 91 | 93 | 93 |
| Dried sugar | | | | | | | | |
| Beet pulp (50 g/kg) | 87 | 90 | 96 | 95 | 94 | 95 | 96 | 96 |
| Beet pulp (100 g/kg) | 81 | 86 | 92 | 89 | 93 | 94 | 94 | 94 |
| Beet pulp (150 g/kg) | 72 | 81 | 90 | 88 | 92 | 92 | 94 | 94 |

(Data from Dierick *et al.*, 1983)

The present level of knowledge makes it very difficult to make predictions about the effects of specific types of dietary fibre on nutrient absorption in the small intestine of pigs, though it is evident that some important changes may occur. At present nothing is known about the effects of dietary fibre on vitamin absorption and little is known of the effects on carbohydrate and lipid absorption. However, enough is known from studies in man and rat to indicate that inclusion of supplementary dietary fibre tends to reduce the rate or the amount of apparent absorption of all nutrient types, to a degree that is determined in part by the source and level used. Such effects may vary with the age of the pig and the level of feeding applied. Clearly some of the effects may be considerable and current views on the nutrient requirements or responses of pigs may need modification when more detailed information is available. This may be especially so for amino acids because any amino acids which are not absorbed by the end of the small intestine appear to play no further role in the nutrition of the animal (Zebrowska, 1973).

## Large intestine

Striking as the effects of dietary fibre may be on the function of the small intestine, it is in the large intestine that it becomes an identifiable nutrient source, rather than a medium eliciting diverse physiological responses. The transit time of digesta is much longer through the large intestine (generally 20–40 h) than through the stomach and small intestine (generally in the range 2–16 h). These conditions allow considerable net absorption of water: for example mean values for 40 kg pigs of 3152 g for a cereal-based diet and 986 g for a semi-purified (low dietary fibre) diet were found during 24 h periods by Low, Partridge and Sambrook (1978). Addition of cellulose to the diet caused a major increase in the volume of water passing into and out of the large intestine (Partridge, 1978).

A major consequence of the slow passage of digesta through the large intestine is that it encourages prolific bacterial growth: up to $10^{11}$ bacteria of both obligate anaerobic and aerobic species have been found per gram of fresh digesta. Although the population is generally thought to be stable, antibiotic treatment may cause big disturbances in the microbial balance. A detailed review of microbial fermentation in the pig alimentary tract was published by Cranwell (1968), but relatively few bacteriological studies have been published since, the main interest in the pig now being focused on the products of the bacterial activity and their potential nutritive value.

It has been recognized for many years that the weight and volume of the gut of pigs tends to increase when diets of a high dietary fibre content are used (for example Coey and Robinson, 1954). Kass *et al.* (1980) found in growing pigs that increasing the dietary fibre content of the diet by means of alfalfa meal reduced the stomach weight, as a proportion of the weight of the whole gut, while the weights of the small intestine, caecum and colon increased, as a proportion of body weight (*Table 11.8*) (all expressed on an empty gut weight basis). The percentage dry matter content of the digesta decreased significantly in all regions of the gut except the caecum as the amount of alfalfa meal in the diet increased. The effect of a high dietary fibre diet on 120–130 kg pigs was to increase the volume of the gut, and in particular that of the stomach and caecum (Horszczaruk, 1962).

**Table 11.8** THE EFFECT OF ALFALFA MEAL ON EMPTY GUT WEIGHT IN GROWING PIGS (EXPRESSED AS % OF BODY WEIGHT)

| *Level of dietary alfalfa meal (%)* | *Stomach* | *Small intestine* | *Caecum* | *Colon* |
|---|---|---|---|---|
| 0 | 0.75 | 1.98 | 0.18 | 1.40 |
| 20 | 0.79 | 2.09 | 0.17 | 1.60 |
| 40 | 0.76 | 2.33 | 0.20 | 1.79 |
| 60 | 0.77 | 2.57 | 0.22 | 2.02 |

(Data from Kass *et al.*, 1980)

The decomposition of dietary fibre in the caecum and colon of pigs was measured using the nylon bag technique by Horszczaruk and Sljivovacki (1971). Four 18-month-old pigs fitted with caecal and colonic cannulas were used. Nylon bags containing ground raw cellulose or lucerne meal were placed on short threads through the cannulas for two or four days. Cellulose was digested faster than lucerne meal: 70 and 30 per cent in two days and 95 and 35 per cent in four days respectively. It was noted that these latter values compared with values of 0 and 17 per cent in young growing pigs.

The rate of flow of digesta through different regions of the large intestine is generally accelerated by the addition of dietary fibre to the diet (Horszczaruk, 1962; Kass *et al.*, 1980). Recent studies at Shinfield indicate that the water phase of digesta moves more rapidly through the caecum and ascending colon than the dry matter, irrespective of the source of dietary fibre (supplementary bran, lactulose and pectin were used) but the effect is much less for pectin which has the greatest water-binding capacity of the sources tested. However the time of arrival of the liquid and solid phases in the faeces did not differ between types of dietary fibre.

The role of the caecum in pigs has been investigated by comparing the digestibility of diets with low or high dietary fibre content in intact and caecectomized pigs by Lloyd, Dale and Crampton (1958) and Gargallo and Zimmerman (1981). In neither case did caecectomy significantly affect the performance of the pigs or the digestibility of crude fibre.

## Volatile fatty acids (VFA)

Large concentrations (150–250 mM of VFA are found throughout the large intestine: smaller amounts (5–40 mM) are found in the stomach and small intestine, as shown by Argenzio and Southworth (1975) and Clemens, Stevens and Southworth (1975). The VFA derived by microbial activity from dietary fibre are predominantly found in the large intestine where cellulolytic bacteria are abundant: although such bacteria may be found in the ileum their importance in this part of the gut is comparatively minor. VFA can be produced from all the components of dietary fibre, as well as undigested starch, lipid and proteins which may enter the large intestine. Acetic acid tends to predominate in the digesta, with smaller amounts of propionic acid and butyric acid, the proportions varying with the type of dietary fibre fed and the particular site. In general as the dietary fibre content of the diet increases, so the proportion of acetate to the other VFA rises, as shown for alfalfa by Kass *et al* (1980) and for sunflower hulls by Gargallo and Zimmerman (1981b). Volatile fatty acids are readily absorbed and their concentrations in blood of pigs were first measured by Barcroft, McAnally and Phillipson (1944).

The rates of VFA production have been measured by a variety of methods in order to calculate the degree to which VFA may contribute to the energy supply of the pig. Among the earliest studies was a comparison of portal-venous VFA concentration differences in 30 kg pigs by Friend, Nicholson and Cunningham (1964) who calculated that 15–28 per cent of the maintenance energy requirements might be met by VFA. However, these values do not allow for hepatic production of VFA, which was demonstrated by Imoto and Namioka (1978a), and are thus open to doubt. A more recent attempt to calculate the amounts of VFA which may be formed in the large intestine of pigs fed diets containing 0, 200, 400 and 600 g alfalfa meal was made by Kass *et al.* (1980) by use of regression equations for each acid after measurement in the caecum and colon 2, 4, 8 or 12 h after feeding. Based on gross caloric values of 3.40, 4.96 and 5.95 kcal/g for acetic, propionic and butyric acids, respectively, the energy values of VFA disappearing from the caecum and colon (assuming complete absorption) were calculated to be 79, 147, 227 and 155 kcal/day for 48 kg pigs and 47, 231, 285 and 245 kcal/day for 89 kg pigs, the values corresponding to diets containing 0, 200, 400 or 600 g alfalfa meal/kg, respectively. The net maintenance requirement of the 48 and 89 kg pigs was calculated to be 1296 and 2028 kcal/kg $W^{0.75}$, so VFA could provide 6.9, 11.3, 12.5 and 12.0 per cent of the energy needed for maintenance in 48 kg pigs and 4.8, 11.4, 14.0 and 12.9 per cent in 89 kg pigs given the diets containing 0, 200, 400 or 600 g alfalfa meal/kg, respectively. These figures are based on the difference between VFA production and absorption and are underestimates of the actual production and absorption rates, to an unknown degree.

Kennelly, Aherne and Sauer (1981) measured VFA production in the caecum by continuous isotope infusion. In the case of a barley–soya diet, net VFA production rates indicated that caecal fermentation could provide 19.7 per cent maintenance energy requirements of growing pigs fed hourly. When alfalfa supplements were provided and the pigs were fed three times daily, it was calculated that VFA from the caecum could provide 10.1, 15.5 or 11.1 per cent of maintenance energy requirements when the diet contained 0, 27.3 or 52.0 per cent alfalfa. The problem with this method is that although accurate values may be obtained, if corrected for VFA interconversions, the isotope is not contained within the caecum and thus a pool of unknown size is obtained and sampled from.

Imoto and Namioka (1978a) attempted to measure VFA production in pigs by examining VFA in blood entering and leaving both the liver and gut. Extensive hepatic production of acetate was seen and this was the major component of circulating acetate. Furthermore extensive metabolism of absorbed VFA was found to occur in the large intestinal wall.

When short-term (30 or 60 min) incubation of caecal contents of pigs was employed by Farrell and Johnson (1972) to measure VFA production, it was concluded that VFA from the caecum provided 2.7 or 1.9 per cent of the apparently digestible energy of pigs given diets with 80 or 260 g cellulose/kg (equivalent to 5.5 or 3.9 per cent of maintenance energy requirements). In a similar way (but using caecal and colonic contents), Imoto and Namioka (1978b) calculated that VFA production from the whole of the large intestine could provide 9.6–11.6 per cent of maintenance energy requirements of growing pigs. Gargallo and Zimmerman (1981b) also used the latter approach and calculated that VFA could provide 6.2, 5.6 and 5.0 per cent of the maintenance energy requirements of 95 kg pigs given diets with 20, 100 and 200 g sunflower hulls/kg, respectively. More recently Argenzio (1982) suggested that VFA could provide 19–25 per cent of daily maintenance energy requirements, using *in vitro* incubation data. Measurements of fermentation by *in vitro* methods rely on steady state assumptions, and must be related to the entire pool of caecal or colonic contents, neither of which can be measured accurately.

There are thus problems of such magnitude that presently available estimates of the contribution of VFA from the large intestine to meeting the energy requirements of the pig must be viewed with a great deal of caution. The different techniques, diets and assumptions used in the experiments reported above make a comparative view almost meaningless. Nevertheless the results available do indicate that further research to quantitate the contribution of VFA to metabolism is warranted.

## VOLATILE FATTY ACID METABOLISM IN PIGS

Although it is believed that a large proportion of absorbed volatile fatty acids may be metabolized in the gut wall, substantial amounts do enter the blood as indicated earlier. Acetate introduced orally (in the form of 1-$^{14}$C-acetate) into newborn piglets was metabolized very rapidly to $^{14}CO_2$ (within minutes of dosing) and 89–92 per cent was recovered within 12 h in this form, while less than 3 per cent appeared in urine and faeces (Mohme, Molnar and Lenkheit, 1970). Small amounts of radioactivity were found in most tissues. More recently Latymer and Woodley (1984) found that $^{14}$C from U-$^{14}$C-acetate injected at physiological levels into the caecum of 22–28 kg pigs was also rapidly absorbed and peak blood concentrations were already observed 30 min later. From this time until 5 h after the injection $^{14}$C was found in all the major lipid classes (including free cholesterol and cholesteryl esters), plasma proteins and other water-soluble compounds. In a second study in two of the same pigs (when they weighed 70 and 78 kg) Latymer and Low (1984) infused U-$^{14}$C acetate into the caecum and collected all urine and faeces for 96 h. The pigs were then killed and the mean distribution of radioactivity found throughout the body is shown in *Table 11.9*. It can be seen that a substantial portion of the dose was retained in the carcass (mainly in the subcutaneous fat). This contrasts with the situation in the newborn piglet which has minimal energy

**Table 11.9** MEAN PERCENTAGE OF U $^{14}C$ FROM U $^{14}C$ ACETATE INJECTED INTO THE CAECUM OF TWO PIGS RECOVERED AFTER 96 h

| | |
|---|---|
| Small intestine wall and contents | 0.8 |
| Large intestine contents | 0.1 |
| Large intestine wall | 1.5 |
| Liver | 0.6 |
| Kidney | 0.1 |
| Blood | 0.1 |
| Carcass | 23.6 |
| Urine | 2.7 |
| Faeces | 5.2 |
| Total | 34.7 |

Losses, assumed to be as $^{14}CO_2$, by difference 65.3
(Data from Latymer and Low, 1984)

reserves and thus used oral acetate as an immediate energy source (Mohme, Molnar and Lenkheit, 1970). During studies *in vitro* using tissue slices from various parts of growing pigs Huang and Kummerow (1976) demonstrated the incorporation of 1-$^{14}C$ acetate or U-$^{14}C$ acetate into fatty acids and cholesterol; the rate of incorporation was highest in adipose tissue.

The nutritive value of acetate was estimated to be 59 per cent in terms of the percentage use of metabolizable energy, in growing pigs by Jentsch, Schiemann and Hoffman (1968): these authors observed that 2100 calories were deposited per gram of supplementary dietary acetate. This value corresponds well with the values of 56–59 per cent for growing pigs obtained by Imoto and Namioko (1983a), who gave acetate orally as triacetin. In another part of the same studies, Imoto and Namioka (1983b) observed a reciprocal relationship between the metabolism of acetate and glucose, depending on the time of day; 12 h after feeding glucose was the dominant blood energy source, replacing acetate which was present at higher concentrations following feeding.

These results indicate that the principal product of fibre digestion is not only absorbed but is metabolized, to an efficiency which is approximately three-quarters of that of glucose, when measured under conditions of growth. Energetic efficiency is generally higher for maintenance than for growth (in other species) but so far no data on the efficiency of acetate use in pigs under the former conditions are available. Furthermore, data on the metabolism of propionic and butyric acids in pigs do not appear to exist, either in growing or maintenance conditions.

The apparently digested energy from fermentation of dietary fibre in the large intestine, in terms of its potential value to the animal, is less than that obtained from the enzymic digestion of starch because some of the apparently absorbed energy is lost, not as volatile fatty acids, but as heat of bacterial fermentation and a further amount is lost as methane. The latter has been calculated to be 14–17 per cent of the apparently digested energy arising from fermentation (Agricultural Research Council, 1981).

## Influence of dietary fibre on nutrient absorption in the large intestine

Although quite a large number of comparisons between ileal and faecal apparent digestibility of nutrients have been made in pigs, the effects of dietary fibre have generally been made using diets differing in both its source and amount. Increasing the fibre content of the diet has been shown to reduce both the ileal and faecal digestibility of nitrogen and organic matter, as shown in the examples in *Tables 11.3, 11.4* and *11.7*. In other reports, similar effects of rye straw meal (Zebrowska, 1982), whole vs. dehulled barley (Just *et al.*, 1980), ground barley straw (Just, 1982a) and wheat and oat brain (Just, 1982b) have been shown. The results are characterized by an increasing proportion of the energy digested in the large intestine as the dietary fibre content is increased, with corresponding reductions in nitrogen and Stoldt fat absorption, as exemplified in *Table 11.4*. In addition, net synthesis of certain amino acids between the ileum and faeces was seen as the barley straw content of the diet was increased (Just, 1982a). The amounts of dietary fibre (measured as crude fibre) which disappeared in the large intestine were not significantly altered when barley straw meal was added to diets (Just, 1982a) or wheat and oat bran (Just, 1982b). *Tables 11.5* and *11.6* show the disappearance of dietary fibre and of minerals respectively during digesta passage through the large intestine. Detailed results on the effects of cereal fibre on the apparent disappearance of fatty acids in the large intestine shows that this generally corresponds with that of crude fat (Just, Andersen and Jorgensen, 1980). Varying the level of feeding of growing pigs between 70 and 100 per cent of the current Danish standard had no effect on the apparent digestibility of the main nutrients either measured in the ileum or faeces (Just, Jorgensen and Fernandez, 1983).

## Influence of dietary fibre on nutrient digestibility measured overall

Although a considerable literature exists on this theme, some of which has already been cited, it is difficult to draw firm conclusions because of the great variety of diets, pigs, sources of dietary fibre, feeding levels and levels of dietary fibre inclusion used. The following examples have been chosen to demonstrate typical effects.

### NITROGEN

Although ground wood cellulose supplements have tended to reduce nitrogen digestibility (Horszczaruk, 1962; Kirchgessner, Roth-Maier and Roth, 1975; Partridge, Keal and Mitchell, 1982), oat feed, by contrast, had no such effect in the studies of Potkins, Lawrence and Tomlinson (1984) although it did depress dry matter and energy digestibility. The soluble galactomannan guar gum did not significantly alter the digestibility or retention of nitrogen in growing pigs (though nitrogen digestibility and retention tended to rise) (Low and Keal, 1981). Alkali-treated straw, which contains a considerable quantity of soluble carbohydrates, tends to reduce nitrogen digestibility as in the work of Farrell (1973) and Bergner, Simon and Bergner (1980). While corn cobs (150 g/kg diet) had little effect on nitrogen digestibility, they depressed energy and dietary fibre digestibility: however the weights of neutral detergent fibre, acid detergent fibre and hemicellulose digested increased by 55, 15 and 90 per cent respectively (Frank, Aherne and Jensen, 1983). Oat hulls (Kennelly and Aherne, 1980) and barley hulls (Bell, Shires and Keith, 1983) had little effect on nitrogen digestibility, but energy and dietary fibre digestibility fell. Pectin reduced nitrogen digestibility in the experiments of Albers and Henkel (1979) and Mosenthin and Henkel (1983), but nitrogen balance was unaffected because nitrogen losses in urine fell.

ENERGY

The effect of rapeseed hulls was to depress energy and dietary fibre digestibility in trials by Bell and Shires (1982). Similar effects of supplementary lucerne leaf meal were observed by Kuan, Stanogias and Dunkin (1983); although nitrogen digestibility fell, daily nitrogen retention increased. Increases in the length and weight of various sections of the gut were also found as the level of dietary lucerne leaf meal increased, in the same studies. Kass *et al.* (1980) measured the effects of alfalfa on energy and dietary fibre digestibility (*Table 11.10*). The apparent

**Table 11.10** EFFECT OF LEVEL OF DIETARY ALFALFA MEAL ON OVERALL APPARENT DIGESTIBILITY (%) OF DRY MATTER, CELL WALL, ACID DETERGENT FIBRE, HEMICELLULOSE, CELLULOSE AND NITROGEN IN PIGS

| | *% alfalfa meal in diet* | | | |
|---|---|---|---|---|
| | *0* | *20* | *40* | *60* |
| Dry matter | 77 | 61 | 52 | 28 |
| Cell wall | 62 | 34 | 27 | 8 |
| Acid detergent fibre | 56 | 10 | 11 | 1 |
| Hemicellulose | 67 | 54 | 49 | 22 |
| Cellulose | 58 | 20 | 9 | 7 |
| Nitrogen | 70 | 52 | 41 | 41 |

(Data from Kass *et al.*, 1980)

digestibility of wood cellulose has often been found to be between 20 and 30 per cent, but wide variation has been noted by several authors. Some explanation for this comes from the work of Cunningham, Friend and Nicholson (1962) who found that the digestibility was 29.1 per cent when pigs were fed at a maintenance level. The same pigs had previously been fed at a growing pig level and only digested 5.0 per cent of the cellulose. In a later growing phase the digestibility was 18.3 per cent. This indicates that age and physiological state may be important determinants of the digestibility of dietary fibre. Straw cellulose is highly digestible (80–95 per cent) after lignin removal (Woodman and Evans, 1947; Forbes and Hamilton, 1952). The reduction in energy digestibility resulting from feeding diets with six levels of wood cellulose was found to be linear, up to a level of 175 g crude fibre/kg diet by Tullis and Whittemore (1981). The digestibility of the dietary fibre in various types of forages was found to be remarkably high by Yoshimoto and Matsubara (1983): some of their results are summarized in *Table 11.11*. Morgan, Whittemore and

**Table 11.11** APPARENT DIGESTIBILITY (%) OF FORAGES MEASURED OVERALL IN PIGS

| | *Organic matter* | *Nitrogen* | *Neutral detergent fibre* | *Cellulose* | *Hemicellulose* |
|---|---|---|---|---|---|
| Shimofusa turnip | 70 | 80 | 66 | 71 | 79 |
| Cabbage leaf | 74 | 69 | 86 | 85 | 94 |
| Alfalfa | 42 | 58 | 38 | 44 | 57 |
| Ladino clover | 59 | 67 | 57 | 61 | 70 |
| Tall fescue grass | 28 | 43 | 32 | 29 | 50 |
| Italian rye grass | 42 | 63 | 44 | 42 | 50 |

Each forage source formed 50% of the diet by weight, except turnip and cabbage (40%). Values are corrected for contributions by the basal diets.
(Data from Yoshimoto and Matsubara, 1983)

Cockburn (1984) examined the effects of straw, oatfeed, rice bran and beet pulp on the energy value of compound pig feeds and concluded that prediction of the digestible energy content of diets was best when based on their neutral detergent fibre content.

MINERALS

Although the retention of phosphorus and calcium was unaffected by oat hulls in the studies of Moser *et al.* (1982a) the apparent digestibility of phosphorus fell. In contrast, wood cellulose increased phosphorus retention and bone breaking strength (Moser *et al.*, 1982b); no explanation for this effect is immediately apparent. Supplementary wheat bran has been shown to decrease zinc absorption in pig diets by Newton, Hale and Plank (1983).

## Influence of dietary fibre on overall transit time

The results of recently published studies on the effect of dietary fibre on transit time are summarized in *Table 11.12*. It is evident that when graded levels of a single type of dietary fibre, or different types of fibre were added to semi-purified or milk-based diets (i.e. free of dietary fibre), variable and often large effects were seen. When bran was added to a cereal-based diet, no effect was seen (Canguilhem and Labie, 1977); this has also been found in recent studies in which supplements of bran, pectin and lactulose were given to growing pigs by Latymer (unpublished). This lack of effect of supplementary dietary fibre in reducing transit time of cereal-based diets (which have a high dietary fibre content) may imply that there is

**Table 11.12** EFFECTS OF DIETARY FIBRE ON OVERALL MEAN TRANSIT TIME IN PIGS

| *Source and level in diet* | (g/kg) | *Diet type* | *Initial pig weight* (kg) | *Time* (h) | *Reference* |
|---|---|---|---|---|---|
| Lucerne leaf meal | 50 | Semi-purified | 44 | 43.7 | 1 |
| Lucerne leaf meal | 100 | Semi-purified | 44 | 41.6 | |
| Lucerne leaf meal | 150 | Semi-purified | 44 | 29.7 | |
| Lucerne leaf meal | 200 | Semi-purified | 44 | 28.4 | |
| Coarse bran | 312 | Semi-purified | 70 | 51.6 | 2 |
| Fine bran | 472 | Semi-purified | 70 | 49.7 | |
| Lucerne meal | 308 | Semi-purified | 70 | 36.0 | |
| Solka floc | 150 | Semi-purified | 70 | 71.0 | |
| Bran | 170 | Milk powder | 50 | 66.0 | 3 |
| No dietary fibre | | Milk powder | 50 | 120.0 | |
| Bran (100 g/day) | | Milk | 90 | 64.3 | 4 |
| No dietary fibre | | Milk | 90 | 98.6 | |
| Bran (100 g or 200 g/day) | | Cereal | 30 | 52.0 | 5 |
| No extra dietary fibre | | Cereal | 30 | 49.0 | |
| Bran (100 g/day) | | Milk replacer | 30 | 79.0 | |
| No extra dietary fibre | | Milk replacer | 30 | 107.0 | |

Data from 1. Kuan, Stanogias and Dunkin (1983)
2. Ehle *et al.* (1982)
3. Fioramonti and Bueno (1980)
4. Bardon and Fioramonti (1983)
5. Canguilhem and Labie (1977)

**Table 11.13** EFFECT OF BRAN SUPPLEMENTATION OF A WEANER DIET FOR PIGLETS OF 9.5 kg INITIAL WEIGHT AT WEANING ON MEAN TRANSIT TIME (80% RECOVERY OF MARKER)

| *% crude fibre (as bran)* | *Mean transit time* (h) *Weeks post weaning* | | | |
|---|---|---|---|---|
| | *1* | *2* | *3* | *4* |
| 0 | 192 | 361 | 215 | 141 |
| 2.1 | 155 | 113 | 111 | 103 |
| 3.1 | 133 | 118 | 118 | 92 |
| 5.5 | 117 | 98 | 63 | 78 |

(Data from Schnabel, Bolduan and Guldenpenning, 1983)

a minimum transit time in growing pigs irrespective of the dietary fibre content of the diet.

Very little information exists on the effects of dietary fibre on transit time in weaner pigs. Schnabel, Bolduan and Guldenpenning (1983) demonstrated that bran supplements accelerated transit through the gut and also that transit time decreased over four weeks after weaning (*Table 11.13*). A crude fibre content of 50–60 g/kg diet was recommended for piglet starter diets by these authors.

## Effects of dietary fibre on the whole animal

### VOLUNTARY FEED INTAKE

It is well known that additional dietary fibre tends to increase voluntary feed intake of pigs. This topic was reviewed in detail by the Agricultural Research Council (1967), which concluded that every 1 per cent increase in the dietary fibre content of the diet is accompanied by an increase of approximately 3 per cent in feed intake. At the same time it was observed that additional dietary fibre reduced the growth rate despite the increased intake, which did not appear to compensate fully for the lower digestible energy content of the diet. The data available did not allow an estimate to be made of the plateau of intake imposed by dietary fibre. A feature of the data reviewed was the great variability, which was probably related to such factors as the age of the pig, the particular botanical type of dietary fibre used and the way in which it had been processed.

Since that time a number of studies have provided additional insight into the role that fibre can play in voluntary feed intake. For example, Owen and Ridgman (1967) found that intakes of high dietary-fibre diets (barley-based, with sawdust and oat feed) were consistently higher than those of low dietary-fibre content only when the pigs were in the finishing phase of growth. The time spent eating corresponded with the weights of feed eaten. In the early phases of growth (27–50 kg) reduced digestible energy intakes were found but these were compensated by higher intakes, when diets of high dietary fibre content were given in the 50–118 kg finishing phase. In a second report Owen and Ridgman (1968) pointed out that the adaptive response to diets of higher dietary fibre content takes a long time to occur, especially in young pigs. The quality of the carcasses was not significantly improved by feeding the high dietary fibre diet.

In experiments with growing-finishing pigs, Baker *et al.* (1968) fed diets with 0, 100, 200 or 400 g/kg added cellulose on an *ad libitum* basis; corresponding feed intakes of 2.63, 2.42, 2.00 and 1.50 kg/day and daily gains of 0.76, 0.68, 0.48 and 0.25 kg/day respectively were obtained. Contrasting daily voluntary feed intakes of lactating sows given oat husks (400 g/kg diet) or straw (300 g/kg diet) of 7.79 and 5.80 kg dry matter (85.0 and 6.04 MJ digestible energy) were found by Zoiopoulos, English and Topps (1982). Another interesting example of the complexity of voluntary feed intake mechanisms is provided by Taverner, Campbell and Biden (1984) who found intakes of growing pigs fell from 2.18 to 1.92 kg/day as the digestible energy content rose from 11.8 to 14.4 MJ/kg diet; all diets contained 120 g acid detergent fibre/kg and supplementary fat provided the increases in energy density. Although there was a 7 per cent increase in digestible energy intake (as the fat content of the diets was increased) the maximum daily digestible energy intake was 27 MJ; intakes of 34 MJ had been found by the same group when similar but low fibre diets had been given (Campbell, Taverner and Curic, 1983).

A recent study by Zoiopoulos, English and Topps (1983) showed that pigs growing between 55 and 87 kg ate different amounts of dry matter per day, when fed a semi-*ad libitum* basis (as much as could be eaten during 1 h in the morning and 1 h in the evening); malt culms and dark grains depressed intake to 1.81 and 1.94 kg/day respectively compared with a control intake of 2.30 kg/day and a high weatings intake of 2.05 kg/day.

The bacterial population of the large intestine hydrolyses undigested proteins to a wide range of products, including tyramine and tryptamine, amine derivatives of tyrosine and tryptophan respectively. These can saturate the hypothalamus and reduce feed intake. Inhibition of the formation of these compounds can be achieved by lowering the pH of the caecal and colonic contents below the high pH requirements of amine-producing bacteria. Such an effect may be caused by volatile fatty acids, produced by degradation of dietary fibre. This mechanism may explain how dietary fibre could influence voluntary feed intake (Bergner, 1981).

It is thus apparent that different types of dietary fibre influence voluntary feed intake in different ways and it is also evident that pigs do not eat to maintain a strictly controlled energy intake. It seems that further studies on the interactions between fibre source, other energy sources and voluntary feed intake are merited.

## EFFECTS OF DIETARY FIBRE ON GROWTH AND FEED:GAIN

The Agricultural Research Council (1967) concluded after a thorough review of the literature that increasing the percentage of crude fibre in the diet depressed the growth of pigs: for every 1 per cent additional crude fibre in the diet, a 2 per cent decrease in growth could be expected. However, the data were very variable. Crude fibre addition to the diet similarly worsened feed:gain ratios; for every 1 per cent increase in crude fibre, a 3 per cent increase in feed required per kg gain was shown. Again, the results were very variable. Although many new publications describing responses to dietary fibre have followed since publication by the Agricultural Research Council (1967) it is doubtful whether these conclusions can be modified or improved, largely because the types of fibre used in experiments have not been well characterized.

The effect of purified cellulose on growth and body composition of growing pigs fed on an *ad libitum* basis by Cunningham, Friend and Nicholson (1961) was to

**Table 11.14** EFFECTS OF CELLULOSE ADDITION TO AN ENERGY-DEFICIENT BASAL DIET FOR GROWING PIGS. ALL VALUES ARE RATIOS OF INTAKE (12 PIGS/TREATMENT)

| | *Basal regime* | *Basal + cellulose* | |
|---|---|---|---|
| 35 kg liveweight: | | | |
| Energy digestibility | 0.81 | 0.69 | *** |
| Energy retention | 0.79 | 0.67 | *** |
| N digestibility | 0.82 | 0.76 | *** |
| N retention | 0.48 | 0.46 | NS |
| ADF digestibility | 0.41 | 0.35 | NS |
| NDF digestibility | 0.53 | 0.46 | NS |
| 65 kg liveweight: | | | |
| Energy digestibility | 0.83 | 0.71 | *** |
| Energy retention | 0.80 | 0.69 | *** |
| N digestibility | 0.85 | 0.78 | *** |
| N retention | 0.51 | 0.48 | NS |
| ADF digestibility | 0.39 | 0.35 | NS |
| NDF digestibility | 0.56 | 0.47 | ** |

Significance of differences: NS, $P < 0.05$; **, $P < 0.01$; ***, $P < 0.001$.
(Data from Partridge, Keal and Mitchell, 1982)

decrease the dressing percentage (and increase the iodine number of the fat). No net gain was obtained from energy derived from cellulose, a conclusion which was also drawn by De Goey and Ewan (1975). Kupke and Henkel (1977) compared straw and wood cellulose and found the former led to fatter carcasses, while wood cellulose reduced nitrogen digestibility; usually nitrogen output in the urine fell so nitrogen balance was not affected. Partridge, Keal and Mitchell (1982) measured the energy value of cellulose (150 g/kg diet) added to barley–soya diets, fed at a restricted and energy-limiting level to growing pigs. The results are shown in *Table 11.14*. Growth rates and nitrogen retention were unaffected by cellulose addition, while dressing percentage fell. The amount of energy digested and absorbed (MJ/day) was the same for both diets (16.59 and 16.60 at 35 kg and 29.81 and 29.91 at 65 kg for control and cellulose-supplemented diets respectively). The results indicate that either the energy from cellulose was not absorbed (i.e. it was used in bacterial metabolism or lost as methane), or it was used in the gut wall, or the energy absorbed from dietary fibre was offset by reduced energy absorption from other sources. In a balance study, including respiration measurements, in sows, Muller and Kirchgessner (1983) measured the energy value of cellulose by subtraction, using data from control and cellulose-supplemented diets. Under these conditions, 29 per cent of the cellulose was digested; 95 and 68 per cent of the absorbed cellulose energy was metabolizable and retained respectively. Nitrogen balance was improved, urinary nitrogen output falling while faecal nitrogen output increased. Thus sows appeared to make better use of supplementary cellulose than growing pigs.

Recently it has been shown that, for every 1 per cent increase in the crude fibre content of the diet by barley straw (Just, 1982a), wheat or oat bran (Just, 1982b) gross energy digestibility fell by 2.1 or 3.5 units respectively and the efficiency of use of ME fell by 0.7 units in each case. These decreases corresponded to increases in the proportion of the dietary energy being digested in the large intestine. A general linear relationship between the percentage of dietary energy disappearing from the large intestine (X) and the net energy value (as a percentage of metabolizable energy in the diet) (Y) was expressed by Just, Fernandez and Jorgensen (1983) as:

$$Y = 74.5 + 0.49X$$

## EFFECTS OF DIETARY FIBRE ON BODY COMPOSITION

From the discussion in several sections of this review, it can be seen that increasing the dietary fibre content of diets results in trends towards greater faecal nitrogen loss largely through increased output of bacteria and endogenous matter (Mason, Kragelund and Eggum, 1982). Conversely the urinary output of nitrogen is usually reduced, with a greater amount of urea excretion into the large intestine, providing a substrate for the bacteria (Mosenthin and Henkel, 1983). The net effect is thus for nitrogen retention and therefore carcass lean or protein content to be relatively unaffected by increasing the dietary fibre content of the diet, although instances of both increases and decreases can be found, but the reasons for these effects remain unclear.

The effect of an increased percentage of dietary fibre in the diet on fat deposition is almost invariably to decrease it in growing pigs: a reduction of approximately 0.5 mm in backfat thickness for every 1 per cent increase in the crude fibre content was calculated by Elsley (1969) from the literature, with a corresponding increase in the days of growth before slaughter. More recent information supports these conclusions. Concomitant reductions in carcass weight are consistently found as a result of increasing the dietary fibre content of the diet in many reports.

## DIETARY FIBRE AND SOWS

Though breeding pigs have often been fed diets with a high content of dietary fibre, few detailed studies have been made on this topic. Problems in the interpretation of some published work include lack of definition of the types of fibre used and failure to allow for the often considerable consumption of straw bedding. Højgaard-Olsen and Nielsen (1966) observed that sows given supplementary straw gave birth to significantly heavier piglets, which consumed more milk and creep feed and thus had a greater weight at weaning. Münchow *et al.* (1982) found that partially hydrolysed straw gave rise to improved reproductive performance compared with unhydrolysed straw; in particular 1.5 more piglets were born per litter, for as yet unknown reasons. It was concluded that partially hydrolysed straw could replace about 45 per cent of the concentrates in sow diets without detrimental effects.

When sows were given either a control diet on a restricted basis, or the same diet with substitution by oat husks (400 g/kg) or barley straw (300 g/kg), both fed on an *ad libitum* basis, daily digestible energy intakes were 70.1, 85.0 and 60.4 MJ respectively (Zoiopoulos, English and Topps, 1982). Corresponding nitrogen balances (g/day) were 18.3, 25.5 and 3.5. The liveweight changes during lactation were −9.8, 5.1 and −16.8 kg. These results indicate that there is potential for *ad libitum* feeding of sows on diets with a high content of dietary fibre provided that the cost is sufficiently low: they also indicate that different sources of dietary fibre have very different effects. Further studies on the effects of barley straw on heat production and energy use in sows have been made by Müller and Kirchgessner (1983).

Several studies on gestating sows have suggested that the digestion of dietary fibre increases during gestation (for example Zivkovic and Bowland, 1970). Hemicellulose digestion improved while cellulose digestion remained constant in sows given alfalfa or tall wheatgrass during gestation by Pollman, Danielson and Peo (1979). However, Zivkovic and Bowland (1970) found reduced gestation weight gains in pigs fed diets with a high dietary fibre content.

There appears to be potential for the use of fibre in improving the satiety and behaviour of sows. Significant increases in the time spent eating (from 15.8 to 52.0 min) and lying down (monitored by video recording) were found by Mroz, Partridge, Broom and Mitchell when a barley–soya diet was supplemented with oat hulls (personal communication).

## INTERACTIONS BETWEEN DIETARY FIBRE AND ANTIBIOTICS

At present the effects of antibiotics on the nutritional value of diets with a high content of dietary fibre is only partly understood. Bohmann, Hunter and McCormick (1955) showed that the addition of alfalfa to young pig diets led to decreased daily gain, but this effect was reversed by aureomycin. However, Powley *et al.* (1981) found no consistent improvement in the use of high-alfalfa diets in pigs after antibiotic supplementation. On the other hand Sherry, Harrison and Fahey (1981) observed that supplementation of maize–soya diets with cellulose (80 g/kg) and antibiotic resulted in a significant fibre × antibiotic interaction, which depressed the resting metabolic rate of weanling pigs, i.e. heat production fell. This was accompanied by higher growth and improved feed:gain. It appeared that the microbial population was on the one hand enhanced by the additional fibre and on the other hand suppressed by the antibiotic, leading to changes in its size or composition, or predominant metabolic pattern. This work suggests that more knowledge of the interactions between the microbial population and dietary fibre could lead to the development of effective practical systems of using feedstuffs with an increased content of dietary fibre.

## DIETARY FIBRE AND WEANING

The role of dietary fibre in the weaning phase of piglets remains uncertain. Creep feeds are usually of relatively low dietary fibre content, but intakes are both variable and often low. At weaning diets of rather higher fibre content are offered, but appetites are often poor, and diarrhoea is a frequent problem. Drochner *et al.* (1978) suggested that a supplement of wood cellulose can depress bacterial activity and help reduce diarrhoea. Diets which combine such potentially beneficial effects with high digestibility and palatability are needed in order for maximal growth potential to be achieved.

## POSSIBLE BREED EFFECTS ON USE OF DIETARY FIBRE

The possibility that some breeds of pigs may use dietary fibre more efficiently than others deserves more investigation; Laurentowska (1959) noted higher cellulose and lignin digestibility in Pulawy than in Large White Pigs. Pekas, Yen and Pond (1983) found that lean and obese genotypes grew at the same rate and efficiency on low or high dietary fibre diets, but gut dimensions differed.

### PROCESSING OF FEEDSTUFFS

Processing of dietary fibre for pigs has received little attention. Pelleting of timothy–red clover diet (Cameron, 1960), oats (Seerley, 1962), lucerne and bran (Kracht *et al.*, 1975) led to improved feed intake and performance. Particle size may also be important: Nutzback, Pollmann and Behnke (1984) found that gravid pigs digested finely-ground (6.25 mm) alfalfa-containing diets better than normally ground diets (12.5 mm). Cellulose digestion was also improved by pelleting.

Chemical treatment of straw has been practised for many decades but Bergner (1981) has refined this procedure and has thoroughly investigated its nutritional and physiological effects in growing and also in breeding pigs, for which it seems suitable. Chemical treatment of other feedstuffs for pig nutrition is certainly merited.

## Conclusions

A wide variety of fibrous feedstuffs is currently available for pig nutrition (Van Es, 1981), and in particular 'bulky' feeds, popular in the past, such as potatoes, fodder beet, brassicas, young grass, grass silage dried beet pulp, and the entire maize plant have attractions (Thomke, 1981; Livingstone, 1983). Future research priorities in this field should include much more detailed knowledge of the chemical and physical composition of each of these sources of dietary fibre in relation to physiological studies on their mode of action, processing methods and their nutritive value at all stages of life in pigs. Until this information is available the use of feedstuffs with a high content of dietary fibre cannot be made on a sound scientific basis. The question of how to express practical responses of pigs to such feedstuffs is very difficult because all types of dietary fibre probably exert both indirect effects on other nutrients and direct effects as nutrient sources in their own right. Herein lies a complex and challenging problem for the animal nutritionist.

## References

AGRICULTURAL RESEARCH COUNCIL (1967). In *The Nutrient Requirements of Farm Livestock No 3: Pigs*, pp. 56–63. Agricultural Research Council; London

AGRICULTURAL RESEARCH COUNCIL (1981). In *The Nutrient Requirements of Pigs*, pp. 41–44. Commonwealth Agricultural Bureaux; Farnham Royal

ALBERS, N. and HENKEL, H. (1979). *Zeitschrift für Tierphysiologie, Tierernährung und Futtermittelkunde*, **42**, 113–121

ARGENZIO, R.A. (1982). *Les Colloques de l'INRA*, **12**, 207–215

ARGENZIO, R.A. and SOUTHWORTH, M. (1975). *American Journal of Physiology*, **228**, 454–460

ASP, N.G., JOHANSSON, C.G., HALLMER, H. and SILJESTROM, M. (1983). *Journal of Agricultural and Food Chemistry*, **31**, 476–482

BAKER, D.H., BECKER, D.E., JENSEN,A.H. and HARMON, B.G. (1968). *Journal of Animal Science*, **27**, 1332–1335

BARCROFT, J., MCANALLY, R.A. and PHILLIPSON, A.T. (1944). *Journal of Experimental Biology*, **20**, 120–129

BARDON, T. and FIORAMONTI, J. (1983). *British Journal of Nutrition*, **50**, 685–690

BEHM, G. (1954). *Archives für Tierernährung*, **4**, 197–218

BELL, J.M. and SHIRES, A. (1982). *Canadian Journal of Animal Science*, **62**, 557–565

BELL, J.M., SHIRES, A. and KEITH, M.O. (1983). *Canadian Journal of Animal Science*, **63**, 201–212

BERGNER, H. (1981). *Pig News and Information*, **2**, 135–140

BERGNER, H., SIMON, O. and BERGNER, U. (1980). In *Protein Metabolism and Nutrition*, pp. 198–204. Ed. H.J. Oslage and K. Rohr. European Association of Animal Production, Braunschweig

BOHMANN, V.R., HUNTER, J.E. and MCCORMICK, J. (1955). *Journal of Animal Science*, **14**, 499–506

CAMERON, C.D.T. (1960). *Canadian Journal of Animal Science*, **40**, 126–133

CAMPBELL, R.G., TAVERNER, M.R. and CURIC, D.M. (1983). *Animal Production*, **36**, 193–199

CANGUILHEM, R. and LABIE, C. (1977). *Revue de Medécine Vétérinaire*, **128**, 1669–1681

CLEMENS, E.T., STEVENS, C.E. and SOUTHWORTH, M. (1975). *Journal of Nutrition*, **105**, 759–768

COOPER, P.H. and TYLER, C. (1959). *Journal of Agricultural Science (Cambridge)*, **52**, 332–339

COEY, W.E. and ROBINSON, K.L. (1954). *Journal of Agricultural Science (Cambridge)*, **45**, 41–47

CRANWELL, P.D. (1968). *Nutrition Abstracts and Reviews*, **38**, 721–730

CUNNINGHAM, H.M., FRIEND, D.W. and NICHOLSON, J.W.G. (1961). *Canadian Journal of Animal Science*, **41**, 120–125

CUNNINGHAM, H.M., FRIEND, D.W. and NICHOLSON, J.W.G. (1962). *Canadian Journal of Animal Science*, **42**, 167–175

DE GOEY, L.W. and EWAN, R.C. (1975). *Journal of Animal Science*, **40**, 1045–1057

DIERICK, N., VERVAEKE, I., DECUYPERE, J. and HENDERICKX, H.K. (1983). *Revue de l'Agriculture (Brussels)*, **36**, 1691–1712

DROCHNER, W., HAZEM, A.S., MEYER, H. and RENSMANN, F.W. (1978). *Fortschritte der Veterinärmedizin*, **28**, 220–225

EHLE, F.R., JERACI, J.L., ROBERTSON, J.B. and VAN SOEST, P.J. (1982). *Journal of Animal Science*, **55**, 1071–1080

ELSLEY, F.W.H. (1969). In *Third Nutrition Conference for Feed Manufacturers*, pp. 126–152. J. and A. Churchill; London

ENGLYST, H.N., ANDERSON, V. and CUMMINGS, J.H. (1983). *Journal of the Science of Food and Agriculture*, **34**, 1434–1440

ENGLYST, H.N., WIGGINS, H.S. and CUMMINGS, J.H. (1982). *Analyst (London)*, **107**, 307–318

ES, A.J.H. VAN (1981). *Agriculture and Environment*, **6**, 195–204

FARRELL, D.J. (1973). *Animal Production*, **16**, 43–47

FARRELL, D.J. and JOHNSON, K.A. (1972). *Animal Production*, **14**, 209–217

FIORAMONTI, J. and BUENO, L. (1980). *British Journal of Nutrition*, **43**, 155–162

FORBES, R.M. and HAMILTON, T.S. (1952). *Journal of Animal Science*, **11**, 480–490

FRANK, G.R., AHERNE, F.X. and JENSEN, A.H. (1983). *Journal of Animal Science*, **57**, 645–654

FRIEND, D.W., NICHOLSON, J.W.G. and CUNNINGHAM, H.M. (1964). *Canadian Journal of Animal Science*, **44**, 303–308

GARGALLO, J. and ZIMMERMAN, D.R. (1981a). *Journal of Animal Science*, **53**, 395–402

GARGALLO, J. and ZIMMERMAN, D.R. (1981b). *Journal of Animal Science*, **53**, 1286–1291

HENRY, Y. (1970). *Annales de Zootechnie*, **19**, 117–141

HØJGAARD-OLSEN, N.J. and NIELSEN, H.E. (1966). In *Forsøgslaboratoriets Aarbog*, pp. 12–15. National Institute of Animal Science; Copenhagen

HORSZCZARUK, F. (1962a). *Roczniki Nauk Rolniczych*, **80B2**, 115–125

HORSZCZARUK, F. (1962b). *Roczniki Nauk Rolniczych*, **80B2**, 5–22

HORSZCZARUK, F. and SLJIVOVACKI, K. (1971). *Rocznini Nauk Rolniczych*, **93B**, 143–147

HUANG, W.Y. and KUMMEROW, F.A. (1976). *Lipids*, **11**, 34–41

IMOTO, S. and NAMIOKA, S. (1978a). *Journal of Animal Science*, **47**, 479–487

IMOTO, S. and NAMIOKA, S. (1978b). *Journal of Animal Science*, **47**, 467–478

IMOTO, S. and NAMIOKA, S. (1983a). *Journal of Animal Science*, **56**, 858–866

IMOTO, S. and NAMIOKA, S. (1983b). *Journal of Animal Science*, **56**, 867–875

JAMES, W.P.T. and THEANDER, O. (Ed.) (1981). *The Analysis of Dietary Fiber in Food*. Marcel Dekker; New York

JENTSCH, W., SCHIEMANN, R. and HOFFMANN, L. (1968). *Archives für Tierernährung*, **18**, 352–357

JUST, A. (1982a). *Livestock Production Science*, **9**, 717–729

JUST, A. (1982b). *Livestock Production Science*, **9**, 569–580

JUST, A., ANDERSEN, J.O. and JØRGENSEN, H. (1980). *Zeitschrift für Tierphysiologie, Tierernährung und Futtermittelkunde*, **44**, 82–90

JUST, A., FERNANDEZ, J.A. and JØRGENSEN, H. (1983). *Livestock Production Science*, **10**, 171–186

JUST, A., JØRGENSEN, H. and FERNANDEZ, J.A. (1983). *Livestock Production Science*, **10**, 487–506

JUST, A., SAUER, W.C., BECH-ANDERSEN, S., JØRGENSEN, H. and EGGUM, B.O. (1980). *Zeitschrift für Tierphysiologie, Tierernährung und Futtermittelkunde,* **43**, 83–91

KASS, M.L., VAN SOEST, P.J., POND, W.G., LEWIS, B. and MCDOWELL, R.E. (1980). *Journal of Animal Science*, **50**, 175–191

KENNELLY, J.J. and AHERNE, F.X. (1980). *Canadian Journal of Animal Science*, **60**, 717–726

KENNELLY, J.J., AHERNE, F.X. and SAUER, W.C. (1981). *Canadian Journal of Animal Science*, **61**, 349–362

KEYS, J.E. and DEBARTHE, J.V. (1974). *Journal of Animal Science*, **39**, 53–57

KIRCHGESSNER, M., ROTH-MAIER, D.A. and ROTH, F. (1975). *Zuchtungskunde*, **47**, 96–103

KRACHT, W., SCHRODER, H., RINNE, W. and FRANKE, M. (1975). In *Tierernährung und Futterung-Erfahrungen, Ergebnisse, Entwicklungen*–**9**, pp. 250–259. VEB Deutscher Landwirtschaftsverlag; Berlin

KUAN, K.K., STANOGIAS, G. and DUNKIN, A.C. (1983). *Animal Production*, **36**, 201–209

KUPKE, B. and HENKEL, H. (1977). *Zeitschrift für Tierphysiologie Tierernährung und Futtermittelkunde*, **38**, 330

LATYMER, E.A. and LOW, A.G. (1984). *Proceedings of the Nutrition Society*, **43**, 12A

LATYMER, E.A. and WOODLEY, S.C. (1984). *Proceedings of the Nutrition Society*, **43**, 22A
LAURENTOWSKA, C. (1959). *Roczniki Nauk Rolniczych*, **74B**, 567–578
LIVINGSTONE, R.M. (1983). *Pig Farming*, **31**, 61–62, 65
LLOYD, L.E., DALE, D.G. and CRAMPTON, E.W. (1958). *Journal of Animal Science*, **17**, 684–692
LOW, A.G. and KEAL, H.D. (1981). *12th International Congress of Nutrition, San Diego*, p. 56
LOW, A.G., PARTRIDGE, I.G. and SAMBROOK, I.E. (1978). *British Journal of Nutrition*, **39**, 515–526
LOW, A.G. and RAINBIRD, A.L. (1984). *British Journal of Nutrition*, **52**, 499–505
MASON, V.C., KRAGELUND, Z. and EGGUM, B.O. (1982). *Zeitschrift für Tierphysiologie, Tierernährung und Futtermittelkunde*, **48**, 241–252
MOHME, H., MOLNAR, S. and LENKHEIT, W. (1970). *Zeitschrift für Tierphysiologie, Tierernährung und Futtermittelkunde*, **21**, 138–146
MORGAN, C.A., WHITTEMORE, C.T. and COCKBURN, J.H.S. (1984). *Animal Feed Science and Technology*, **11**, 11–34
MOSENTHIN, R. and HENKEL, H. (1983). *Les Colloques de l'INRA*, **16**, 447–450
MOSER, R.L., PEO, E.R., MOSER, B.D. and LEWIS, A.J. (1982a). *Journal of Animal Science*, **54**, 800–805
MOSER, R.L., PEO, E.R., MOSER, B.D. and LEWIS, A.J. (1982b). *Journal of Animal Science*, **54**, 1181–1195
MÜLLER, H.L. and KIRCHGESSNER, M. (1983). *Zeitschrift für Tierphysiologie, Tierernährung und Futtermittelkunde*, **49**, 127–133
MUNCHOW, H., BERGNER, H., SEIFERT, H., SCHONMUTH, G. and BRABAND, E. (1982). *Archiv für Tierernährung*, **32**, 483–491
MURRAY, A.G., FULLER, M.F. and PIRIE, A.R. (1977). *Animal Production*, **24**, 139
NEWTON, G.L., HALE, O.M. and PLANK, C.O. (1983). *Canadian Journal of Animal Science*, **63**, 399–408
NUZBACK, L.J., POLLMANN, D.S. and BEHNKE, K.C. (1984). *Journal of Animal Science*, **58**, 378–385
OWEN, J.B. and RIDGMAN, W.J. (1967). *Animal Production*, **9**, 107–113
OWEN, J.B. and RIDGMAN, W.J. (1968). *Animal Production*, **10**, 85–91
PARTRIDGE, I.G. (1978). *British Journal of Nutrition*, **39**, 539–545
PARTRIDGE, I.G., KEAL, H.D. and MITCHELL, K.G. (1982). *Animal Production*, **35**, 209–214
PEKAS, J.C., YEN, J.T. and POND, W.G. (1983). *Nutrition Reports International*, **27**, 259–270
POLLMANN, D.S., DANIELSON, D.M. and PEO, E.R. JR. (1979). *Journal of Animal Science*, **48**, 1385–1393
POTKINS, Z.V., LAWRENCE, T.L.J. and THOMLINSON, J.R. (1984). *Animal Production*, **38**, 534
POWLEY, J.S., CHEEKE, P.R., ENGLAND, D.C., DAVIDSON, T.P. and KENNICK, W.H. (1981). *Journal of Animal Science*, **53**, 308–316
RAINBIRD, A.L. and LOW, A.G. (1983). *Proceedings of the Nutrition Society*, **42**, 88A
RAINBIRD, A.L., LOW, A.G. and ZEBROWSKA, T. (1983). *British Journal of Nutrition*, **52**, 489–498
RAINBIRD, A.L., LOW, A.G. and SAMBROOK, I.E. (1984). *Proceedings of the Nutrition Society*, **43**, 28A
REEDS, P.J., CADENHEAD, A., FULLER, M.F., LOBLEY, G.E. and MCDONALD, J.D. (1980). *British Journal of Nutrition*, **43**, 445–455

RERAT, A., VAISSADE, P. and VAUGELADE, P. (1979). *Annals de Biologie Animale Biochimie Biophysique*, **19**, 739–747

SAMBROOK, I.E. (1981). *Journal of the Science of Food and Agriculture*, **32**, 781–791

SAMBROOK, I.E., RAINBIRD, A.L. and LOW, A.G. (1982). In *Fibre in Human and Animal Nutrition* (Abstract). Royal Society of New Zealand; Wellington

SAUER, W.C., STOTHERS, S.C. and PARKER, R.J. (1977). *Canadian Journal of Animal Science*, **57**, 775–784

SCHNABEL, E., BOLDUAN, G. and GULDENPENNING, A. (1983). *Archives für Tierernährung*, **33**, 371–378

SEERLEY, R.W. (1962). *Dissertation Abstracts*, **22**, 4143

SHERRY, P.A., HARRISON, P.C. and FAHEY, G.C. JR. (1981). *Journal of Animal Science*, **53**, 1309–1315

SIMON, O., MÜNCHMEYER, R., BERGNER, H. and ZEBROWSKA, T. (1978). *British Journal of Nutrition*, **40**, 243–252

SOEST, P.J. VAN (1963a). *Journal of Association of Official Agricultural Chemists*, **46**, 825–828

SOEST, P.J. VAN (1963b). *Journal of Association of Official Agricultural Chemists*, **46**, 829–835

SOEST, P.J. VAN and MCQUEEN, R.W. (1973). *Proceedings of the Nutrition Society*, **32**, 123–130

SOUTHGATE, D.A.T. (1969). *Journal of the Science of Food and Agriculture*, **20**, 331–335

TAVERNER, M.R., CAMPBELL, R.G. and BIDEN, S. (1984). *Proceedings of the Australian Society of Animal Production*, **15**, 757

TAVERNER, M.R., HUME, I.D. and FARRELL, D.J. (1981). *British Journal of Nutrition*, **46**, 149–158

THOMKE, S. (1981). *Livestock Production Science*, **8**, 188–189

TROWELL, H., SOUTHGATE, D.A.T., WOLEVER, T.M.S., LEEDS, A.R., GASSULL, M.A. and JENKINS, D.J.A. (1976). *Lancet*, **1**, 967

TULLIS, J.B. and WHITTEMORE, C.T. (1981). *Animal Production*, **32**, 395

WHITING, F. and BEZEAU, L.M. (1957). *Canadian Journal of Animal Science*, **37**, 106–113

WOODMAN, H.E. and EVANS, R.E. (1947). *Journal of Agricultural Science (Cambridge)*, **37**, 202–210

YOSHIMOTO, T. and MATSUBARA, N. (1983). *Japanese Journal of Zootechnical Science*, **54**, 748–754

ZEBROWSKA, T. (1973). *Roczniki Nauk Rolniczych*, **95B1**, 115–123

ZEBROWSKA, T. (1982). *Les Colloques de l'INRA*, **12**, 225–236

ZEBROWSKA, T., LOW, A.G. and ZEBROWSKA, H. (1983). *British Journal of Nutrition*, **49**, 401–410

ZEBROWSKA, T., SIMON, O., MUNCHMEYER, R., WOLF, E., BERGNER, H. and ZEBROWSKA, H. (1982). *Archives für Tierernährung*, **32**, 431–444

ZIVKOVIC, S. and BOWLAND, J.P. (1970). *Canadian Journal of Animal Science*, **50**, 177–184

ZOIOPOULOS, P.E., ENGLISH, P.R. and TOPPS, J.H. (1982). *Animal Production*, **35**, 25–33

ZOIOPOULOS, P.E., ENGLISH, P.R. and TOPPS, J.H. (1983a). *Zeitschrift für Tierphysiologie, Tierernährung und Futtermittelkunde*, **49**, 210–218

ZOIOPOULOS, P.E., TOPPS, J.H. and ENGLISH, P.R. (1983b). *Zeitschrift für Tierphysiologie, Tierernährung und Futtermittelkunde*, **49**, 219–228

12

# PHOSPHORUS AVAILABILITY AND REQUIREMENTS IN PIGS

A.W. JONGBLOED, H. EVERTS and P.A. KEMME
*Research Institute for Livestock Feeding and Nutrition, Lelystad, The Netherlands*

## Introduction

Phosphorus (P) is an essential element in the animal's body. In addition to its vital participation in the development and maintenance of skeletal tissue, it plays an important role in many other metabolic functions (NRC, 1980). Usually, the amounts of P available to the animal from feeds of plant origin are insufficient to obtain high levels of performance (Jongbloed, 1987). Therefore, additional inorganic phosphorus is supplied to diets, particularly for pigs and poultry.

It has, however, been recognized that excretion of P can lead to environmental problems (Gerritse and Zugec, 1977), particularly in areas with large numbers of pigs and poultry per unit of land. Application of manure in large quantities in certain areas leads to accumulation of P in the soil, together with leaching and run-off. The effect is eutrophication of ground water and fresh water sources. To minimize environmental pollution, the excretion of P by animals should be reduced as much as possible.

As a first step in this process the supply of P in the feed should be in accordance with the animal's requirement. Adequate knowledge is therefore required of the digestibility of P in the feedstuffs used and of the animal's requirement for it at any stage and type of production. Secondly, P excretion can be reduced by enhancement of the digestibility of P in feeds by means of plant or microbial phytase (Simons *et al.*, 1990). Another possibility for decreasing excretion of P is to choose those feedstuffs in the mixed feed which have a low total P concentration together with a high P digestibility. Throughout this chapter the term 'digestible P' will be used for apparently absorbable P; only where the literature specifies it will the term 'available P' be used.

In this chapter information is provided on the P digestibility of various feedstuffs from plant and animal origin and from feed phosphates. Until recently it has usually been assumed that P availability in feedstuffs of plant origin is about 30–35%, but current evidence indicates a wide range in P digestibility between feedstuffs. The availability of P in feedstuffs of animal origin (DLG, 1987) and in feed phosphates (NRC, 1988) is often assumed to be around 95–100%. However, even these feedstuffs have recently been shown to have a wider range of P digestibility.

Recent experiments conducted at Lelystad have investigated the use of microbial phytase. The results are very promising and are reported here.

In the literature (ARC, 1981; NRC, 1988) there is a consensus about the requirement of available P for maintenance and production, but a problem arises when the requirement is expressed as a dietary concentration. To be able to balance P concentrations in the diet according to requirements, the availability or digestibility of P within the feed used must be known. Until recently this has not been possible, but this problem can now be overcome, by using the same term to express phosphorus requirements of pigs and its concentration in the diet. More detailed information on the requirements of P and Ca is given by Jongbloed and Everts (1992).

## Digestible phosphorus in feedstuffs

Until recently knowledge of the digestibility and availability of P from materials of plant origin has been scant. Most experiments have focused on maize, barley, wheat, wheat bran and soya bean meal. Only one or two measurements have been reported for about 10 other feedstuffs. The results of these experiments have been extensively discussed by Jongbloed (1987) and Jongbloed and Kemme (1990b).

Two main methods have been used for assessing the availability of P in feedstuffs. The first one is the slope ratio technique where animals are given graded amounts of P from feeds of unknown availability. The slope of the relationship between P intake and bone ash content or bone breaking strength is compared with that of a standard form of P (mostly monosodium phosphate). The availability of monosodium phosphate is assumed to be 100 (Cromwell, 1983). The second approach is a balance technique in which the apparent digestibility can be measured by the difference between P intake and its faecal excretion.

The results obtained by the two techniques should be considered in combination because they can provide complementary information. From recent studies (Dellaert *et al.*, 1990) it is suggested that the availability figures obtained by the slope ratio technique (with monosodium phosphate as a reference) should be multiplied by 0.9 to obtain apparent digestibility coefficients. This correction factor was applied to the availability figures cited in the literature and the recalculated results are presented in Table 12.1.

Average P digestibility in maize was calculated by leaving out the value of 48% obtained by Fourdin, Fontaine and Pointillart (1986). Ensiled moist maize has a much higher P digestibility than dry maize. This can be explained by the hydrolysis of phytate by microbial phytase during the ensiling process. Wheat and wheat by-

**Table 12.1** RECALCULATED APPARENT DIGESTIBILITY OF P IN SOME FEEDSTUFFS

| *Feedstuff* | *n* | *Digestibility (%)* (mean and SD) | *Range* |
|---|---|---|---|
| Barley | 6 | 32±14 | 16–51 |
| Maize | 9 | 17± 9 | 8–29 |
| Maize (moist, ensiled) | 4 | 42± 7 | 36–52 |
| Wheat | 5 | 44± 4 | 36–46 |
| Wheat bran/middlings | 6 | 41±11 | 32–50 |
| Soya bean meal | 6 | 24± 8 | 15–34 |

(Jongbloed and Kemme, 1990b)

products have a considerably higher P digestibility than maize, barley and soya bean meal. This can be explained by the presence of wheat phytase (Pointillart, Fontaine and Thomasset, 1984). Feedstuffs of animal origin are usually rich in P, which is almost all in the form of inorganic phosphates. Therefore, a high availability and a substantial contribution to the supply of P for pigs can be expected in feeds containing such ingredients. Phosphorus availability for these products ranged from 76 to 102% (Jongbloed and Kemme, 1990b).

Knowledge of the availability of P in feed phosphates is necessary to compare the nutritive value of P in different sources. Neither the recommendations of NRC (1988) nor in those of DLG (1987) indicate any discrimination in availability of P between different sources; P availabilities of 100% and 95%, respectively, were suggested. However, the results of recent experiments suggest significant differences in P digestibility for feed phosphates of different types and brands.

### FEEDSTUFFS OF PLANT ORIGIN

Details of experimental procedures employed to estimate P digestibility within various feeds have been described fully by Jongbloed (1987) and Jongbloed and Kemme (1990a,b). Therefore, only relevant aspects of these trials will be discussed here. Most trials were performed according to a Latin square design. Altogether more than 100 batches of feedstuffs have been tested for apparent digestibility of P. The mean apparent digestibility coefficients presented for feedstuffs of plant origin were derived from at least three batches. In each trial, four animals were used in the weight range of 45–110 kg. In all trials a basal feed was used except for barley, maize and wheat, which were tested as a single feed. For the first 29 trials, barley served as the basal feed, but due to the presence of some phytase in barley which might affect the P digestibility, it was later substituted by maize. The procedure adopted was designed to ensure that 50% or more of total P came from the test material provided that the estimated concentration of digestible P in the diet did not exceed 1.6 g/kg. This is regarded as the minimum P requirement for growing pigs (Jongbloed, 1987). To balance the feeds, all vitamins and essential trace elements were added. The Ca/total P ratio in the diet was maintained at 1.3:1 by the addition of limestone, as long as the concentration of Ca was above 5.0 g/kg. Since 1988, all the experimental diets have been formulated to contain just 6.0 g Ca/kg. In all trials using the balance technique, the amount of energy supplied in the diet was 2.3 times maintenance (maintenance = 292 kJ $NE_f$ or 418 kJ ME/$kg^{3/4}$). Just before feeding, about 2.5 litres water was added per kg of diet, and the diet was offered twice daily in similar amounts. The procedures for sampling, weighing and analytical techniques have been described by Jongbloed and Kemme (1990a). Digestibility of P in the tested feedstuffs was calculated by difference assuming that the digestibility of P in the basal diet was constant and equal to the results of the trial with basal feed only.

The results of the chemical analyses of the feedstuffs of plant origin and the P digestibility are presented in Table 12.2. The mean standard deviation of P digestibility in the total ration (basal diet + tested feedstuff) over all trials was 3.8%, while the mean standard deviation of P digestibility in feedstuffs added to a basal diet was 6.4%. This was higher than that of the total ration because usually only 50–80% of total digestible P originated from the feedstuff tested. Concentration of P differed substantially between the feedstuffs, ranging from 1.4 g/kg DM (tapioca meal), to 17.0 g P/kg DM (rice bran). The concentration of phytate P as a

**Table 12.2** Ca, P AND PHYTATE P CONCENTRATION OF FEEDSTUFFS TESTED AND APPARENT P DIGESTIBILITY COEFFICIENTS (MEAN AND SD)

| *Feedstuff* | *Number of trials* | *Content in feedstuff* *Ca* | *P* (g/kg DM) | *Phytate P* | *P digestibility(%)* *Mean* | *SD* | *Range* |
|---|---|---|---|---|---|---|---|
| Barley | 5 | 1.3±0.4 | 4.4±0.3 | 2.8±0.2 | 39 | 4 | 34–44 |
| Maize | 7 | 0.6±0.3 | 3.2±0.4 | 2.1±0.3 | 17 | 5 | 12–26 |
| Wheat | 5 | 0.6±0.2 | 4.1±0.4 | 2.9±0.2 | 47 | 2 | 45–51 |
| Beans (*Phaseolus* spp.) | 3 | 2.1±0.5 | 5.2±0.4 | 1.7±0.4 | 38 | 10 | 29–48 |
| Peas | 4 | 1.9±1.4 | 4.8±0.9 | 2.4±0.7 | 45 | 4 | 42–51 |
| Hominy feed | 7 | 2.9±3.3 | 7.3±1.5 | 4.9±2.0 | 19 | 8 | 10–34 |
| Rice bran (<3% husks) | 4 | 2.7±2.7 | 17.0±2.7 | 13.8±3.3 | 12 | 2 | 9–13 |
| Wheat middlings | 6 | 1.3±0.2 | 12.0±0.8 | 9.6±0.9 | 28 | 6 | 18–35 |
| Maize gluten feed | 10 | 0.8±0.5 | 9.8±1.6 | 6.3±1.6 | 20 | 6 | 12–32 |
| Tapioca meal | 3 | 2.8±0.4 | 1.4±0.3 | 0.4±0.2 | 10 | 12 | 1–24 |
| Coconut expeller | 5 | 1.3±0.5 | 5.8±0.2 | 2.6±0.5 | 34 | 8 | 25–43 |
| Maize meal solvent extract | 4 | 0.4±0.3 | 7.4±0.8 | 5.4±0.3 | 20 | 9 | 11–31 |
| Soya bean meal solvent extract (XF>7%) | 3 | 3.7±0.7 | 6.6±0.5 | 4.0±0.4 | 37 | 1 | 36–38 |
| Soya bean meal solvent extract (XF<3.5%; dehulled) | 3 | 3.7±0.6 | 7.3±0.6 | 4.2±0.9 | 38 | 4 | 33–41 |
| Sunflower seed meal solvent extract | 4 | 3.9±0.3 | 11.6±2.1 | 8.9±2.1 | 16 | 1 | 14–17 |

proportion of total P ranged from 30% (in tapioca and beans) to 80% (in rice bran and wheat middlings). About two-thirds of total P in barley, wheat and maize was present as phytate P. In some other feedstuffs such as hominy feed and maize gluten feed, phytate P concentration varied markedly, presumably due to differences in processing.

Relatively large differences in P digestibility were observed between feedstuffs. The lowest P digestibility was noted for tapioca meal and rice bran (10 and 12%, respectively). The extremely low digestibility coefficient for tapioca meal (10%) was presumably due to the small amount of P in this feedstuff, but might also be affected by a relatively high proportion of endogenous P in total intake of P. The highest values were obtained for wheat and peas (47 and 45%, respectively). The apparent digestibility of P in wheat has also been reported to be relatively high (44%) in the literature. This may be due to a high phytase activity, as suggested by Fourdin, Fontaine and Pointillart (1986). In the present trials it was found that wheat contained about 1000 phytase units per kg (1 unit = release of 1 μmol inorganic P/min), which is rather high. Peas have a rather high P digestibility which can partly be explained by a relatively low concentration of phytate P. The apparent digestibility of P in wheat bran/middlings was much lower (28%) than suggested by the literature (41%). Phytase activity measured in some batches of wheat middlings ranged from 3700 to 5100 units. Also, there was a relatively large degree of variation in P digestibility in wheat bran. Recent trials have suggested that, apart from differences in phytase activity, a high calcium concentration in the diet significantly diminished P absorption (Jongbloed and Kemme, 1990c).

For maize and maize by-products, such as hominy feed, maize gluten feed and solvent extracted maize meal, the P digestibility was low varying from 17% to 20%. Although maize by-products contained two to three times more phosphorus than maize itself, the apparent digestibility was very similar (17 and 20%, respectively).

Digestibility of phosphorus from barley used in the present trials (39%) was not significantly different from that cited in the literature (32%). Part of the differences may be due to differences in phytase activity, as it has been shown that summer varieties have a higher phytase activity than winter varieties, 630 and 350 units/kg, respectively (A.W. Jongbloed, unpublished data). Furthermore, it was noted that there was a tendency for a higher P digestibility when the proportion of phytate P was lower. P digestibility in solvent extracted soya bean meal was 38%. This value is higher than that in the literature, and it is assumed that the value in the literature underestimates P digestibility. According to NRC (1988), the availability of P in soya bean meal and dehulled soya bean meal is 38% and 25%, respectively, whereas the present studies indicate that both feedstuffs have a similar P digestibility (37% and 38%).

Based on the phytate P content in the plant feedstuffs, and assuming that non-phytate P in the feedstuffs is 80% absorbed, it was calculated that no P bound to phytate complexes was absorbed from most of the feedstuffs. However, from wheat, wheat middlings and barley, which contain phytase, some P from phytate complexes was absorbed. Similar conclusions were drawn by Pointillart, Fontaine and Thomasset (1984) and Fourdin *et al.* (1986, 1988).

### FEEDSTUFFS OF ANIMAL ORIGIN

In eight trials with animal products maize was used as a basal diet, and this was supplemented with different animal products at a constant total Ca content of 6.0 g/kg feed. In addition, two trials were carried out using the slope ratio technique to evaluate batches of meat and bone meal, and bone precipitate (Dellaert *et al.*, 1990).

Concentrations of Ca and P in all the tested products of animal origin, and the apparent digestibility coefficients of phosphorus are presented in Table 12.3. The observed Ca:P ratio was nearly 2:1 for meat meal, bone meal and meat+bone meal. For fish meal and skimmed milk powder it ranged from 1.2:1 to 1.3:1 and in whey powder (poor in lactose) the Ca:P ratio was 0.7:1. The apparent P digestibility ranged from 68% to 91%. The values are quite close to those obtained

**Table 12.3** Ca and P CONCENTRATION IN FEEDSTUFFS OF ANIMAL ORIGIN AND APPARENT P DIGESTIBILITY COEFFICIENTS

| *Feedstuff* | *Technique* | *Number trials* | *Content in feedstuff* Ca (g/kg DM) | P (g/kg DM) | *P digestibility (%)* |
|---|---|---|---|---|---|
| Meat meal | B | 1 | 48.6 | 22.5 | 74 |
| Meat meal | B | 1 | 66.3 | 33.1 | 85 |
| Bone meal | B | 1 | 174.2 | 85.7 | 68 |
| Fish meal | B | 2 | 33.1 | 25.2 | 86 |
| Feather meal (hydrolysed) | B | 1 | 3.6 | 1.6 | 75 |
| Skimmed milk powder | B | 1 | 12.7 | 10.6 | 91 |
| Whey powder (poor in lactose) | B | 1 | 10.6 | 14.8 | 82 |
| Bone precipitate | S | 1 | 229.0 | 176.3 | 87 |
| Meat and bone meal | S | 1 | 182.2 | 87.6 | 80 |

Jongbloed and Kemme (1990b)
(B = balance technique; S = slope ratio technique)

for feed grade phosphates (Dellaert *et al.*, 1990). Due to the high concentration of P in bone meal it seems that the balance technique is less suitable for experiments on this feedstuff with growing-finishing pigs. To obtain a level of less than 1.6 g digestible P in the feed, only a small amount of bone meal can be added to the basal diet and therefore the analytical error is relatively high. In the literature the P digestibilities in fish meal, meat+bone meal and steamed bone meal (availability coefficients × 0.9) are slightly higher than those presented in Table 12.3, except for dried whey. Variability of P content between different batches of the same feedstuff might be due to differences in the manufacturing process, subsequently affecting the digestibility of P. It seems likely that P digestibility in bone precipitate is higher than in steamed bone meal, which may arise from differences in physicochemical structure of the products.

### FEED PHOSPHATES

Various experiments were also performed to determine the nutritional value of feed phosphates of different types and brands. Piglets weaned at about 4½ weeks of age were used for this work between 5 and 10 weeks of age. In all experiments the slope ratio technique was used, sometimes with small adaptations. Detailed descriptions of the experiments have been given by Jongbloed (1987) and Dellaert *et al.* (1990). In order to calculate digestibility, chromic oxide was used as an indigestible marker. In the third and fifth week, but sometimes also in the fourth week after weaning, faeces were collected for 2 days. As well as digestibility, other parameters such as blood P levels and several bone parameters were taken into account.

Results of the apparent digestibility of P in the feed phosphates are given in Table 12.4. Significant differences were found between the various types of feed phosphates. Disodium phosphate, used as a reference, had the highest P digestibility. Monocalcium phosphates generally had a higher P digestibility than dicalcium phosphates. There were even significant differences between the various sources of monocalcium phosphate. The values across experiments for a particular type of feed phosphate showed a high repeatability. These values show a very good degree of agreement with those of Grimbergen, Cornelissen and Stappers (1985).

**Table 12.4** THE APPARENT DIGESTIBILITY OF PHOSPHORUS (%) OF FEED PHOSPHATES

| *Experiment* | *I* | | *II* | | *III* | | | *IV* | | *Mean* |
|---|---|---|---|---|---|---|---|---|---|---|
| *Week of determination* | *3* | *5* | *3* | *5* | *3* | *4* | *5* | *3* | *4* | |
| | *P digestibility (%)* | | | | | | | | | |
| *P source* | | | | | | | | | | |
| Dicalcium phosphate (anhydrous) | 65[b] | 64[b] | 69[c] | 63[c] | – | – | – | – | – | 65±3 |
| Dicalcium phosphate (dihydrate) | – | – | 71[bc] | 67[bc] | – | – | – | – | – | 69 |
| Monocalcium phosphate (A) | 76[a] | 79[a] | 77[b] | 74[ab] | 72[a] | 74[a] | 71[a] | 77 | 75 | 75±3 |
| Monocalcium phosphate (B) | 78[a] | 83[ac] | 85[a] | 80[a] | 84[b] | 84[b] | 84[b] | – | – | 83±3 |
| Monocalcium phosphate (C) | 83[a] | 84[ac] | – | – | – | – | – | – | – | 84 |
| Disodium phosphate | 96[c] | 90[c] | 90[c] | 85[a] | – | – | – | – | – | 90±4 |
| Calcium sodium phosphate (1) | – | – | – | – | 85[b] | 86[b] | 81[b] | – | – | 84 |
| Calcium sodium phosphate (2) | – | – | – | – | 86[b] | 87[b] | 86[c] | – | – | 86 |

Means in the same column with superscript letters are significantly different ($P<0.05$)

In their experiments Dellaert *et al.* (1990) showed that for feed phosphates, digestibility of P had a higher correlation with P retention than blood or bone related parameters. This means that P digestibility is a reliable estimate for the nutritive value of P in feedstuffs.

### CONCLUSIONS ON DIGESTIBILITY OF P IN FEEDSTUFFS

The results presented indicate that the digestibility of P in various feedstuffs of plant origin varies substantially. Factors which cause variation in P digestibility between feedstuffs are the origin of the feedstuff, the concentrations of phytate P and of total P and the presence of phytase. The digestibility of P in maize, maize by-products, rice bran and sunflower seed meal is 20% or lower, and thus substantially lower than the generally accepted value of between 30% and 35%. However, in the case of wheat (phytase-rich) and peas these values are largely underestimated.

The apparent digestibility of P in feedstuffs of animal origin is high and ranges from 70 to 90%. These feedstuffs can supply large amounts of digestible P in pig diets. More observations are needed for each animal product to justify whether the P digestibility is below or above 80%. From the results presented it can be concluded that around 80% of P from meat meal, bone meal, and meat and bone meal is digested.

**Table 12.5** COMPARISON OF NUTRITIVE VALUE OF P IN SOME COMMON FEEDSTUFFS

| | *P content* (g/kg DM) | | *CVB (1990)* | *NRC (1988)* | *DLG (1987)* |
|---|---|---|---|---|---|
| | *Total P* | *Phytate P* | *Apparent digestibility* (%) | *Availability* (%) | *Intestinal availability* (%) |
| Barley | 3.3 | 2.1 | 37 | 31 | 55 |
| Maize | 2.5 | 1.6 | 16 | 15 | 50 |
| Wheat | 3.3 | 2.2 | 46 | 50 | 55 |
| Rice bran | 13.3 | 10.0 | 13 | 25 | 50 |
| Soya bean meal | 6.5 | 3.9 | 38 | 38 | 55 |
| Dicalcium phosphate $0H_2O$ | 200 | – | 65 | 100 | 95 |
| Dicalcium phosphate $2H_2O$ | 182 | – | 69 | 100 | 95 |
| Monocalcium phosphate | 225 | – | 80 | 100 | 95 |

There are marked differences in digestibility of P between types and origin of feed phosphates tested, ranging from 65% to 90%. The repeatability of digestibility of P within a type of feed phosphate is high.

Table 12.5 provides a comparison of the apparent digestibility of P (CVB, 1990), the availability of P (NRC, 1988) and the intestinal availability of P (DLG, 1987). It can be seen that, in general, for feedstuffs of plant origin there is good agreement between the values of CVB (1990) and NRC (1988), but the estimations of availability of P according to DLG (1987) are much higher. However, NRC (1988) and DLG (1987) give much higher availability figures for feed phosphates than CVB (1990).

### ENHANCING DIGESTIBILITY OF P BY PLANT AND MICROBIAL PHYTASE

Several authors have described a positive effect of phytase on the absorption of phytate phosphorus (Nelson, 1967; Pointillart, Fontaine and Thomasset, 1984;

Williams and Taylor, 1985; Jongbloed, 1987). As well as being present in feedstuffs such as wheat, barley and rye, phytases are produced by the microbial flora in the intestine and are probably also secreted into the lumen of the intestinal tract (alkaline phosphatase) from the brush border of the small intestine (Davies and Flett, 1978; Cooper and Gowing, 1983). However, Pointillart, Fontaine and Thomasset (1984) and Williams and Taylor (1985) concluded that, for pigs, intestinal phytase does not seem to be of great significance for the hydrolysis of phytate. This section discusses first the effect of plant phytase and then that of microbial phytase on the digestibility of P.

*Effect of plant phytase*

It has already been shown that the availability of P from wheat or wheat by-products is quite high (see Tables 12.1 and 12.2). This is attributed to their high phytase activity (Pointillart, Fontaine and Thomasset, 1984; Cromwell, Stahly and Moneque, 1985; Scheuerman, Lantsch and Menke, 1988). In a few experiments at Lelystad attempts have been made to quantify the effect on P digestibility of intrinsic phytase in individual feedstuffs of plant origin and in diets of pigs. Parts of these experiments have been described by Jongbloed and Kemme (1990c), who compared the influence of phytase present in wheat and wheat middlings on P digestibility. Part of the batch of the product was inactivated by heating at temperatures above 80°C. The normal and phytase inactivated products were tested for their P digestibility in separate experiments or added to a basal diet to see whether there was an additional effect of plant phytase on P digestibility of the basal diet. The results of these studies are presented in Table 12.6. The presence of phytase enhanced the P digestibility in wheat from 27% to 50%. Also the P digestibility in wheat bran was almost doubled due to its phytase. In Experiment I, the presence of wheat phytase in the diet enhanced the P digestibility of the maize/soya bean meal diet from 31% to 49%. However, in the feed with wheat middlings a smaller increase was noted. These results show that the digestibility of P can be increased substantially by the presence of plant phytase. However, when feeds are steam pelleted at temperatures around 80°C the phytase is inactivated and no beneficial effect on digestibility of P was obtained (Jongbloed and Kemme, 1990a).

**Table 12.6** EFFECT OF WHEAT PHYTASE ON THE DIGESTIBILITY OF P

| *Experiment* | *Feed* | *Digestibility of P (%)* | |
|---|---|---|---|
| | | *With phytase* | *Without phytase* |
| 1+2 | Wheat | 50 | 27 |
| 1 | Maize + soya bean meal | – | 29 |
| 1 | Maize + soya bean meal + 40% wheat | 49 | 31 |
| 2 | Wheat middlings | 33 | 19 |
| 2 | Maize + soya bean meal | – | 28 |
| 2 | Maize + soya bean meal & wheat Middlings (20%) | 40 | 32 |

(Adapted from Jongbloed and Kemme, 1990c)

*Effect of microbial phytase*

Nelson *et al.* (1968) were the first to add phytase, produced by a culture of *Aspergillus ficuum*, to liquid soya bean meal. The feed was incubated for 2–24 h at 50°C and after drying it was fed to 1-day-old chicks. The birds showed a considerable increase in bone ash percentage. Until recently, comparable results with diets for pigs containing added phytase were lacking (Cromwell and Stahly, 1978; Chapple, Yen and Veum, 1979; Shurson, Ku and Miller, 1984).

The first promising results were recently reported by Simons *et al.* (1990), who also obtained the phytase from a strain of *Aspergillus ficuum*. In an experiment with pigs in the liveweight range 35–70 kg, two diets were used. The first diet was based on maize and solvent extracted soya bean meal; the second diet was more like a practical diet, as used in the Netherlands. To one part of both diets 1000 units phytase/kg diet were added. It was shown that this enzyme increased the digestibility of P from 20% to 46% for the maize/soya bean meal diet and from 34% to 56% for the practical diet. Studies with cannulated pigs indicated that this hydrolysis of phytic acid by microbial phytase took place mainly in the stomach (Jongbloed, Kemme and Mroz, 1990). Further studies have shown that, in piglets of the liveweight range 10–30 kg, microbial phytase not only enhanced the digestibility of P more than 20%, but also improved growth rate and feed conversion ratio (Beers and Koorn, 1990). This was mainly due to a higher feed intake.

In the most recent experiments performed at Lelystad a phytase preparation from a genetically modified organism has been used. Again, this product substantially enhanced the digestibility of P. The prospect of using microbial phytase in pig feeds appears very promising. It may mean that in effect feeds for growing-finishing pigs and for pregnant sows may need little or no supplementary feed phosphate.

## The requirement for phosphorus

To reduce the excretion of phosphorus data are needed on the requirement for P by the animals at various physiological stages as well as knowledge of the digestibility of P in feedstuffs. Current recommendations for P requirements of pigs vary widely in different countries. The reason for this is likely to be due to differences in environment, genotype, level of feeding, major ingredients used in the diets and criteria of adequacy.

The methods used to estimate P requirement of pigs have been fully discussed (ARC, 1981; Guéguen and Perez, 1981; NRC, 1988). Two approaches for estimating P requirements have been described, the empirical method and the factorial method. The advantages and disadvantages of both methods have been extensively discussed (Jongbloed, 1987), and scientifically the factorial method is considered to be advantageous for several reasons, not least that it allows a more accurate approach because it can be applied to various systems of production.

The factorial method relies on the amount of endogenous P excreted in faeces and urine (maintenance) as well as knowledge of the P requirement for production of meat, maternal and fetal tissue or milk. The maintenance requirement of P is expressed as g P/kg live weight/d. The requirement of P for production is expressed as g P/kg growth for growing pigs, as g P/d for pregnant and lactating sows.

MAINTENANCE REQUIREMENT FOR P

The maintenance requirement for P is determined by the loss of endogenous P in faeces and the loss of P in urine. Jongbloed (1987) undertook an extensive survey of the estimates of endogenous excretion of P in faeces, from which it was concluded that the amount of faecal endogenous loss of P was low and to some extent depended on the level of supply (Table 12.7). It can be seen that the amount of endogenous P loss is lower at deficient dietary levels of P. It can be assumed that, for growing pigs at a dietary supply of P above 5.8 g/kg DM, 9 mg of endogenous P/kg live weight/d is excreted in faeces. For breeding sows at a live weight above 140 kg the mean daily endogenous loss of P in faeces was 4.5 mg/kg live weight (Table 12.7). However, only three observations were obtained for sows on P free diets and only two observations were available for lactating sows. Therefore, data for breeding sows are not well established.

**Table 12.7** DAILY LOSS OF ENDOGENOUS P IN FAECES OF PIGS

| *Liveweight range* (kg) | *Number of experiments* | *Range of P contents in diet* (g/kg DM) | *Mean (and range) of endogenous faecal P loss* (mg/kg live weight) |
|---|---|---|---|
| 15–80 | 11 | 0–3.3 | 2.9 (1.3–4.8) |
| 14–80 | 8 | 5.8–8.3 | 8.8 (1.8–17.6) |
| >140 | 8 | 0–6.3 | 4.5 (3.0–7.1) |

(After Jongbloed, 1987)

Excretion of P in urine is highly correlated with its dietary supply due to homeostatic mechanisms. Jongbloed (1987) showed that, at low dietary P levels and Ca:P ratios ranging from 1.0:1 and 2.0:1, 20 mg P/d was excreted in the urine of pigs weighing 30 kg, whereas 40 mg P/d was recorded for pigs weighing 100 kg. This amounts to between 0.7 and 0.4 mg P/kg live weight/d. Similar values were also reported in studies of Grimbergen, Cornelissen and Stappers (1985) and Pointillart *et al.* (1985). In breeding sows, urinary losses of P are also low. In experiments at Lelystad they ranged from 0.10 to 0.15 g P/d when low dietary P levels were used up to 0.6 g/d with high levels of P in the diets. At low levels of dietary P, the urinary excretion of P in the urine amounts to about 1 mg P/kg live weight/d.

It can be concluded therefore that a maintenance requirement for P of growing pigs of 10 mg/kg live weight/d can be adopted for diets containing adequate levels of P. A similar value was also suggested by Guéguen and Perez (1981). For diets containing low levels of P, the maintenance requirement will be about 4 mg P/kg live weight/d.

By expressing the requirement in terms of apparent digestible P it is clear that the contribution of endogenous faecal excretion is already taken into account. Theoretically, the maintenance requirement of apparent digestible P is equal to the urinary loss of 1 mg P/kg live weight. However, it is known from the literature (Vemmer, 1982; Jongbloed, 1987), that the absorption coefficient of P is lower when a surplus of P is offered relative to requirement. This means that in practice, where animals are mostly fed above their requirement, the digestibility of P will be somewhat lower than that obtained in our experiments. In these, with growing pigs used to assess the digestibility of P, the concentration of P in the diets was mostly between 4 and 6 g/kg DM, which is supposed to supply a minimum amount of digestible P. To compensate for the lower digestibility of P, when animals are fed above their requirement, it is suggested that instead of lowering the digestibility of P in feedstuffs, a value of 6 mg/kg live weight is used as 'endogenous loss' both for growing pigs and breeding sows. This value for faecal endogenous loss is based on the data in Table 12.7.

### THE REQUIREMENT OF P FOR GROWTH

The P requirement for growth can be best estimated by data obtained using the slaughter technique, because in most balance experiments the retention of P is overestimated (Jongbloed, 1987). From our earlier studies it was concluded by means of an allometric function that the requirement for growth is on average 5.1 g P/kg liveweight gain (Jongbloed, 1987). There was a small non-significant quadratic effect. The diets used for the calculations contained a high concentration of P and Ca, so it may be assumed that almost maximal bone calcification occurred, although for optimal performance this is not required (Guéguen and Perez, 1981; NRC, 1988). The value of 5.4 g P/kg liveweight gain for pigs from 10 to 20 kg is markedly lower than those proposed by Guéguen and Perez (1981) and ARC (1981). However, in more recent studies at the IVVO with piglets weaned at 25 and 31 days of age, a value of 4.9 to 5.0 g P/kg liveweight gain was again found (Dellaert *et al.*, 1990). Differences between these values for young pigs and those adopted by Guéguen and Perez (1981) and ARC (1981) may be attributable in part to genetic factors, but there is also likely to be a large element of overestimation of amount of P in this type of animal. However, for pigs from 25 kg live weight onwards there are no major differences between our values and those in the literature, except for the values of ARC (1981) for pigs over 80 kg.

**Table 12.8** CALCULATED REQUIREMENT OF P FOR GROWTH IN PIGS (g P/kg BODYWEIGHT)

| *Pig type* | *Live weight* (kg) | | | | | | |
|---|---|---|---|---|---|---|---|
| | *10* | *30* | *50* | *60* | *70* | *90* | *110* |
| Normal | – | 5.10 | 4.94 | 4.87 | 4.80 | 4.70 | 4.60 |
| Very lean | 5.45 | 5.35 | 5.19 | 5.12 | 5.05 | 4.95 | 4.85 |

In studying the P requirement for growth, two types of growth were distinguished, normal and very lean (Jongbloed, 1987). Although there was a non-significant quadratic effect it was decided to take the quadratic effect into account when calculating requirements, because a quadratic effect is more in agreement with the concept of growth models. The function with the quadratic term for live weight was; $\ln P = 1.494 + 1.108 \ln W - 0.018 (\ln W)^2$. The results of these calculations are presented in Table 12.8, and indicate higher requirements (g/kg bodyweight basis) for very lean than for normal genotypes.

### THE REQUIREMENT OF P FOR FETAL GROWTH AND MATERNAL TISSUE

In the first 2 months of gestation the requirement of pregnancy for P is low. In that period P is mainly used for restoring body reserves lost during the last lactation period. Detailed information on deposition of fat and protein to replenish body reserves, together with restoration of the P status of bone tissue has been described by Jongbloed and Everts (1992). Data on the deposition of P in the placenta, uterus, allantoic fluids, and udder during pregnancy are based on Den Hartog *et al.* (1988) and other studies at Lelystad. Mineralization in fetuses predominantly takes place in the last month of pregnancy (Becker, 1976; Den Hartog *et al.*, 1988). The latter authors concluded from their slaughter experiments that the live weight of a fetus $y$ could be estimated by Equation (12.1).

$$y\,(\mathrm{g}) = 8.58 \times 10^{-6} \times t^4 \qquad (12.1)$$

where $t$ = days of pregnancy

Together with Equation (12.2) which is used for estimating the concentration of P/kg fetus the amount of P can be estimated. These appropriate values for different stages of pregnancy are given in Figure 12.1.

$$\mathrm{P}\,(\mathrm{g/kg}) = 0.82 + 0.058 \times t \qquad (12.2)$$

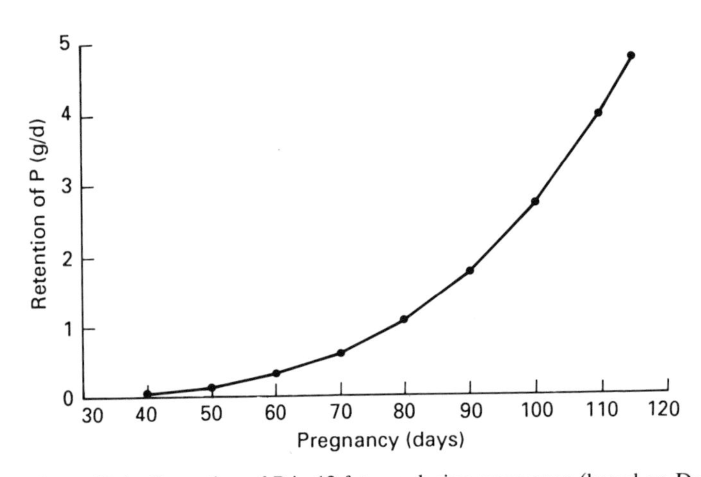


**Figure 12.1** Retention of P in 12 fetuses during pregnancy (based on Den Hartog *et al.*, 1988)

### THE REQUIREMENT OF P FOR MILK PRODUCTION

Milk production in the sow is the major determinant of the P requirement for lactating sows. To date, the requirement for P has been derived from the amount of milk produced and its P content. However, measuring the amount of milk produced is difficult and several assumptions have to be made (Goerke, 1979; Den Hartog *et al.*, 1984, 1987; Noblet and Etienne, 1986). A further complication is to obtain a good estimate of the P concentration in sow's milk. This may vary from 1.2 to 1.7 g/kg, although in the literature a value of 1.55 g/kg is generally adopted

(Freese, 1958; Guéguen and Salmon-Legagneur, 1959). An alternative approach is to base estimates of requirements for lactation on the quantity of P in piglets. To obtain a concentration of 5.45 g P/kg liveweight gain (Table 12.8) and a growth rate of 250 g/day, the total requirement both for maintenance and growth by a piglet of 5 kg live weight can be calculated. The apparent digestibility of P in milk (skimmed milk powder) is assumed to be 91% (Table 12.3). This results in a supply of 1.52 g P in sow's milk to the piglet. To excrete this amount of P in milk the sow can, apart from P from the feed, rely on catabolism of stored P in the uterus and maternal body. When a total of 1.75 kg protein is catabolized, this will result in 17 g available P. Resorption of P from bones may also take place during lactation. However, this contribution is supposed to be minor, and has therefore not been taken into account as a source of P supply in the milk.

### SYNTHESIS OF P REQUIREMENTS AND APPLICATION IN PRACTICE

Based on the knowledge presented in the preceding sections, requirement for P can be formulated for animals in various physiological stages. Details of the calculations have been given by Jongbloed and Everts (1992). In this section only summarized results of these calculations are presented (Table 12.9 for growing pigs and Table 12.10 for breeding sows).

The data shown in Table 12.9 indicate that the dietary concentration of digestible P decreases from 3.6 to 1.5 g/kg as the live weight of the animal increases from 10

**Table 12.9** ESTIMATED REQUIREMENT OF APPARENT DIGESTIBLE P FOR GROWING PIGS

| *Live weight* (kg) | *Growth rate* (g/d) | *Requirement for apparent digestible P* *Maintenance* (g/d) | *Growth*[a] (g/d) | *Total* (g/d) | *Feed intake* (kg/d) | *P content of feed* (g dig. P/kg) |
|---|---|---|---|---|---|---|
| 10 | 300 | 0.07 | 1.64 | 1.71 | 0.48 | 3.6 |
| 30 | 570 | 0.21 | 2.91 | 3.12 | 1.35 | 2.3 |
| 50 | 820 | 0.35 | 4.05 | 4.40 | 2.15 | 2.0 |
| 80 | 870 | 0.56 | 4.13 | 4.69 | 2.95 | 1.6 |
| 100 | 850 | 0.70 | 3.91 | 4.61 | 3.10 | 1.5 |

[a]Based on the composition of growth for a normal type of pig (Table 12.8)

**Table 12.10** ESTIMATED P REQUIREMENT FOR PREGNANT AND LACTATING SOWS (200 kg LIVE WEIGHT; 12 PIGLETS)

| | *Requirement for apparent digestible P* 'Maintenance (g/d) | *Body stores* (g/d) | *In fetuses+annexa/ milk* (g/d) | *Total* (g/d) | *Feed intake* (kg/d) | *P content of feed* (g dig. P/kg) |
|---|---|---|---|---|---|---|
| Pregnancy | | | | | | |
| Day 0 | 1.4 | 1.6 | 0 | 2.6 | 2.5 | 1.0 |
| Day 90 | 1.4 | 1.3 | 1.9 | 4.6 | 3.0 | 1.5 |
| Day 110 | 1.4 | 1.1 | 4.5 | 7.0 | 3.0 | 2.3 |
| Day 115 | 1.4 | 1.1 | 5.3 | 7.8 | 3.0 | 2.6 |
| Lactating | 1.4 | −0.5 | 18.2 | 19.1 | 7.2 | 2.6 |

to 100 kg live weight. The calculations for pigs from 25 to 110 kg are based on a mean growth rate of 780 g/d and a feed conversion ratio of 3.0 (CVB, 1990). Comparison of these estimates with recommended P levels in feeds for growing-finishing pigs indicates that they are in fairly good agreement with those of NRC (1988) but substantially lower than those of ARC (1981) and DLG (1987).

The estimated P requirements for pregnant and lactating sows are based on a litter of 12 piglets and a sow weighing 200 kg. The concentration of digestible P in the feed for pregnant sows increases from 1.0 g/kg at the start of pregnancy to 2.6 g/kg at term. Especially during the last week of pregnancy there is a substantial increase in the required dietary concentration. However, 1 week before farrowing pregnant sows are moved to the farrowing house and the feed for lactating sows will be offered. This means that a dietary content of 2.2 g digestible P/kg will suffice for pregnant sows. The concentration of digestible P in feeds for lactating sows should be much higher than that for pregnant animals. A problem which does arise is that in many cases a high feed intake, such as that recommended by CVB (1990), is not achieved, especially with primiparous sows. When feed intake is 6.5 kg/day instead of 7.2 kg/day, then the recommended level in the feed should be 2.9 g digestible P/kg. For safety reasons this may be increased slightly. Some of the P requirements of the piglet can be contributed by the creep feed, although intake of creep feed is low (0.2–0.5 kg/piglet) for piglets weaned at 5 weeks of age.

The P levels recommended by NRC (1988) in feeds for sows during pregnancy are very much higher, while those recommended by DLG (1987) are also substantially higher than the values recommended in this chapter. The NRC (1988) and DLG (1987) recommended levels for lactating sows are also substantially higher. Part of these differences may be due to dietary energy concentrations and feed intake levels, but are also due to differences in the assumed availability of P in the feed. In this respect limited evidence from unpublished studies at Lelystad suggested that the P digestibility coefficients determined in growing pigs are slightly higher than those for multiparous breeding sows. If lower P digestibility coefficients in breeding sows are confirmed in future experiments, the recommendations based on digestible P for these categories of animals will need to be corrected.

Since introduction of digestible P for pigs as a criterion for estimating the nutritive value of P in the Netherlands, the excretion of P in growing-finishing pigs has decreased substantially from 1200 to 810 g/pig over the entire growing-finishing period. This reduction of more than 30% results in less environmental pollution per pig (Coppoolse *et al.*, 1990).

## Conclusions

The apparent digestibility of P in various feedstuffs of plant origin varies from 10 to 50%. This is due to origin of the feedstuff, the proportion of phytate P and the presence of intrinsic phytase. The digestibility of P in feedstuffs of animal origin and feed phosphates is high, ranging from 65 to 90%. Differences may be due to technologies of manufacturing and physicochemical structure of the products.

Estimates of P digestibility with feed phosphates show a high degree of repeatability within a type of feed phosphate. Furthermore digestibility of P gave a reliable estimate for the nutritive value of P and had a higher correlation with retention of P than blood and bone related parameters.

Both intrinsic plant and microbial phytases enhance the digestibility of phytate P in feedstuffs of plant origin. Using 1000 units of microbial phytase in feeds for pigs enhanced the digestibility of P more than 20%. The consequence is that much less supplementary P from feed phosphate will be used for pig feeds in the near future.

Large differences in recommendations of P in feeds exist between those reported in the literature and those reported in this chapter. Digestible P is much better than total P as a recommendation for P requirement. Further research is necessary to explain the differences in recommendations, especially those for breeding sows. The better knowledge of supply and requirement of digestible P has already led to a considerable reduction of the excretion of P by pigs in the Netherlands.

## References

Agricultural Research Council (1981). Commonwealth Agricultural Bureaux, Slough

Becker, K. (1976). *Übersicht zur Tierernährung*, **4**, 167–195

Beers, S. and Koorn, T. (1990). Report IVVO no. 223, Lelystad, 24 pp

Chapple, R.P., Yen, J.T. and Veum, T.L. (1979). *Journal of Animal Science*, **49** (supplement 1), 99–100

Cooper, J.S. and Gowing, H.S. (1983). *British Journal of Nutrition*, **50**, 673–678

Coppoolse, J., Vuuren, A.M. van, Huisman, J., Jansen, W.M.M.A., Jongbloed, A.W., Lenis, N.P. and Simons, P.C.M. (1990). Dienst Landbouwkundig Onderzoek, Wageningen

Cromwell, G.L. and Stahly, T.S. (1978). *Feedstuffs*, **50** (14), 14

Cromwell, G.L. (1983). *Proceedings of the National Feed Ingredient Association*. Mineral Nutrition Institute, Chicago, IL; NFIA, Des Moines, IA

Cromwell, G.L., Stahly, T.S. and Moneque, H.J. (1985). *Journal of Animal Science*, **61** (supplement 1), 320

CVB (1990). Verkorte tabel. Centraal Veevoederbureau in Nederland, Lelystad

Davies, N.T. and Flett, A.A. (1978). *British Journal of Nutrition*, **39**, 307–316

Dellaert, B.M., van der Peet, G.F.V., Jongbloed, A.W. and Beers, S. (1990). *Netherlands Journal of Agricultural Science*, **38**, 555–566

Deutsche Landwirtschafts-Gesellschaft (1987). DLG-Verlag GmbH, Frankfurt

Fourdin, A., Fontaine, N. and Pointillart, A. (1986). *Journées Recherches Porcines en France*, **18**, 83–90

Fourdin, A., Camus, P., Cayron, B., Colin, C. and Pointillart, A. (1988). *Journées Recherches Porcines en France*, **20**, 327–332

Freese, H.H. (1958). *Archiv für Tierernährung*, **8**, 330–392

Gerritse, R.G. and Zugec, I. (1977). *Journal of Agricultural Science (Cambridge)*, **88**, 101–109

Goerke, R. (1979). Dissertation, Göttingen

Grimbergen, A.H.M., Cornelissen, J.P. and Stappers, H.P. (1985). *Animal Feed Science Technology*, **13**, 117–130

Guéguen, L. and Perez, J.M. (1981). *Proceedings of the Nutrition Society*, **40**, 273–278

Guéguen, L. and Salmon-Legagneur, E. (1959). *Comptes rendus des séances de l'Académie des Sciences, Paris*, **249**, 784–786

Hartog, L.A. den, Verstegen, M.W.A., Hermans, H.A.T.M., Noorderwier, G.J.

and Kempen, G.J.M. van (1984). *Zeitschrift für Tierphysiologie Tierernährung und Futtermittelkunde*, **51**, 148–157
Hartog, L.A. den, Boer, H., Bosch, M.W., Klaassen, G.J., Steen, H.A.M. van der (1987). *Journal of Animal Physiology and Animal Nutrition*, **58**, 253–261
Hartog, L.A. den, Zandstra, T., Kemp, B., Verstegen, M.W.A. (1988). *Journal of Animal Physiology and Animal Nutrition*, **60**, 4–7
Jongbloed, A.W. (1987). Thesis, Agricultural University Wageningen (Rapport IVVO no. 179, Lelystad) 343 pp
Jongbloed, A.W. and Kemme, P.A. (1990a). *Animal Feed Science and Technology*, **28**, 233–242
Jongbloed, A.W. and Kemme, P.A. (1990b). *Netherlands Journal of Agricultural Science*, **38**, 567–575
Jongbloed, A.W. and Kemme, P.A. (1990c). In: *Mestproblematiek: aanpak via de voeding van varkens en pluimvee.* Verslag van de themadag Veevoeding en Milieu, Lelystad, 19 April 1990, Lelystad: IVVO pp. 33–42
Jongbloed, A.W., Kemme, P.A. and Mroz, Z. (1990). Rapport IVVO no. 221, Lelystad 28 pp
Jongbloed, A.W. and Everts, H. (1992). *Netherlands Journal of Agricultural Science* 40: 123–136
Nelson, T.S. (1967). *Poultry Science*, **46**, 862–871
Nelson, T.S., Shieh, T.R., Wodzinski, R.J. and Ware, J.H. (1968). *Poultry Science*, **47**, 1289–1293
Noblet, J. and Etienne, M. (1986). *Journal of Animal Science*, **63**, 1888–1896
NRC (1980) National Academy of Sciences Washington, DC
NRC (1988) Ninth revised edition. National Academy Press, Washington, DC
Pointillart, A., Fontaine, N. and Thomasset, M. (1984). *Nutrition Reports International*, **29**, 473–483
Pointillart, A., Fontaine, N., Thomasset, M. and Jay, M.E. (1985). *Nutrition Reports International*, **32**, 155–167
Scheuermann, S.E., Lantsch, H.J. and Menke, K.H. (1988). *Journal of Animal Physiology and Animal Nutrition*, **60**, 64–75
Shurson, G.C., Ku, P.K. and Miller, E.R. (1984). *Journal of Animal Science*, **59** (supplement 1), 106
Simons, P.C.M., Versteegh, H.A.J., Jongbloed, A.W., Kemme, P.A., Slump, P., Bos, K.D., Wolters, M.G.E., Beudeker, R.F. and Verschoor, G.J. (1990). *British Journal of Nutrition*, **64**, 525–540
Vemmer, H. (1982). *Zeitschrift Tierphysiologie, Tierernährung und Futtermittelkunde*, **47**, 220–230
Williams, P.J. and Taylor, T.G. (1985). *British Journal of Nutrition*, **54**, 429–435

13

# THE WATER REQUIREMENT OF GROWING-FINISHING PIGS – THEORETICAL AND PRACTICAL CONSIDERATIONS

P. H. BROOKS and J. L. CARPENTER
*Seale-Hayne Faculty of Agriculture, Food and Land Use, Polytechnic South West, Newton Abbot, Devon, UK*

## Introduction

Water is often referred to as the 'forgotten nutrient' and it is certainly true that it has received less attention than virtually any other nutrient. This is remarkable when one considers that not only is water the single nutrient required in the greatest quantity by animals but also the most essential. Water is intimately involved in virtually all metabolic functions as well as comprising almost 70% of the adult animal's body mass. In addition Adolph (1933) pointed out that water ranks far above every other substance in the body as regards rate of turnover.

'The vital role of water in the body is indicated by Rubner's observation that the body can lose practically all its fat and over half its protein and yet live, while a loss of one-tenth of its water results in death' (Maynard *et al.*, 1979). When the Agricultural Research Council (1981) reviewed the literature on water they listed only 28 references on which they based their recommendations. They concluded that:

'From the various reports considered it is apparent that in conditions of free access to water there are wide variations in individual consumption. Generally, it is not possible to decide whether these represent unimportant idiosyncrasies or physiological needs which should be met if possible. Excluding suckling sows and their offspring, the evidence suggests that the water requirement of pigs is about two parts of water for each one part of feed (by weight), the ratio being perhaps somewhat wider for recently weaned pigs and narrower for older animals.'

This view is considerably at variance with that of Chew (1965) who in his comprehensive review of the water metabolism of mammals concluded:

'It is necessary to recognize that there is no single water requirement for a species or individual; the amount drunk depends upon factors such as: temperature, humidity, diet, frequency with which water is provided, conditions of caging and stresses of the environment. The evaluation of published data is very uncertain unless such experimental conditions are described.'

If water was only a nutrient it would be relatively easy to measure input/output relationships and derive 'requirement' values which would satisfy specified production objectives. Indeed, this is the approach that a number of reviewers have taken. Unfortunately, in determining the animal's 'need' for water, and in attempting to identify the manner in which water should be provided, such an approach is too simplistic.

In addition to the requirements for tissue maintenance and growth, for reproduction and lactation, the animal needs water in order to fulfil a number of other physiologically and biochemically significant functions, namely:

(1) for the adjustment of body temperature,
(2) for the maintenance of mineral homeostasis,
(3) for the excretion of the end products of digestion (particularly urea),
(4) for the excretion of antinutritional factors ingested with the diet,
(5) for the excretion of drugs and drug residues,
(6) for the achievement of satiety (gut fill),
(7) for the satisfaction of behavioural drives.

Four important points need to be made in the context of this list of needs. First, that the individual functions listed above may be additive and furthermore that they may be additional to the requirements for maintenance and production. Thus, the classic requirement in conventional nutritional terms represents a physiological minimum which will be operative only if the requirements for the other functions have fortuitously been met within a given situation.

Second, some of the functions listed above have a higher priority than production. Hence, in a situation where the water supply is not adequate to support all the animal's needs, the prioritization of these functions will result in an undersupply of water to support production functions and therefore result in a reduction in animal performance.

Third, under extreme circumstances the animal may have to satisfy a need, for example to reduce body temperature or to detoxify the body by greatly increasing its water intake. Such circumstances may result in a water intake so great that feed intake is depressed, i.e. the animal's capacity for total volumetric fill is exceeded before its capacity for voluntary feed intake has been satisfied.

Fourth, as will be discussed later in this chapter, the timeliness and manner of water presentation may affect not only voluntary feed intake but also the digestibility of feed. The significance that recent information on this subject may have for the interpretation of the results of studies on other nutrients and diets is the focus of this chapter.

## Water input/output relationships in the pig

The pig obtains water from three sources:

(1) from water ingested with its food (generally water which is intimately bound into the feed ingredients);
(2) from water consumed directly through drinking;
(3) from water formed during metabolism by the oxidation of hydrogen-containing foods (so-called 'metabolic water').

The pig uses water for growth and in the case of reproducing animals for the products of conception and for milk production. In addition the pig loses body water via four routes:

(1) in expired air during respiration;

(2) from the skin surface through perspiration and insensible water loss;
(3) in faeces;
(4) in urine.

The relative contribution of these different inputs and losses is extremely variable. The interactions between them produced by differences in health status, nutrition and environment are considerable and complex. Consequently, factorial estimation of water requirement is neither a reliable nor a practical proposition. Table 13.1 illustrates the relative quantitative contribution of the different inputs and losses for a 60-kg pig fed *ad libitum* on a well balanced compound diet containing no mineral excesses. It assumes that the pig is in good health and is maintained in a thermoneutral environment. Given these conditions the water demand is probably close to the minimum per unit of food consumed and/or per unit of gain.

Some of the more important factors influencing the relative proportions of water inputs and losses are discussed below.

**Table 13.1** EXAMPLE OF WATER BALANCE IN A 60 kg LIVEWEIGHT PIG FED A COMPOUND DIET AND GAINING 700 g/day IN A THERMONEUTRAL ENVIRONMENT

| *Water used/lost* | (ml) | (%) | *Water consumed/formed* | (ml) | (%) |
|---|---|---|---|---|---|
| Growth[a] | 469 | 8.2 | Food water[e] | 380 | 6.6 |
| Respiration[b] | 580 | 10.1 | Food oxidation[f] | 1015 | 17.7 |
| Skin[c] | 420 | 7.3 | | | |
| Faeces[d] | 742 | 12.9 | | | |
| Urine[h] | 3536 | 61.5 | Water consumed[g] | 4352 | 75.7 |
| Total | 5747 | | | 5747 | |

[a] Growth (700 g/day) assumed to be 67% water
[b] Respiration loss assumed to be 0.58 litres/day (Holmes and Mount, 1967)
[c] Insensible moisture loss from skin assumes 13.4 g/m$^2$ per h at thermoneutral temperature and 70% RH as obtained by Morrison *et al.* (1967). Surface area = $0.10W^{0.63}$ (Brody, 1964)
[d] *Ad libitum* fed pig of 60 kg liveweight assumed to eat 2.72 kg food (2.23 kg DM). DM digestibility assumed to be 82% and faecal DM 35%
[e] Compound diet assumed to be 14% moisture
[f] The diet is assumed to contain per 1000 g fresh weight, fat 70 g, carbohydrate 590 g and protein 180 g. The protein is assumed to have a biological value of 70 therefore 54 g of the protein would not be used in protein growth and would be deaminated. Thus the yield of metabolic water per kg feed would be

| | (g/kg) | *Water yield/g** | *Total* |
|---|---|---|---|
| Fat | 70 | 1.10 | 77 |
| Carbohydrate | 590 | 0.060 | 354 |
| Protein | 54 | 0.44 | 24 |
| Total | | | 455 |

* See text

[g] Water intake assumed to be 1.6 kg per kg feed which was the lowest ratio recorded by Yang *et al.* (1981) for pigs fed *ad libitum*
[h] Urine volume derived by difference

## Factors affecting the relative contribution of different water inputs

### WATER CONSUMED IN FOOD

The pig is obliged to drink a certain amount of water in its food. Pigs fed compound, cereal based, 'dry', diets obtain comparatively little water by this route. In the example a 60-kg pig fed *ad libitum* is assumed to consume around 2.72 kg feed (ARC, 1981). Typically this feed would contain about 14% moisture giving a food-water intake of 0.38 litres/day). The contribution of food to the supply of water may be considerably greater under other feeding regimens. For example, systems may require the pig to obtain significant portions of its nutrient intake from grass or roots (e.g. fodder beet), from a variety of ensiled products or from a range of liquid by-products such as human food waste (swill or garbage feeding) milk by-products (skim or whey) or liquid processing wastes (brewery washings, pot ale syrup, starch manufacturing residues). Consequently a 60-kg pig consuming 2 kg of dry matter from these different sources could ingest as little as 0.22 litres of water or as much as 4.7 litres water in association with its feed.

Liquid feed systems represent a particular and unusual case, for in these the more normal dry compound feed components are suspended in water to enable the feed to be distributed to the pig through a pipeline. Typically these systems would use a liquid to dry matter (water to feed) ratio in the range 2.5–3.5:1. Thus at the extreme the same 60-kg pig receiving 2 kg food dry matter could receive this in association with 7 litres of water.

### WATER PRODUCED BY FOOD OXIDATION

Metabolic water is one of the end products of oxidation of organic dietary nutrients, fat depots and tissue protein. The precise quantity of water produced depends upon the molecular weight of the particular substrate and the amount of water produced by the reaction. Two examples are given below:

*Glucose*

$$C_6H_{12}O_6 + 6O_2 \rightarrow 6CO_2 + 6H_2O$$

Molecular weights 180 + 192 → 264 + 108

Metabolic water/g carbohydrate oxidized = 108/180 = 0.6 ml

*Stearic acid*

$$CH_3(CH_2)_{16}COOH + 26O_2 \rightarrow 18CO_2 + 18H_2O$$

Molecular weights 284 + 832 → 792 + 324

Metabolic water/g fat oxidized = 324/284 = 1.14 ml

In general carbohydrates yield approximately 60 ml water per 100 g of carbohydrate oxidized. Fats are interesting in that they yield a greater weight of water than the weight of fat oxidized. A figure of 110 ml water per 100 g fat oxidized is a reasonable approximation for the yield of water from a mixed fat sample. Similarly, the catabolism of amino acids, which is more complex than fatty acid and monosaccharide metabolism, yields approximately 44 ml of water per g of protein oxidized. According to Yang *et al.* (1984) every kg of air-dry feed eaten will contribute between 0.38 and 0.48 litres of metabolic water. However, simple calculations based on the apparent composition of the diet fed may considerably over-estimate the yield of metabolic water for the following reasons.

(1) The digestibility of individual diet components needs to be taken into account.
(2) All fat digested is unlikely to be oxidized for energy. A proportion of the fat absorbed will be deposited intact in the adipocytes and will not be oxidized.
(3) A large proportion of the protein absorbed will be utilized in protein metabolism and not oxidized. The higher the biological value of the protein and the more accurately protein requirements are fulfilled the less protein will be subjected to oxidation.

As Lloyd *et al.* (1978) have pointed out, although oxidation yields metabolic water this process actually results in a net *demand* for water, as the water required for dissipation of the heat produced and the water required to excrete the end products of the process exceed that yielded by the reaction (Table 13.2). In

**Table 13.2** WATER NEEDED FOR METABOLISM OF EACH 100 kcal FURNISHED BY VARIOUS FOODS (g)

| *Food* | *Preformed water* | *Metabolic water formed* | *Water lost in dissipation of heat* | *Water lost in excreting end products* | *Net deficit of water* |
|---|---|---|---|---|---|
| Protein | 0 | 10.3 | 60 | 300 | 350 |
| Starch ($CH_2O$) | 0 | 13.9 | 60 | 0 | 46 |
| Fat | 0 | 11.9 | 60 | 0 | 48 |
| Beef | 25 | 11.3 | 60 | 119 | 143 |
| Eggs | 47 | 11.1 | 60 | 154 | 156 |
| Milk | 127 | 12.5 | 60 | 123 | 43 |
| Bread | 14 | 13.2 | 60 | 69 | 102 |
| Apples | 150 | 13.9 | 60 | 56 | –48 |

(After Lloyd *et al.*, 1978)

particular the provision of excess and non 'ideal' protein will increase the amount of urea which has to be excreted and, in turn, will increase the demand for water. In this context it is interesting to note that milk, a product which is itself almost 90% water, actually creates a water deficit because it is a high protein and mineral material. This has particular implications in liquid feeding systems for growing/finishing pigs, in supplementary liquid feeding systems for baby pigs, for orphan pigs fed on milk substitute and for suckling pigs where no separate water supply has been provided.

WATER CONSUMED

The model in Table 13.1 represents production conditions which would minimize the pig's demand for water. Interestingly, although the absolute water intake of the pig is positively correlated with feed intake, the pig appears to minimize its demand for water per unit of feed dry matter when it is fed *ad libitum* (Yang *et al.*, 1981). It would appear that when offered unrestricted access to food and water the pig maximizes the proportion of food that it consumes within its volumetric limit consistent with consuming adequate water to maintain its homeostatic balance. Conversely, when feed intake is reduced below the level producing physical satiety the pig increases its water intake. This increase in water intake, appears to satisfy the pig's requirement for volumetric (gut) fill. From their studies Yang *et al.* (1981) suggested that the young pig may have a requirement for total volumetric intake equivalent to 19% of bodyweight. However, this cannot be regarded as a 'norm' as it has not been possible to validate this conclusion in studies at Seale-Hayne or by reference to the results of other published studies.

Nevertheless it is apparent that the pig will imbibe additional water to compensate for low dietary bulk. In subsequent studies Yang *et al.* (1984) found that polydipsia occurred when the daily dry matter intake of growing pigs was reduced below 30 g/kg bodyweight per day. It is interesting to note that if this value also applied to reproducing sows a daily feed intake of 2.5 kg/day (2.15 kg DM) would only be sufficient to prevent polydipsia in pregnant sows weighing less than 72 kg! Interestingly, polydipsia is frequently seen in sows confined in stalls or tethers when it is usually associated with other stereotypic behaviours (Cronin *et al.*, 1986). Under these circumstances the additional activity associated with the continual performance of these stereotypic behaviours has a significant energy cost utilizing up to 23% of the sow's dietary energy intake.

Recently, Appleby and Lawrence (1987) have reported that stereotypic behaviour in sows appears to be a result of inadequate feed intake, i.e. the stereotypes result from stress induced by hunger rather than confinement. Therefore, it would seem reasonable to postulate that the polydipsia observed in confined sows might also be a response to inadequate gut fill.

## Factors affecting the relative contribution of different water losses/uses

WATER FOR GROWTH

A detailed consideration of the nature and composition of tissue growth is outside the scope of this chapter. Suffice it to say that water contributes significantly to the development of liveweight gain. The water content of the animal approaches 70% and that of lean tissue exceeds it. Consequently, the demand for water for growth will depend upon the animal's lean growth potential, and its stage of development. The latter affects not only the relative proportion of lean and fat in daily gain but also the extent (rate) of body water and protein turnover and the hydration of tissues. As a general rule water turnover, body and individual tissue water content tend to reduce as the animal proceeds towards mature body size (Whittemore and Elsley, 1979).

WATER LOST THROUGH RESPIRATION

Moisture is continually lost from the respiratory tract during the normal process of breathing. Incoming air is both warmed and wetted as it passes over the moist lining of the respiratory tract and is expired at around 90% saturation (Roubicek, 1969).

The rate of water loss is a function of four factors:

(1) ambient temperature,
(2) the vapour pressure gradient between the incoming air and the expired air,
(3) respiration rate,
(4) the tidal volume of the lungs.

For pigs in a thermoneutral environment (20°C) respiratory water loss was estimated to be 0.29 and 0.58 litres/pig for pigs of 20 and 60 kg (Holmes and Mount, 1987). Similarly Randall (1983) estimated a daily loss of the order 0.43 litres/pig for growing pigs.

Both temperature and humidity affect respiratory moisture loss. At a constant dew point of 10°C, moisture loss increased from 0.40 to 1.25 litres/pig day as temperature increased from 15.6 to 24.4°C whilst at a constant temperature of 29.4°C (dry bulb) increasing the relative humidity from 50 to 90% decreased moisture loss from 1.61 to 0.59 litres/pig day.

WATER LOST THROUGH THE SKIN

Sweating and insensible moisture loss from the skin are not major sources of water loss in pigs. Apart from small areas of the body surface the pig's sweat glands are dormant. Histological examinations have shown the glands to be blocked by plugs of keratin (Ingram, 1967). Within the thermoneutral range the rate of moisture loss has been estimated to be between 12 and 16 g/m$^2$ (Moritz and Henriques, 1947; Ingram, 1964; Morrison *et al.*, 1967). Increasing temperature from −5 to 30°C (dry bulb) increased water loss from 7 to 32 g/m$^2$ (Ingram, 1964). It might be expected that relative humidity would also affect the rate of moisture loss. However, Morrison *et al.* (1967) found no effect on moisture loss when the relative humidity was increased from 50 to 90%.

WATER LOSS IN FAECES

The amount of faeces produced and the volume of water lost through this route depend largely upon nutrient digestibility and feed intake. Faecal moisture percentage is little affected by the amount of water ingested. The results of four independent studies on the effects of feed moisture content on the dry matter of faeces are summarized in Table 13.3. These show only minor differences in percentage faecal moisture across a wide range of water to meal intake ratios. However, faecal moisture content is significantly increased by high levels of dietary fibre (Cooper and Tyler, 1959a,b). This increase in moisture content may result from the hygroscopic nature of the fibre, from the activity of hind gut micro-organisms or from the increased rate of passage through the alimentary tract reducing the time available for reabsorption of water in the hind gut.

**Table 13.3** THE EFFECTS OF FEED MOISTURE CONTENT ON THE DRY MATTER CONTENT OF FAECES PRODUCED BY GROWING PIGS

| *Moisture content of feed (%)* | *Dry matter content of faeces (%)* | *Water provision*[a] | *Reference no.*[b] |
|---|---|---|---|
| 10 | 32.8–36.6 | A | 3 |
| 10 | 34.0–34.6 | A | 4 |
| 40 | 34.2 | A | 4 |
| 50 | 33.5–35.2 | A | 3 |
| 60 | 23.7 | N | 1 |
| 60 | 28.9 | N | 1 |
| 70 | 21.2–24.9 | N | 1 |
| 73 | 28.9 | N | 2 |
| 75 | 23.5 | N | 1 |
| 77 | 31.5 | N | 2 |
| 79 | 22.2 | N | 1 |
| 85 | 33.2–38.1 | A | 4 |
| 95 | 32.8 | N | 2 |

[a] Supplementary water not provided (N); supplementary water was provided *ad libitum* (A)
[b] 1, Castle and Castle (1957); 2, Cooper and Tyler (1959); 3, Kornegay and Graber (1968); 4, Kornegay and Van der Noot (1968)

## WATER LOSS IN URINE

Water is the main constituent of urine, generally contributing around 95% of the volume. It is the vehicle by which excess minerals, toxins and the end products of catabolism, filtered from the blood by the kidneys, are removed from the body.

The minimum daily water requirement for eliminating these various components is extremely difficult to assess and virtually impossible to predict. Although the renal system can excrete large quantities of ingested water, when the water supply is severely restricted there is a limit to the amount that water can be economized through renal reabsorption. This is because there is a finite limit beyond which the kidneys cannot increase urine concentration. In pigs the maximum concentration is 1 osmol/litre (McFarlane, 1976).

The renal pathway responds very quickly to short-term fluctuations in water intake. The volume of urinary water loss is thus largely dependent on the amount of water ingested. Whilst the other inputs and losses of water may be regarded as obligatory, some part at least of ingested water intake and hence urinary loss may be seen as voluntary. Water intake, controlled by the thirst mechanisms may be regarded as the 'coarse adjustment' in maintaining homeostasis while the rapid and much more sensitive mechanism of renal clearance provides the 'fine tuning'. As a consequence it will be an extremely rare event for the animal to consume the absolute minimum quantity of water required to fulfil all the indefinable physiological functions. The norm will be for it to overconsume water to some degree and use increased renal clearance to correct the oversupply and maintain homeostasis.

## Factors resulting in an increased demand for water

### NUTRITIONAL FACTORS

Our knowledge of the influence of nutritional factors on the demand for water is extremely fragmentary. This is because water demand has rarely been studied as an integral part of nutritional studies. Very few workers have incorporated measurements of water intake in their studies. Consequently, little is known about the effects of specific dietary components, raw materials or feed manufacturing processes on the water demand of pigs. The two major nutritional factors known to increase water demand are:

(1) the quantity and quality of protein in the diet,
(2) the mineral content of the diet particularly the sodium and potassium levels.

The significance of protein has been discussed earlier in this chapter. The more the amino acid supply deviates from 'ideal' the greater the demand for water. As protein supplied to the pig is rarely 'ideal', water demand will increase in proportion to the crude protein content of the diet. Thus, pigs fed 12 or 16% crude protein diets between 3 and 6 weeks of age consumed 3.90 and 5.26 litres water per pig day respectively (Wahlstrom *et al.*, 1970).

Although everyone is conscious that increased salt intake increases the demand for water surprisingly few studies have been conducted on the subject. One of the few studies that has been published (Table 13.4) showed that a 450% increase in the salt content of the diet increased the water demand per kg diet by 16%. Potassium has a more dramatic effect (Table 13.5). In two studies at Seale-Hayne, doubling the potassium content of the diet increased water demand by 14 and 25%.

High mineral levels can be tolerated by the pig providing that it has adequate water available to detoxify itself. The toxic level of 2% NaCl suggested by ARC (1981) is an extremely conservative figure based on the stated need to protect pigs on fixed water to feed ratios (fed through liquid feeding systems) where no alternative water supply is available to them. However, in studies where water was freely available a salt level of 100 g/kg diet (140 g/kg DM) did not produce any adverse effects (Gyrd-Hansen, 1972).

**Table 13.4** EFFECT OF DIETARY SALT ON PERFORMANCE AND WATER DEMAND

| | *NaCl in diet (%)* | | | |
|---|---|---|---|---|
| | *0.06* | *0.13* | *0.20* | *0.27* |
| Daily feed (kg) | 1.32 | 1.43 | 1.51 | 1.45 |
| Daily gain (g) | 410* | 560 | 630 | 610 |
| FCR | 3.22* | 2.56 | 2.40 | 2.37 |
| Daily water (p) | 4.2 | 4.7 | 5.1 | 5.3 |
| kg water per kg feed | 3.2 | 3.3 | 3.4 | 3.7 |

(Calculated from Hagsten & Perry, 1976)
Pigs initial wt, 17 kg *ad libitum* fed, 28-day trial

**Table 13.5** EFFECT OF DIETARY POTASSIUM ON WATER DEMAND OF *AD LIBITUM* AND SCALE FED PIGS

| | *Potassium* (g/kg diet) | | | |
|---|---|---|---|---|
| | *8* | *11* | *14* | *17* |
| *Scale fed* | | | | |
| Daily feed (kg) | 1.45 | 1.40 | 1.40 | 1.39 |
| Daily water (litre) | 4.07 | 4.42 | 4.81 | 5.04[a] |
| Water per kg feed | 2.91 | 3.26 | 3.46 | 3.64 |
| *Ad libitum* fed | | | | |
| Daily feed (kg) | 1.58 | 1.60 | 1.63 | 1.58 |
| Daily water (litre) | 4.98 | 5.41 | 5.91 | 5.51[a] |
| Water per kg feed | 3.21 | 3.42 | 3.68 | 3.67 |

(After Gill, 1989)
[a] Initial wt 25 kg, 12-week trial scale fed pigs received 700 g/kg $W^{0.75}$

So-called 'salt poisoning' is a misnomer as the condition is not generally a result of a toxic level of salt intake *per se* but of a disruption of the pig's water balance. Thus, any circumstance which results in an increase in water demand, which cannot be satisfied, or results in a disruption in water supply, can produce the condition called 'salt poisoning'. Thus, pigs may die of this condition as a result of circumstances as varied as gastrointestinal disturbances resulting in diarrhoea when water supply is limited (Taylor, 1979) or the temporary loss of water due to frozen pipes or blocked watering equipment. Therefore, it is a misnomer to describe this condition as 'salt poisoning'. The term 'water starvation' would be a much more accurate descriptor and one which could help to focus attention on the real cause(s) of the problem.

In addition to excess, unutilized protein and minerals in the diet, the addition of antibiotics may also affect water balance. The type of antibiotic and the circumstances in which it is fed may produce differing results. Consequently, responses have ranged from a reduction in water demand when penicillin was added to diets (Robinson *et al.*, 1953) through little or no effect when penicillin or aureomycin were used (Holmes and Robinson, 1965; Braude and Johnson, 1953) to a 13.7% increase in demand when pigs were fed 20 mg/tonne oxytetracycline (Pieterse, 1963). In the absence of definitive information it is hypothesized that the effect of antibiotics on water demand will depend upon the relative extent to which water loss is reduced by the control of gastrointestinal disruption (scour) and water demand is increased to enable renal clearance of the antibiotic or its residues.

Very little is known about the effect of feed processing or feed form on water consumption. However, a recent study (Brooks, Parkins and Russell, unpublished data) suggests that more attention needs to be given to this subject. Weaned pigs (21–35 days of age) had a water to feed ratio of 2.19 for a diet fed as meal and 3.13 when the same diet was fed in pellet form. Clearly there is a need to establish the effect of processing and feed form on the water demand of the pig.

HIGH AMBIENT TEMPERATURE

Increased water consumption coupled with increased urinary water loss is an effective mechanism by which the pig can lose body heat. The extent of losses by this route depend upon

(1) the temperature of water consumed,
(2) the quantity of water consumed.

Much of the available information on the relationship between ambient temperature and water intake comes from studies using calorimetry. The results of a number of studies using growing-finishing pigs is summarized in Table 13.6.

Increasing ambient temperature from 12–15°C to 30–35°C gave an increase in water consumption per kg liveweight of approximately 57% in pigs of 33.5 kg liveweight (Mount *et al.*, 1971) and of approximately 63% in pigs of 90 kg liveweight (Straub *et al.*, 1976). The percentage increase in water to feed ratio is generally much greater as the pig, fed *ad libitum*, reduces food intake under high temperature conditions as a means of avoiding the embarrassment of ridding itself of metabolic heat.

Information obtained in calorimetric studies may not prove to be helpful in anticipating increased water demand under commercial conditions for a number of reasons. First, ambient temperatures generally show a diurnal pattern. Conse-

**Table 13.6** THE EFFECTS OF AMBIENT TEMPERATURE ON WATER USAGE OF GROWING PIGS REPORTED IN VARIOUS PUBLISHED STUDIES

| *Mean liveweight* (kg) | *Feeding regimen* (g/kg LW/day) | *Ambient temperature* (°C) | *Water intake* (litre/p/day) | (litre/kg/feed) | (litre/kg W) | *Reference no.* |
|---|---|---|---|---|---|---|
| 25 | 39–52 g | 7 | | 2.88 | 0.14 | 1 |
| 38 | 39–52 g | 12 | | 2.76 | 0.12 | |
| 35 | 39–52 g | 20 | | 2.74 | 0.12 | |
| 35 | 39–52 g | 30 | | 4.28 | 0.18 | |
| 73 | *Ad libitum* (46) | 20 | 7.36 | 2.18 | | 2 |
| 37 | *Ad libitum* (63) | 22 | 4.98 | 2.12 | | |
| 50 | *Ad libitum* (34) | 33 | 8.45 | 5.00 | | |
| 33.5 | 42–45 g | 7–12 | | 2.60 | 0.12 | 3 |
| 33.5 | 42–52 | 20 | | 2.60 | 0.13–0.12 | |
| 33.5 | 42–45 | 30 | | 4.20 | 0.18 | |
| 59 | 1.83 kg/day (31) | 20 | | 2.2 | 0.09 | |
| 59 | 1.83 kg/day (31) | 30 | | 2.8 | 0.12 | |
| 73 | *Ad libitum* (46) | 20 | | 2.2 | 0.1 | |
| 50 | *Ad libitum* (34) | 33 | | 5.00 | 0.17 | |
| 37 | *Ad libitum* (03) | 22 | | 2.1 | 0.14 | |
| 23 | *Ad libitum* | 5 | 4.10 | | | 4 |
| | | 35 | 7.33 | | | |
| 30 | 1.5 kg/pig day | 27 | 4.60 | 3.10 | | 5 |
| | | 33 | 7.50 | 5.00 | | |
| | *Ad libitum* | 27 | 4.00 | 1.60 | | |
| | | 33 | 6.10 | 2.40 | | |
| 90 | *Ad libitum* (32) | 15 | | 2.2 | 0.07 | 6 |
| 90 | *Ad libitum* (21) | 35 | | 5.4 | 0.12 | |

1, Close *et al.* (1971); 2, Morrison and Mount (1971); 3, Mount *et al.* (1971); 4, Nienaber and Hahn (19084); 5, Nang *et al.*(1981); 6, Straub *et al.* (1976)

quently the pig is not subjeced to high temperatures throughout the 24-h period. In such circumstances, the pig tends to alter its feeding behaviour and consume more of its food in the cooler parts of the day when dissipation of heat is not such a problem (Feddes *et al.*, 1989). Second, feed intake may be maintained by reformulation of the diet to provide more of the dietary energy from fat. Fat has lower heat increment than cereals consequently less metabolic heat has to be dissipated when high fat diets are used (Stahly, 1983).

Third, in commercial situations the pig may use (or be encouraged to use) evaporative heat loss to reduce body heat. Although the pig does not have effective sweat glands it can achieve evaporative cooling by wallowing. Similarly, under commercial conditions the provision of water sprays or drips, which wet the pig, can provide the same advantages.

As a consequence of these interacting factors water demand may be very variable. The data of Deinum (cited by CSIRO, 1987) demonstrated the variability which occurred under high temperature conditions in South Australia (Table 13.7).

**Table 13.7** DAILY WATER USAGE (LITRE/PIG) OF PIGS KEPT UNDER CONFINED AND DRY FEEDING CONDITIONS IN A SOUTH AUSTRALIAN PIGGERY OF 2000 SOWS

| | *Month* | | |
|---|---|---|---|
| | *Nov 1981* | *Dec 1981* | *Jan 1982* |
| Mean max. temperature (°C) | 25 ± 7 | 26 ± 5 | 29 ± 7 |
| Mean daily water intake (litre/pig) | | | |
| Weaners | – | 9 ± 1 | 9 ± 1 |
| Growers | 11 ± 1 | 12 ± 1 | 12 ± 1 |
| Finishers | 14 ± 2 | 14 ± 2 | 14 ± 1 |
| Range max. temperature (°C) | 26 – 38 | 19 – 37 | 21 – 43 |
| Range daily water intake (litre/pig) | | | |
| Weaners | – | 8 – 11 | 8 – 10 |
| Growers | 9 – 14 | 10 – 14 | 10 – 14 |
| Finishers | 11 – 18 | 12 – 21 | 13 – 15 |

After Deinum in CSIRO (1987)

As mentioned previously the extent that increased water intake will contribute to cooling depends on water temperature. Water temperature will also affect the amount of water consumed. Thus at low ambient temperatures high water temperatures will encourage drinking behaviour whereas at high ambient temperatures the reverse is the case (Table 13.8).

## Factors affecting the supply of water to the pig

It is clear from the foregoing discussion that in situations where there is no limitation to the water supply of the pig it can cope adequately with a wide range of nutritional and environmental conditions. However, in practice the supply of water to the pig may pose a limitation to the pig's ability to maintain its water balance and hence adversely affect its performance.

**Table 13.8** EFFECT OF AMBIENT TEMPERATURE AND DRINKING WATER TEMPERATURE ON WATER INTAKE (litres) OF PIGS FROM 45–90 kg

| *Water temperature* | *House temperature* | |
|---|---|---|
| | *Cool 22°C* | *Hot 35/25°C* |
| 11°C | 3.3 (1.5)[b] | 10.5 (6.0) |
| 30°C | 3.9 (2.3) | 6.6 (3.8) |

After Vajrabukka *et al.* (1981) cited in CSIRO (1987)
[a] Temperatures fluctuated every 12 h
[b] Values in parenthesis are litre/kg feed consumed

It is convenient to group those situations in which water supply has, or may, become inadequate into two categories.

(1) Those systems or production practices which *by design* impose limitations on *how much* water the pig may drink and *when* they may drink it.
(2) Those systems in which there *appears* to be an unrestricted water provision but where physical, environmental or social factors render the supply inadequate.

## INTENTIONAL RESTRICTION OF WATER SUPPLY

There are several different feeding systems which either restrict water intake to certain periods of the day, or alternatively provide the pig with its daily water input mixed with its food (liquid feeding). Some systems ostensibly allow the pig to select its own, preferred, water to feed ratio by operating a water valve in the feed trough. Many systems of this latter type deny the pig a supply of water uncontaminated by feed and as a consequence can impose is imposed by liquid feeding systems where no alternative water supply is provided. Such systems fail to recognize that a major function of water is the maintenance of mineral and temperature homeostasis. Because of the individual biological variation and the additive nature of the factors which determine water demand described above, it is impossible to anticipate demand accurately. Consequently, it is not possible to specify a water to feed ratio that can be guaranteed to satisfy the pig's needs. Therefore, pigs should always have free access to an unrestricted supply of water.

## UNINTENTIONAL RESTRICTION OF WATER SUPPLY

When water is deliberately withheld or restricted, as in the cases above, it is easy to see how changes in the pig's diet or health might increase the demand for water and change a previously adequate supply into an inadequate one. However, when presented with pigs which are apparently supplied with water a*d libitum* through bowls or drinkers it is more difficult to accept that the water supply may be a limiting factor. In just the same way that a chemical analysis of lysine does not tell us how much of that amino acid is available to the animal, so the mere presence of a watering device in a pen tells us nothing about the availability of water to the pig. It is generally assumed that voluntary intake is a satisfactory guide to the quantity of fluid the body needs and that thirst is a sufficiently strong drive to motivate an animal to obtain the water it requires if a source is available to it. This is not always the case.

For example, humans suffering from water deprivation, as in heat exhaustion or when sweating rapidly due to intense activity, have to be persuaded to drink sufficient water to balance water losses from the body (Lloyd *et al.*, 1978). Sheep and, more particularly, goats will only drink water at certain temperatures and from some utensils. If such a supply is not available they will refuse water to the point of death. The pig is not usually as extreme in its reactions and will generally consume sufficient water to keep it alive. However, the pig cannot be relied upon to consume enough water to maximize biological performance.

Two factors appear to be particularly important in influencing the pig's willingness to drink, namely, water quality and water delivery rate. Unfortunately, we are only just beginning to understand how these factors influence water consumption and pig performance and how they interact. Nevertheless a few points are worth noting at this time.

### *Water quality*

Although the pig gives the impression that it is not too fastidious in its eating and drinking habits this is somewhat illusory. When given the choice between drinking from a clean water bowl or a nipple drinker pigs preferred to drink from a bowl (Figure 13.1). However, as soon as the bowl became contaminated with food from

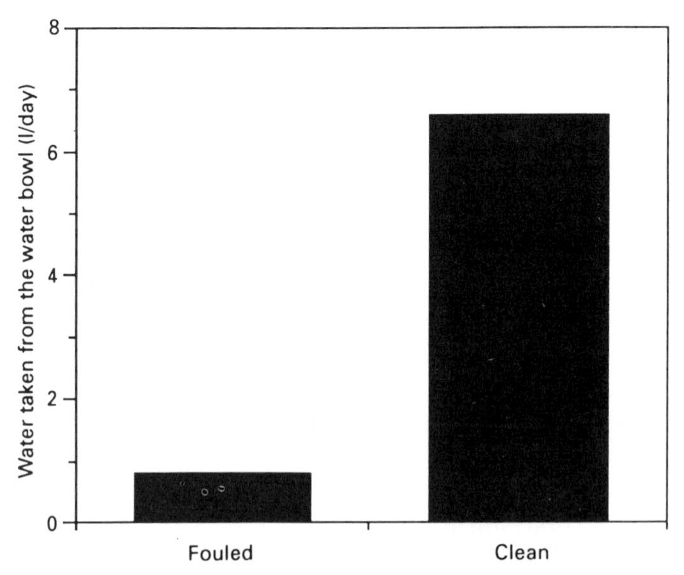


**Figure 13.1** Effect of fouling of water bowl on water consumption (Carpenter and Brooks, 1989, unpublished data): in his study pigs were given the choice of drinking from a nipple waterer or from a bowl drinker. The data show the extent the bowl drinker was discriminated against as soon as it became fouled. In this study fouling with feed resulted in as much discrimination against the bowl drinker as did contamination with faeces or urine

their mouths they preferred to use the nipple drinker. It has been well established that water with a high mineral content or containing toxic substances either of mineral or biological origin (e.g. toxins from blue green algae) will reduce pig performance (NAC, 1974). In the case of heavily contaminated water (e.g. very saline water) a vicious circle is set up in which the pig consumes more water in an attempt to detoxify itself and eventually builds up the offending material to a toxic level. Thus, contaminants, depending on what they are, can both depress and increase water intake.

Recommended limits to the concentration of minerals in water are given in Table 13.9. Ideally water for pigs should be clean, unpolluted, and with just enough mineral content to give it a good taste.

**Table 13.9** RECOMMENDED LIMITS OF CONCENTRATION (mg/ml) OF SOME POTENTIALLY TOXIC SUBSTANCES IN WATER

| *Item* | *CSIRO* (1987) | *NAC* (1974) | *Item* | *CSIRO* (1987) | *NAC* (1974) |
|---|---|---|---|---|---|
| Total dissolved salt | 6000 | 5000 | Chromium | 1.0 | 1.0 |
| Bicarbonate ($HCO_3^-$) | 1000 | – | Cobalt | – | 1.0 |
| Calcium ($Ca^{2+}$) | 1000 | – | Copper | 0.5 | 0.1 |
| Fluoride ($F^-$) | 2 | 2 | Lead | 0.5 | 0.5 |
| Magnesium ($Mg^{2+}$) | 400 | – | Mercury | 0.002 | 0.01 |
| Nitrate ($NO_3^-$) | 500 | 100 | Nickel | 5.0 | 1.0 |
| Nitrite ($NO_2^-$) | 100 | 10 | Vanadium | – | 0.1 |
| Sulphate ($SO_4^{2-}$) | 1000 | – | Zinc | 20.0 | 25.0 |
| Arsenic | 0.5 | 0.2 | Selenium | 0.02 | – |
| Cadmium | 0.01 | 0.05 | | | |

The pH of the water appears to have little direct effect on consumption. To date studies at Seale-Hayne have shown no reduction in water intake as a result of the inclusion of lactic acid in water (to a pH of 3.23) (Table 13.10). Nevertheless, caution must be exercised when adding acids or antimicrobials to water supplies as they can result in the detachment of organic matter from the delivery system, thereby producing complete or partial blockage of drinkers which if unnoticed can lead to 'water starvation'.

**Table 13.10** EFFECT OF LACTIC ACID AND LACTIC ACID PLUS FLAVOUR ON WATER INTAKE OF WEANERS FROM 21 TO 35 DAYS OF AGE

| | *Control* | *Lactic acid* | *Lactic acid + flavour* | $SE_D$ |
|---|---|---|---|---|
| Weaning wt (kg) | 6.32 | 6.01 | 5.72 | 0.31 |
| Daily feed intake (g) | 271 | 280 | 269 | 8.99 |
| Daily gain (g) | 176 | 178 | 178 | 10.04 |
| Daily water intake (g) | 728 | 767 | 666 | 55.0 |
| FCR | 1.59 | 1.73 | 1.69 | 0.10 |
| g water per g feed | 2.69 | 2.74 | 2.48 | – |
| Scouring days per pig | 0.19 | 0.17 | 0.29 | – |

Brooks, Parkins and Russell, unpublished data

### *Water delivery rate*

The other factor which is proving to have a considerable influence on water consumption is water delivery rate. It has been assumed in the past that the pig consumes food and then drinks sufficient water to metabolize its food and detoxify itself. The data in Tables 13.11 and 13.12 suggest that the reverse may be the case, namely, that the availability of water influences the amount of water the pig consumes and that this in turn affects its voluntary feed intake and its subsequent performance. In the Seale-Hayne study (Table 13.11) the pigs were in their thermoneutral zone and performance improved with increasing water intake. A feature of the trial was the very short time for which pigs were prepared to drink each day. Clearly the pigs on the most restricted water delivery rate were not prepared to extend their drinking time in order to obtain a greater water intake. It is interesting to speculate whether the time that they are prepared to spend drinking is conditioned by the time they have been spending nursing the sow preweaning and whether their intake would be increased if they were all able to drink at the same time. These possibilities are currently under investigation.

**Table 13.11** EFFECT OF WATER DELIVERY RATE ON WATER INTAKE AND PERFORMANCE OF WEANED PIGS FROM 3–6 WEEKS OF AGE

| | *Water delivery rate* (cm$^3$/min) | | | |
|---|---|---|---|---|
| | *175* | *350* | *450* | *700* |
| Water intake (litre/day) | 0.78$^d$ | 1.04$^c$ | 1.32$^b$ | 1.63$^a$ |
| Feed intake (g/day) | 303$^c$ | 323$^b$ | 341$^a$ | 347$^a$ |
| Daily gain | 210$^c$ | 235$^b$ | 250$^a$ | 247$^a$ |
| FCR (kg feed/kg liveweight gain) | 1.48 | 1.39 | 1.37 | 1.42 |
| Apparent time spent drinking (min/day) | 4.46$^b$ | 2.97$^a$ | 2.93$^a$ | 2.32$^c$ |
| Time required to consume 1.63 litre (min) | 9.31 | 4.66 | 3.62 | 2.32 |

(Barber *et al.*, 1989)
Pigs housed at 28°C
Means with different superscript letters are significantly ($P$<0.005) different

**Table 13.12** EFFECT OF WATER DELIVERY RATE AND ENVIRONMENTAL TEMPERATURE ON WATER INTAKE AND PERFORMANCE OF WEANED PIGS FROM 10–14 WEEKS OF AGE

| *House temperature* (°C) | | *5* | | | *35* | |
|---|---|---|---|---|---|---|
| Water delivery rate (ml/min) | 100 | 600 | 1100 | 100 | 600 | 1100 |
| Water intake (litre/day)$^a$ | 3.26 | 4.43 | 4.62 | 3.13 | 8.02 | 10.83 |
| Feed intake (kg/day)$^b$ | 2.24 | 2.04 | 2.18 | 0.74 | 1.12 | 1.09 |
| Daily gain (g)$^c$ | 855 | 774 | 730 | 278 | 384 | 466 |
| FCR (kg feed/kg liveweight/gain) | 2.62 | 2.64 | 2.99 | 2.66 | 2.92 | 2.34 |
| Drinking time (min/day)$^a$ | 32.6 | 7.4 | 4.2 | 31.3 | 13.4 | 9.9 |

After Nienaber and Hahn (1974)
[a] Difference due to house temperature and water delivery rate significant at $P < 0.05$
[b] Difference due to house temperature significant at $P < 0.01$

In the study by Nienaber and Hahn (1974), which involved somewhat older pigs, drinking times were extended. Despite this and high ambient temperatures they still would not compensate for the restricted intake by increasing their drinking time. At the low environmental temperature (5°C) water intake was greatest at the highest water delivery rate. These animals had depressed performance due to reduced food intake and the need for energy to raise the temperature of the water consumed to body temperature. These two studies suggested that in newly weaned pigs the mechanisms controlling water balance and thereby influencing water intake are not fully developed. Consequently the pig's behavioural characteristics and the design and operation of the water delivery system can become the limiting factor to performance. If dietary factors increase the water demand of the pig beyond its ability or willingness to obtain the required water intake, food intake will be depressed and performance will suffer.

It would appear that not only is the quantity of water the pig consumes per day important but also when the pig receives it. In a trial with scale fed growing pigs (32 kg) it was found that providing one or two drinkers per pen of eight pigs had no effect on their water consumption, while increasing the water delivery rate from 300 to 900 $cm^3$/min increased water intake by 80% (Table 13.13). In this study more water was 'used' but it is not clear whether this represented additional consumption. Certainly the apparent increase in water intake had no beneficial effect on pig performance. This is in marked contrast to a study conducted with liquid fed pigs (Table 13.14).

In this study pigs were given different water to feed ratios and were provided with an additional supply of water via drinkers. As the water to feed ratio increased so too did the biological performance of the pig. In a subsequent digestibility study (Table 13.15) it was found that dry matter digestibility increased as the water to feed ratio was increased. These findings suggest that dilution of the digesta has a favourable effect on food utilization. It has generally been assumed that the advantages in feed economy claimed for liquid feeding systems were derived from reduction of waste. However, these data suggest that they may also come from improved food utilization. What is not yet clear is whether the performance of pigs on a dry feed system would be improved if they took the same amount of water immediately after feeding or indeed whether they could be persuaded to consume such a large quantity of water immediately around the time of food ingestion.

**Table 13.13** EFFECT OF DELIVERY RATE AND DRINKER NUMBER ON WATER USE BY GROWING PIGS

| *Water delivery rate* ($cm^3$/min) | *300* | | *900* | |
|---|---|---|---|---|
| Number of drinkers | 1 | 2 | 1 | 2 |
| Water intake (litre/day)[a] | 2.0 | 1.7 | 3.6 | 4.0 |
| Feed intake (kg/day) | 1.69 | 1.67 | 1.68 | 1.67 |
| Daily gain (g) | 724 | 696 | 701 | 695 |
| FCR (kg feed/kg liveweight/gain) | 2.0 | 2.4 | 2.4 | 2.4 |
| Drinking time (min/day)[b] | 6.7 | 5.7 | 5.0 | 4.4 |

(Barber *et al.*, 1988)
[a] Differences between delivery rates significant at $P < 0.01$
[b] Differences between delivery rates significant at $P < 0.01$

**Table 13.14** VOLUNTARY WATER USE AND PERFORMANCE OF GROWING PIGS OFFERED LIQUID DIETS AT DIFFERENT WATER TO MEAL RATIOS

| | *Water to meal ratio* | | | |
|---|---|---|---|---|
| | *2:1* | *2.5:1* | *3:1* | *3.5:1* |
| Meal intake (kg/day) | 1.48 | 1.49 | 1.46 | 1.47 |
| Voluntary water use (kg/day) | 1.26[a] | 0.78[b] | 0.44[c] | 0.24[d] |
| Total water use (kg/day) | 4.23[a] | 4.51[b] | 4.86[c] | 5.40[d] |
| Daily gain (kg/day) | 0.73[a] | 0.74[a] | 0.75[a,b] | 0.77[b] |
| FCR (kg feed/kg liveweight/gain) | 2.01 | 2.00 | 1.95 | 1.90 |
| Water to feed ratio (w/w) | 2.97 | 3.12 | 3.36 | 3.68 |

(Gill, Brooks and Carpenter 1986)
(Means with the same superscript do not differ significantly at $P < 0.05$)

**Table 13.15** EFFECT OF WATER TO FEED RATIO ON DIET DIGESTIBILITY

| | *Water to feed ratio* | | | |
|---|---|---|---|---|
| | *2:1* | *2.67:1* | *3.33:1* | *4:1* |
| Dry matter digestibility (%) | 79.12[a] | 77.78[a] | 80.34[a,b] | 82.93[b] |
| Estimate DE (MJ/kg/DM) | 15.16 | 14.96 | 15.41 | 15.80 |
| Nitrogen retention (g/kg $W^{0.75}$/day) | 1.49 | 1.40 | 1.63 | 1.74 |

(Barber, Carpenter and Brooks, 1987, unpublished)
[a,b] Means with the same superscript do not differ significantly at $P < 0.05$

## *Limitations imposed by water delivery systems*

It is clear from the foregoing discussion that:

(1) water delivery rate has a considerable influence on performance,
(2) pigs will invest relatively little time in drinking behaviour.

It follows that any delivery system which fails to ensure that pigs can obtain water in the quantity they want at the time that they want it are likely to have a deleterious effect on pig performance.

Three problems occur frequently in the UK.

(1) Inadequate delivery rates in weaner accommodation due to a combination of low head (pressure), small diameter delivery pipes and small drinker orifice sizes.
(2) A combination of both excessive and inadequate delivery rates in finishing houses. This problem stems from the provision of long pipe runs in large houses. Drinkers in pens near the header tank frequently have excessive delivery rates producing high levels of spillage, excessive effluent production

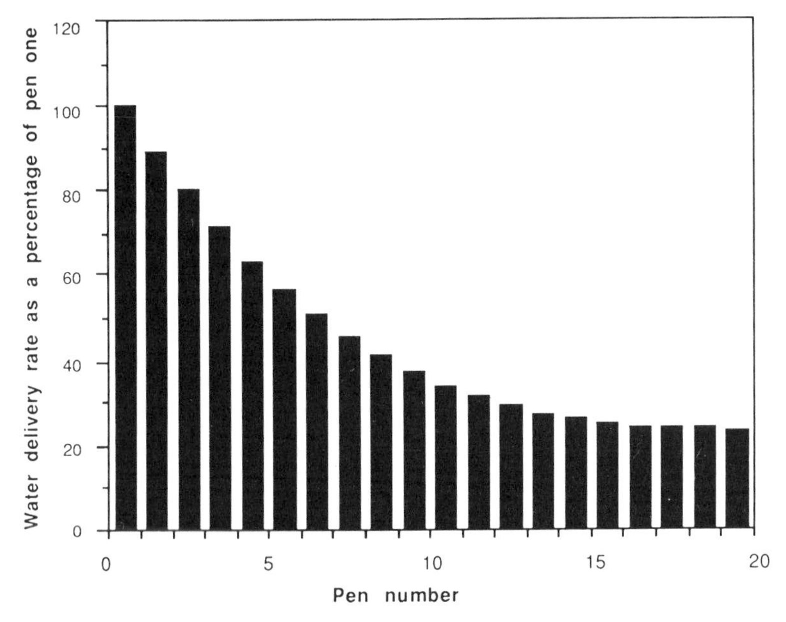


**Figure 13.2** Loss of water delivery rate down the length of a hose due to friction losses in the pipe (Barber, 1989, personal communication)

and wet pig environments. In pens furthest from the drinkers flows are often inadequate due to the loss of effective head due to friction losses in the delivery pipe and 'upstream' demand (Figure 13.2).

(3) Contamination of pipes/drinkers with solid material. This solid matter has a variety of origins. In some cases, particularly where bore holes and streams are the source of supply the material enters with the mains supply. In other cases the material results from ingress of dust and other solid matter through the header tank. Finally, the material may develop in the supply system. In hard water areas, pipes and drinkers may become mineralized. Galvanized pipes rust and the resultant material clogs filter gauzes on drinkers. When supplies are not cleaned regularly, bacteria, algae and slime moulds develop. Sheets of this material may detach and block drinker orifices.

In countries where there is no legal requirement for an air break between the pig and the incoming mains supply, water pressure is often high. This causes its own problems as drinkers frequently produce a jet of water of such velocity that it creates discomfort for the pig attempting to drink such that drinking is inhibited.

Irrespective of whether the water supply is high or low pressure, few units have a routine for cleaning water supply systems. The importance of water hygiene is not generally recognized and is certainly not acted upon. One suspects that many intransigent and difficult to diagnose problems of ill-health may have their genesis in contaminated water supply systems.

To overcome the problems listed above and to provide a controlled delivery rate at every drinker point requires a complete rethink of the plumbing system. This has been achieved in a new delivery system which has recently been introduced onto the market. The system (known as ZERH$_2$O PIPE) replaces the conventional header tank with a small sealed valve unit which provides an air break on the incoming main but prevents any ingress of solid material by the provision of a breather pipe with a filter. Large bore delivery pipes take over the water storage function and ensure that there are no friction losses even on very long pipe runs. Connections from the main delivery pipe to individual drinkers are via a push fit flexible down pipe which enables the easy insertion of a suitable water proportioner when the medication of an individual pen is indicated.

Using this system it is possible to select a drinker, and to achieve the desired delivery rate in a single pen and be confident that an equivalent delivery rate will be achieved in all other pens on the same line if similarly equipped.

## Conclusions

(1) The pig uses water for a variety of different functions. Because all of these functions cannot be anticipated and because many may be additive it is almost impossible to specify a water requirement in any given circumstance.

(2) It is clear that the nature of water provision can significantly affect performance. When water is withheld the animal loses its ability to detoxify itself. When water and feed are provided in a fixed ratio, as in liquid feed systems, performance will only be adequate if the food component creates a water demand less than that provided by the water component of the diet. In general, the water to feed ratios used in practice are inadequate to ensure this. Practical constraints on production units are such that ratios can never be set at a level which will ensure that the needs of all pigs within the unit are catered for. In order to safeguard the welfare of pigs all pigs should be provided with a water delivery system which provides an unrestricted supply of water at all times.

(3) Recent results indicate that considerable improvements in performance could be achieved on many commercial units by giving more attention to water management. They also highlight the important interaction between food and water and in so doing bring into question the validity of many nutritional studies where pigs have not been provided with an unrestricted supply of water. In particular they cast doubt on the results of numerous digestibility studies where pigs have been subjected to fixed (and on the basis of recent evidence) inadequate water supplies.

(4) There is now evidence that in many circumstances the pig does not, of its own volition, consume sufficient water to maximize its biological performance. Consequently, it is vital to engineer water delivery systems which do not contribute to an underconsumption of water and to develop management practices which encourage the pig to consume sufficient water.

(5) It is apparent that the inclusion of antimicrobials, to improve the microbiological quality of water, may be beneficial. Water is not a single commodity. The flavour and acceptability of waters, even of an appropriate microbiological quality, differ greatly. Consequently, there are considerable opportunities for improving the consumption of some water by the use of flavours and flavour enhancers to mask less pleasant water flavours. For this reason alone, if for no other, any moves by EC legislators to ban the addition of materials to water must be strenuously opposed. The proper use of water additives to improve water quality, to remove natural limitations to

consumption and to provide a vehicle for the provision of micronutrients and therapeutic drugs should be seen to have the potential to make a major contribution to pig health and welfare.

## References


Adolph, E. F. (1933). *Physiological Reviews,* **13**, 336–371

Agricultural Research Council (1981). *The Nutrient Requirements of Pigs.* Commonwealth Agricultural Bureaux, Farnham Royal, Slough

Appleby, M. C. and Lawrence, A. (1987). *Animal Production,* **45**, 103–110

Barber, J., Brooks, P. H. and Carpenter, J. L. (1988). *Animal Production,* **46**, 521 (abstract)

Barber, J., Brooks, P. H. and Carpenter, J. L. (1989). In *The Voluntary Food Intake of Pigs*, pp. 103–104. Ed. Forbes, J. M., Varley, M. A. and Lawrence, T. L. J. BSAP, Occasional Publication No. 13

Braude, R. and Johnson, B. C. (1953). *Journal of Nutrition,* **49**, 505–512

Brody, S. F. (1964). *Bioenergetics and Growth.* Hunter, New York

Castle, E. J. and Castle, M. E. (1957). *Journal of Agricultural Science, Cambridge,* **49**, 106–112

Chew, R. M. (1965). In *Physiological Mammalogy. Volume II Mammalian Reaction to Stressful Environments*, pp. 44–149. Ed. Mayer, M. V. and Van Gelder, R. G. Academic Press, New York

Close, W. H., Mount, L. E. and Start, I. B. (1971). *Animal Production,* **13**, 285–294

CSIRO (1987). *Feeding Standards for Australian Livestock: Pigs*, Commonwealth Scientific and Industrial Research Organisation, East Melbourne

Cooper, P. H. and Tyler, C. (1959a). *Journal of Agricultural Science, Cambridge,* **52**, 332–347

Cooper, P. H. and Tyler, C. (1959b). *Journal of Agricultural Science, Cambridge,* **52**, 348–351

Cronin, G. M., Var Tartwijk, J. M., Vander Hel, W. and Verstegen, M. W. A. (1986). *Animal Production,* **42**, 257–268

Feddes, J. J. R., Young, B. A. and DeShazer, J. A. (1989). *Applied Animal Behaviour Science,* **23**, 215–222

Gill, B. P. (1989). *Water use by pigs managed under various conditions of housing, feeding and nutrition.* PhD Thesis, Polytechnic South West, Plymouth

Gill, B. P., Brooks, P. H. and Carpenter, J. L. (1986). *Proceedings BSAP Occasional Meeting, Pig Housing and the Environment*, Stoneleigh

Gyrd-Hansen, N. (1972). *Medlemsblad Danske Dyrlaegeforening,* **114**, 631–635

Hagsten, I. and Perry, T. W. (1976). *Journal of Animal Science,* **42**, 1187–1190

Holme, D. W. and Robinson, K. L. (1965). *Animal Production,* **7**, 377–384

Holmes, C. W. and Mount, L. E. (1967). *Animal Production,* **9**, 435–452

Ingram, D. L. (1964). *Research in Veterinary Science,* **5**, 357–364

Ingram, D. L. (1967). *Journal of Comparative Pathology,* **77**, 93–98

Kornegay, E. T. and Graber, G. (1968). *Journal of Animal Science,* **27**, 1591–1595

Kornegay, E. T. and Van der Noot, G. W. (1968). *Journal of Animal Science,* **27**, 1307–1312

Lloyd, L. E., McDonald, B. E. and Crampton, E. W. (1978). *Fundamentals of Nutrition*, 2nd Edn. W. H. Freeman & Co., San Francisco

McFarlane, W. V. (1976). In *Veterinary Physiology*, pp. 463–539. Ed. Phillis, J. W. Wright-Scientechnia, Bristol
Maynard, L. A., Loosli, J. K., Mints, H. F. and Warner, R. G. (1979). *Animal Nutrition*, 7th Edn. McGraw Hill, New York
Moritz, A. R. and Henriques, F. C. (1947). *American Journal of Pathology,* **22**, 695–720
Morrison, S. R. and Mount, L. E. (1971). *Animal Production,* **13**, 51–57
Morrison, S. R., Bond, T. E. and Heitman, H. (1967). *Transactions of American Society of Agricultural Engineers,* **10**, 691–696
Mount, L. E., Holmes, C. W., Close, W. H., Morrison, S. R. and Start, I. B. (1971). *Animal Production,* **13**, 561–563
NAC (1974). *Nutrients and Toxic Substances in Water for Livestock and Poultry.* National Academy of Sciences, Washington DC
Nienaber, J. A. and Hahn, G. L. (1984). *Journal of Animal Science,* **59**, 1423–1429
Pieterse, P. J. S. (1963). *South African Journal of Agricultural Science,* **6**, 47–54
Randall, J. M. (1983). *Journal of Agricultural Engineering Research,* **28**, 451–461
Robinson, K. L., Coey, W. E. and Burnett, G. S. (1953). *Chemistry and Industry,* **1**, 18–23
Roubicek, C. B. (1969). In *Animal Growth and Nutrition*, pp. 353–373. Ed. Hafez, E. S. E. and Dyer, I. A. Lea and Febiger, Philadelphia
Stahly, T. S. (1983). In *Fats in Animal Nutrition*, pp. 313–332. Ed. Wiseman, J. and Cole, D. J. A. Butterworths, England
Straub, G., Weniger, J. H., Tawfik, E. S. and Steinhauf, D. (1976). *Livestock Production Science,* **3**, 65–74
Taylor, D. J. (1979). *Pig Diseases*, 3rd Edn. Burlington Press (Cambridge) Ltd, Foxton, Cambridge
Wahlstron, C., Taylor, A. R. and Seerley, R. W. (1970). *Journal of Animal Science,* **30**, 368–373
Whittemore, C. T. and Elsley, F. W. H. (1979). *Practical Pig Nutrition*, 2nd Edn. Farming Press Ltd, Ipswich
Yang, T. S., Howard, B. and McFarlane, W. V. (1981). *Applied Animal Ethology,* **7**, 259–270
Yang, T. S., Price, M. A. and Aherne, F. X. (1984). *Applied Animal Behaviour Science,* **12**, 103–109

14

# WATER FOR PIGLETS AND LACTATING SOWS: QUANTITY, QUALITY AND QUANDARIES

D. FRASER[1], J. F. PATIENCE[2], P. A. PHILLIPS[1] and J. M. McLEESE[2]
[1]*Animal Research Centre, Agriculture Canada, Ottawa K1A 0C6 and* [2]*Prairie Swine Centre, Department of Animal and Poultry Science, University of Saskatchewan, Saskatoon S7N 0W0, Canada*

## Introduction

As a nutrient for domestic animals, water presents something of a paradox. On the one hand, it fulfills a broad range of physiological and chemical functions for which it is uniquely suited. As a solvent, water is the body's major transportation medium for nutrients, metabolites and waste products, as well as hormones and other chemical messengers. Having a high specific heat, water plays an important role in thermostasis. Water is a lubricant, and a component of such basic chemical reactions as oxidation, hydrolysis, and acid-base balance. In short, few nutrients play such a varied and fundamental role in the processes of life.

On the other hand, our knowledge of animal water requirements is surprisingly limited, and the subject continues to stimulate little research. Perhaps this is because water, as 'a clear, colourless, nearly odourless and tasteless liquid' (Morris, 1970) is too mundane to excite much curiosity. More plausibly, water is generally abundant, inexpensive, and not traded commercially; hence, some of the usual incentives for research are missing. Here the prospect for change is quite strong. Water is playing an increasing role in the delivery of nutrients and medications, and the cost of storing and handling waste water is becoming a serious concern. In sum, economically significant effects of water on animal production are being more widely recognized.

These impacts involve both the quality of drinking water and the quantity consumed. This chapter begins with an overview of water quality as it relates to pigs, and of the difficulties in establishing quantitative water requirements. This is followed by a review of research on lactating sows, suckling piglets, and newly-weaned piglets, emphasizing throughout the fragmentary nature of our knowledge.

## Establishing water requirements: quality

Of the 1400 million km$^3$ of water in the world, 97% is sea water; although unsatisfactory for immediate consumption, this represents an enormous reservoir fundamental to the global water ecosystem. A further 2.2% of the world's water supply is frozen in glaciers and polar ice caps, leaving only 0.8% available as ground or surface water (Price, 1985). In spite of these statistics, water supplies worldwide would be of much less concern if ground and surface water were distributed in proportion to the human population and if water quality were less variable.

The quality of ground and surface water (i.e. the degree of chemical, microbiological and other impurities) is highly variable, depending on many factors including aquifer depth, soil/rock formations, and climate (Price, 1985; Viessman and Hammer, 1985). Human influence is becoming an increasing concern, as the changing nature of water quality and quantity can often be directly attributed to human intervention at various points in the water cycle.

Water quality is usually evaluated by microbiological, physical and chemical criteria; within these categories, health and aesthetic standards may apply, with the latter being more stringent. Specific water quality standards, based primarily on safety and health, have been established for livestock. They provide useful guidance, but rest on relatively little definitive research.

The standard of water quality acceptable for livestock depends on many factors. The class of livestock is particularly important, with cattle generally less susceptible than poultry and pigs (NRC, 1974). An exception is susceptibility to nitrate poisoning, as cattle convert nitrates to the more toxic nitrites in the rumen. Age is also important, since younger animals require higher quality water than adults (NRC, 1974). Feed composition should be considered within the context of water quality; diets which protect against scouring, for example in weanling pigs, are also helpful if poor quality water is being used (McLeese, unpublished data). Ambient temperature is also a factor. For example, increased water consumption associated with thermal stress will concurrently increase the intake of water-borne minerals and may thereby encourage gastrointestinal distress. Most published standards either ignore the influence of species and age (Table 14.1), or provide very broad classifications (Table 14.2). Published standards for human consumption often include aesthetic considerations that are relevant only to humans; such standards are not generally applicable to livestock.

Notwithstanding these limitations, published guidelines serve a useful purpose if applied with appropriate discretion. Like traditional nutrient requirement tables, they provide a standard basis for discussion. However, it is our hope that recognition of their limitations will stimulate further research on the subject.

### MICROBIOLOGY

Water may contain a variety of microorganisms, including both bacteria and viruses. Of the former, *Salmonella* spp, *Vibrio cholera, Leptospira* spp and *Escherichia coli* are the most commonly encountered. Water can also carry pathogenic protozoa as well as eggs or cysts of intestinal worms. Consequently, contaminated water can represent a potential health risk, especially to young or otherwise susceptible pigs.

**Table 14.1** CANADIAN WATER QUALITY GUIDELINES FOR LIVESTOCK

| *Item* | *Maximum recommended limit* (ppm) |
|---|---|
| *Major ions* | |
| Calcium | 1000 |
| Nitrate + nitrite | 100 |
| Nitrite alone | 10 |
| Sulphate | 1000 |
| TDS | 3000 |
| *Heavy metals and trace ions* | |
| Aluminium | 5.0 |
| Arsenic | 0.5[a] |
| Beryllium | 0.1[b] |
| Boron | 5.0 |
| Cadmium | 0.0 |
| Chromium | 1.0 |
| Cobalt | 1.0 |
| Copper (swine) | 5.0 |
| Fluoride | 2.0[c] |
| Iron | no guideline |
| Lead | 0.1 |
| Manganese | no guideline |
| Mercury | 0.003 |
| Molybdenum | 0.5 |
| Nickel | 1.0 |
| Selenium | 0.05 |
| Uranium | 0.2 |
| Vanadium | 0.1 |
| Zinc | 50.0 |

Source: Task Force on Water Quality Guidelines (1987)
[a] 5.0 if not added to feed
[b] Tentative guideline
[c] 1.0 if fluoride present in feed

**Table 14.2** EVALUATION OF WATER QUALITY FOR PIGS BASED ON TOTAL DISSOLVED SOLIDS

| *Total dissolved solids* | | *Comments* |
|---|---|---|
| <1000 | Safe | No risk to pigs |
| 1–2999 | Satisfactory | Mild diarrhoea may occur in pigs not adapted to it |
| 3000–4999 | Satisfactory | May cause temporary refusal of water |
| 5000–6999 | Reasonable | Higher levels should be avoided by pregnant or lactating pigs |
| 7000–10 000 | Unfit | Risky for pregnant, lactating or young pigs, or those exposed to heat stress or water loss |
| >10 000 | Not recommended | |

Adapted from NRC (1974)

Iron bacteria can be most troubling. They use inorganic ferrous iron as an energy source, producing a reddish slime that plugs filters and can reduce well flow rates to barely perceptible levels. The foul odours associated with iron bacteria are believed to arise from the decomposition of dead bacteria.

### PHYSICAL STANDARDS

Standards for the physical properties of water often relate more to aesthetics than to health or safety, but some physical criteria can reflect quality. Colour is not itself a concern, but it may indicate contamination. For example, organic acids may impart a yellowish tinge. Turbidity, resulting from suspended colloidal solids may be due to harmless clay or silt particles, or it may result from waste material of human, animal or industrial origin, in which case the water may not be safe to drink. Likewise, flavours and odours signify contamination (organic or inorganic) since pure water is inert in this regard.

### CHEMICAL STANDARDS

Dissolved minerals and other chemical properties of water are of greatest relevance for evaluating water quality for livestock. The analytical procedures are generally straightforward, but establishing safe levels remains a problem.

Chemical evaluation of farm water typically begins with several primary tests (total dissolved solids, pH, iron, hardness, and nitrates/nitrites), capable of identifying, at least in general terms, the presence or absence of quality problems. If the results of these tests prove satisfactory further testing is unnecessary. If total dissolved solids are elevated, more detailed analyses (sulphate, chloride, sodium, potassium, calcium, magnesium and manganese) will specify the nature of the contamination.

The chemical composition of water from a given well or water body is quite consistent, and except in the case of inadvertent contamination or excessive pumping, will change very little over time (Van der Kamp, 1989). However, wells draining small reserves are more susceptible to change than those charged by large, deep aquifers.

Removal of water from the well and movement through the delivery system also have minimal effect. A survey of Saskatchewan pig farms, in which concurrent water samples were collected at both the well-head and in the barn, identified no major differences between the two sites (Patience *et al.*, 1989). The one exception in this study was iron, which presumably existed in the soluble ferrous ($Fe^{2+}$) state in the ground water but precipitated as the much less soluble ferric ($Fe^{3+}$) ion when pumped and thus exposed to oxygen. Madec (1987) has reported that when water high in nitrate remained in a feed trough for 12–15 h, nitrites and ultimately ammonium were formed. The potential for change therefore exists, but for the most part, these changes relate to specific circumstances (high iron, high nitrate, etc.).

#### *Total dissolved solids (TDS)*

TDS measures the sum of inorganic matter dissolved in a sample of water. Standards of water for pigs, based on TDS, appear in Table 14.2. Water is considered safe if the TDS is less than 1000 ppm, and unfit at levels of TDS greater than 7000 ppm (roughly one-fifth the level found in sea water). Between these two extremes, the effect on pigs is unpredictable. Some farmers and veterinarians report economically relevant losses at levels well below 7000 ppm, while other farms experience transient or minor inconvenience at worst (McLeese *et al.*, 1989). Some of this variation is probably due to the wide range of minerals which contribute to TDS, as different minerals have very different physiological effects on the pig. This underlines the need for more detailed analysis if TDS is greater than 1000 ppm. Water with high levels of TDS can impair pig performance, often in association with diarrhoea, as discussed below. Conductivity (the capacity of the water to conduct an electrical current) can be used to estimate TDS, but the exact mathematical conversion varies depending on water type.

### *pH*

The pH of ground water is rarely a problem in itself, since it normally falls within the acceptable range 6.5–8.5. However, it can have an impact on chemical reactions involved in the treatment of water; high pH impairs the efficiency of chlorination, and low pH may cause precipitation of some antibacterial agents delivered via the water system. Sulphonamides are particularly at risk (Russell, 1985); this could possibly lead to carcass sulpha residues, since precipitated medication in the water lines may leach back into the water after medication has been terminated.

### *Alkalinity*

Alkalinity not considered undesirable by itself, is a measure of the water's capacity to absorb hydrogen ions (acid) without a major change in pH. Although expressed in terms of $CaCO_3$, it is the sum of bicarbonate, carbonate and hydroxide.

### *Hardness*

Hardness is caused by multivalent metal cations, usually calcium and magnesium. Although it has no effect on safety, hardness does impair the cleansing ability of water and also results in the accumulation of scale, as $Mg(OH)_2$ and $CaCO_3$, which can impair water delivery and treatment equipment. Hardness is generally calculated as the sum of calcium plus magnesium, expressed as $CaCO_3$ equivalents. Water is considered soft if hardness is below 50 ppm, and very hard if above 300 ppm.

### *Chloride*

Chloride is generally a less common anion in groundwater than sulphate, although local conditions may result in elevated levels. At concentrations above 250–500 ppm, a brackish taste may develop.

### *Iron*

Iron is not itself a safety problem, but it can encourage the growth of iron bacteria, as discussed above. If precipitated in the oxidized ferric state it may plug fine screens as in drinkers. Ferric iron poses no health or safety risk to the pig, but is visually unappealing.

*Sulphates*

These are of particular concern in pig production. Magnesium sulphate (Epsom salts) and sodium sulphate (Glauber's salts) are laxative and cathartic agents, and may give rise to diarrhoea. The problem is most acute in the newly weaned pig, which is already susceptible to diarrhoea, although cases of adult diarrhoea associated with sulphate waters have been reported.

*Manganese*

Manganese poses no health or safety risk to the pig, but like iron, will precipitate upon oxidation, possibly leading to problems in the water delivery system. Unlike iron, a pH greater than 9.5 favours oxidation, and few water supplies will have such a high pH.

*Nitrates and nitrites*

These have given more concern in pig production than they probably deserve. They are often a man-made problem, formed by the decomposition of organic material. Although nitrates are a major concern for human infants and for ruminant livestock, pigs appear to be much less susceptible. Indeed, many experiments investigating the effect of nitrates on pigs have found little problem, even at levels much higher than normally observed in groundwater supplies (Garrison *et al.*, 1966; Anderson and Stothers, 1978). The results of one such experiment are summarized in Table 14.3.

**Table 14.3** EFFECT ON PIGS OF NITRATES AND NITRITES IN WATER

| | *Nitrates* (ppm) | | | | *Nitrites* (ppm) | | | |
|---|---|---|---|---|---|---|---|---|
| | *0* | *750* | *1500* | *3000* | *0* | *200* | *400* | *800* |
| Average gain (g/day) | 730 | 748 | 649 | 630 | 730 | 549 | 576 | 499 |
| Feed:gain | 3.56 | 3.59 | 3.58 | 3.82 | 3.53 | 3.43 | 3.71 | 3.53 |
| Serum vitamin A (μg/100 ml) | 39.5 | 47.1 | 40.2 | 22.2 | 24.0 | 21.3 | 21.3 | 17.7 |
| Liver vitamin A (μg/g) | 21.8 | 11.8 | 11.8 | 5.8 | 30.9 | 16.2 | 9.6 | 8.7 |

Source: Garrison *et al.* (1966)

STRATEGIES FOR POOR QUALITY WATER

Only some water quality problems can be resolved by water treatment. For example, water can be treated to remove iron and reduce hardness, but no cost-effective methods are available to remove inorganic sulphate.

Water softeners of many types are available, the most common being the ion-exchange unit. The cation resin bed replaces calcium and magnesium with sodium, thereby reducing the hardness of the water, but clearly having no effect on TDS. Because of sodium's role in the process, water softened in this manner will have elevated levels of sodium.

Chlorination is used to disinfect water, with bacteria controlled more effectively than protozoa and enteroviruses. Chlorination of water results in the formation of hypochlorous acid (HOCl), an effective disinfecting agent. Unfortunately, hypochlorous acid dissociates to form hypochlorite ion, a less effective disinfectant. This reaction is heavily influenced by the pH of the water as shown in Figure 14.1.

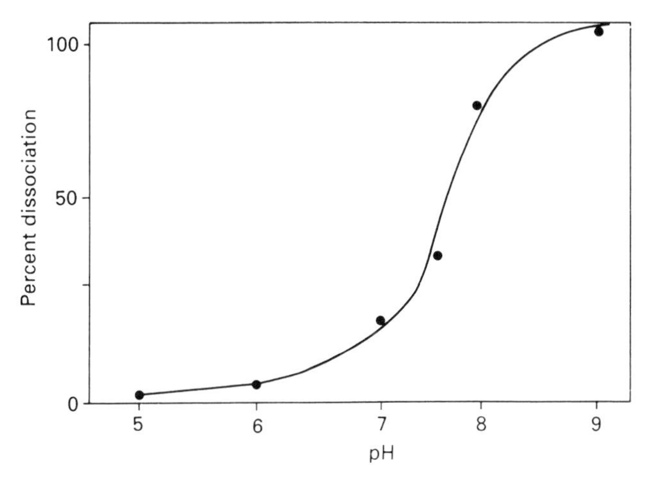


**Figure 14.1** Influence of pH on the dissociation of hypochlorous acid to hypochlorate ion. As pH rises, the proportion of undissociated acid declines

The pH of groundwater often lies on the steepest part of this dissociation curve; hence small differences in pH can exert a large influence on the effectiveness of chlorination. Other factors that influence the antibacterial effect of chlorination include the presence of organic matter in the water (which impairs disinfection), water temperature, and the time of contact between the microorganism and the chlorine.

'Shock chlorination' has also been practised, especially as a means of controlling iron bacteria in farm wells where flow rates have been severely restricted by microbial growth. A high concentration of chlorine, such as in bleach, is introduced into the well and delivery system and left to stand overnight; the entire system is then flushed thoroughly to remove residual chlorine. Useful in some instances, shock chlorination has often proven ineffective in the more serious cases of bacterial loads.

Chlorine dioxide may also be used for disinfection. It has a number of advantages over chlorine, including less dependence on pH and less reactivity with ammonia and other nitrogenous compounds. It is a strong bactericide and viricide. However, it is costly and tends to form chlorate and chlorite residuals which are considered toxins. Chlorine dioxide is available in commercial disinfectants.

The removal of sulphates from water cannot be accomplished by ion exchange. The options available are distillation, electrodialysis or reverse osmosis. At present, none of these is economical under our agricultural conditions. Reverse osmosis has been evaluated, but the capital costs combined with operating costs prove to be prohibitive; in Canada, the total of fixed and variable costs of reverse osmosis is about Cdn $20 (£10 sterling) per pig sold.

Diet adjustment is a common response to poor quality water. The most common approach is removing the salt (NaCl). However, caution is advised. Most saline water contains small quantities of chloride, compared with sodium, so that salt removal may result in a primary chloride deficiency which in turn impairs appetite. Hence it is not surprising that this method generally reduces the incidence of scouring, but at the cost of impairing growth rate. Other nutritionists have reduced the nutrient density of the diet, since high-energy, low-fibre diets tend to generate loose stools in young pigs. Here too, the solution may have a greater effect on performance than the original problem, and therefore is only advised in the most serious cases.

As noted above, certain features of poor quality water can impair the water delivery system. Hardness, iron, iron bacteria and perhaps other factors, can plug nipple drinkers, filters, meters and other components of the water system. Consequently, careful maintenance of the delivery system is particularly important if water quality is poor. Although hardness is often cited as a potential concern, it is probably an exaggerated one; other aspects of water quality, including iron and iron bacteria, pose a much greater challenge for maintaining adequate flow rates.

## Establishing water requirements: quantity

Our poor understanding of the pig's quantitative water requirements is due in part to the fact that the conventional methods of establishing nutrient requirements are not easily applied, or not often applied, to water.

For other nutrients, we often estimate requirements by feeding different levels and monitoring the animal's performance. This is rarely done with water, partly because an animal's water requirements vary widely depending on such factors as environmental temperature and the quantity and quality of feed consumed.

A second approach involves identifying a level of intake adequate to prevent deficiency symptoms. This approach has been used to diagnose extremes of water imbalance such as dehydration of lactating sows (Loje and Bing, 1951) and inadequate clearance of salt from the body (Smith, 1975). However, the approach offers little promise in general because the body will resort to physiologically costly mechanisms in order to prevent dehydration. For example, a lactating sow with limited access to water will evidently reduce her feed intake and catabolize body reserves to produce milk (Liebbrandt, 1989). This prevents dehydration, but at a price.

A third approach is to provide a choice of nutrients and determine the animal's preferred, voluntary level of intake. Many studies have used this approach implicitly by assuming that voluntary intake corresponds to requirements. This is particularly problematic in the case of water because, depending on circumstances, animals will often drink water in 'luxury' amounts, well beyond physiological need. Examples include hunger-induced drinking (Yang *et al.*, 1981) and 'schedule-induced polydipsia' (Stephens *et al.*, 1983). A further concern, related to research methods, involves the use of indwelling bladder (Foley) catheters. Although preferred for urine collection because of improved precision and sample quality, catheters have been shown to increase urine excretion and water intake (Patience *et al.*, 1987).

Pigs seem prone to excessive drinking in other, less specialized contexts. Sows in tether stalls often toy with the nipple drinker and may drink far more water than they require. In one study, the provision of straw bedding to tethered sows – which made them less restless and provided some recreation – greatly reduced the frequency of urination, probably because it reduced excessive drinking (Fraser, 1975). In another study, the simple addition of bulk (oat hulls) to a concentrated diet reduced sows' average daily urine output from about 17 kg to about 6 kg (Mroz *et al.*, 1986). Since little of this reduction could be attributed to increased faecal moisture loss, it presumably resulted from a reduction in excessive drinking.

These special cases of excessive drinking are probably related to some degree of frustration or mild stress (Falk, 1971; Stephens *et al.*, 1983; Mittleman *et al.*, 1986). However, in other contexts as well, animals often drink in the absence of apparent need (Epstein, 1982). Such drinking may be motivated by pleasant taste or temperature sensations of the liquid; it may occur from habit, in certain circumstances, or at certain times of day; or drinking may occur while the animal eats, depending on the sensory properties of the food. In all these cases, there is ample room for voluntary intake to become uncoupled from the homeostatic mechanisms regulating water balance.

A fourth approach for determining requirements is to provide water in a manner which is nominally *ad libitum* but with intake varying from animal to animal, depending on water quality, the dispensing apparatus, and other aspects of the physical and social environment. The resulting intakes can then be measured and correlated with aspects of animal health and performance or, alternatively, performance can be compared for different treatments. This approach, used in a number of recent studies, brings us closer to identifying real benefits, but there are obvious drawbacks. In particular, correlation studies are notoriously difficult to interpret, and simple, empirical comparisons of, for example, biological effects of two different water dispensing systems, may give results that apply in one situation but not another, for reasons that cannot readily be identified.

## Lactating sows

A half century ago, Garner and Sanders (1937) questioned whether some cases of milk shortage in lactating sows could be ascribed to inadequate water intake. It is embarrassing to admit that this apparently simple issue is still far from settled.

Most studies of water consumption by lactating sows have done little more than establish voluntary intake levels, but the results, summarized from 12 studies in Table 14.4, have yielded a number of insights. The lowest mean intake of 8.1 litres/day (Friend, 1971) was associated with unusually low dry matter intake in a feed selection experiment. The two highest values (25.1 and 19.9 litres/day) are based on the very limited studies of Riley (1978) and Bauer (1982). Most of the remaining means fall in the range of 13–19 litres/day, values lower than those cited in some advisory literature (Agriculture Canada, 1988; Midwest Plan Service, 1982).

The various studies showed that water intake is influenced by several factors. Water intake appears to increase with increasing solid food intake (Friend, 1971; Bauer, 1982) and with an increasing percentage of protein in the diet (Mahan, 1969). Water intake may (Mahan, 1969; Friend, 1971) or may not (Garner and Sanders, 1937) increase with advancing lactation, perhaps depending on trends in food intake. Environmental temperature appears to exert only a modest influence; Garner and Sanders (1937) and Lightfoot (1978) reported only slightly higher

**Table 14.4** MEAN WATER INTAKE (litres/day) BY LACTATING SOWS IN 12 STUDIES

| *Mean intake* (litre/day) | *Number of sows* | *Period of lactation studied* | *Reference* |
|---|---|---|---|
| 19.4 | 37 | Weeks 1–6 | Garner and Sanders, 1937 |
| 17–18[a] | 2 | Not clear | Kozhakhmetova, 1966 |
| 12.7 | 17[b] | Weeks 2–4 | Mahan, 1969 |
| 8.1 | 19[c] | Weeks 1–4 | Friend, 1971 |
| 18 | 40 | Weeks 1–3 | Lightfoot, 1978 |
| 25.1 | 4 | Not clear | Riley, 1978 |
| 15[d] | 12 | Weeks 1–3 | Friend and Wolynetz, 1981 |
| 19.9 | 2 | Weeks 1–4 | Bauer, 1982 |
| 17.7 | 24 | Weeks 1–3 | Lightfoot and Armsby, 1984 |
| 12.7[e] | 9 | Not clear | Diblik, 1986 |
| 14[f] | 40 | Days 4–14 | Fraser and Phillips, 1989 |
| 14[f] | 36 | Days 4–14 | Phillips *et al.*, 1990 |

[a] Includes water consumed in wet mash feed
[b] Average of 12 first-parity and five second-parity animals
[c] 12 gilts, seven of which were studied in a second lactation
[d] Includes a small amount of water consumed as 0.5% NaCl solution. Eight sows had supplementary NaCl in the diet and four did not
[e] Average over four types of water dispensers
[f] Lower values were obtained on days 1–3 of lactation

intake in summer than in winter, and Phillips *et al.* (1990) found only 17% greater intake at 28°C than at 20°C. The type of water dispenser can have a significant influence. Diblik (1986) reported the highest mean intakes from a bowl or open water surface (13.6 and 14.1 litres/day), less from a nipple drinker (12.4 litres/day) and least from a 'drinking straw' (10.8 litres/day). However the very high value of 25.1 litres/day reported by Riley (1978) was from a 'drinking straw', and the low mean value of 8.1 litres/day (Friend, 1971) was from a bowl.

While these systematic differences doubtless play a role, individual differences among animals account for a large part of the variation. In the study by Mahan (1969), the average daily intake of individual sows ranged from 9.3 to 21.5 litres/day; sows studied by Fraser and Phillips (1989) ranged from 6.1 to 21.7 litres/day; and Garner and Sanders (1937) reported the standard deviation for between-sow variation as 36% of the mean. Some of this variation appears to be driven by litter size and, hence, the piglets' presumed demand for milk (Mahan, 1969; Lightfoot and Armsby, 1984), but large differences between sows remain when litter size is relatively uniform (Fraser and Phillips, 1989). But do the lowest values mean that some sows under some conditions drink inadequate water to support a substantial lactation?

Figure 14.2 shows some relevant findings on water intake and piglet performance for 37 sows in farrowing crates equipped with nipple drinkers. After an average intake of 10 litres/day during the 4 days before farrowing, intake dropped precipitously on the day after farrowing, then gradually increased over 3–5 days to an average of about 14 litres/day which was maintained to the end of the study on day 14 of lactation. Water intake in lactation could thus be divided into two periods: the first few days after farrowing when intakes were low but increasing, and the later period of relatively stable intakes in the established lactation.

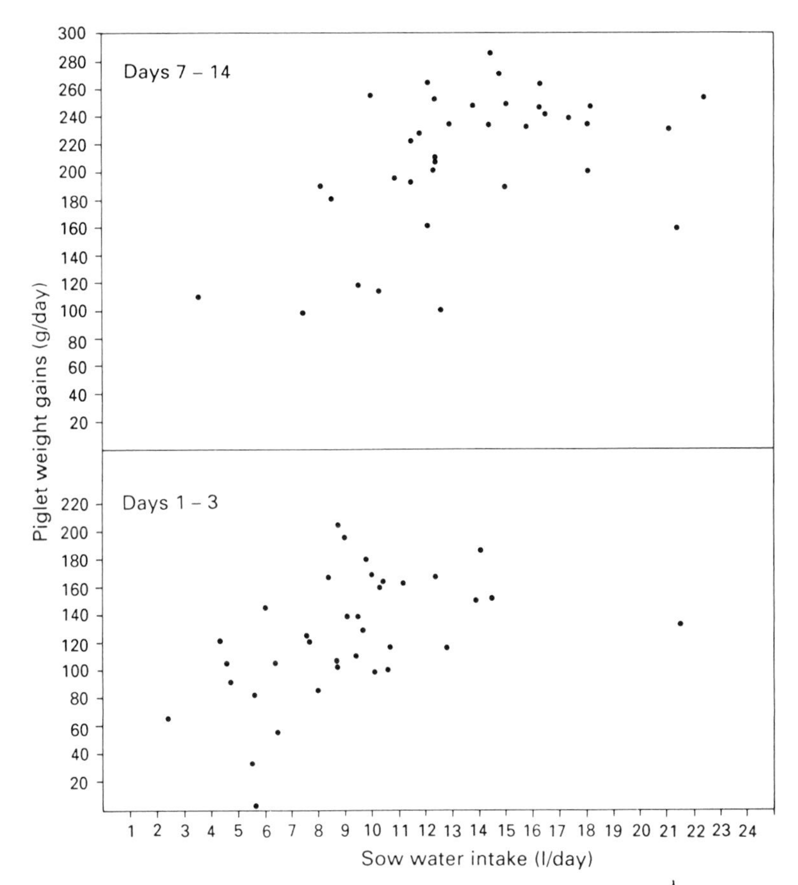


**Figure 14.2** Mean daily weight gains of 37 litters (g/day) in relation to daily water intake of the sow, shown for days 1–3 and days 7–14 of lactation (based on Fraser and Phillips, 1989)

In the second week, with lactation well established, there was little evidence that the sow's water intake was limiting lactation (Figure 14.2). Most of the sows averaged 10–20 litres/day, and most of the litters averaged a daily piglet weight gain of 200 g/day or more over this range of sow water intake. Five litters had distinctly inferior gains of 100–120 g/day, and their respective dams drank less than average, but in three of the five cases, the sow's water intake overlapped the range that supported normal gains in other litters.

Early lactation gave a different picture. During the first 3 days, many sows averaged less than 10 litres/day, and litter weight gains were significantly correlated with sow water intake within this range (see Fraser and Phillips, 1989, for details). This correlation could mean (i) that small or weak litters demanded little milk and stimulated little drinking by the sow, (ii) that both low water intake and low milk production were caused by a third factor, such as sow illness, but were not causally related, or (iii) that low water intake contributed to low milk production. The first two possibilities are unlikely to be a complete explanation because many of the sows actually drank less in early lactation than they had done before farrowing, and the low intakes were not strongly associated with illness in the sow or the size and weight of the litter. Furthermore, the litters with the lowest gains were clearly malnourished, with many piglets losing weight on at least one of the days.

Our interpretation of the data in Figure 14.2 is that once lactation was established, most sows achieved a water intake that met or exceeded their minimum requirement. In early lactation, however, some of these animals had a low water intake which was at least symptomatic of, and probably contributed to, inferior milk production.

What accounted for these low intakes? In this experiment, sows obtained water from a nipple drinker which delivered 0.7 litres/min and which, allowing for some spillage, required the sow to stand or sit for about 20 min/day in order to drink 10 litres. The sows with the lowest water intake on days 1–3 tended to be very inactive during this time, but even the least active animals assumed a standing or sitting position for 40–60 min/day, suggesting that water intake was not simply limited by the amount of time the sow spent active.

A suggested hypothesis is that water intake is linked to activity in a more subtle way. During pregnancy, confined sows with ample access to water probably drink 'luxury' amounts, mainly in two contexts: when eating, and as one of the few types of natural behaviour that can help to fill unoccupied active time. Just after farrowing, sows are naturally less active, and some may seem quite listless and lacking in appetite, especially in a warm environment. With a low food intake and low activity level, a sow is missing both of the major contexts that promoted drinking during gestation, and the animals may incur some degree of water deficit before adopting drinking habits suited to the new demands of lactation.

Ensuring that sows have easy access to water may help to avoid such problems, but sound recommendations are difficult to make, even for such elementary points as nipple drinker flow rate. In one study 36 sows were given a nipple drinker adjusted to deliver 0.6 litres/min, or a higher flow rate of about 2 litres/min, or a 'deluxe' treatment with two high-flow-rate nipples which allowed the sow to drink while lying as well as while standing (Phillips *et al.*, 1990). The three treatments had no impact on average water intake, although spillage was increased, especially in a hot environment, by easier access to water. Even drastic restriction of water flow rate sometimes has only modest effects on lactating sow performance. Hoppe *et al.* (1987a) housed 16 sows in crates with nipple drinkers delivering about 70 or 700 ml/min during a 3-week lactation. Sows with the lower flow rate had slightly (but not significantly) lower voluntary feed intake and greater weight loss in lactation, but there was no apparent effect on piglet survival or weight gain. In a more extensive study, Leibbrandt (1989) assigned 72 sows to either 70 or 700 ml/min. Sows with the low flow rate ate about 15% less feed and lost about three times more weight during lactation, but again survival and weight gain of piglets were seemingly unaffected.

Advisory literature often suggests a flow rate of 2 litres/min for lactating sows. Such a high target value may well help encourage producers to avoid harmfully low flow rates, and may be suitable if spillage and management of waste water is not a problem. In reality, however, research has not shown any greater water intake at 2 litres/min than at 0.6 litres/min, many sows appear to cope with even lower rates, and for specific sows with inadequate intake, it is questionable whether a very high flow rate will provide a solution. Thus, pending further research, flow rates of 0.6–1 litres/min would be difficult to fault.

Nonetheless, we must expect to see occasional baffling exceptional cases. Loje and Bing (1951) described a condition in lactating sows involving loss of appetite, reduced lactation and often death, which they attributed to inadequate water intake by sows in the established lactation, but no reasons for this inadequate intake were given. Another case study reported a herd where slow-flowing nipple drinkers were blamed for massive piglet mortality (Anonymous, 1987). In France, low water intake in pregnant sows appeared to be a contributing factor to urinary disorders (Tillon and Madec, 1984) and a water intake of 15 litres/day during pregnancy was advocated (Madec, 1984, 1985). In practice, therefore, it may be found that water intake interacts with other husbandry factors such as water quality and sanitary conditions to produce greater-than-average water requirements in specific cases.

## Unweaned piglets

Some pig producers provide supplementary water for piglets from birth, others begin supplying water at a later age, and others provide no supplementary water before weaning (but see Figure 14.3). This variety of practices reflects the lack of consensus on whether supplementary water is of benefit to suckling piglets.

One common belief is that provision of water to older piglets encourages the intake of solid 'creep' feed. Friend and Cunningham (1966) showed this to be true, but the difference was small until the fifth week of lactation. With 3- or 4-week weaning, the effect of water on creep-feed intake has yet to be shown clearly.

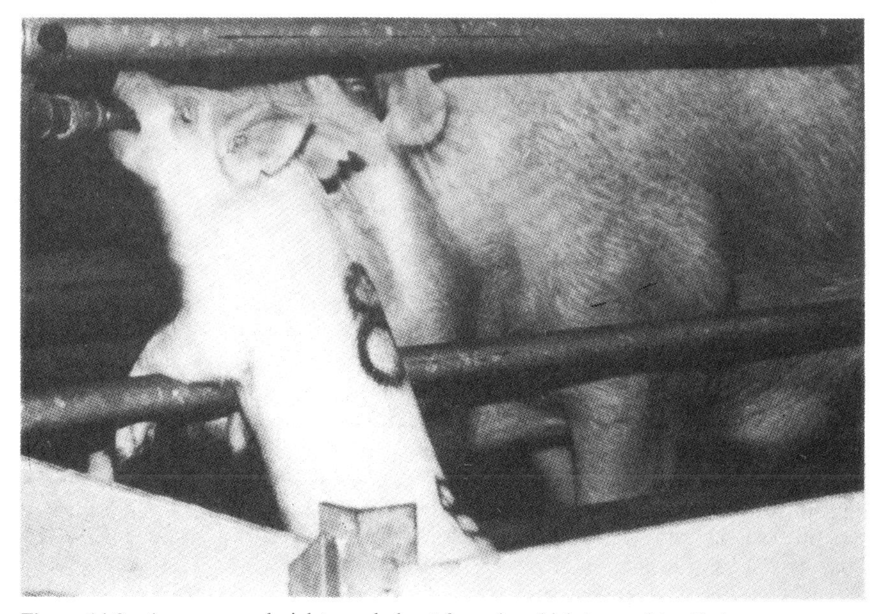

**Figure 14.3** An unweaned piglet aged about 3 weeks which learned to climb up the farrowing crate and drink from the sow's nipple drinker 0.6 m above the floor. Although effective on occasion, this is not a recommended method of providing water to piglets

Even less clear is the importance of water for very young piglets. It seems logical that a piglet on an all-milk diet will have no need for additional water, and in most cases this is true. If, however, a piglet is uncompetitive and fails to establish ownership of a teat, or if the sow produces little milk in the first days after farrowing, then young piglets are likely to be severely malnourished. The question is whether dehydration may result and whether provision of water will help to prevent death.

BEHAVIOURAL STUDIES

Studies reporting water intake by very young piglets show an interesting trend (Table 14.5), with little or no water intake seen in the earlier studies, but appreciable intakes in more recent work. We suspect that this trend is due to modern improvements in the farrowing environment. In recent decades, much emphasis has been placed on providing warmth for young piglets to reduce chilling and hypoglycaemia. In thus countering the effects of energy loss, have we encouraged greater moisture loss, to the extent that poorly-fed piglets would benefit from extra water?

**Table 14.5** AVERAGE WATER USE PER PIGLET IN THE FIRST WEEK (OR PART THEREOF) AFTER BIRTH IN SEVEN STUDIES

| *Mean water use per piglet* (ml/day) | *Number of litters* | *Period studied* | *Reference* |
|---|---|---|---|
| 9 | 33 | Days 1–7 | Aumaitre, 1964 |
| 12 | 9 | Days 1–7 | Friend and Cunningham, 1966[a] |
| 0 | 14 | Days 1–7 | Bekaert and Daelemans, 1970 |
| 0 | 20 | Days 1–5 | Wojcik *et al.*, 1978 |
| 12 | 20 | Days 5–10 | Wojcik *et al.*, 1978 |
| 40 | 36 | Days 1–7 | Svendsen and Andreasson, 1981 |
| 35 | 51 | Days 1–4 | Fraser *et al.*, 1988[b] |
| 130 | 32 | Days 1–2 | Phillips and Fraser, 1989[b, c] |

(From Fraser, 1990)
[a] Water provided 9 h/day
[b] After correction for evaporation and estimated spillage
[c] Two water dispensers (one conventional and one modified) provided per pen

In recent work several things have been learned about water consumption by piglets in their first days after birth. First, consumption is strongly influenced by ambient temperature, as illustrated in Figure 14.4.

Second, in addition to the large differences between litters, individual litter-mates differ greatly in their intake. For example, an intensive study of six litters, whose drinking was monitored continuously for 2 days after birth, showed individual intakes ranging from nearly zero to several hundred g/day within the same litters (P. A. Phillips and D. Fraser, unpublished).

Third, early drinking is influenced by social and exploratory factors. In one study, young piglets drank more from a bowl in which a stream of air bubbles created sound and movement at the water surface (Phillips and Fraser, 1989). In another test, 1-day-old litters drank about twice as much from a wide dispenser,

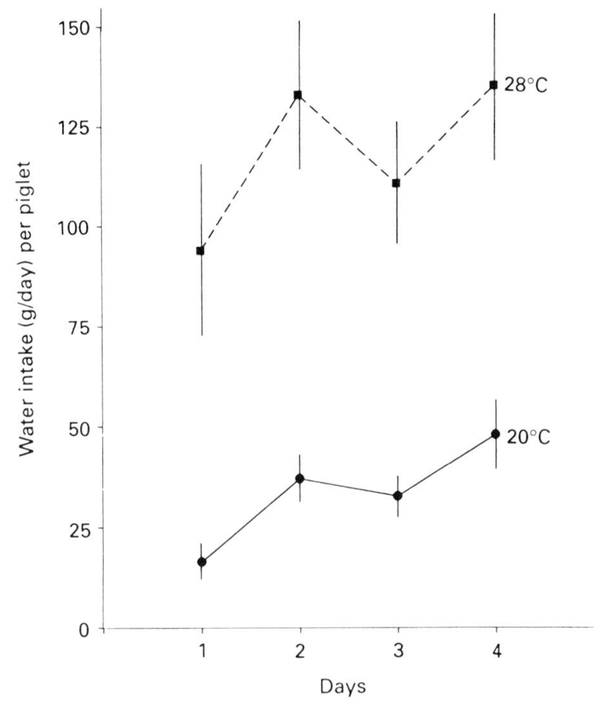


**Figure 14.4** Mean water intake per piglet (g/day) on the first 4 days after birth in a room held at either 28 (±1)°C or 20 (±1)°C for 18 litters studied at each temperature. Water was provided in a bowl, and apparent intakes were adjusted for spillage and evaporation. Data were expressed as litter means (in g/day per piglet), then averaged (±SEM) over the 18 litters in each treatment. (Previously unpublished data, collected in conjunction with the study by Phillips *et al.*, 1990)

which allowed up to three piglets to drink simultaneously, than from a conventional one (Phillips and Fraser, 1990).

Fourth, litters of higher birth weight tend to drink more in their first days than litters of light piglets (Fraser *et al.*, 1988; Phillips and Fraser, 1989). This may simply reflect a tendency for larger, vigorous piglets to be more active in exploring and using features in their environment.

Finally, despite enormous variation among litters, there is a clear tendency for litters that are gaining poorly during their first 1–2 days to drink more water at that time. In one study, the 10 litters (out of 51) with the lowest gains on days 1 and 2 showed a rapid increase in water intake to a mean of about 110 g/day per piglet on day 2, while the 10 litters with the greatest weight gains drank only about 40 g/day per piglet (Figure 14.5). In a further study (Phillips and Fraser, 1989) a significant negative correlation between water intake and weight gain on day 1 but not day 2 was found.

It is concluded from these data that although many factors influence early water intake, and although there is great variation among animals, litters that obtain relatively little milk from the sow in the early lactation (as reflected by very poor gains) show an unusually strong interest in water.

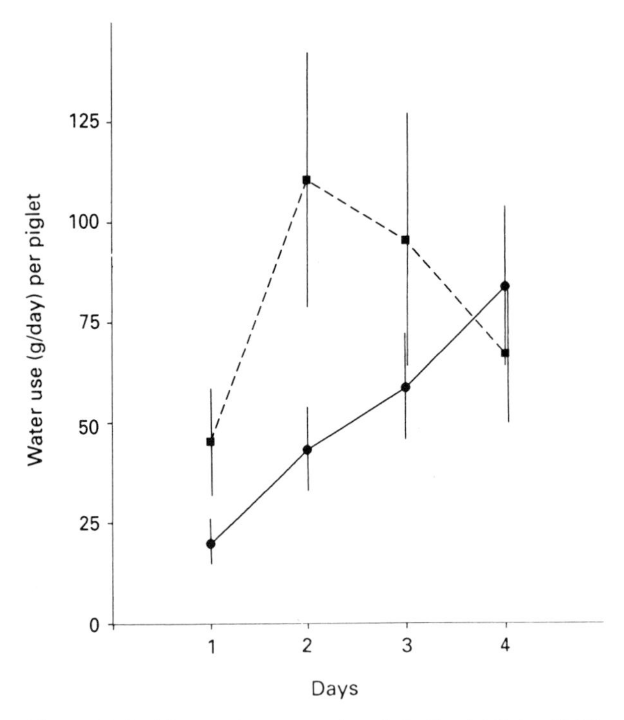


**Figure 14.5** Mean water use on days 1–4 after birth by 10 litters with low average weight gain (litter averages ranging from a loss of 33 g/day, up to gain of 70 g/day per piglet: broken line) and 10 litters with high average weight gain (average gains of 155–217 g/day per piglet: solid line) on days 1 and 2. Data were expressed as litter means (in g/day per piglet), then averaged (±SEM) over the 10 litters in each group. (From Fraser *et al.*, 1988)

## IS EARLY DRINKING BENEFICIAL?

Clearly, physiological studies are needed to determine whether water will be beneficial to piglets under particular circumstances. Lacking the necessary data, however, the following observations are offered in an attempt to indicate the conditions under which dehydration could possibly develop.

Apart from water required to support the growth of muscle tissue and to clear wastes from the body, young animals require water to replace that lost by evaporation and respiration (Hey and Katz, 1969). Therefore, a piglet's water requirements will depend greatly on the effective environmental temperature. Most farm animals show a roughly linear increase in the proportion of heat loss due to evaporation as environmental temperature increases from about 20°C to body temperature (Curtis, 1983). Piglets appear to be exceptional. Calorimetry studies indicated little evaporative heat loss from piglets as ambient temperature increases to 34°C, followed by either a sharp increase (Stombaugh *et al.*, 1973) or only a small increase (Mount, 1962), as ambient temperature increases to 38°C and above. At about 38°C, evaporation accounts for roughly 50% of total heat loss but with very large differences among individuals (Mount, 1962). Thus, it appears that piglet water requirements will be modest at most temperatures but may increase considerably (with large individual differences) under very warm conditions.

Sows' milk typically contains about 7% fat, 5% lactose and 6% protein (Klobasa *et al.*, 1987) which, if completely metabolized, generate about 1.2 kcal/g (5.0 kJ/g) (Kleiber, 1961; Noblet and Etienne, 1986). The same 1 g of milk also contains about 0.8 g water which, at a latent heat of evaporation of 0.6 kcal/g (2.5 kJ/g), would dissipate 0.48 kcal (2 kJ) on evaporation. Thus, evaporation of all the available water would dissipate only about 40% of the heat generated by metabolism of all available nutrients. Comparable values would be about 70% for human milk and cows' milk (based on data from Kleiber, 1961).

Under most circumstances, this amount of water would be more than enough, partly because of the piglet's limited evaporative moisture loss at most environmental temperatures, and partly because a well-fed piglet will not catabolize all of the fat, sugar and protein in its diet and will thus generate less heat per g of milk than the calculations imply. However, the situation may be quite different for an underfed piglet in a very warm environment. Based on data from Mount (1959) and Studzinski (1972), it is estimated that a poorly-nourished 1–2 kg piglet at its minimum metabolic rate and dissipating half of its heat by evaporation would require 100–200 g of milk per day simply to replace evaporative water loss. Although pig producers do not purposely house piglets in excessively warm environments, effective environmental temperature in a farrowing pen is difficult to estimate in practice, especially when powerful radiant heat sources are used. In our own studies, highly variable rates of non-excretory body weight loss (largely evaporative moisture loss) by piglets under radiant heaters have been seen, with values of 3–10 g/h per kg of bodyweight being relatively common. These seem to provide ample potential for dehydration in a piglet receiving little milk from the sow.

Even if some dehydration does occur, however, it may be only a secondary problem whose resolution will do little to improve survival. In a series of rather drastic experiments (Morrill, 1952; Morrill and Sampson, 1952), piglets fasted in an environment at 15°C showed almost complete exhaustion of body energy reserves in about 1 day. In an environment at 31°C, the animals survived 3–4 days, by which time their energy reserves were depleted and signs of dehydration were also apparent. Provision of drinking water reduced the rate of bodyweight loss and other symptoms of dehydration, but the depletion of energy reserves remained critical (Morrill and Sampson, 1952). In this case, preventing the dehydration did little to improve survival. It is suspected that in even warmer environments, dehydration would develop more quickly and depletion of energy reserves more slowly, with the possible result that dehydration would become the more life-threatening problem. The point is that studies of piglet dehydration need to consider the animal's energy balance as well as its water balance.

It also needs to be asked whether piglets may require water to offset a loss of body moisture from diarrhoea, as suggested by Ehlert *et al.* (1981). In calves, diarrhoea can lead to large faecal water loss and a reduction in total body water (Phillips *et al.*, 1971, but see also Fayet, 1971). In newborn piglets, both experimentally-induced and spontaneous diarrhoea caused physiological disturbance (Andren and Persson, 1983). In these studies, those animals which subsequently died experienced severe weight loss, and developed elevated haematocrit and serum proteins, indicative of dehydration. Infected piglets which survived were apparently able to adjust to fluid loss, since no changes in haematocrit or serum proteins were observed. However, bodyweight loss was still evident. Balsbaugh *et al.* (1986) found that experimentally-induced diarrhoea in the piglet caused an increase in the haematocrit (presumably reflecting altered partitioning of water in intra- and extracellular fractions), but no net decline in body water concentration. However, they also observed bodyweight loss. Like the study of Andren and Persson (1983) these data demonstrated that the young piglet has considerable capability to maintain water homeostasis.

Finally, it needs to be asked whether the provision of drinking water to very young piglets may do harm by, for example, reducing their motivation to suckle. Phillips *et al.* (unpublished data) studied 44 litters, 22 of which had access to a water bowl from birth while the remainder had access to the water bowl at the end of their first week. Water was considered unnecessary for these litters as room temperature averaged only 23°C and almost all litters had reasonable weight gains. Weight gain and mortality were not different between the treatments, suggesting that provision of water to litters that did not require it brought no harm to the animals.

## Newly-weaned litters

At weaning, along with the many other changes that it experiences, the young pig must suddenly consume water as its major fluid source. However, it is not altogether clear how well the newly-weaned pig can regulate water metabolism, or select a daily intake appropriate to its needs.

Several studies have demonstrated that restricted nipple-drinker flow rates can impair water intake and, under severe restriction, growth rate as well. Barber *et al.* (1988) reported that nipple drinker flow rates below 700 ml/min restricted water intake and below 450 ml/min restricted feed intake, and thus growth rate. Shurson (1989) reported significant improvements in growth rate during the first and third weeks when pigs received water at 700 *versus* 70 ml/min. Nienaber and Hahn (1984) reported a similar effect, using 10-week-old piglets, and also noted that optimum flow rates rose with increased environmental temperatures. In contrast, Hoppe *et al.* (1987b) found no difference in feed intake or body weight gain between newly-weaned piglets with nipple drinkers flowing at 700 or 70 ml/min. Brumm and Shelton (1983) reported that two nipple drinkers per pen of 16 pigs supported better growth than a single drinker, but flow rates were not reported.

A further concern is whether newly-weaned pigs consume water appropriate to their needs. In three separate experiments, pigs weaned at 4 weeks of age followed unexpected water consumption patterns – relatively large quantities on the first 1–2 days after weaning, and considerably lower intakes for the next 2–3 days before rising again on day 5 (Figures 14.6 and 14.7). This pattern was the same with fresh water or water containing more than 4000 ppm of total dissolved solids (Figure 14.6). Significantly, scouring was continually rising during this period, possibly causing greater water requirement on those days when water intake remained low. At present it is impossible to say whether the high intake on days 1 and 2 were excessive, perhaps driven by a desire for satiety, whether the low intakes around days 3–4 were inadequate, or whether gastrointestinal water absorption was impaired by some post-weaning disturbance. Clearly, the mechanisms regulating water consumption in the newly-weaned pig need further study.

The weanling pig is more susceptible to poor water quality than other classes of pig. Pork producers with poor quality water most commonly report an increased

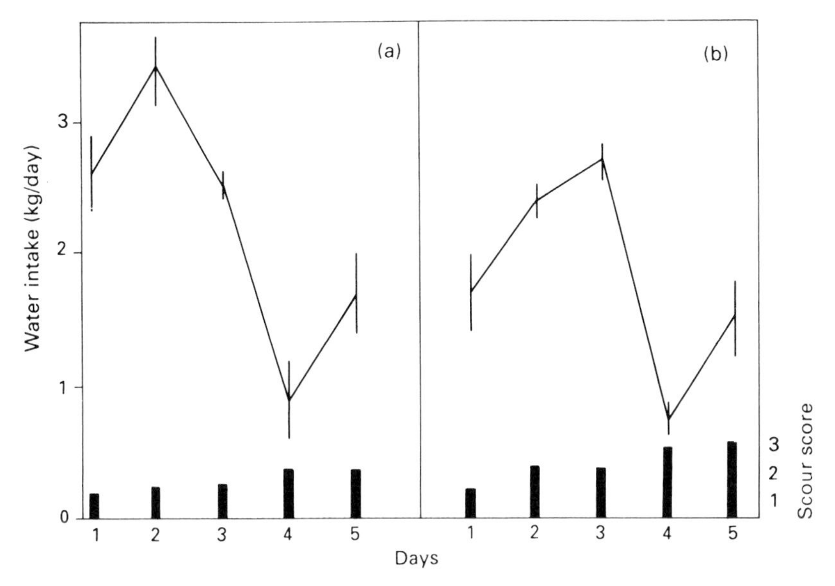


**Figure 14.6** Water consumption patterns and average scour scores for pigs fed (a) fresh (217 ppm TDS) or (b) saline (4100 ppm TDS) water for the first 5 days after weaning. Scour score is based on a scale reflecting decreasing stool consistency from 1 = firm to 3 = loose). Pigs were housed individually at 25°C following weaning at 4 weeks of age. (Previously unpublished data from J. M. McLeese and J. F. Patience)

incidence of diarrhoea although slow growth and, in some cases, death have also been seen (McLeese *et al.*, 1989). However, research has not always been able to corroborate field observations, and many experiments have found pigs to grow quite well on poor quality water. For example, Young (1986) found that pigs fed water containing up to 2172 ppm of sulphates grew as well as or better than pigs offered deionized water. Anderson and Stothers (1978) reported the results of three experiments, in which pigs were fed water containing about 6000 ppm TDS, supplied either as sulphates or chloride. In only one of their studies did the pigs on the fresh water outperform the others in growth rate, feed intake and feed conversion. Our studies have provided similarly contradictory findings. For example, in one experiment, pigs fed a medicated feed suffered no ill effects due to the consumption of water containing greater than 4000 ppm TDS (Table 14.6). However, a second study using unmedicated feed revealed a clear growth depression attributable to the same water (Table 14.7).

The effect of poor quality water on the pig is no doubt influenced by other factors. We have observed that increasing the intake of high sulphate water, even in pigs previously 'adapted' to it, leads to diarrhoea. Therefore, research must focus on daily intake of sulphates, rather than their concentration in the water.

Because of its known laxative effects, sulphate is often assumed to be the major deleterious factor in the water; this has never been confirmed and requires further study. For example, given the different metabolism of sodium and magnesium, it is possible that one cation in association with sulphate is more serious than the other.

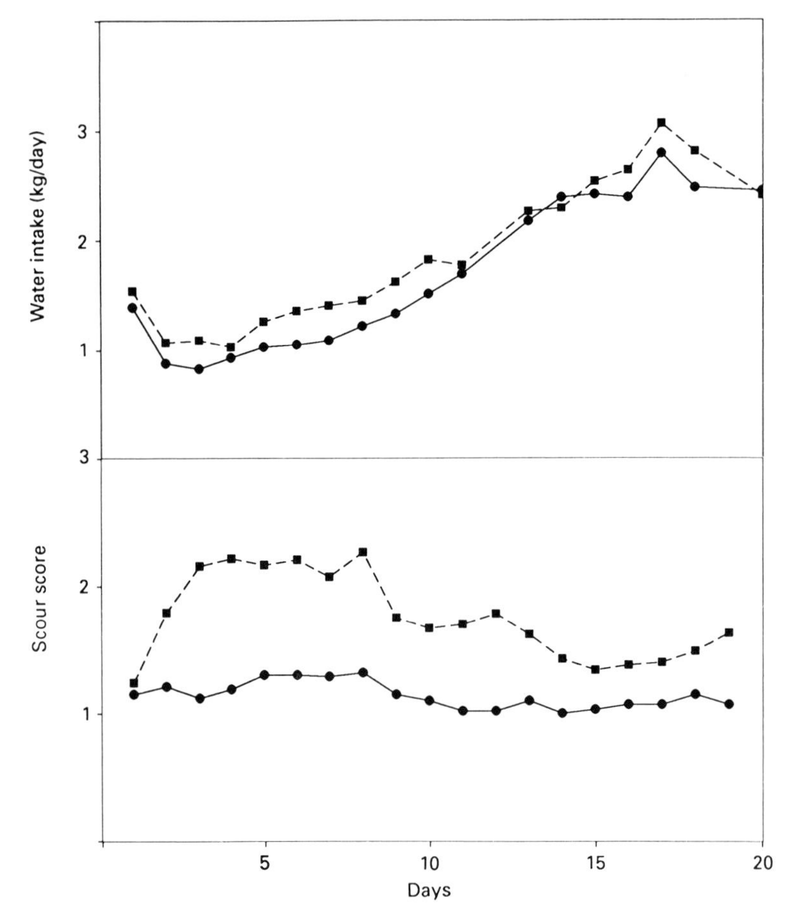


**Figure 14.7** Water consumption and average scour score of pigs fed low (150 ppm) or high (2300 ppm) sulphate water for the 20 days immediately after weaning. (- - -) High sulphate; (—) low sulphate. Data summarize the results of two independent experiments. (Refer to Figure 14.6 for other details.) M. L. Tremblay, J. M. McLeese, G. I. Christison, and J. F. Patience, (unpublished data)

## Conclusions

Based on this review, it is concluded that the water requirements of pigs need more practical attention and more research effort in at least three areas: for sows in the first days after farrowing when occasional low water intake correlates with, and may contribute to, poor milk production; for piglets in the first days after birth where malnutrition combined with high environmental temperature may cause dehydration in some animals; and for newly-weaned piglets which must suddenly regulate their water balance by drinking, and whose susceptibility to diarrhoea may be compounded by excessive mineral content in the water. Progress in these areas will require experimentation to replace correlation studies, and physiological work

**Table 14.6** EFFECT OF WATER AT THREE LEVELS OF TOTAL DISSOLVED SOLIDS (TDS) ON THE PERFORMANCE OF PIGS FED A DIET CONTAINING ANTIBIOTICS[a]

| *Variable* | *TDS* (ppm) | | |
|---|---|---|---|
| | *217* | *2350* | *4390* |
| *Water quality* (ppm) | | | |
| Sulphates | 83 | 1280 | 2650 |
| Calcium | 24 | 184 | 288 |
| Chloride | 8 | 34 | 70 |
| Magnesium | 15 | 74 | 88 |
| Sodium | 24 | 446 | 947 |
| Hardness | 124 | 767 | 1080 |
| pH | 8.4 | 8.1 | 8.0 |
| *Performance*[b] | | | |
| Average gain (kg/day) | 0.43 | 0.43 | 0.44 |
| Average feed intake (kg/day) | 0.55 | 0.56 | 0.57 |
| Feed:gain | 1.28 | 1.31 | 1.30 |
| Average water intake (kg/day)[c] | 1.60 | 1.84 | 1.81 |
| Scour score[c] | 1.07 | 1.30 | 1.46 |

Source: M. L. Tremblay, J. M. McLeese, G. I. Christison and J. F. Patience, unpublished data
[a] Tylosin phosphate plus nitrofurazolidone
[b] Performance measures were averaged over two water treatments (with or without Acid-Pak 4-Way, Alltech, Inc., Nicholasville, KY) which had no significant effect on most of the measures
[c] Effect of water source significant, $P < 0.05$

**Table 14.7** EFFECT OF WATER AT TWO LEVELS OF TOTAL DISSOLVED SOLIDS (TDS) ON THE PERFORMANCE OF PIGS FED A DIET FREE OF ANTIBIOTICS

| *Item* | *TDS* (ppm) | | *P*[a] |
|---|---|---|---|
| | *217* | *4390* | |
| Average gain (kg/day) | | | |
| Day 0–4 | 0.083 | 0.069 | NS |
| Day 5–8 | 0.226 | 0.166 | 0.08 |
| Day 9–15 | 0.568 | 0.481 | 0.11 |
| Day 16–21 | 0.581 | 0.542 | NS |
| Day 0–21 | 0.418 | 0.360 | 0.04 |
| Average feed intake (kg/day) | | | |
| Day 0–21 | 0.530 | 0.521 | NS |
| Feed:gain | | | |
| Day 0–21 | 1.33 | 1.47 | 0.01 |

Source: J. M. McLeese, J. F. Patience and G. I. Christison, unpublished data
[a] Statistical significance of difference

to supplement simple water intake measures. In particular, work is needed on basic principles so that superficial empirical findings can be extrapolated beyond the exact conditions under which they were obtained.

Finally, research on water requirements needs to look well beyond 'average' values. The 'average' piglet does not become dehydrated; the 'average' sow may have no serious depression in water intake after farrowing; yet this does not mean that there are no problems. The difference between good and excellent management may rest on how well the system caters for the atypical needs of the more vulnerable animals.

## Acknowledgements

We are grateful to Dr B. K. Thompson, Dr G. I. Christison and Ms Monique Tremblay for continuing collaboration in these studies and to Mr E. A. Pajor for Figure 3. Our ideas on water requirements owe much to valuable discussion with Drs M. Etienne and S. E. Curtis, although the views expressed may not reflect theirs. Dr A. Lawrence kindly provided the references on excessive drinking by pigs, and Dr M. C. Appleby made many useful comments on the manuscript. Financial support from Agriculture Canada, Saskatchewan Agriculture and Food, Hillcrest Farms, and Alltech, Inc., is gratefully acknowledged.

## References

Agriculture Canada (1988). *Canadian Farm Buildings Handbook.* Research Branch, Agriculture Canada, Ottawa, 155 pp

Anderson, D. M. and Stothers, S. C. (1978). *Journal of Animal Science,* **47**, 900–907

Andren, B. and Persson, S. (1983). *Veterinaria Scandinavica,* **24**, 84–98

Anonymous (1987). *National Hog Farmer,* **32**(11), 32–39

Aumaitre, A. (1964). *Annales Zootechnologie,* **13**, 183–198

Balsbaugh, R. K., Curtis, S. E. and Meyer, R. C. (1986). *Journal of Animal Science,* **62**, 307–314

Bauer, W. (1982). *Archiv für Experimentelle Veterinärmedizin,* Leipzig, **36**, 823–827

Barber, J., Brooks, P. H. and Carpenter, J. L. (1988). *Proceedings of the British Society of Animal Production Occasional Meeting*, Voluntary Feed Intake of Pigs, Leeds

Bekaert, H. and Daelemans, J. (1970). *Revue de l'Agriculture*, (no. 6–7), 935–949

Brumm, M. C. and Shelton, D. P. (1986). *Nebraska Swine Report.* Nebraska Co-operative Extension Service EC-86-219, pp. 5–6

Curtis, S. E. *Environmental Management in Animal Agriculture.* Iowa State University Press, Ames

Diblik, T. (1986). *Živočišna Výroba,* **31**, 1029–1036

Ehlert, D., Peters, F. and Heinze, C. (1981). *Tierzucht,* **35**, 461–462

Epstein, A. N. (1982). In *The Physiological Mechanisms of Motivation*, pp. 165–214. Ed. Pfaff, D. W. Springer, New York

Falk, J. L. (1971). *Physiology and Behaviour,* **6**, 577–588

Fayet, J. C. (1971). *British Veterinary Journal,* **127**, 37–44
Fraser, D. (1975). *Animal Production,* **21**, 59–68
Fraser, D. (1990). *Journal of Reproduction and Fertility*, (in press)
Fraser, D. and Phillips, P. A. (1989). *Applied Animal Behaviour Science,* **24**, 13–22
Fraser, D., Phillips, P. A., Thompson, B. K. and Peeters Weem, W. B. (1988). *Canadian Journal of Animal Sciences,* **68**, 603–610
Friend, D. W. (1971). *Journal of Animal Science,* **32**, 658–666
Friend, D. W. and Cunningham, H. M. (1966). *Canadian Journal of Animal Science,* **46**, 203–209
Friend, D. W. and Wolynetz, M. S. (1981). *Canadian Journal of Animal Sciences,* **61**, 429–438
Garrison, G. W., Weed, R. D., Chaney, C. H. and Waddill, D. G. (1966). *Kentucky Animal Science Research Report,* pp. 85–87, University of Kentucky, Lexington
Garner, F. H. and Sanders, H. D. (1937). *Journal of Agricultural Science, Cambridge,* **27**, 638–643
Hey, E. N. and Katz, G. (1969). *Journal of Physiology, London,* **200**, 605–619
Hoppe, M. K., Libal, G. W. and Wahlstrom, R. C. (1987a). *South Dakota State University Swine Day Report*, 43–45
Hoppe, M. K., Libal, G. W. and Wahlstrom, R. C. (1987b). *South Dakota State University Swine Day Report*, 46–48
Kleiber, M. (1961). *The Fire of Life.* New York, John Wiley & Sons
Klobasa, F., Werhahn, E. and Butler, J. E. (1987). *Journal of Animal Science,* **64**, 1458–1466
Kozhakhmetova, A. Y. (1966). *Trudy Instituta fiziologii, Akademiya nauk, Kazakhskoi USSR,* **10**, 190–194. (English translation available from Agriculture Canada Library, Ottawa)
Leibbrandt, V. (1989). *Swine Report 1989,* pp. 14–17. University of Wisconsin – Madison
Lightfoot, A. L. (1978). *Animal Production,* **26**, 386 (abstract)
Lightfoot, A. L. and Armsby, A. W. (1984). *Animal Production,* **38**, 541 (abstract)
Loje, K. and Bing, J. (1951). *Nordisk Veterinaermedicin,* **3**, 247–254
McLeese, J. M., Patience, J. F. and Christison, G. I. (1989). *Prairie Swine Centre Annual Report,* pp. 31–34, University of Saskatchewan
Madec, F. (1984). *Pig News & Information,* **5**(2), 89–93
Madec, F. (1985). *Annales de Zootechnologie,* **34**(3), 373
Madec, F. (1987). *Le Point Vétérinaire,* **19**, 611–617
Mahan, D. C. (1969). *Nitrogen and water metabolism in the lactating sow.* PhD Thesis, University Illinois, Urbana, IL
Midwest Plan Service (1982). *Swine Housing and Equipment Handbook.* Swine Housing Subcommittee, Midwest Plan Service, Ames, Iowa, 112 pp
Mittleman, G., Castaneda, E., Robinson, T. E. and Valenstein, E. S. (1986). *Behavioural Neuroscience,* **100**, 213–220
Morrill, C. C. (1952). *American Journal of Veterinary Research,* **13**, 322–324
Morrill, C. C. and Sampson, J. (1952). *American Journal of Veterinary Research,* **13**, 327–329
Morris, W. (1970). *The American Heritage Dictionary of the English Language.* American Heritage Publ. Co., Inc., Boston
Mount, L. E. (1959). *Journal of Physiology, London,* **147**, 333–345
Mount, L. E. (1962). *Journal of Physiology, London,* **164**, 274–281

Mroz, Z., Partridge, I. G., Mitchell, G. and Keal, H. D. (1986). *Journal of the Science of Food and Agriculture,* **37**, 239–247
Nienaber, J. A. and Hahn, G. (1984). *Journal of Animal Science,* **59**, 1423–1429
Noblet, J. and Etienne, M. (1986). *Journal of Animal Science,* **63**, 1888–1896
NRC (1974) *Nutrients and toxic substances in water for livestock and poultry.* National Academy of Sciences, Washington, DC, 93 pp
Patience, J. F., McLeese, J. and Tremblay, M. L. (1989). *Proceedings of the Western Nutrition Conferences,* pp. 113–138, University of Saskatchewan, Saskatoon
Patience, J. F., Wolynetz, M. S., Friend, D. W. and Hartin, K. E. (1987). *Canadian Journal of Animal Science,* **67**, 859–863
Phillips, P. A. and Fraser, D. (1989). *Canadian Agricultural Engineering,* **31**, 175–177
Phillips, P. A. and Fraser, D. (1990). *Applied Engineering in Agriculture,* (in press)
Phillips, P. A., Fraser, D. and Thompson, B. K. (1990). *Applied Engineering in Agriculture,* (in press)
Phillips, R. W., Lewis, L. D. and Knox, K. L. (1971). *Annals New York Academy of Sciences,* **176**, 231–243
Price, M. (1985). *Introducing Groundwater.* George Allen & Unwin, London, **195**
Riley, J. E. (1978). *Animal Production,* **26**, 386 (abstract)
Russell, I. D. (1985). *Poultry Digest,* **44**(524), 422
Shurson, G. C. (1989). *Ohio Swine Research and Industry Report,* pp. 37–39, The Ohio State University, Columbus
Smith, D. L. T. (1975). *Diseases of Swine,* 4th edition, pp. 854–860. Ed. Dunne, H. W. and Leman, A. D. Iowa State University Press, Ames
Stephens, D. B., Ingram, D. L. and Sharman, D. F. (1983). *Quarterly Journal of Experimental Physiology,* **68**, 653–660
Stombaugh, D. P., Roller, W. L., Adams, T. and Teague, H. S. (1973). *American Journal of Physiology,* **225**, 1192–1198
Studzinski, T. (1972). *Journal of Physiology, London,* **224**, 305–316
Svendsen, J. and Andreasson, B. (1981). *Investigations into the supplying of liquids to piglets and to weaned pigs: liquid consumption and production results.* Swedish University of Agricultural Sciences, Department of Farm Buildings, Report 14
Task Force on Water Quality Guidelines (1987). *Canadian Water Quality Guidelines,* prepared for the Canadian Council of Resource and Environment Ministers. Environment Canada, Ottawa
Tillon, J. P. and Madec, F. (1984). *Annales Recherches Vétérinaires,* **15**, 195–199
Van der Kamp, G. (1989). *Proceedings of the Western Nutrition Conference,* pp. 105–111, University of Saskatchewan, Saskatoon
Viessman, W., Jr and Hammer, M. J. (1985). *Water Supply and Pollution Control,* 4th edn. Harper & Row, Publications, New York
Wojcik, S., Widenski, K. and Mroz, Z. (1978). *Medycyna Weterynaryjna,* **34**, 161–164
Yang, T. S., Howard, B. and Macfarlane, W. V. (1981). *Applied Animal Ethology,* **7**, 259–270
Young, L. G. (1986). *Swine Research Update,* Centralia College of Agricultural Technology, Centralia, Ontario

15

# THE PHYSIOLOGICAL BASIS OF ELECTROLYTES IN ANIMAL NUTRITION

J. F. PATIENCE
*Prairie Swine Centre, University of Saskatchewan, Saskatoon, Saskatchewan, Canada*

## Introduction

From a purely chemical point of view, an electrolyte is any substance that when added to pure water produces a conducting solution (Sienko and Plane, 1979). Thus, any chemical that dissociates into its constituent ions is an electrolyte. A strong electrolyte dissociates completely, or almost completely, while a weak electrolyte dissociates to only a limited extent. Thus, a solution of strong electrolytes will consist mainly of ions and only a few or almost no undissociated molecules in a solution of weak electrolytes, the undissociated molecules will be more common. Table 15.1 lists common substances that are strong or weak electrolytes, or non-electrolytes.

**Table 15.1** CLASSIFICATION OF ELECTROLYTES

| *Electrolytes* | | *Non-electrolytes* |
|---|---|---|
| *Strong* | *Weak* | |
| Hydrogen chloride | Hydrogen fluoride | Glucose |
| Sodium chloride | Ammonia | Sucrose |
| Sodium hydroxide | Acetic acid | Ethanol |
| Potassium fluoride | Mercuric chloride | Oxygen |
| | Acetone | |

In nutrition and physiology, it is common to consider specific electrolytes, i.e. ions found in the various body fluids (plasma, interstitial fluid, intracellular fluid, urine, etc.). For example, sodium, bicarbonate and chloride are the predominant electrolytes in extracellular fluid, with much smaller quantities of potassium, calcium, magnesium, phosphate, organic acids and protein also present. In the intracellular fluid, such electrolytes as potassium, phosphate, magnesium and protein dominate. Sodium, calcium and bicarbonate are less common species. The ionic composition of typical intracellular and extracellular body fluids appears in Table 15.2.

**Table 15.2** IONIC COMPOSITION OF BODY FLUIDS (mEq/l)

| *Electrolyte* | *Plasma* | *Interstitial fluid* | *Intracellular fluid* |
|---|---|---|---|
| *Cations* | | | |
| Sodium | 142 | 145 | 12 |
| Potassium | 4 | 4 | 150 |
| Magnesium | 1 | 1 | 34 |
| Calcium | 3 | 2 | 4 |
| *Anions* | | | |
| Chloride | 104 | 117 | 4 |
| Bicarbonate | 24 | 27 | 12 |
| Phosphate | 2 | 2 | 40 |
| Protein | 14 | 0 | 54 |

Adapted from Rose (1977)

The concentration of electrolytes is often expressed as milliequivalents (mEq) or milliosmoles (mOsm), rather than milligrams (mg), since the electrical and osmotic properties of these ions are of particular interest. Values expressed in milligrams can be converted to milliequivalents by dividing by the molecular weight and multiplying by the valency. Thus, 100 mg of sodium equals 4.3 mEq [(100/23) × 1] and 100 mg of calcium is the same as 5 mEq [(100/40.08) × 2)].

No discussion of electrolytes would be complete without a brief summary of water metabolism, sinced a major function of electrolytes is to assist in the maintenance of water balance within the body. Indeed, it is impossible to develop a functional discussion of one without the other.

## Water

As part of the evolutionary process some 400 million years ago, when living organisms moved from an aquatic environment to dry land, they developed a layer of skin that provided a barrier between the fluid environment required by the cells and the surrounding air. At the same time, mechanisms evolved to regulate the composition of fluid inside the cell as well as that surrounding it (Maloiy, MacFarlane and Shkolnik, 1979). In essence, these organisms developed their own 'aquatic environment.'

Water is required to transport nutrients, gases, waste products and hormones about the body. It lubricates many moving parts and is essential in helping to maintain the delicate balance of acids and bases. It provides the medium in which the body's metabolic processes may occur and it assists in dissipating heat. Movement of digesta through the gastrointestinal tract could not occur without water, nor could the absorption of nutrients acorss the gut wall. Indeed, it is difficult to think of a single body function that does not require water, at least indirectly.

Not surprisingly, the chemical properties of water make it ideally suited to fulfil its central role in life. For example, water has a very high heat capacity, supporting its role in thermal homeostasis. It maximizes its electrolyte activities at 37°C which

**Table 15.3** TYPICAL DISTRIBUTION OF BODY WATER

| *Compartment* | *As a percentage of total water* | *As a percentage of total body* |
|---|---|---|
| Total body water | 100 | 60 |
| Extracellular | 40 | 24 |
| Interstitial | 19 | 11.5 |
| Plasma | 7.5 | 4.5 |
| Connective tissue | 7.5 | 4.5 |
| Bone | 5.0 | 3.0 |
| Intracellular | 60 | 36 |

Adapted from Rose (1977)

is about body temperature and has very low viscosity. Furthermore, it has high surface tension, supporting capillary movement in the vascular system (Quinton, 1979).

Typically, total body water is divided into two major compartments: extracellular, including both interstitial fluid and blood plasma, and intracellular. A typical breakdown is illustrated in Table 15.3.

However, many factors can and do influence the proportion of the body that is water; for example, as the animal matures, body fat increases and body water content decreases. The reason for this is clear; adipose tissue contains only 7% water, while muscle is three-quarters water (Georgievskii, 1982a). Thus, as an animal matures, water content will vary in concert with changes in the proportions of fat and muscle.

Water is derived by the body from three sources, the primary one, of course, being drinking water. Common feed ingredients never contain 100% dry matter, so that eating food also provides some free water; most foodstuffs contain at least 8% water, and some much more. Finally, metabolic processes generate water as shown in equations 11.1, 11.2 and 11.3.

Carbohydrate (glucose)

$$C_6H_{12}O_6 + 6O_2 \rightarrow 6H_2O + 6CO_2 \qquad (15.1)$$

Amino acid (methionine)

$$2C_5O_2NH_{11}S + 15O_2 \rightarrow 7H_2O + 9CO_2 + (NH_3)_2CO + 2H_2SO_4 \qquad (15.2)$$

Fat (tripalmitin)

$$2C_{51}H_{98}O_6 + 145O_2 \rightarrow 98H_2O + 102CO_2 \qquad (15.3)$$

It can be seen that oxidation of 100 g carbohydrate (equation 15.1) yields about 60 g water; a similar quantity of protein (equation 15.2) would yield about 44 g water, depending on the amino acid profile. In the case of methionine illustrated above, oxidation of 100 g would yield 42.3 g of water.

Interestingly, oxidation of fat (equation 15.3) yields more than its own weight in water, a fact exploited by the camel. It stores fat (not water) in its humps and then utilizes these deposits to generate water when required. The calculated water balance in a more typical farm animal, the growing pig, appears in Table 15.4.

**Table 15.4** TYPICAL WATER BALANCE IN THE GROWING PIG

| *Intake* | *ml* | *Output* | *ml* |
|---|---|---|---|
| Drinking water | 4000 | Urine | 2930 |
| Food water | 200 | Lungs, etc | 1530 |
| Water of oxidation | 990 | Faeces | 250 |
| | | Growth | 480 |
| Total | 5190 | | 5190 |

Patience (unpublished data). This example pertains to a young pig, growing at the rate of about 700 g/day, eating 2000 g/day of a typical maize-soyabean meal diet in a thermoneutral environment

Obviously, urine is the primary vehicle for elimination of water, and is certainly the one that can be regulated most effectively according to physiological need. Fundamentally, urine volume is determined by the quantity of water requiring removal from the body. The composition of urine varies tremendously, being influenced by the quantity of solutes eliminated and by the quantity of water consumed relative to need. A complete consideration of water balance must also include losses due to sweat and respiration, which tend to be influenced in the first instance not by a need to maintain water balance, but rather to achieve some other homeostatic objective.

The major physiological regulators of water balance relate to intake (thirst) and excretion. Excretion is under the influence of hormones, including antidiurectic hormone (ADH), the renin-angiotensin-aldosterone complex and possibly prolactin. If plasma osmolality falls below 'osmotic threshold' ADH secretion is suppressed to very low levels and urine output rises; conversely if osmolality rises, ADH secretion increases. Thus, release of ADH results in the excretion of decreasing amounts of a urine that is increasing in concentration. The regulation of ADH is so sensitive that a change in osmolality of only 1% is sufficient to elicit a measurable change in ADH and subsequent adjustment in urinary excretion (Zerbe and Robertson, 1983). ADH release may also be influenced by haemodynamic factors, such as blood volume and pressure, which stimulate baroreceptors located in the heart, aorta and carotid artery (Robertson, 1977). The baroregulatory system is less sensitive to small changes than the osmotic system, but responds in a more dramatic fashion to large changes. Other factors, such as nausea and hypoglycaemia influence ADH release (Baylis, Zerbe and Robertson, 1981) but their practical application remains obscure.

Aldosterone is involved in water retention indirectly via its sodium-sparing action (Legros, 1979). Angiotensin and renin may act directly on the pituitary to stimulate ADH release. The role of prolactin in water balance is minimal, if it exists at all (Legros, 1979).

Thirst is an important primary regulator of water balance. Interestingly, thirst is stimulated by many of the same factors that increase the secretion of ADH, such as increased plasma osmolality (Thornton, Baldwin and Purdew, 1985). This is a sensitive response; only a 2–3% increase in osmolality will induce thirst (Fitzsimmons, 1972). Hypovolaemia will also initiate the sensation of thirst involving, in ways which remain unclear, the same stretch receptors involved in

ADH release (Thrasher, Keil and Ramsay, 1982). Thus, thirst and ADH appear to be closely linked.

However, it must also be noted that thirst can be stimulated by many factors unrelated to water requirement; important examples include hunger (Patience *et al.*, 1987d) and behaviour (Madec, Cariolet and Dantzer, 1986).

## Electrolyte metabolism

### SODIUM

Sodium plays a primary role in the regulation of water balance and, in particular, in extracellular fluid volume (Alcantara, Hanson and Smith, 1980). It also plays a critical role in cellular transport systems, due to its participation in the sodium-potassium ATPase. This system is responsible, in part, for the electrochemical gradient across cell membranes and thus is critical not only for both active and passive transport function but also for the creation of the electrical potential required for nervous system activity and muscle contraction. Clearly, the body's emphasis on sodium homeostasis is well founded.

The total amount of sodium in the body is divided into exchangeable and non-exchangeable compartments. About 43% is in bone, of which 68% is bound within the crystalline structure and is therefore unavailable for other metabolic use (Edelmon and Leibman, 1959). The remainder of bone sodium, plus that found in the various fluid compartments is considered to be exchangeable. In total, about 70% of total body sodium exists in the exchangeable pool (Rudd, Pailthorp and Nelp, 1972).

Sodium balance is monitored by a variety of systems. Receptors in various tissues detect changes in fluid colume, blood pressure or other phenomena which are related to sodium concentration (Gardenswartz and Schrier, 1982). Numerous hormones, including the renin-angiotensin system, aldosterone, some prostaglandins, the kallikrein-kinin system and natriuretic hormone are all involved to a lesser or greater extent. Clearly, the body appears to place a very high priority on the regulation of body sodium content.

Under normal circumstances, about 90% of the sodium in the diet will be absorbed; even when intake rises 4- or 5-fold, the proportion absorbed remains largely unchanged (Patience, Austic and Boyd, 1987c). Thus, the kidney remains the primary organ of sodium homeostasis, with the intestinal tract playing a relatively minor role.

Sodium metabolism can be influenced by a wide variety of nutritional circumstances, some of which might not be readily considered. For example, in ruminants, ingestion of a high fibre diet stimulates parotid secretion of sodium bicarbonate, impairing normal sodium and water circulation to the rumen. This generates a short-term acidaemia, hyperhydration, renal sodium retention and urinary acidification (Andersson *et al.*, 1986). Although such changes are transient, they underline the need to obtain blood samples for acid-base measurements with a full understanding of their relationship to such factors as time of feeding.

### POTASSIUM

Potassium is the principal cation of the intracellular environment (Table 15.2). Almost 90% of total body potassium can be found in the intracellular spaces. The majority of extracellular potassium is present in bone (Tannen, 1986). Unlike sodium, potassium is largely an intracellular electrolyte, although exceptions do exist. Maintenance of constant extracellular potassium is absolutely critical for life. Hypokalaemia is defined by serum potassium $<2.4$ mEq/l and hyperkalaemia by serum levels $>7.0$ mEq/l. Hyperkalaemia appears to be particularly serious, due to the apparent vulnerability of cardiac tissue (Rose, 1977).

Potassium is readily absorbed from the gastrointestinal tract, although in the pig, the extent of absorption tends to be somewhat less than that for sodium or chloride (Patience *et al.*, 1987c). Potassium is involved in a large number of functions. Like sodium, it is part of the Na-K ATPase, mentioned above. Furthermore, it is a cofactor in a number of enzymes, participates in the maintenance of acid-base balance, gas transport by haemoglobin, cardiac function and protein synthesis (Fregly, 1981; Georgievskii, 1982a).

As in the case of sodium, potassium homeostasis is the responsibility of the renal tissue (Mason and Scott, 1972). However, the actual mechanisms in some cases differ substantially. Potassium is filtered by the glomerulii, but is largely reabsorbed in the proximal tubule. Thus, renal regulation of potassium is dependent on its secretion into the more distal portions of the tubules, largely under the influence of aldosterone (Brobst, 1986). Interestingly, glucocorticoids (Bastl, Binder and Hayslett, 1980) and insulin (Bia and DeFronzo, 1981) also play a role. Sodium concentration in the urine and acid-base status of the animal are also known to influence the extent of renal potassium secretion (Mason and Scott, 1972; Kem and Trachewsky, 1983). Because of the relationship between potassium excretion, sodium excretion and mineralocorticoid levels, urine and salivary sodium:potassium ratios are sometimes used as an indicator of mineralocorticoid effect (Kem and Trachewsky, 1983).

The role of insulin is an interesting one; the fact that such a critical component of the endocrine system responds to extracellular potassium concentration underlines the importance placed on its homeostasis. Increases in extracellular potassium stimulate insulin release resulting in enhanced potassium transport into the cell. The relationship between glucose and potassium, mediated via insulin must be understood by clinicians since any disturbance in carbohydrate metabolism may influence potassium, and vice versa. For example, infusion of glucose can cause hypokalemia (Rose, 1977) and similarly, potassium infusion could lead to a potentially lethal hypoglycaemia.

Acid-base disturbance has an interesting effect on potassium metabolism. Acidosis results in its translocation from within the cell to the extracellular fluid; alkalosis has the opposite effect (Adrogue and Madias, 1981). The basis of this effect is the buffering of excess hydrogen ions by the intracellular fluids; as the protons enter the cells, potassium leaves in order to maintain electrical neutrality. Burnell *et al.* (1956) have suggested that for each 0.1 unit decrease in extracellular pH, serum potassium would rise by 0.5 to 1.2 mEq/l. Thsi relationship is not universally valid, since organic (lactic) acidoses do not appear to apply (Lindeman and Pederson, 1983).

Another interesting phenomenon is the accumulation of basic amino acids in skeletal tissue of potassium deficient animals (Eckel, Norris and Pope, 1958). This response to potassium depletion is unique to the basic amino acids and the magnitude of change is substantial (Arnauld and Lachance, 1980a). A number of hypotheses have been proposed to explain this observation, including the action of basic amino acids as intracellular cations or the impairment of amino acid transport systems. However, experimental data refute both explanations.

CHLORIDE

Chloride tends to be overlooked in many discussions on electrolyte metabolism. This is quite surprising, since chloride represents approximately 65% of the total anions in the extracellular fluids; consequently, the regulation of total body chloride is also closely linked to extracellular water balance.

The close association between chloride and sodium has meant that it has traditionally been considered as part of sodium metabolism. More recently, unique aspects of chloride transport, for example, have received considerable attention (Alperin *et al.,* 1985).

Chloride participates in many critical metabolic functions. For example, as hydrochloric acid, it plays a major role in gastric digestion. In the blood, its transfer between the erythrocytes and surrounding plasma, in what is referred to as the 'chloride shift' helps to support more efficient transport of carbon dioxide from the cells to the lungs.

OTHER ELECTROLYTES

There are other important electrolytes in the body. However, from a nutritional perspective, those of greatest interest have been discussed above; sodium, potassium and chloride. The other inorganic electrolytes which are present in the body in substantial quantity include calcium, magnesium and phosphate.

## Electrolyte disturbance

VOMITING

Vomiting can have a major effect on electrolyte balance. Vomitus generally contains more chloride than sodium, since it is rich in HCl, and a relatively large amount of potassium. Vomiting generally is associated with metabolic alkalosis, depletion of chloride reserves, potassium deficiency and dehydration. The most profound effects are associated with the hypochloraemia and hypokalaemia. Clinically, patients suffering from gastric fluid losses experience muscle weakness and cardiac arrhythmia, muscle tetany, respiratory depression and generalized weakness (Weinberg, 1986).

DIARRHOEA

Diarrhoea can have a profound influence on water and electrolyte metabolism. Whereas gastric contents are rich in acid, intestinal contents are highly alkaline, being rich in bicarbonate. Thus, whereas emesis results in alkalosis, diarrhoea is associated with metabolic acidosis. Diarrhoea also results in the abnormal loss of water, sodium, potassium and nitrogen (Krehl, 1966).

Diarrhoeas maybe categorized according to cause: osmotic, secretory or primary motility. Osmotic diarrhoea results from the ingestion of poorly absorbable solutes, such as magnesium sulphate, oxide or hydroxide, sodium phosphate, sulphate or citrate or mannitol. It may also be created by malabsorption or poor digestion of carbohydrates (fructose, lactose, etc.). Osmotic diarrhoea secondary to carbohydrate malabsorption can be differentiated from that due to poorly absorbed mineral salts by measuring faecal pH. The former is usually associated with low pH due to microbial fermentation, whereas the latter creates normal (7.0) or high pH.

Secretory diarrhoeas are due to depressed absorption of fluids secreted higher in the gastrointestinal tract, or due to chemicals that actually stimulate secretory activity in the gut. For example, ricinoleic acid, the active component in castor oil is an effective laxative because it stimulates secretory activity in the colon through a cAMP-dependent mechanism (Ammon, Thomas and Phillips, 1974). Previously, castor oil was thought to exert its effect by stimulating gut motility.

Oral rehydration therapy was a concept introduced into human medicine about two decades ago. The high cost and technical demands of conventional intravenous therapy limited its utility to cholera victims in remote areas. The development of oral treatment that was safe, simple and effective was a tremendous advancement. The key to success was the inclusion of glucose or similar carbohydrate that exploited the intestinal sodium:glucose cotransport system to increase sodium and thus water absorption (Cash, 1983). Nalin *et al.* (1970) clearly demonstrated the benefits of such oral therapy in reducing the requirement for intravenous fluids. Some typical rehydration formulae are shown in Table 15.5. Cash (1983) offered a

**Table 15.5** ORAL ELECTROLYTE SOLUTIONS FOR CHOLERA PATIENTS

| | *Solution* (mM) | |
|---|---|---|
| *Component* | *a* | *b* |
| Sodium | 100 | 96 |
| Potassium | 10 | 25 |
| Chloride | 70 | 72 |
| Bicarbonate | 40 | 24 |
| Citrate | – | 25 |
| Glucose | 120 | 111 |

[a] Pierce *et al.* (1969)
[b] Palmer *et al.* (1977)

specific formula comprising (per litre of distilled water): 8 g glucose, 4 g sodium chloride, 6.5 g sodium acetate (or 5.4 g sodium lactate) and 1 g potassium chloride. Starch has been suggested as a replacement for glucose, since it will be gradually hydrolysed to glucose, and thus reduce the osmotic load on the intestine (Jelliffe *et al.*, 1987). Nalin *et al.* (1970) have proposed that combining a neutral amino acid with glucose will provide an even more effective rehydration solution.

Oral rehydration therapy represents an excellent example of applying fundamental knowledge on physiology to resolve a practical problem.

## ENDOCRINE DISTURBANCE

Polyuria (excessive urine output) can result from endocrine disturbance, the two most common causes in humans being diabetes mellitus and diabetes insipidus. In the former, excessive glucose in the urine generates what is called a solute diuresis.

Thus, one can differentiate between the two forms of diabetes by measuring daily urine solute excretion. If it exceeds 1500 mOsm/kg, the polyuria is probably due to solute diuresis; an assay for glucose will determine if it is the problem substance. Conversely, if solute excretion is less than 1500 mOsm/kg, then the polyuria is probably due to water diuresis and is called diabetes insipidus.

Diabetes insipidus can be the result of one of three pathological situations: inadequate production of ADH, renal insensitivity to circulating ADH or reduced release of ADH due to high fluid intake (Vokes, 1987). In all cases, the final result is polyuria or excessive urine output (Vick, 1984).

## STARVATION AND REFEEDING

Starvation has profound effects on water and electrolyte retention. During the early phases of fasting, renal excretion of sodium rises (Bloom and Mitchell, 1960). Extended fasting is associated with sodium conservation and subsequent refeeding is accompanied by sodium accumulation (Boulter, Hoffman and Arky, 1973). Previous salt restriction will reduce starvation-induced natriuresis, but will not eliminate it entirely (Schloeder and Stinebaugh, 1966).

The relationship with body weight gain or loss is interesting, and has been exploited in some human diets to effect rapid weight loss (unfortunately of water, not of actual body mass). During the early phases of fasting, increased sodium losses lead to reduced fluid osmolality with concomitant suppression of ADH; urine excretion increases and body weight loss is rapid. Refeeding increases body sodium reserves which in turn affects osmolality, stimulates ADH release and urine output falls with an accompanying rise in body water and thus weight gain (Vokes, 1987).

Starvation also results in kaliuresis; part of the loss is due to the loss of body mass and associated breakdown of muscle cells. However, there is also an impairment of renal conservation of potassium (Drenick *et al.*, 1966). Although other minerals, such as calcium, phosphorus and magnesium are also excreted in increasing amounts during fasting, serum levels rarely fall below normal and no adverse effects are reported. In all cases, including the sodium and potassium deficits noted above, oral supplements are most effective in reversing the problem (Weisner, 1971).

## STRESS

The use of supplemental electrolytes during 'stressful' situations remains a somewhat confused topic, no doubt due in part to the difficulty in defining and quantifying 'stress'. There is some physiological basis for the practice, given that elevation of ACTH and/or glucocorticoids, a phenomenon associated with stress, has been shown to impair the utilization of many minerals including potassium (Kem and Trachewsky, 1983) and calcium (Kenny, 1981). For example, glucocorticoids have been shown to inhibit calcium absorption from the intestinal tract in the pig (Fox, Care and Marshall, 1978), chicken (Corradino, 1979) and humans (Klein *et al.*, 1977). Rude and Singer (1982) provided a thorough review of the role of adrenal corticoids in mineral metabolism.

Heroux (1981) discussed how cold stress increases renal excretion of magnesium and potassium, but not sodium. Apparently, the effects lessen over time, presumably as the animal acclimatizes to the environment. However, the challenge remains to determine the extent to which electrolyte requirements might be affected by environmental stress and whether normal diet composition meets the animal's needs under practical conditions.

## Diagnosis of electrolyte disturbance

Diagnosis of mineral disturbance under clinical circumstances is quite difficult. Many overt symptoms of deficiency or imbalance are not specific to one mineral, so that differential diagnosis is particularly trying. Conversely, deficiencies of single minerals may occur, but often, inadequate dietary supplementation leads to multiple symptoms. Also, observable disturbance of dietary inadequacy is rarely acute, and chronic pathologies develop over extended periods of time. Clearly, more definitive means of evaluating electrolyte status than gross observation are required. The following is a list of some of the approaches that may be applied to evaluating the electrolyte status of farm animals. It is not intended to be inclusive, but rather provide a brief overview of some possible approaches.

### HAEMATOCRIT

Haematocrit is a measure of the proportion of the blood represented by red blood cells. Thus, hypovolaemia (loss of water) or hypervolaemia (water accumulation) may be reflected not only in altered haematocrit, but in plasma protein values as well. In fact, plasma osmolality is a more sensitive indicator of water status than is haematocrit.

### PLASMA SODIUM

This is generally reduced in primary sodium depletion. However, if sodium loss is accompanied by dehydration, changes in the concentration of sodium may not reflect the absolute loss of sodium from the body. Plasma sodium concentration may also be misleading when osmotically active solutes accumulate in the blood, drawing water from the cells into the plasma and diluting sodium concentration.

### PLASMA POTASSIUM

This is not a good indicator of potassium status since most of the body potassium is found in the intracellular space. Depression of plasma potassium requires a severe depletion of potassium or a disturbance of water balance, in which case plasma potassium may reflect water, not potassium, disturbance.

### PLASMA BICARBONATE AND CHLORIDE

These are useful indicators if evaluated in concert with plasma sodium and potassium. The anion gap can be calculated as shown in equation 15.4 (Cohen,

1984), identifying the nature of a disturbance or the success of treatment. By measuring known ion content, and assuming electroneutrality of gross body fluids, one can estimate the presence of other counterbalancing ions, such as lactate, sulphate, phosphate, ketoacids, etc.

$$\text{Anion gap} = Na^+ + K^+ - HCO_3^- - Cl^- \tag{15.4}$$

BLOOD pH, $pCO_2$ AND $HCO_3$

These define the acid-base status of the animal and therefore assist in determining if the disturbance includes an acidaemia or alkalaemia.

TISSUE CONTENT

Depending on the nature of the mineral and the suspected disturbance, tissue biopsy may prove useful. However, standard values are difficult to obtain and many non-pathological circumstances are known to alter analytical values.

## New approaches

DIETARY UNDETERMINED ANION

The balance of macromineral cations and anions has recently become a topic of greater interest among animal nutritionists. The reports of Mongin (1981) in poultry, Leibholz *et al.* (1966), Yen, Pond and Prior (1981) in pigs, Thacker (1959) in rabbits, Chui, Austic and Rumsey (1984) in fish and Block (1984) in cattle demonstrate the importance of considering dietary electrolyte balance as an entity quite distinct from that involving the individual ions. All of the above reports revealed a close relationship between the balance of dietary cations and anions and the performance of the respective species. This broad application across so many species reflects the fundamental nature of the phenomenon.

Two estimates of electrolyte balance have been proposed (Austic and Patience, 1988); one, the more comprehensive is called dietary undetermined anion (dUA) and is calculated as:

$$dUA = (Na^+ + K^+ + Ca^{2+} + Mg^{2+}) - (Cl^- + H_2PO_4^- + HPO_4^{2-} + SO_4^{2-}) \tag{15.5}$$

Dietary electrolyte balance (dEB) considers only the monovalent ions and is calculated as:

$$dEB = Na^+ + K^+ - Cl^- \tag{15.6}$$

The relative advantages of each calculation have been discussed in detail elsewhere (Austic and Patience, 1988); suffice it to say that dUA is more comprehensive, considering all of the relevant fixed cations and anions, but is encumbered by the need for seven individual analyses in order to characterize a given diet or feedstuff. Dietary electrolyte balance is more convenient, requiring only three analyses but it ignores the potential contributions of the polyvalent ions.

The use of either must recognize the fact that other components of the diet, notably amino acid oxidation, can and do influence acid-base contributions (Brosnan and Brosnan, 1982; Chan, 1981). Also, adjustment for the relevant differences in digestibility is required (Lennon, Lemann and Litzow, 1966).

In essence, dUA and dEB are estimates of the dietary content of metabolizable anions and cations that consume or generate acid upon metabolism (Patience, Austic and Boyd, 1987b). Since feedstuffs are electrically neutral, any excess of positive or negative charges calculated from equations 15.5 or 15.6 above must be indicative of the presence of counterbalancing metabolizable anions or cations. In essence, dUA or dEB employ the assumption of electrical neutrality to quantify elements in the diet which would otherwise be difficult to analyse. These anions and cations generate base or acid, respectively, upon metabolism. Thus, a diet containing a relative excess of mineral cations will be more alkalinogenic than a diet containing a relative excess of mineral anions.

This phenomenon is independent of specific ion effects (Patience and Wolynetz, 1987). This was clearly demonstrated in an experiment in which pig performance was monitored in the presence of increasing dietary chloride (by replacing calcium carbonate with calcium chloride, so that calcium remained unchanged) at constant or varied dUA. To maintain constant dUA at elevated chloride, sodium and potassium, in a 2:1 molar ratio were added as their bicarbonates to the diet. Increasing dietary chloride at constant dUA had no effect on any of the variables measured; however, chloride-mediated changes in dUA depressed growth rate, feed intake and feed efficiency and also generated an apparent metabolic acidosis (Table 15.6).

Clearly, although most considerations of electrolyte nutrition deal with the function or metabolism of specific ions, this concept is more 'generic' and represents a combined effect. This occurs because the calculation of dUA and dEB is really a calculation of charges, and uses the macrominerals only as vehicles of this determination.

Alterations in dUA have been shown to influence the metabolism of a number of nutrients, including amino acids (Scott and Austic, 1978), vitamin D (Reddy *et al.*,

**Table 15.6** EFFECT OF ELEVATED CHLORIDE AT CONSTANT OR VARIED DIETARY UNDETERMINED ANION

| | | | | | |
|---|---|---|---|---|---|
| Diet content, assayed | | | | | |
| Sodium (g/kg) | 2.1 | 2.0 | 6.5 | 2.4 | 7.3 |
| Potassium (g/kg) | 7.5 | 7.6 | 10.8 | 7.6 | 11.4 |
| Chloride (g/kg) | 2.7 | 10.5 | 10.5 | 12.7 | 12.7 |
| dUA (mEq/kg) | 388 | 172 | 412 | 98 | 431 |
| Performance | | | | | |
| Growth rate (kg/day) | 0.70 | 0.65 | 0.72 | 0.56 | 0.71 |
| Feed intake (kg/day) | 1.54 | 1.47 | 1.55 | 1.30 | 1.54 |
| Feed conversion ratio (kg feed/kg liveweight gain) | 0.46 | 0.44 | 0.47 | 0.43 | 0.46 |
| Blood parameters | | | | | |
| pH | 7.21 | 7.14 | 7.19 | 7.11 | 7.21 |
| Bicarbonate (mmol/l) | 29 | 26 | 29 | 23 | 28 |
| Base excess (mmol/l) | 0.6 | −3.6 | −0.2 | −6.9 | −0.5 |

Patience and Wolynetz (1987). Effect of dUA but not chloride significant ($P<0.05$) for all variables shown above

1982) and calcium (Block, 1984) as well as affecting growth in all species considered: poultry (Mongin, 1981), pigs (Yen *et al.*, 1981) and fish (Chui *et al.*, 1984). Furthermore, alteration in dUA has been exploited under practical conditions to alleviate the adverse effects of heat stress in poultry (Teeter *et al.*, 1985).

An example of dietary electrolyte balance influencing amino acid metabolism occurs in the lysine-arginine antagonism in the chick, which can be exacerbated or ameliorated by electrolyte additions to the diet (O'Dell and Savage, 1966; Savage, 1972). It has been suggested that alkaline salts stimulate lysine oxidation and thus reduce the effect of the antagonism (Scott and Austic, 1978).

Johnson and Farrell (1985) reported that electrolyte balance interacted with energy utilization in young growing broiler chicks. They concluded that the electrolyte balance affected the bird's maintenance energy requirements.

Specific ion effects are different from the more generalized dUA and dEB. For example, Patience, Austic and Boyd (1986) reported that alkaline salts of potassium but not sodium depressed the digestibility of energy in growing pigs (Table 15.7). In this experiment, pigs were fitted with simple T cannulae proximal

**Table 15.7** EFFECT OF SODIUM AND POTASSIUM SUPPLEMENTATION ON APPARENT NUTRIENT DIGESTIBILITY IN PIGS

| | | | | |
|---|---|---|---|---|
| Diet content, assayed | | | | |
| Sodium (g/kg) | 2.4 | 7.5 | 10.2 | 2.2 |
| Potassium (g/kg) | 3.3 | 3.3 | 3.3 | 17.7 |
| Chloride (g/kg) | 3.5 | 3.1 | 3.4 | 3.5 |
| Ileal digestibility (%) | | | | |
| Nitrogen | 82.3 | 83.1 | 82.2 | 80.1 |
| Lysine | 85.0 | 87.0 | 86.6 | 82.9 |
| Tryptophan | 66.9 | 65.0 | 65.5 | 62.2 |
| Energy | 76.7 | 77.1 | 76.7 | 74.0 |
| Faecal digestibility (%) | | | | |
| Nitrogen | 84.9 | 83.1 | 82.2 | 80.1 |
| Lysine | 73.6 | 72.7 | 73.6 | 71.1 |
| Tryptophan | 78.7 | 77.0 | 77.8 | 73.9 |
| Energy | 85.2 | 84.8 | 85.1 | 82.7 |

Patience, Austic and Boyd (1986). The effect of potassium supplementation on apparent faecal energy digestibility significant ($P<0.05$)

to the ileo-caecal junction and fed diets based on maize and maize-gluten meal. Sodium or potassium were supplemented with their respective bicarbonate salts. There tended to be a generalized depression in nutrient absorption in the presence of excess potassium, but only the effect on energy was statistically significant ($P<0.05$). Since equivalent quantities of sodium and potassium had differing effects, it is clear that this was not related to electrolyte balance.

The relationship betwen dietary electrolytes and egg shell quality in the laying hen is well known (Hamilton and Thompson, 1980). Mongin (1981) has reviewed the subject thoroughly.

In dairy cattle, Block (1984) has confirmed the earlier observations of Dishington (1975) that electrolyte levels in the diet can have a profound effect on the incidence

**Table 15.8** EFFECT OF ELECTROLYTE BALANCE IN CATTLE DIET ON THE INCIDENCE OF MILK FEVER

| *No. cows* | *Ration* | *Milk yield* (kg) | *Milk fever* (%) |
|---|---|---|---|
| 19 | Anion | 7142 | 0 |
| 19 | Cation | 6656 | 47 |

Source: Block (1984)

of milk fever. His results, summarized in Table 15.8, demonstrate that a diet rich in anions reduces the incidence of milk fever, while one rich in cations has the opposite effect.

Although electrolyte balance in the diet has been shown to influence growth, its application in evaluating the quality of 'novel' feed ingredients has been sketchy. Miller (1970) demonstrated that the balance of electrolytes in the diet could affect the perceived quality and therefore value of fish meal in poultry diets. Adjustment of mineral mixture composition used in chick bioassays of fish meal significantly altered growth performance.

It is well established that bone metabolism responds to chronic acid-base disturbance (Lemann, Litzow and Lennon, 1966) and further that dietary electrolyte balance can effect such changes in acid-base status (Patience, Austic and Boyd, 1987a). Consequently, researchers have investigated a direct link between dietary electrolyte levels and skeletal integrity. In poultry, it is now accepted that adjustment in dietary electrolyte balance will influence the incidence of tibial dischondroplasia (Sauveur, 1984). However, attempts to find such a relationship in pigs has met with limited success (Van der Wal *et al.*, 1986).

### TAIL-BITING IN PIGS

Tail-biting and other cannibalistic activities of pigs has been a frustrating subject for many years. Although many environmental factors are known to encourage tail biting, a reasoned explanation has been lacking. Recently, however, Fraser (1987b) developed a working hypothesis that helps to focus much of the vague reasoning associated with tail-biting. By developing an experimental model for his research, involving pseudo-tails made from 1.3-cm thick sash cord, and studying the activity of pigs very intensively (one experiment alone employed over 211 000 observations), he has formulated an interesting description of the phenomenon.

He suggests in his model that tail-biting is a two-phase event. Confinement housing, devoid of straw and other chewable objects, causes the pig to direct his inate chewing behaviour to whatever is at hand, such as another pig's tail. Large pen size or overcrowding increases the opportunity for such behaviour. Other stressors may also contribute to the problem. Circumstances such as poor health, which creates a general apathy on the part of the victim, or poor environment, which causes general discomfort and associated hyperactivity, may lead to greater chewing activity.

Once the tail is wounded, and blood appears, phase two begins. Attraction to blood may be increased by what is often referred to as salt appetite, a phenomenon observed in many mammalian species (Mackay, 1979; Denton, 1982). An unbalanced diet, lack of salt or dietary monotony may contribute to the problem.

This model explains why many researchers have been unable to initiate tail-biting by feeding a low-salt diet, since phase one is required to create an outbreak, and salt deficiency only becomes a factor in phase two. Furthermore, it encompasses the many factors suggested to predispose pigs to tail-biting, such as overcrowding, poor ventilation, lack of feeder or waterer space, etc. (van Putten, 1969; Jericho and Church, 1972; Ewbank, 1973; Kelley, McGlone and Gaskins, 1980; Penny, Walters and Tredget, 1981; Lohr, 1983; Fraser, 1987a).

Fraser (1987b) proposes that stress may also contribute to the problem. For example, it has been shown that elevation of circulating ACTH, which can occur under the influence of 'stress', increases sodium appetite in other species such as sheep (Weisinger *et al.*, 1980) and rats (Weisinger *et al.*, 1978). Fraser hypothesizes that stress in pigs may also increase sodium appetite and thus exacerbate the attraction to blood as described above.

## References

ADROGUE, H.J. and MADIAS, N.E. (1981). *American Journal of Medicine,* **71**, 456–467

ALCANTARA, P.F., HANSON, L.E. and SMITH, J.D. (1980). *Journal of Animal Science,* **50**, 1092–1101

ALPERIN, R.J., HOWLIN, J.J., PREISIG, P.A. and WONG, K.R. (1985). *Journal of Clinical Investigation,* **76**, 1360–1366

AMMON, H.V., THOMAS, P.J. and PHILLIPS, S.F. (1974). *Journal of Clinical Investigation,* **53**, 374–379

ANDERSSON, B., ANDERSSON, H., AUGUSTINSSON, O., FORSGREN, M., HOLST, H. and JONASSON, H. (1986). *Acta Physiologica Scandinavica,* **126**, 9–14

ARNAUD, J. and LACHANCE, P.A. (1980). *Journal of Nutrition,* **110**, 2480–2489

AUSTIC, R.E. and PATIENCE, J.F. (1988). *Critical Reviews in Poultry Biology*, **1**, 315–345

BASTL, C.P., BINDER, H.J. and HAYSLETT, J.P. (1980). *American Journal of Physiology,* **238**, F181–F186

BAYLIS, P.H., ZERBE, R.L. and ROBERTSON, G.L. (1981). *Journal of Clinical Endocrinology and Metabolism,* **53**, 935–940

BIA, M.J. and DeFRONZO, R.A. (1981). *American Journal of Physiology,* **240**, F257–F268

BLOCK, E. (1984). *Journal of Dairy Science,* **67**, 2939–2948

BLOOM, W.L. and MITCHELL, W. JR (1960). *Archives of Internal Medicine,* **106**, 321–326

BOULTER, P.R., HOFFMAN, R.S. and ARKY, R.A. (1973). *Metabolism,* **22**, 675–683

BROBST, D. (1986). *Journal of the American Veterinary Medical Association,* **188**, 1019–1025

BROSNAN, J.T. and BROSNAN, M.E. (1982). In *Advances in Nutrition Research*, pp. 77–105. Ed. Draper, H.H. Plenum Press, New York

BURNELL, J.M., VILLAMIL, M.F., UYENO, B.T. and SCRIBNER, B.H. (1956). *Journal of Clinical Investigation,* **35**, 935–939

CASH, R.A. (1983). In *Diarrhea and Malnutrition*, pp. 203–222. Ed. Chen, L.C. and Scrimshaw, N.S. Plenum Press, New York

CHAN, J.C.M. (1981). *Federation Proceedings,* **40**, 2423–2428

CHUI, Y.N., AUSTIC, R.E. and RUMSEY, G.L. (1984). *Comparative Biochemistry and Physiology,* **78B**, 777–783

COHEN, R.D. (1984). In *Clinical Physiology*, pp. 1–40. Ed. Campbell, E.J.M., Dickinson, C.J., Slater, J.D.H., Edwards, C.R.W. and Sikora, K. Blackwell Scientific, Oxford

CORRADINO, R.A. (1979). *Archives of Biochemistry and Biophysics,* **192**, 302–310

COX, M., STERNS, R.H. and SINGER, I. (1978). *New England Journal of Medicine,* **299**, 525–532

DENTON, D. (1982). *The Hunger for Salt.* Springer, Berlin

DISHINGTON, I.W. (1975). *Acta Veterinaria Scandinavica,* **16**, 503–512

DRENICK, E.J., BLAHD, W.H., SINGER, F.R. and LEDERER, M. (1966). *American Journal of Clinical Nutrition,* **18**, 278–285

ECKEL, R.E., NORRIS, J.E. and POPE, C.E. II (1958). *American Journal of Physiology,* **193**, 644–652

EDELMAN, I.S. and LEIBMAN, J. (1959). *American Journal of Medicine,* **27**, 256–277

EWBANK, R. (1973). *British Veterinary Journal,* **129**, 366–369

FIELD, M.J., STANTON, B.A. and GIEBISCH, G.H. (1984). *Kidney International,* **25**, 502–511

FITZSIMMONS, J.T. (1972). *Physiological Reviews,* **52**, 468–561

FOX, J., CARE, A.D. and MARSHALL, D.H. (1978). *Journal of Endocrinology,* **78**, 187–194

FRASER, D. (1987a). *Applied Animal Behaviour Science,* **17**, 61–68

FRASER, D. (1987b). *Canadian Journal of Animal Science,* **67**, 909–918

FREGLY, M.J. (1981). *Annual Review of Nutrition,* **1**, 69–93

GARDENSWARTZ, M.H. and SCHRIER, R.W. (1982). In *Sodium: its Biological Significance*, pp. 19–71. Ed. Papper, S. CRC Press, Inc., Boca Raton

GEORGIEVSKII, V.I. (1982a). In *Mineral Nutrition of Animals*, pp. 91–170. Ed. Georgievskii, V.I., Annenkov, B.N. and Samokhin, V.T. Butterworths, London

GEORGIEVSKII, V.I. (1982b). In *Mineral Nutrition of Animals*, pp. 79–89. Ed. Georgievskii, V.I., Annenkov, B.N. and Samokhin, V.T. Butterworths, London

HAMILTON, R.M.G. and THOMPSON, B.K. (1980). *Poultry Science,* **59**, 1294–1303

HEROUX, O. (1981). In *Handbook of Nutritional Requirements in a Functional Context*, pp. 523–542. Ed. Rechcigl, M. CRC Press, Inc., Boca Raton

JELLIFFE, E.F.P., JELLIFFE, D.B., FELDIN, K. and NGOKWEY, N. (1987). *World Review of Nutrition and Dietetics,* **53**, 218–295

JERICHO, K.W.F. and CHURCH, T.L. (1972). *Canadian Veterinary Journal,* **13**, 156–159

JOHNSON, R.J. and FARRELL, D.J. (1985). In *Energy Metabolism of Farm Animals*, pp. 102–105. Ed. Moe, P.W., Tyrrell, H.F. and Reynolds, P.J. European Assoc. for Animal Production

KELLY, K. W., McGLONE, J.J. and GASKINS, C.T. (1980). *Journal of Animal Science,* **50**, 336–341

KEM, D.C. and TRACHEWSKY, D. (1983). In *Potassium: Its Biological Significance*, pp. 25–36. Ed. Whang, R. CRC Press, Inc., Boca Raton

KENNY, A.D. (1981). *Intestinal Calcium Absorption and its Regulation.* CRC Press, Inc., Boca Raton

KLEIN, R.G., ARNAUD, S.B., GALLAGHER, J.C., DeLUCA, H.F. and RIGGS, B.L. (1977). *Journal of Clinical Investigation,* **60**, 253–259

KREHL, W.A. (1966). *Nutrition Today,* **1**, 20–23

LEGROS, J.J. (1979). In *Mechanisms of Osmoregulation in Animals*, pp. 611–636. Ed. Gilles, R. John Wiley and Sons, Chichester

LEIBHOLZ, J.M., McCALL, J.T., HAYS, V.W. and SPEER, V.C. (1966). *Journal of Animal Science,* **25**, 37–43

LEMANN, J. JR, LITZOW, J.R. and LENNON, E.J. (1966). *Journal of Clinical Investigation,* **45**, 1608–1614

LENNON, E.J., LEMANN, J. JR, and LITZOW, J.R. (1966). *Journal of Clinical Investigation,* **45**, 1601–1607

LINDEMAN, R.D. and PEDERSON, J.A. (1983). In *Potassium: Its Biological Significance*, pp. 45–75. Ed. Whang, R. CRC Press, Inc., Boca Raton

LOHR, J.E. (1983). *New Zealand Veterinary Journal,* **31**, 205

MacKAY, W.C. (1979). *Comparative Animal Nutrition,* **3**, 80–99

MADEC, F., CARIOLET, R. and DANTZER, R. (1986). *Annales de Recherches Veterinaires,* **17**, 177–184

MALOIY, G.M.O., MacFARLANE, W.V. and SHKOLNIK, A. (1979). In *Comparative Physiology of Osmoregulation in Animals*, pp. 185–209. Ed. Maloiy, G.M.O. New York, Academic Press

MASON, G.D. and SCOTT, D. (1972). *Quarterly Journal of Experimental Physiology,* **57**, 393–403

MILLER, D. (1970). *Poultry Science,* **49**, 1535–1540

MONGIN, P. (1981). *Proceedings of the Nutrition Society,* **40**, 285–294

NALIN, D.R., CASH, R.A., RAHMAN, M. and YUNUS, M.D. (1970). *Gut,* **11**, 768–772

O'DELL, B.L. and SAVAGE, J.E. (1966). *Journal of Nutrition,* **90,** 364–370

PALMER, D.L., KOSTER, F.T., ISLAM, A.F.M.R., RAHMAN, A.S.M.M. and SACK, R.B. (1977). *New England Journal of Medicine,* **297**, 1107–1110

PATIENCE, J.F. and WOLYNETZ, M.S. (1987). *Journal of Animal Science,* **65** (Supp. 1): 303–304

PATIENCE, J.F., AUSTIC, R.E. and BOYD, R.D. (1986). *Nutrition Research,* **6**, 263–273

PATIENCE, J.F., AUSTIC, R.E. and BOYD, R.D. (1987a). *Feedstuffs,* **59**(27), 13 and 15–18

PATIENCE, J.F., AUSTIC, R.E. and BOYD, R.D. (1987b). *Journal of Animal Science,* **64**, 457–466

PATIENCE, J.F., AUSTIC, R.E. and BOYD, R.D. (1987c). *Journal of Animal Science,* **64**, 1079–1085

PATIENCE, J.F., WOLYNETZ, M.S., FRIEND, D.W. and HARTIN, K.E. (1987d). *Canadian Journal of Animal Science,* **67**, 859–863

PENNY, R.H.C., WALTERS, J.R. and TREDGET, S.J. (1981). *The Veterinary Record,* **108**, 35

PIERCE, N.F., SACK, S.B., MITRA, R.C., BANWELL, J.C., BRIGHAM, K.L., FEDSON, D.S. and MONDAL, A. (1969). *Annals of Internal Medicine,* **70**, 1173–1181

POND, W.G. and HOUPT, K.A. (1978). *The Biology of the Pig.* Cornell University Press, Ithaca

QUINTON, P.M. (1979). *Comparative Animal Nutrition,* **3**, 100

REDDY, G.S., JONES, G., KOOH, S.W. and FRASER, D. (1982). *American Journal of Physiology,* **243**, E265–E271

ROBERTSON, G.L. (1977). *Recent Progress in Hormone Research,* **33**, 333–374

ROBERTSON, G.L. and BERL, T. (1985). In *The Kidney*, pp. 385–432. Ed. Brenner, B.M. and Rector, F.C. W. Saunders Co., Philadelphia

ROSE, D.B. (1977). *Clinical Physiology of Acid-Base and Electrolyte Disorders.* McGraw-Hill Book Co., New York

RUDD, T.G., PAILTHORP, K.G. and NELP, W.B. (1972). *Journal of Laboratory and Clinical Medicine,* **80**, 442–448

RUDE, R.K. and SINGER, F.R. (1982). In *Disorders of Mineral Metabolism*, pp. 482–556. Ed. Bonner, F. and Coburn, J.W. Academic Press, New York

SAUVEUR, B. (1984). *World's Poultry Science,* **40**, 195–206

SAVAGE, J.E. (1972). *Poultry Science,* **51**, 35–43

SCHLOEDER, F.X. and STINEBAUGH, B.J. (1966). *Metabolism,* **15**, 838–846

SCOTT, R.L. and AUSTIC, R.E. (1978). *Journal of Nutrition,* **108**, 137–144

SIENKO, M.J. and PLANE, R.A. (1979). *Chemistry Principles and Applications.* McGraw-Hill Book Co., New York

STERN, P.H. (1981). *Calcified Tissue International,* **33**, 1–4

TANNEN, R.L. (1986). In *Fluids and Electrolytes*, pp. 150–228. Ed. Kokko, J.P. and Tannen, R.L. W.B. Saunders Co., Philadelphia

TEETER, R.G., SMITH, M.O., OWENS, F.N., ARP, S.C., SANGIAH, S. and BREAZILE, J.E. (1985). *Poultry Science,* **64**, 1060–1064

THACKER, E.J. (1959). *Journal of Nutrition,* **69**, 28–32

THORNTON, S.N., BLADWIN, B.A. and PURDEW, T. (1985). *Quarterly Journal of Experimental Physiology,* **70**, 549

THRASHER, T.N., KEIL, L.C. and RAMSAY, D.J. (1982). *American Journal of Physiology,* **243**, R354–R362

VAN DER WAL, P.G., HEMMINGA, H., GOEDEGEBUURE, S.A. and VAN DER VALK, P.C. (1986). *Veterinary Quarterly,* **8**, 136–144

VAN PUTTEN, G. (1969). *British Veterinary Journal,* **125**, 511–517

VICK, R.L. (1984). *Contemporary Medical Physiology.* Addison-Wesley Publ. Co., Menlo Park

VOKES, T. (1987). *Annual Review of Nutrition,* **7**, 383–406

WEINBERG, J.M. (1986). In *Fluids and Electrolytes*, pp. 742–759. Ed. Kokko, J.P. and Tannen, R.L. W.B. Saunders Co., Philadelphia

WEISNER, R.L. (1971). *American Journal of Medicine,* **50**, 233–240

WEISINGER, R.S., DENTON, D.A., McKINLEY, M.J. and NELSON, J.F. (1978). *Pharmacology, Biochemistry and Behaviour,* **8**, 339–342

WEISINGER, R.S., COGHLAN, J.P., DENTON, D.A., FAN, J.S.K., HATZIKOSTAS, S., McKINLEY, M.J., NELSON, J.F. and SCOGGINS, B.A. (1980). *American Journal of Physiology,* **239**, E45–E50

YEN, J.T., POND, W.G. and PRIOR, R.L. (1981). *Journal of Animal Science,* **52**, 778–782

ZERBE, R.L. and ROBERTSON, G.L. (1983). *American Journal of Physiology,* **224**, E607–E614

16

# MANIPULATION OF THE GUT ENVIRONMENT OF PIGS

T.L.J. LAWRENCE
*University of Liverpool, UK*

## Introduction

In the first instance an explanation of the title of this chapter is necessary. What is the meaning of the word 'manipulation' and the phrase 'gut environment', and why should it be necessary to consider any sort of manipulation? These are points that need clarifying. 'Gut environment' is not easy to define and neither is it easy to explain in an unambiguous manner the word 'manipulation'. A starting point is to define a baseline. Here it is assumed that the normal physiology and biochemistry of the small and large intestine, reviewed by Rerat (1978, 1984) and Laplace (1984), form the baseline, to be elaborated upon below. Manipulation is any change in that baseline beyond that caused by feeding the type of complete diet, formulated to meet known nutrient requirements and perhaps, certain production targets, used in the majority of cases in practice on a twice daily restricted feeding regimen. Inherent in this approach is the acceptance of the premise that the majority of growing pigs are fed restrictedly, and twice daily, on diets which are complete and which are manufactured within reasonably narrow limits in terms of nutrient composition relative to live weight and level of feeding. The reasons behind the need to manipulate the environment are concerned with the high rates of morbidity and mortality, resultant from bacterial (particularly *Escherichia coli*) infections and gastric abnormalities (e.g., ulceration of the *pars oesophagea*), found in practice and the need to be less dependent on drug therapy for their alleviation and control. The possibilities of dietary alleviation and control rest on tenuous threads of evidence in many instances. Thus, this chapter is one in which the philosophy of such an approach is aired although in certain areas some firm evidence is available to support the hypotheses put forward.

## Normal gut function

The role of the stomach must be regarded as one of storing and mixing food which has been eaten. Food is acidified and the onset of enzymatic and microbial degradation is initiated. Gastric emptying is basically a discontinuous process characterized by successive emissions of volumes of digesta varying normally between 5 ml and 50 ml but which may be as high as 120 ml immediately after the intake of food (Rerat, 1984). Emptying starts soon after food is eaten and usually takes a long time to end (Cuber and Laplace, 1979). For example, with a semi-purified diet based on maize starch, Cuber and Laplace (1979) found that emptying started whilst the act of eating was still in progress, with about 20–30% of the dry matter intake leaving the stomach in approximately 30 min. Subsequently, emptying was slower and more regular with virtually all of the dry matter ingested leaving the stomach within 24 h of the start of feeding. Where the interval between feeding times is regular and of approximately 12 h duration, the residues of the former meal (amounting to 20–25% of the intake) are emptied first, but mixed with the first fractions of the new meal (10–15%), within approximately 1 h of the new food being offered (Cuber and Laplace, 1979). In addition, the finest and most acid-soluble particles are emptied from the stomach first (Hill, Noakes and Lowe, 1970).

In the small intestine the migrating motor complex, the myoelectric migrating complex, gut motility and peristalsis and the passage of food along the digestive tract are all to a greater or lesser extent interlinked (Laplace, 1984). The myoelectric migrating complex is characterized by periods of regular and irregular movement (spiking activity) as well as by a quiescent phase (Laplace, 1984). When given one or two feeds daily there is a continuous spiking activity in the proximal small intestine which lasts for 2–3 h (Ruckebusch and Bueno, 1976; Laplace, 1978). Peristaltic waves propel abrupt gushes of digesta along the tract, the digesta then spreading distally over the surface of the mucosa. When the spiking activity arrives at the segment where the digesta lie, they are gathered together to form a bolus which is propelled in front of the spikes. Therefore, apart from a post-prandial interruption, the flow of digesta occurs in batches and the rate of flow along the whole intestine depends upon the rate of migration of the myoelectric migrating complex which, according to Laplace (1981), acts as a 'cruise control mechanism'.

In the duodenum digesta are mixed with bile and pancreatic juice. There is an increase in bile salt secretion immediately after feeding and within 24 h of feeding a 45 kg pig may secrete about 2100 ml of bile containing 37 g of bile salts (Juste, Corring and Breaut, 1979). In addition, there is a further secretion of digestive juices, for hydrolysis of digesta, and this causes the dry matter of the digesta to decrease. The proximal small intestine is the primary site of protein digestion and the soluble nitrogen fraction in this part of the digestive tract contains a high proportion of α-amino nitrogen (57–71%) together with peptides (27%) and proteins (20%). The overall effect is to increase pH but cyclical variation is evident within 2.5 h of feeding and this corresponds to changes in spiking activity (Bueno and Fioramonti, 1982). The amplitude of this variation exceeds 2.5 units for between 8 and 14 h after feeding but after 24 h of fasting does not exceed 1 pH unit.

The amount of nitrogen reaching the ileo–caecal junction and the proportions of exogenous nitrogen vary according to the protein ingested and the rest of the diet. In the proximal part of the gut the apparent digestibility of nitrogen varies from 60 to 90% (Zebrowska, 1973; Keys and de Barthe, 1974b). Therefore differing quantities of amino acids and nitrogen enter the large intestine.

The ileal and rectal digestibility of fibre varies according to its botanical origin (Keys and de Barthe, 1974a) but there is a marked degradation in the distal part of the small bowel (Sambrook, 1979). Most fibre is broken down in the hind-gut but again there is variation according to source (Keys and de Barthe, 1974a; Sambrook, 1979). In the large intestine motor activity is present with approximately 1500 colonic spike bursts occurring daily on a standard cereal diet, the retention time of about 40 h (Fioramonti and Bueno, 1980a) representing approximately 80% of the total transit time (Keys and de Barthe, 1974b).

## Possible manipulations

There are many ways in which the 'baseline environment' within the tract, as described above, may be manipulated and modified. A survey of the literature in both the human and domesticated animal fields indicates that six areas of possible influence can be identified although basically these fall under two main headings, namely alterations in the diet on the one hand and in the method of feeding on the other. The possible mechanisms which may be involved, many of which may be interlinked are: water and bile salt adsorption, gel filtration and formation, gastric emptying rate, transit time through the various parts of and the total gastrointestinal tract, fermentation levels in the hind-gut and gastric digesta pH changes. The six areas are:

(1) Changes in feeding method. For example, restricted *vs ad libitum* feeding; wet *vs* dry feeding.
(2) Increasing the nutrient content of the diet either totally or individually. For example, increasing the crude protein level or the level of one specific amino acid by using a synthetic isomer.
(3) Increasing the overall crude fibre level of the diet.
(4) Adding chemicals which influence the pH of the digesta (e.g. organic acids).
(5) Modifying the physical form of either the cereal component or the complete diet.
(6) Using probiotics in the diet.

### FEEDING METHOD

Although, as postulated, most pigs are given their food twice daily, some fed restrictedly are given their food once daily and others are given their food *ad libitum*, at least for part, if not the whole, of their growing period. With *ad libitum* feeding food is offered dry; with restricted feeding the food is offered either dry or mixed with varying quantities of water. The ratio of water to dry food may vary between 2:1 and 5:1 with the advent of pipeline feeding systems necessitating a ratio nearer the latter rather than the former.

Pigs fed *ad libitum* still have a myoelectric migrating complex (Laplace, 1978) and 20 complexes within 24 h have been recorded in the duodenum but with only 40% of these migrating along the entire length of the small intestine (Laplace, 1984). Thus, there is a response to *ad libitum* feeding but a different pattern is elicited compared with the situation on twice daily feeding as described previously from the work of Ruckebusch and Bueno (1976). This disruption of the normal pattern may reflect an increase in the volume of intestinal contents (Laplace, 1980).

Stomach emptying may also be influenced by *ad libitum* feeding and Rerat (1984) concluded that the amount of food in the stomach is the most important factor in this context. This view is supported by the findings of Hunt (1980) in studies in the human and therefore may be linked to the changes in intestinal content and the effects on the myoelectric migrating complex discussed above. Possibly linked to these are the changes which may be induced in the pH of the gastric contents (Lawrence, 1972 and Laplace, 1974) and the subsequent effect which this may have on gastric emptying (Rerat, Corring and Laplace, 1976), the characteristics of the digesta and their movement in the other parts of the intestinal tract.

## DIETARY CHANGES

### *Density changes*

During the past decade there has been a progressive tendency to give pigs diets which have become increasingly dense in nutrient content. The increased density has been effected in various ways by increasing the inclusion levels of fats, sugars, protein rich materials and synthetic isomers of certain amino acids. It would appear feasible that such modifications can have a big part to play in influencing the environment of the gastrointestinal tract. For example Rerat, Corring and Laplace (1976) suggest that the lower the protein content of the diet the faster the rate of gastric emptying but that free amino acids added to the diet will not modify gastric emptying unless given in large quantities. They also imply that lipid content of the diet can modify gastric emptying rate. Hunt (1980) suggests that there is a strong relationship between food intake and gastric emptying: the greater the concentration of energy in the gastric contents the slower the volume rate of transfer per unit of time to the duodenum. In the small intestine Armstrong and Cline (1977) found that, in fistulated pigs, high levels of crude protein (220 g/kg cf. 160 g/kg) produced high levels of intestinal fluid.

### *Crude fibre*

Although used widely the term crude fibre is of course essentially a term which is definable by chemical analysis but which because of its all-embracing nature gives no indication of the types and proportions of polysaccharides present in a particular food or diet. In the context of the considerations here it has a very limited value and it is very important to consider those polysaccharides which occur in the greatest amount and with the greatest frequency in plant materials that are used for feeding. These are cellulose, hemicellulose, pectin, lignin and the mucilages. In addition to the indigenous sources of these polysaccharides 'synthetic' forms are available (e.g. as solka floc). In the *Gramineae*, hemicellulose generally accounts for a higher proportion of the total polysaccharides than cellulose. However, in wheat bran, crude fibre accounts for only about 25% of the total cell wall material present (Van Soest and McQueen, 1973). Pectin, a polyuronic acid polymer, is found in the primary cell walls and intercellular layers. Details of analytical procedures which may be used to separate various polysaccharides in foods have been given by Southgate, Hudson and Englyst (1978). Certain analytical methods for determining the various fractions of crude fibre which are used commonly, indicate the different proportions of some of the various polysaccharides:

cellulose = difference between modified acid detergent fibre (MADF) and acid detergent lignin (ADL);

hemicellulose = difference between neutral detergent fibre (NDF) and MADF.

The effects of differing types of polysaccharide on the gut environment vary. However, increasing crude fibre *per se* may also modify the function of and environment within the digestive tract. Fioramonti and Bueno (1980b) found that adding 170 g bran/kg to a standard cereal diet increased retention time from 44 to 60 h (cf. 120 h on a milk-based diet) but decreased the number of long, and increased the number of short, spike-bursts recorded daily. They pointed out that one of the factors controlling colonic motility is the volume of the dietary residue entering the colon. The increased volume of digesta referred to here perhaps indicates the prime way in which fibre can affect the gut environment, that is through its adsorptive properties.

There are a number of different ways in which water is adsorbed on to the surface of fibre in the digestive tract but the end result is reasonably similar whichever method predominates: a semi-rigid, jelly-like mass is formed which holds all the liquid present. When a polysaccharide forms a gel, a gel-filtration system may develop. In the case of bile, in the upper small intestine, the conjugated bile acids (taurine and glycine conjugates of cholic and chenodeoxycholic acid) are weakly adsorbed on fibre. In the lower gut, however, the unconjugated acids are strongly adsorbed. The strong adsorptive properties in this region, where adsorption is greatest at an acid pH, reflect the hydrolysis which has taken place by bacteria, of the water-soluble conjugated acids, and the physical transformation accompanied by the formation of bacterial metabolites which has ensued (Eastwood, 1973).

The magnitude of the response elicited and the extent to which the physical characteristics outlined above become apparent depend on the type of fibre. McConnell, Eastwood and Mitchell (1974) investigated as sources of fibre, *in vitro*, 26 different raw materials which had been dried to powdered forms after various treatments had been imposed. These included cooking, soaking and heating to various temperatures. They found that none of these treatments affected water-holding capacity but that grinding materials, including bran, to fine powders in a rotary mill, decreased this capacity, although not greatly. There were tenfold differences at the extremes in water-holding capacity between the materials studied but the authors pointed out that the results *in vitro* may have little bearing on the results which might be obtained *in vivo* because no account could be taken of bacterial action in the caecum. In terms of water adsorption, lignin adsorbs water poorly but hemicellulose adsorbs it well (Eastwood, 1974) whilst purified cellulose is considerably less efficient than the indigenous cellulose and other polysaccharides of cereals (Partridge, 1978). In the case of cereals this is related to the fact that cellulose is only at best a poor gel formant, whereas hemicellulose, methyl cellulose and pectin form gels readily with water. The extremely hydrophilic nature of pectin, enabling it to adsorb up to 50 times its own weight as water, means that it has a very much stronger ability to adsorb water than does bran (Stephen and Cummings, 1979). Polysaccharides which are hydrophilic and which adsorb water readily tend to adsorb bile acids poorly compared with those which are hydrophobic. Thus, lignin adsorbs water poorly but adsorbs on to its surface, with ease, bile salts.

The overall effect of changes such as these is to affect the bulk of the digesta in the tract and this in turn can affect the rate of passage. Furthermore, the utilization of other nutrients in the diet may be affected if the overall level of fibre is increased. In terms of rate of passage of digesta, the general effect is that the increased bulk from the non-assimilable material has an effect on transit time so that there is less time for the processes of digestion and absorption to take place, the physicochemical effects from the increased bulk reducing the diffusion of digestion products towards the absorptive mucosal surfaces (Southgate, 1973). The bulk laxative effect is reflected in increases in stool weight and changes in dry matter content. Eastwood (1973) quotes the following increasing order of effectiveness in plant sources in increasing stool weight: cellulose, lucerne leaf meal, wheat bran, maize germ, sugar beet pulp and agar agar. A further effect, not necessarily significantly deleterious, is that fibre has a considerable cation binding capacity with the result that minerals are less available (Partridge, 1978). Also manipulation of gastric pH may ensue (Lawrence, 1972).

### *Chemicals*

A variety of chemicals are added to components of pig food. For example organic acids, in particular propionic acid, are used as preservatives of cereals containing more than 160 g/kg of moisture and which are stored aerobically. The possibilities of manipulating gastric pH by including organic acids in the drinking water has been demonstrated (Thomlinson and Lawrence, 1981). There are possibilities that this effect may influence events in other parts of the digestive tract. Lactic acid would appear to be the organic acid of choice and several workers have studied its effects (e.g. White *et al.*, 1969). It is possible that chemicals other than acids may be able to exert a similar influence on gastric pH, for example sodium hydrogen diacetate, but evidence is currently not available. If gastric pH can be influenced in this way then presumably gastric emptying may also be influenced (Hill, Noakes and Lowe, 1970).

Other chemicals can be added with a very different aim in mind. For example, various hydrocolloids added to the diet can affect the physical form of digesta and its rate of passage through the gut. Various gums, such as agar, carageenan and carob bean gum, sodium alginate and sodium carboxymethyl cellulose affect the adsorption of water in similar ways to fibre and accelerate the rate of passage of digesta (Gohl, 1977).

### *Physical form changes*

There are many possibilities. In particular the particle size distribution of cereals may vary widely according to such factors as the screen size used in milling the cereal and the moisture content of the cereal itself. Alternatively the diet may be in a meal or pelleted form when fed to the animal. Also fibrous materials may be presented to the digestive juices of the gut as different sized particles.

With cereals, different sized particles have been shown to modify the gut environment via differences reflected in rate of passage and stool dry matter content (Lawrence, 1970a). Also, gastric pH was affected (Lawrence, 1970b). In such cases it is to be presumed that the differences induced are consequential to differential adsorptive properties in the gastrointestinal tract as a whole. Diets offered in pelleted forms may also induce changes, particularly as pellets are likely to contain higher percentages of smaller sized particles than meals. Meyer (1980) points to the fact that there is a relationship between food particle size and gastric emptying in the sense that the stomach retains particles until they are fragmented below 0.5 mm diameter.

*Probiotics*

The term probiotic is used to describe lactic acid producing additives. Cultures of certain lactobacilli have this capacity and several workers have shown that lactobacilli can suppress haemolytic coliforms (e.g. Mitchell and Kenworthy, 1976) and improve growth and food conversion efficiency (e.g. Hale and Newton, 1979). Possible modes of action in reducing *Escheria coli* numbers may be associated with the production of antibiotic-type substances and lactic acid. In such cases there could be a concomitant reduction in intestinal pH and an influence on adhesion to and colonization of the digestive tract by bacteria and thus, ultimately, a prevention of toxic amine synthesis.

## Effects of manipulation on bacterial activity and gastric abnormalities

It would appear, from recent evidence from a wide variety of sources, that gastric ulcers occur frequently in pigs and are the cause of much concern. Other lesions, which may predispose ulcers, also appear to occur frequently and these are of varying severity and may culminate ultimately in frank ulcers which bleed profusely or perforate. Undoubtedly there are many factors which may be involved and a comprehensive list has been compiled by Smith (1980). It would appear that one of the main predisposing factors, if not a factor *per se*, is the particle size of the cereal component of the diet, assuming that, and as is the usual case, cereal forms a high proportion of the total diet. There are numerous reports of American-type diets, based on maize in particular, but also sorghum and wheat to a lesser extent, ground too finely causing gastric lesions of varying severity in the *pars oesophagea* region of the stomach (e.g. Reimann *et al.*, 1968). There is evidence now that UK barley, which is quantitatively the most important cereal used in pig feeding in this country, may elicit the same response if ground too finely (Lawrence, Thomlinson and Whitney, 1980). If this is so, and apart from consistently producing an optimal grist size, what preventive steps can be taken?

Adding materials rich in fibre to the diet can undoubtedly help in this context. However, the results obtained would appear to depend on the source of fibre and its physical form in the diet. Henry (1970) found that purified cellulose was effective in giving protection but that if ground too finely the ameliorating effect was diminished. Thus 80 g/kg of powdered cellulose was needed to give the same protection afforded by 50 g/kg of the same material in a coarser form. Grass meal as a source of fibre, and perhaps because it is usually very finely ground, may not be an effective protective agent (Bjorklund and Pettersson, 1976). Seether *et al.* (1971) point out that oat hulls *per se* do not prevent ulcers, the polar (ether extract) fraction of the hulls being the responsible protectant.

In work at this centre (Potkins, Lawrence and Thomlinson, unpublished) some detailed studies have been made on grist sizes of barley based diets, additions of fibrous materials to these diets and the incidence of gastric lesions. This work has shown consistently that lesions of varying type and severity can be induced by grinding barley through 1.5 mm screens to produce a modulus of fineness of grinding between 1.5 and 1.9 approximately. The experiments have lasted for about 100 days and have involved growing pigs from approximately 20 kg live weight to weights around those usually associated with marketing bacon pigs. Substituting 50 g/kg of barley with an equivalent amount of broad bran, either ground or unground, reduced the lesion inducing effect of the finely ground barley with a slightly greater effect from the unground material. In comparison, substitution of either 50 g/kg or 100 g/kg of barley with equivalent amounts of oatfeed had a much greater ameliorating effect although the two substitution levels induced the same effect. Subsequent studies showed that many of the lesions had probably developed after 50 to 60 days although their severity was less at this stage. In contrast, grinding barley through a 4.5 mm screen was associated with few lesions in pigs to which it was fed. The modulus of fineness of grinding in this case was approximately 2.6. However, when such a diet was pelleted, the incidence of gastric lesions was as high as with the finely ground barley. In all of these experiments it was deliberately planned that pigs from different sources should be used and the results indicate that there were distinct differences between sources. Accordingly, care needs to be taken in assessing the apparent differences between bran and oatfeed referred to above. This finding indicates that husbandry and feeding factors in early life may have a part to play and point to the need for further work in this area. The possible transient nature of some of the lesions cannot be overlooked (Tournut and Labie, 1970). However, overall there is other confirmatory work associating finely ground barley with gastric lesions and indicating that certain blood parameters may be affected also (Crabo, Björklund and Simonsson, 1973). Another factor which may need further investigation is the part which micro-organisms play in developing the lesions. In the work at this centre referred to above, certain species of *Lactobacillus* and yeasts have been detected and it is possible that yeasts may invade the gastric epithelium when the surface layer of *Lactobacilli* has been displaced (Tannock and Smith, 1970).

There is evidence in the rat and the dog that gastric ulceration causes a disruption of the myoelectric migrating complex (Fioramonti and Bueno, 1980b). Bile staining is a common feature in many cases where lesions are found (Reed and Kidder, 1970) but to what extent a refluxing of bile is linked to such a disruption is uncertain. However, hexosamine levels have been found to be higher in the mucosal layers in animals that have lesions and invariably the gastric digesta are very fluid. Thus it has been postulated that the fluid nature of the digesta, present from feeding a finely ground diet, is more frequently brought into contact with the unprotected *pars oesophagea* region and causes bile juices to be regurgitated readily. These and other gastric chemicals then precipitate the lesions which can culminate in ulcers. On the other hand, if this is the case then the work of Bunn *et al.* (1981) does not indicate that gastrin *per se* is a causative agent. The importance of the adsorptive properties of fibre in such situations is not difficult to appreciate. The adsorption of bile to a variety of fibrous sources, including bran, is well established (e.g. Calvert and Yeates, 1982) and if the digesta are rendered less fluid when fibre, particularly coarse as opposed to fine, is added, then a possible mechanism of prevention or alleviation becomes apparent. Recent evidence adds a further note to this conjecture. Sambrook (1981) reported that bile flow into the duodenum was affected by fibre type, being higher for a barley based diet in which bran was included than for a purified diet to which solka floc had been added.

It is feasible that effects such as those discussed above may be linked to suppressing the activity of certain bacteria, particularly *Escherichia coli*. For example, it is possible that the decrease in gastric pH from feeding coarsely ground barley-based diets, restrictedly and/or with bran included (Lawrence, 1970b; 1972), could have an effect on this activity. Furthermore, the changes in faecal dry matter and digesta transit time, relative to bile secretion and volatile fatty acid production, could have an integral part to play. In a similar manner, additions of organic acids and probiotics and the feeding of nutrient-dense diets could influence the sequence of events. What is the evidence on these points?

There is evidence, though not conclusive, that lactic acid added to the diet and/or drinking water, may control or prevent diarrhoea associated with some bacterial infections. Diarrhoea is preceded by gastric stasis (White *et al.*, 1969) and the rapid flow of excess digesta may disorganize the myoelectric migrating complex (Dardillat and Marrero, 1977; Defilippi and Valenzuela, 1981) and accentuate the condition. Lactic acid added to the diet can cause a reduction in pH of one unit (White *et al.*, 1969) and although in this case the duration of the gastric stasis was not altered, the severity of the scour, the bacterial population and the loss of condition were all reduced. Their later work (White *et al.*, 1972) showed little effect on gastric malfunction but the previously cited work of Hunt (1980) and Meyer (1980) on gastric emptying and that of Armstrong and Cline (1977) with high dietary levels of crude protein relative to fluid production, are clearly important relative to these concepts and the situations found in practice.

Bran can also reduce gastric pH (Lawrence, 1972) and Thomlinson and Lawrence (1981) found that the addition of lactic acid to the drinking water, or bran to the creep food, significantly reduced gastric pH and that in naturally occurring oedema disease and *Escherichia coli* enteritis the multiplication of *Escherichia coli* 0.141:K85(B) was delayed and mortality correspondingly reduced. Recently Rainbird and Low (1983) have demonstrated that 40 g/kg of bran substituted into a diet based on casein, maize starch, soyabean oil, tallow and cellulose, significantly slowed the emptying of dry matter from the stomach after 4 h had elapsed from the time the pigs were fed. Clearly, this indicates a method of manipulating the environment in the stomach and therefore, in the intestine, but to what extent, if at all, such events can be related to the change in pH discussed above, is uncertain.

The effects on the response of two strains (K1261 and K1362) of *Escherichia coli* of dietary protein level, energy source and the addition of lactose to the diet were studied by Armstrong and Cline (1976) using intestinal loop ligation techniques. Fluid accumulation from the two strains indicated their enteropathogenic nature but none of the dietary manipulations affected this response. However, when 200 g/kg of oats were added to the basal (control) diet, less fluid accumulated, especially in the absence of the two *Escherichia coli* strains. In further work (Armstrong and Cline, 1977) the crude protein level of the diet (160 cf. 220 g/kg), the environmental temperature (about 4 °C cf. 27 °C) and the proportion of oats in the diet (0 cf. 200 mg/kg) were all varied. The high level of protein (as previously stated) and the high level of oats gave, respectively, the highest and lowest intestinal fluid volumes. Also the lower of the two temperatures gave higher fluid accumulation and more diarrhoea. The maize/soya bean diet which was used as the control was associated with lower *Escherichia coli* numbers in the anterior small intestine and this was postulated as having been due to a slower gastric emptying time which exposed the ingested bacteria to a low pH for a longer period of time. The corollary was that there were higher numbers of *Escherichia coli* in the posterior portion of the small intestine, possibly because of the greater proliferation of *Escherichia coli* in this part of the tract.

The use of commercially available probiotics has received some attention (Pollman, Danielson and Peo, 1980a; 1980b; Pollman *et al.*, 1980). They used a probiotic containing *Lactobacillus acidophilus* and found that this was as effective as several different antibiotics in improving the performance of their pigs. However, in these particular studies *Escherichia coli* numbers were not suppressed although the work of Mitchell and Kenworthy (1976) indicated improvements in this direction. It would appear that further studies to investigate the growth of *Lactobacillus* bacteria on different types of substrate would be worthwhile.

## Effects of manipulation on nutritive value of diet and pig performance

It is pertinent to consider what effects might be induced in diet utilization and consequent pig performance from the dietary manipulations discussed above. Obviously it is desirable that the effects should be minimal.

It is outside the scope of this chapter to deal in detail with the effects of different feeding methods, physical form of the diet and diet density on pig response. Extensive reviews of the two former are those of Braude (1972), Vanschoubroek, Coucke and Van Spaendonck (1971) and Lawrence (1976), whilst the advantages of nutrient-dense diets on performance have been recorded by several workers (e.g. Lawrence, 1977).

The effects of the addition of fibre to the diet are best set against the data of Keys and De Barthe (1974a,b). These data show that approximately 80% of hemicellulose and 100% of cellulose digestion occurs in the large intestine. If this is the case it is appropriate to ask to what extent additions of fibre from natural and commonly used feedstuffs affect the normal digestion process, particularly in the small intestine. An unambiguous answer to this is difficult to find from the literature. Kennelly and Aherne (1980) reviewed published evidence on the effects of crude fibre on the apparent digestibility of dietary dry matter, nitrogen and gross energy and found the evidence to be extremely variable. The following quotation from their paper epitomizes the situation: 'Despite its aesthetic appeal, the temptation to ascribe a general cause and effect relationship between crude fibre and digestibility coefficients must be resisted. The present results indicate that the dietary model (the model took into consideration whether or not the diets were of equal energy and nitrogen content and the source of fibre) selected in diets with similar crude fibre levels is an important variable which can confound the results obtained'. In the context of the relatively small additions or substitutions of fibre from natural and commonly used feedstuffs considered here, it would appear, from various prediction equations available, very unlikely that there would be any major effect on dietary nutritive value or pig performance (e.g. Just, 1982). However, much may depend on the type of fibre used. For example Henry (1976) found that hemicellulose was more closely associated with reductions in digestible energy than cellulose. This was confirmed by Taverner and Farrell (1981) who found that neutral detergent fibre (of which hemicellulose is a major component) was a better predictor of ileal protein digestibility than acid detergent fibre (cellulose and lignin). Nevertheless the possibility that natural sources of fibre rich in cellulose, hemicellulose and lignin, at inclusion levels no greater than 100 g/kg, may have little effect on dietary nutritive value and pig performance, is supported by the work of Kass *et al.* (1980). They substituted 200, 400 or 600 g/kg of alfalfa meal in an otherwise normal diet; 200 mg/kg had no significant effect on growth rate and food conversion efficiency but the two other levels significantly depressed both parameters. The increases in rate of passage of digesta which were found with increasing levels of substitution were accompanied by progressive decreases in apparent digestibility of dry matter, cellulose, hemicellulose and nitrogen. Also the dry matter content of the digesta in all sections of the digestive tract was negatively correlated with fibre level of the diet.

Other sources of polysaccharides may, however, influence the nutritive value of the diet to a greater extent. For example, Murray, Fuller and Pirie (1977) investigated the effects of various types of fibre on protein digestibility by replacing starch in a control diet, containing starch, barley and soyabean meal, with either 100 g/kg of cellulose or 60 g/kg of one of the two gel formants methyl cellulose and pectin. They found that the gel formants, but not cellulose, gave a significant reduction in apparent nitrogen digestibility at the terminal ileum, the largest depression from 0.76 in the control diet to 0.48 in the substituted diet being with methyl cellulose. Protein-bound lysine digestibility was also reduced with methyl cellulose but free (synthetic) lysine was almost completely digested in all diets. The rate of passage of digesta to the terminal ileum was increased when the methyl cellulose was added. They concluded that the results suggested that bulk *per se* may have little or no effect on digestion of protein but that the hydrolysis of protein, rather than the absorption of the products of digestion, may be impaired when gel-forming polysaccharides are added to the diet. Preliminary work at this centre (Potkins, Lawrence and Thomlinson, unpublished) indicated that up to 50 g/kg of pectin or guar gum had little effect on the nutritive value of the diet.

If fibre levels are increased, then fermentation increases in the hind-gut can be expected and as a result increased quantities of volatile fatty acids will be produced. In terms of the overall energy metabolism of the animal it is unlikely that the increased levels of volatile fatty acids will contribute significantly to the total energy available to the animal (Farrell and Johnson, 1972), although there is evidence that they are absorbed efficiently (Lawrence, 1973). Thus it is very unlikely that additions of organic acids to diets, as gastrointestinal tract manipulators, will affect nutrient utilization and performance. Evidence to support this view is to be found from many studies where propionic acid treated grain (where propionic acid has been used as a preservative for damp grain) has been fed (e.g. Lawrence, 1971).

## References

ARMSTRONG, W.D. and CLINE, T.R. (1976). *J. Anim. Sci.*, **42**, 592–598

ARMSTRONG, W.D. and CLINE, T.R. (1977). *J. Anim. Sci.*, **45**, 1042–1050

BJÖRKLUND, N.E. and PETTERSSON, A. (1976). *Nordisk Veterinaermedicin*, **28**, 33–39

BRAUDE, R. (1972). In *Pig Production*, pp.279–291. Ed. Cole, D.J.A. Butterworths; London

BUENO, L. and FIORAMONTI, J. (1982). In *Motility of the Digestive Tract*. pp. 169–173. Ed. Wienbeck, M. Raven Press; New York

BUNN, C.M., HANSKY, J., KELLY, A. and TITCHEN, D.A. (1981). *Res. Vet. Sci.*, **30**, 376–378

CALVERT, G.D. and YEATES, R.A. (1982). *Br. J. Nutr.*, **47**, 45–52

CRABO, B., BJÖRKLUND, N.E. and SIMONSSON, A. (1973). *Acta Vet. Scand.*, **14**, 263–271

CUBER, J.C. and LAPLACE, J.P. (1979). *Ann. Zootech.*, **28**, 173–184

DARDILLAT, C. and MARRERO, E. (1977). *Ann. Biol. Anim. Biochim. Biophys.*, **17**, 523–530

DEFILIPPI, C. and VALENZUELA, J.E. (1981). *Scand. J. Gastroent.*, **16**, 977–979

EASTWOOD, M.A. (1973). *Proc. Nutr. Soc.*, **32**, 137–143
EASTWOOD, M.A. (1974). *J. Sci. Fd Agric.*, **25**, 1523–1527
FARRELL, D.J. and JOHNSON, K.A. (1972). *Anim. Prod.*, **14**, 209–218
FIORAMONTI, J. and BUENO, L. (1980a). *Br. J. Nutr.*, **43**, 155–162
FIORAMONTI, J. and BUENO, L. (1980b). *Digestive Dis. Sci.*, **25**, 575–580
GOHL, B. (1977). Report Department Animal Husbandry, Agricultural College of Sweden, 65 pp. Uppsala; Sweden
HALE, O.M. and NEWTON, G.L. (1979). *J. Anim. Sci.*, **48**, 770–775
HENRY, Y. (1970). *Ann. Zootech.*, **19**, 117–141
HENRY, Y. (1976). *Proceedings, First International Symposium on Feed Composition, Amino Acid Requirements and Computerization of Diets*, pp. 270–280. Utah State University; Utah
HILL, K.J., NOAKES, D.E. and LOWE, R.A. (1970). In *Physiology of Digestion and Metabolism in the Ruminant*. pp. 166–179. Ed. Phillipson, A.T. Oriel Press; Newcastle-upon-Tyne
HUNT, J.N.A. (1980). *Am. J. Physiol.*, **239**, Part 1. G1–G4
JUST, A. (1982). *Livestock Prodn Sci.*, **9**, 717–729
JUSTE, CATHERINE, CORRING, T. and BREAUT, Ph. (1979). *Ann. Biol. Anim. Biochim. Biophys.*, **19**, 79–90
KASS, MARIE L., VAN SOEST, P.J., POND, W.G., LEWIS, BERTHE and McDOWELL, R.E. (1980). *J. Anim. Sci.*, **50**, 175–191
KENNELLY, J.J. and AHERNE, F.X. (1980). *Can. J. Anim. Sci.*, **60**, 717–726
KEYS, J.E. and DE BARTHE, J.V. (1974a). *J. Anim. Sci.*, **39**, 53–56
KEYS, J.E. and DE BARTHE, J.V. (1974b). *J. Anim. Sci.*, **39**, 57–62
LAPLACE, J-P. (1974). *Recherches Médicin Véterinaire*, **150**, 121–129
LAPLACE, J-P. (1978). *Ann. Zootech.*, **27**, 377–408
LAPLACE, J-P. (1980). In *Current Concepts of Digestion and Absorption in Pigs*. pp. 24–27. Eds Low, A.G. and Partridge, I.G. NIRD Press; Reading
LAPLACE, J-P (1981). In *Nutrition in Health and Disease and International Development*. pp. 847–872. Eds Harper, A.E. and Davis, G.K. Alan Liss; New York
LAPLACE, J-P (1984). In *Function and Dysfunction of the Small Intestine*. pp. 1–20. Eds Batt, R.G. and Lawrence, T.L.J. Liverpool University Press; Liverpool
LAWRENCE, T.L.J. (1970a). *Anim. Prod.*, **12**, 139–150
LAWRENCE, T.L.J. (1970b). *Anim. Prod.*, **12**, 151–163
LAWRENCE, T.L.J. (1971). *J. Sci. Fd Agric.*, **22**, 407–411
LAWRENCE, T.L.J. (1972). *Br. Vet. J.*, **128**, 402–411
LAWRENCE, T.L.J. (1973). *Int. Res. Commun. Syst.*, **(73–3)** 45–6–1
LAWRENCE, T.L.J. (1976). *Proc. Nutr. Soc.*, **35**, 237–243
LAWRENCE, T.L.J. (1977). *Anim. Prod.*, **25**, 261–270
LAWRENCE, T.L.J., THOMLINSON, J.R. and WHITNEY, J.C. (1980). *Anim. Prod.*, **31**, 93–99
McCONNELL, A.A., EASTWOOD, M.A. and MITCHELL, W.D. (1974). *J. Sci. Fd Agric.*, **25**, 1457–1464
MEYER, J.H. (1980). *Am. J. Physiol.*, **239**, Part 2. G133–135
MITCHELL, I. de G. and KENWORTHY, R. (1976). *J. Appl. Bact.*, **41**, 163–174
MURRAY, A.G., FULLER, M.F. and PIRIE, A.R. (1977). *Anim. Prod.*, **24**, 139 (Abstr.)
PARTRIDGE, I.G. (1978). *Br. J. Nutr.*, **39**, 539–545
POLLMAN, D.S., DANIELSON, D.M. and PEO, E.R. (1980a). *J. Anim. Sci.*, **51**, 577–581
POLLMAN, D.S., DANIELSON, D.M. and PEO,E.R. (1980b). *J. Anim. Sci.*, **51**, 638–644

POLLMAN, D.S., DANIELSON, D.M., WREN, W.B., PEO, E.R. and SHAHANI, K.M. (1980). *J. Anim. Sci.*, **51**, 629–637

RAINBIRD, ANNA L. and LOW, A.G. (1983). *Proc. Nutr. Soc.*, **42**, 88A

REED, J.H. and KIDDER, D.E. (1970). *Res. Vet. Sci.*, **11**, 438–440

REIMANN, E.M., MAXWELL, C.V., KOWALZYK, T., BENEVENGA, N.J., GRUMMER, R.H. and HOEKSTRA, W.G. (1968). *J. Anim. Sci.*, **27**, 992–999

RERAT, A. (1978). *J. Anim. Sci.*, **46**, 1808–1837

RERAT, A. (1984). In *Function and Dysfunction of the Small Intestine*. pp. 21–38. Eds Batt, R.G. and Lawrence, T.L.J. Liverpool University Press; Liverpool

RERAT, A., CORRING, T. and LAPLACE, J-P. (1976). In *Protein Metabolism and Nutrition*. pp. 97–138. Eds Cole, D.J.A., Boorman, K.W., Buttery, P.J., Lewis, D., Neale, R.J. and Swan, H. Butterworths; London

RUCKEBUSCH, Y. and BUENO, L. (1976). *Br. J. Nutr.*, **35**, 397–405

SAMBROOK, I.E. (1979). *Br. J. Nutr.*, **42**, 279–287

SAMBROOK, I.E. (1981). *J. Sci. Fd Agric.*, **32**, 781–791

SEETHER, K.A., MIYA, T.S., PERRY, T.W. and BOEHM, P.N. (1971). *J. Anim. Sci.*, **32**, 1160–1163

SMITH, W.J. (1980). *The Pig Veterinary Society Proceedings*, pp. 1–13

SOUTHGATE, D.A.T. (1973). *Proc. Nutr. Soc.*, **32**, 131–136

SOUTHGATE, D.A.T., HUDSON, G.J. and ENGLYST, H. (1978). *J. Sci. Fd Agric.*, **29**, 979–988

STEPHEN, ALISON M. and CUMMINGS, J.H. (1979). *Proc. Nutr. Soc.*, **38**, 55A

TANNOCK, G.W. and SMITH, J.M.B. (1970). *J. Comp. Path.*, **80**, 359–367

TAVERNER, M.R. and FARRELL, D.J. (1981). *Br. J. Nutr.*, **46**, 181–192

THOMLINSON, J.R. and LAWRENCE, T.L.J. (1981). *Vet. Rec.*, **109**, 120–122

TOURNUT, J. and LABIE, C. (1970). *Proceedings of the Symposium on Stress in the Pig—Janssen Pharmaceuticals*. Beerse; Belgium

VANSCHOUBROEK, F., COUCKE, L. and VAN SPAENDONCK, R. (1971). *Nutr. Abst. Rev.*, **41**, 1–9

VAN SOEST, P.J. and McQUEEN, R.W. (1973). *Proc. Nutr. Soc.*, **32**, 123–130

WHITE, F.G., WENHAM, G., ROBERTSON, V.A.W. and RATTRAY, E.A.S. (1972). *Proc. Nutr. Soc.*, **31**, 67–71

WHITE, F., WENHAM, G., SHARMAN, G.A.M., JONES, A.S., RATTRAY, E.A.S. and McDONALD, I. (1969). *Br. J. Nutr.*, **23**, 847–858

ZEBROWSKA, T. (1973). *Roczniki Nauk Rolniczych B.*, **95**, 115–133

# 17

# ACIDIFICATION OF DIETS FOR PIGS

R. A. EASTER
*Department of Animal Sciences, University of Illinois, Urbana, Illinois, USA*

## Introduction

The evolution of a biologically and economically satisfactory strategy for feeding piglets weaned at an early age is not yet complete. Progress is being made on new research providing fundamental information on the functional capabilities, and limitations, of the gastrointestinal tract during the transition from sow's milk to dry, cereal-based diets. A key discovery was the recognition that the young piglet may not be able to maintain appropriate gastric pH during the period. The discussion that follows provides an overview of the research leading to that conclusion and the experimental evidence to establish the basis for the use of acidification in weanling pig diets.

## Physiological difficulties for the early-weaned piglet

Growth failure is a well established phenomenon in early-weaned pigs (Okai, Aherne and Hardin, 1976). Undoubtedly, this is a manifestation of an array of interacting environmental, social and physiological factors, not the least of which is digestive immaturity. The data reported by Etheridge, Seerley and Huber (1984) and presented in *Table 17.1* serve to illustrate this problem. Pigs were either provided cereal-based diets, beginning at day 7, and weaned at 21 days of age or allowed to suckle the sow for 35 days post partum without access to dry feed. Faecal samples which were collected and analysed provided an indication that fermentation in the lower bowel was significantly greater in pigs receiving the cereal-based diet. Faecal osmolarity and volatile fatty acid concentrations were both increased.

These results are not particularly surprising. It has been adequately demonstrated that the weanling pig is ill-prepared, enzymatically (Becker *et al.*, 1954; Corring, Aumaitre and Durand, 1978), to digest the complex carbohydrates found in most cereal-based weaner diets. Attempts to cause precocious maturation of digestive functions by treatment with hydrocortisone and/or ACTH have met with only limited success (Chapple, Cuaron and Easter, 1983). Incomplete digestion results in passage of fermentable substrate into the lower bowel. The osmotic upset observed by Etheridge, Seerley and Huber (1984) supports the notion that there is a relationship between diarrhoea and digestive development. Early-weaned pigs also lack the

**Table 17.1** EFFECT OF DIET-TYPE ON MEASURED OSMOLARITY AND OSMOTICALLY ACTIVE CONSTITUENTS IN FAECAL CONTENTS OF PIGS AT 35 DAYS OF AGE (VALUES, EXCEPT FOR pH, ARE EXPRESSED AS mosmol/litre)

| | *Dietary treatment* | | |
|---|---|---|---|
| | *Maize–soyabean meal* | *Oats–casein* | *Sow's milk* |
| pH | 5.9 | 6.5 | 7.1 |
| Lactic acid$^-$ | 14.1 | 5.5 | 0.3 |
| Total VFA$^-$ | 8.7 | 3.9 | 2.7 |
| $Na^+$ | 13.8 | 10.8 | 2.0 |
| $K^+$ | 15.0 | 10.0 | 20.0 |
| $Cl^-$ | 4.7 | 3.0 | 1.1 |
| $Ca^+$ | 1.4 | 0.5 | 0.3 |
| $R^-$ | 12.2 | 10.3 | 11.3 |
| Measured | 149.8 | 88.5 | 49.0 |

After Etheridge, Seerley and Huber (1984)

capacity to produce gastric acid and it is probable that this has a negative effect on digestion. This chapter will provide a review of the information that leads to the hypothesis that digestion and, consequently, growth efficiency can be enhanced by diet acidification.

The need for dietary acid presumes that there is a deficiency in the pig's ability to maintain proper gastric pH. Mature pigs adjust stomach pH by secretion of hydrochloric acid from the parietal cells. Although there is variation due to diet, time after the meal at which the pH is measured and sampling site, stomach of mature pigs can reach very acid values, i.e. pH 2.0–3.5 (Slivitskii, 1975 as cited by Kidder and Manners, 1978).

The situation in young pigs is quite different. Although a subject of debate for some time, it is now evident that the newborn pig does produce some hydrochloric acid (Forte, Forte and Machen, 1975; Cranwell, Noakes and Hill, 1976). Initially, the production is low but increases with advancing age (Cranwell, 1985). This is apparently the consequence of limited secretory capacity and not lack of stimulation. Cranwell (1985) reported that maximal acid output in response to intravenous betazole hydrochloride infusion averaged 3.4 mmol $H^+$/h for pigs at 9–12 days and increased to 7.6 mmol $H^+$/h when pigs reached an age of 27–38 days.

The suckling pig employs several strategies to solve the problem of limited acid secretion. First, the primary carbohydrate in sow's milk is lactose which can be converted to lactic acid by the *Lactobacillus* bacteria normally resident in the stomach. This, in fact, appears to be the primary method of gastric acidification in suckling pigs. Secondly, nursing pigs reduce the need for momentary secretion of copious amounts of acid by consuming frequent, relatively small, meals (Pond and Maner, 1984). Finally, diets differ greatly in buffering capacity (Manners, 1970). Sow's milk is undoubtedly easier to acidify than is a high-protein weanling diet that is richly supplemented with calcium carbonate. Maner *et al.* (1962) found that stomach pH values dropped below 2.0 within 2 h after feeding a casein–dextrose diet while more than 4 h were required for a similar pH to be reached when pigs were fed a soyabean protein–dextrose diet.

What are the consequences of failure to maintain a low gastric pH? There are two working hypotheses. First, stomach pH has a role in preventing the movement of viable bacteria from the environment into the upper small intestine (Stevens, 1977).

Second, hydrochloric acid is involved in the activation of pepsinogens. Additionally pepsin has two pH optima, one at pH 2.0 and another at pH 3.5 (Rerat, 1981). Thus, it is likely that pigs having an elevated gastric pH also experience a net reduction in efficiency of protein digestion.

## Response of piglets to organic acids

It isn't at all surprising that the first attempts to use acidification in pig farming were directed at the alleviation of post-weaning diarrhoea (*cf.* Kershaw, Luscombe and Cole, 1966). These workers found that 1% lactic acid addition to the drinking water would improve growth rate and feed efficiency. They also reported reductions in the *Escherichia coli* count in the duodenum and jejunum of pigs fed acids in the drinking water. In another experiment, Cole, Beal and Luscombe (1968) found that growth rate and feed efficiency were significantly improved by the addition of 0.8% lactic acid to the drinking water (*Table 17.2*). Moreover, there was a reduction in haemolytic *E. coli* counts in both the duodenum and jejunum.

The hypothesis that acidification may reduce the incidence of scours in young pigs was tested in an experiment by White *et al.* (1969). Pigs were separated from (or left with) the sow 48 h post partum. Those removed from the sow were given a standard rearing diet or that diet with sufficient lactic acid to reduce the diet pH to 4.8. Pigs fed the diets with lactic acid had lower stomach pH values than either those fed the standard diet or those that were allowed to suckle the sow. Acidification provided a prophylaxis against scouring (*Table 17.3*). These results imply an effect of lactic acid therapy on stomach pH values. This was confirmed by Thomlinson and Lawrence (1981). They also reported that the multiplication of *E. coli* 0141:K85(B) was reduced by acidification with a corresponding reduction in piglet mortality.

**Table 17.2** EFFECT ON GROWTH PERFORMANCE OF ADDING SEVERAL MATERIALS AT 0.8% TO THE DRINKING WATER OF POST-WEANLING PIGLETS

| | *Control* | *Lactic acid* | *Propionic acid* | *Calcium propionate* | *Calcium acrylate* |
|---|---|---|---|---|---|
| Feed intake (kg/day) | 1.009 | 1.027 | 0.927 | 0.931 | 0.868 |
| Liveweight gain (kg/day)[a] | 0.372 | 0.409 | 0.354 | 0.345 | 0.310 |
| Gain: feed ratio (kg gain/kg feed) | 0.367 | 0.395 | 0.373 | 0.370 | 0.355 |

After Cole, Beal and Luscombe (1968)
[a]Improvement due to lactic acid ($P < 0.05$)

**Table 17.3** EFFECT OF ACIDIFICATION OF THE DIET ON INCIDENCE OF SCOURING

| | | *Post-weaning diet treatment* | | |
|---|---|---|---|---|
| | *Nursed by sow* | *'Normal' diet* | *Lactic acid diet* | *High casein diet* |
| No. pigs | 6 | 7 | 6 | 6 |
| No. scouring | 4 | 7 | 3 | 6 |
| Mean days scouring/pig | 1.2 | 6.0 | 1.0 | 3.5 |

After White *et al.* (1969)

The ultimate value of organic acids in the prophylaxis of post-weaning diarrhoea is yet to be fully established. It is of passing interest that 'folk' remedies in tropical regions of the world include the use of lime juice to treat scours in young piglets (Costa Rican swine producer, personal communication).

Research attention in the 1970s turned from reduction of diarrhoea to more general effects on growth rate and efficiency of feed utilization. A report by Kirchgessner and Roth (1982) summarized several experiments involving the use of fumaric acid. These studies showed improved liveweight gain, feed intake and feed conversion efficiency when weaner pigs were fed diets supplemented with 1.5–2.0% fumaric acid. They showed that older pigs also responded to fumaric acid but the magnitude of the response was less. The effects were attributed in part to improved digestibilities of nutrients. Nitrogen balance was improved by 5–7% and the metabolizable energy values of the diets were increased by 1.5–2.1% (Kirchgessner and Roth, 1980).

Attempts to replicate the European work in North America have met with mixed success. In an early experiment, Lewis (1981) fed pigs, weaned at four weeks of age, diets containing graded levels of fumaric acid. These diets were based on maize and soyabean meal and contained small amounts of dried whey, fat and fish solubles. Gain was improved but evidence of an improvement in feed intake or feed conversion efficiency was lacking. Moreover, the incidence of scours was unaffected.

Using relatively more complex diets, i.e. containing barley, wheat, oat groats, soyabean meal, fish meal, dried skim milk, tallow, vitamins and minerals, Falkowski and Aherne (1984) reported that grain tended to be improved by 1 or 2% addition of either fumaric acid or citric acid but the effect was not significant. Feed conversion efficiency was improved ($P < 0.05$) by 5–10%, depending on treatment. The response was similar for both fumaric and citric acid. The trend for feed conversion efficiency to respond more consistently than growth has been confirmed by Giesting and Easter (1985).

The response to acid appeared to occur independently of effects due to other growth promotants. Edmonds, Izquierdo and Baker (1985) conducted a series of factorial experiments to evaluate the interaction of citric acid, copper sulphate and antibiotics relative to effects on growth performance of weanling pigs. Acidification improved performance in the presence and absence of both copper and antibiotic. As in previous studies, the feed conversion efficiency response tended to be of greater magnitude than the growth response.

Similar positive benefits were noted when 3.0% citric acid was added to the diets of pigs weaned at ten days of age (Henry, Pickard and Hughes, 1985). In these experiments the response to citric acid was greater than the response to fumaric acid. Interestingly, when pigs were given a choice there was evidence of a selection preference for non-acidified diets.

Giesting and Easter (1985) compared the response of weaner pigs to 2% additions of propionic, fumaric and citric acid to simple diets formulated with maize, soyabean meal, vitamins and minerals. The pigs were weaned at an average age of 30 days and were fed the assigned diets for a four-week period. Addition of each acid improved efficiency of gain but propionic acid caused a reduction in food intake and a depression in gain. In a second experiment, fumaric acid additions of 1, 2 or 3% were made to the diet. There was a linear improvement (*Table 17.4*) in both growth rate and feed conversion efficiency with the maximum response at 3% acid addition.

**Table 17.4** EFFECT OF GRADED LEVELS OF FUMARIC ACID ON GROWTH-PERFORMANCE OF WEANER PIGS (10.0–18.7 kg)

| | *Fumaric acid level* (%) | | | | | |
|---|---|---|---|---|---|---|
| | *0* | *1* | *2* | *3* | *4* | *PSE*[a] |
| Diet pH | 5.96 | 4.77 | 4.33 | 3.98 | 3.80 | |
| Liveweight gain (g/day) | 261 | 261 | 257 | 296 | 297 | 14.6 |
| Feed intake (g/day) | 501 | 484 | 445 | 493 | 493 | 23.4 |
| Gain: feed ratio (kg gain/kg feed) | 0.52 | 0.54 | 0.57 | 0.60 | 0.60 | 0.02 |

After Giesting and Easter (1985)
[a]Pooled standard error

## Relationship of diet type to the acidification response

The fact that Giesting and Easter (1985) were able to obtain a response to higher levels of acid than those used previously led to the hypothesis of a possible interaction between diet type and the acid response. Prior to the work by Giesting and Easter (1985), acid additions had been made to diets containing some milk product. The lactose contained in these diets would have been available for the formation of lactic acid which may have ameliorated the need for dietary acid.

The results from an experiment designed to evaluate the relationship between diet type and the acid response are presented in *Table 17.5* (Giesting, 1986). Diets were formulated with either maize and soyabean meal or maize with 11.95% soyabean meal and 25% dried skim milk. The level of acid addition to each type was 0, 2 or 3%. The maximum response was obtained with 2% acid when the diet containing dried skim milk. In contrast, there was a linear gain and efficiency response with up to 3% acid in the simple, maize–soyabean meal diet.

It has been suggested (Kidder, 1982) that vegetable–protein diets are more difficult for the weaner pig to digest than are diets having milk protein as the supplemental amino acid source. If proteolysis is enhanced by acidification, then it might be expected that pigs fed diets formulated with soya–protein concentrate would exhibit a

**Table 17.5** EFFECT OF FUMARIC ACID ADDITION ON PERFORMANCE OF PIGS FED DIETS FORMULATED WITH SOYABEAN MEAL OR DRIED SKIM MILK[a]

| *Protein source:* | *Soyabean meal* | | | *Dried skim milk* | | | |
|---|---|---|---|---|---|---|---|
| *% Fumaric acid:* | *0* | *2* | *3* | *0* | *2* | *3* | *PSEM*[b] |
| | | | *Weeks 0–2* | | | | |
| Liveweight gain (g/day) | 133 | 152 | 171 | 195 | 223 | 195 | 23.3 |
| Feed intake (g/day) | 306 | 296 | 304 | 312 | 332 | 310 | 22.1 |
| Gain: feed ratio (kg gain/kg feed) | 0.430 | 0.510 | 0.560 | 0.600 | 0.670 | 0.630 | 0.0486 |
| | | | *Weeks 0–4* | | | | |
| Liveweight gain (g/day) | 289 | 320 | 311 | 327 | 359 | 350 | 22.3 |
| Feed intake (g/day) | 540 | 549 | 533 | 532 | 565 | 536 | 33.1 |
| Gain: feed ratio (kg gain/kg feed) | 0.540 | 0.580 | 0.580 | 0.610 | 0.640 | 0.650 | 0.137 |

[a]After Giesting (1986). The average initial weight was 8.2 kg and the duration of the experiment was four weeks
[b]Pooled standard error of the mean

greater response to acid than pigs fed diets prepared with casein. The results of an experiment to test that hypothesis are shown in *Table 17.6* (Giesting, 1986). Zero or 3% fumaric acid was added to diets of similar lysine content. Both liveweight gain and feed conversion efficiency were improved by acid addition. There was, however, an interaction with diet type. Pigs fed the diet formulated with soya–protein concentrate responded more to acidification than did pigs fed the diet containing a large amount of casein.

This experiment included a fifth treatment combination wherein sodium bicarbonate was added to the fumaric acid-supplemented, casein-based diet in an attempt to demonstrate that the effect of the acid could be negated by neutralization of the acid. There was an intriguing response to bicarbonate addition. Pigs fed the diet containing both bicarbonate and fumaric acid grew more efficiently ($P < 0.05$) than did those consuming the other diets. This response to bicarbonate was unexpected, but has been confirmed (Roos, Giesting and Easter, 1987) by subsequent experiments. Giesting (1986) has proposed that the bicarbonate response results from correction of a metabolic acid load, i.e. $H^+$, arising from ingestion of large quantities of fumaric acid. Kirchgessner and Roth (1982) reported that rats fed high-energy, low-protein diets supplemented with fumaric acid had elevated activities of glutamate–oxaloacetate transaminase and glutamate–pyruvate transaminase in the liver, along with reduced serum urea levels. The increased enzyme levels may be an indicator of increased deamination of amino acids to produce ammonia to buffer $H^+$ in the urine.

In an attempt to establish that diet acidification does, indeed, improve protein digestion, Giesting (1986) used surgically modified pigs to obtain ileal samples on which to base digestion estimates. A simple 'T' cannula was installed in the terminal ileum at two weeks of age using a modification of the procedure first described by Funderburke *et al.* (1982). Following surgery, the pigs were returned to the sow and allowed to suckle normally until four weeks of age. They were weaned, placed in individual cages, fed test diets and ileal samples were obtained. The pigs were fed *ad libitum* and digestion coefficients were calculated using chronic oxide as an indigestible marker.

The results of an experiment (Giesting, 1986) designed to examine the effects of diet type, i.e. milk-based versus soya-based, and fumaric acid addition on ileal digestibility values are presented in *Table 17.7*. Not unexpectedly, there was substantial variation in the data. It was clear, however, that both nitrogen and dry matter digestibility increased linearly with age. Both nitrogen and dry matter digestibility were greater for diets containing skim milk than for diets formulated with soyabean meal. Trends for improved digestibility with fumaric acid addition were present but not significant.

These data also suggest that the acid response may decline with maturation of the gastrointestinal tract. To test this hypothesis, Giesting and Easter (1985) fed diets containing fumaric acid to finishing pigs. There was no response in either growth rate or efficiency of feed conversion.

## Alternatives to organic acids

In view of economic considerations, alternatives to the organic acids discussed above have been investigated for their potential as diet acidifiers. Giesting (1986) attempted to demonstrate growth responses to the addition of hydrochloric, phosphoric and sulphuric acids in amounts calculated to provide acidification similar to that obtained with 3% fumaric acid. Concentrated hydrochloric acid addition to weaner-pig diets

**Table 17.6** EFFECT OF FUMARIC ACID ADDITION ON PERFORMANCE OF PIGS FED DIETS BASED ON SOYA-PROTEIN CONCENTRATE OR CASEIN[a,b]

| *Protein source:* | *Soya–protein concentrate* | *Soya–protein concentrate* | *Casein* | *Casein* | *Casein* | |
|---|---|---|---|---|---|---|
| *Fumaric acid:* | − | + | − | + | + | |
| *Sodium bicarbonate:* | − | − | − | − | − | *PSEM*[c] |
| | | *Weeks 0–2* | | | | |
| Liveweight gain (g/day) | 123 | 174 | 148 | 169 | 198 | 13.3 |
| Feed intake (g/day) | 306 | 322 | 330 | 352 | 348 | 16.4 |
| Gain: feed ratio (kg gain/kg feed) | 0.410 | 0.530 | 0.460 | 0.480 | 0.570 | 0.0279 |
| | | *Weeks 0–4* | | | | |
| Liveweight gain (g/day) | 298 | 311 | 292 | 330 | 315 | 11.7 |
| Feed intake (g/day) | 580 | 598 | 578 | 621 | 592 | 19.5 |
| Gain: feed ratio (kg gain/kg feed) | 0.510 | 0.520 | 0.500 | 0.530 | 0.530 | 0.0104 |

[a]After Giesting (1986). The average initial weight was 7.4 kg and the duration of the experiment was four weeks
[b]Fumaric acid addition was 3.0%, sodium bicarbonate addition was 2.74%
[c]Pooled standard error of the mean

**Table 17.7** EFFECT OF DIET TYPE, TIME AFTER WEANING, AND FUMARIC ACID ADDITION ON ILEAL DIGESTIBILITY VALUES FOR NITROGEN AND DRY MATTER[a]

| | *By week* | | | | |
|---|---|---|---|---|---|
| *Item* | *1* | *2* | *3* | *4* | *PSEM*[c] |
| No. of observations | 18 | 19 | 17 | 14 | |
| Liveweight gain (g/day) | 142 | 207 | 418 | 573 | 28.1 |
| DM digestibility (%) | 68.5 | 69.8 | 75.9 | 71.8 | 1.55 |
| N digestibility (%) | 63.5 | 69.7 | 77.8 | 75.2 | 1.90 |
| | *By diet* | | | | |
| Fumaric acid: | – | + | – | + | |
| Protein:[b] | *SBM* | *SBM* | *DSM* | *DSM* | |
| Liveweight gain (g/day) | 231 | 265 | 421 | 423 | 28.1 |
| DM digestibility (%) | 68.8 | 69.4 | 73.8 | 74.0 | 1.55 |
| N digestibility (%) | 67.7 | 70.5 | 74.0 | 74.0 | 1.90 |

[a]After Giesting (1986). Pigs were fitted with ileal cannulae at 14 days of age, allowed to suckle until weaned at day 25 of life. Samples were obtained during the following four weeks
[b]Abbreviations: SBM = simple, maize–soyabean meal diet, DSM = maize diet with 25% dried skim milk
[c]Pooled standard error of the mean

resulted in a severe depression in growth. This is not surprising in view of the fact that this treatment resulted in a diet with about 1.3% $Cl^-$. Dietary electrolyte balance affects animal performance. The calculated index ($Na^+ + K^+ - Cl^-$ expressed in mEq/100 g of diet) was − 6.7. Data from experiments conducted by Patience, Austic and Boyd (1987) provide evidence that an index value in the negative range is consistent with dramatic reductions in growth. Sulphuric acid addition also depressed performance, probably for the same reason.

Of the three inorganic acids tests, only phosphoric did not result in a growth depression. However, there was no indication of improvement in performance. This is particularly disappointing in view of the fact that this acid could serve a dual role as a source of both acidity and inorganic phosphorus.

Other organic acids have also been investigated. For example, Schutte and van Weerden (1986) found positive effects from feeding calcium formate in combination with either fumaric or propionic acid to pigs during a 35-day period beginning at an initial weight of 12.5 kg. The results are presented in *Table 17.8*. This comprehensive experiment also included the use of copper, added as copper sulphate. Growth rate and feed conversion efficiency were significantly improved by calcium formate addition but not by fumaric acid addition. Calcium formate in combination with fumaric acid gave a significant response in feed conversion efficiency but not in growth rate. The best performance was obtained when calcium formate was used in combination with the copper. This agrees nicely with the earlier observation by Edmonds, Izquierdo and Baker (1985) regarding the additivity of the organic acid and copper sulphate responses.

A logical extension of the diet acidification research is the hypothesis that performance of young pigs can be enhanced by the addition of lactic acid-producing miocrobes to the diet on the assumption that this will result in increased formation of lactic acid in the stomach. Pollmann, Danielson and Peo (1980a) fed weaner pigs having an average initial weight of 7 kg, diets supplemented with *Lactobacillus acidophilus* or *Streptococcus faecium*. There was a significant improvement in daily

**Table 17.8** EFFECT OF CALCIUM FORMATE ON GROWTH-PERFORMANCE OF PIGS

| *Diet description* | *Liveweight*[a] *gain* | *Gain: feed ratio* (kg gain/kg feed) | *Feed intake* (g/day) |
|---|---|---|---|
| Control[b] | 16.6 | 0.51 | 927 |
| + 1.5% fumaric acid | 16.4 | 0.52 | 901 |
| + 1.5% calcium formate | 18.6 | 0.53 | 995 |
| + 1.0% fumaric acid + 0.5% calcium formate | 17.6 | 0.53 | 954 |
| + 1.0% calcium formate + 0.5% propionic acid | 19.0 | 0.56 | 976 |
| + 165 ppm copper | 18.9 | 0.56 | 1070 |
| + 165 ppm copper + 1.5% calcium formate | 20.0 | | 1013 |

After Schutte and von Weerden (1986)
[a]Pigs were assigned to treatment at an average weight of 12.5 kg and remained on the test for a total of 35 days
[b]The control diet was of practical composition and contained no antibiotics

gain and feed conversion efficiency in the first experiment with the addition of either organism, but in the second trial only a response to *Lactobacillus* was detected. In the second experiment, lactic acid (DL-lactic acid, 220 mg/kg) was also included and gave a significant (3.09 versus 2.61) improvement in feed conversion ratio.

In a subsequent series of experiments Pollmann, Danielson and Peo (1980b) demonstrated that colonization of the gastrointestinal tract by *Lactobacillus acidophilus* is enhanced by dietary lactose. The utility of lactose in enhancement of growth in weaner pigs has also been demonstrated by Giesting (1986). Pigs were fed diets supplemented with 25% dried skim milk, or diets wherein the carbohydrate and protein components in the first diet were simulated by addition of 13% lactose and 10% casein. Replacement of lactose by either cornstarch or a variety of hydrolysed cornstarch products (*Table 17.9*) depressed performance. This observation is consistent with the notion that gastric acidification in the young pig is mediated through the formation of lactic acid from lactose. The starch or hydrolysed starch diets did not provide the substrate for this function.

A recent experiment at the University of Kentucky (Cromwell and Burnell, 1987) tested the hypothesis that an 'acidification response' could be obtained by feeding a combination of organic acids along with a substantial number of bacterial cells (*Lactobacillus acidophilus* and *Streptococcus faecium*), enzymes and flavouring agents. The results are presented in *Table 17.10*. A growth response was evident, both when

**Table 17.9** EFFECT OF VARYING PROTEIN AND CARBOHYDRATE SOURCES ON PERFORMANCE OF STARTER PIGS (8.09–19.54 kg BODY WEIGHT)[a,b]

| *Item* | *SBM* | *DSM* | *LAC CAS* | *HMS CAS* | *LAC ISP* | *HMS ISP* | *PSEM*[c] |
|---|---|---|---|---|---|---|---|
| Liveweight gain (g/day) | 345 | 454 | 457 | 422 | 406 | 367 | 15.1 |
| Feed intake (g/day) | 618 | 641 | 659 | 645 | 630 | 570 | 19.9 |
| Gain: feed ratio (kg gain/kg feed) | 0.560 | 0.710 | 0.690 | 0.660 | 0.650 | 0.650 | 0.0129 |

After Giesting (1986)
[a]Abbreviations used: SBM = soyabean meal, DSM = dried skim milk, LAC = lactose, CAS = casein, HMS = hydrolysed maize starch, ISP = isolated soya protein
[b]The negative control diet was formulated with maize and soyabean meal. The second diet contained 25% dried skim milk and the remaining diets were formed by substituting LAC or HMS for the carbohydrate in the skim milk and CAS or ISP for the protein.
[c]Pooled standard error of the means

**Table 17.10** PERFORMANCE RESPONSE OF PIGS TO AN ACIDIFYING AGENT[a,b]

| *Basal diet:* | *Maize–soya* | | *Maize–soya–whey* | |
|---|---|---|---|---|
| *Acidifier:* | *0* | *1.0* | *0* | *1.0* |
| Number of pigs | 48 | 48 | 48 | 48 |
| Liveweight gain (g/day) | 291 | 318 | 327 | 341 |
| Feed intake (g/day) | 514 | 527 | 577 | 559 |
| Gain: feed ratio (kg gain/kg feed) | 0.564 | 0.602 | 0.561 | 0.595 |

Cromwell and Burnell (1987)
[a]Data are averaged over two experiments. Pigs weighed 6.8 kg initially and were treated for a 28-day period
[b]The acidifier was a commercial product containing citric acid, sorbic acid and benzoate along with small amounts of *Lactobacillus acidophilus*, *Streptococcus faecium*, enzymes and flavouring agents

pigs were fed simple maize–soyabean meal diets or diets formulated with maize, soyabean meal and dried whey. Additionally, though not shown in the table, the reponse to the acidifier was additive with antibiotic and copper responses. Additional research will undoubtedly be forthcoming in this area.

Diet acidification is not the complete answer to the post-weaning growth check. However, a substantial body of published literature does support the general conclusion that pigs will respond to reduced diet pH in the weeks immediately following weaning. The magnitude of the response is likely related to the nature of the diet with the greatest benefit evident when diets are formulated with cereal grains supplemented with plant proteins and are devoid of lactose.

## References


BECKER, D.E., ULLREY, D.E., TERRILL, S.W. and NOTZOLD, R.A. (1954). *Science*, **120,** 345

CHAPPLE, R.P., CUARON, J.A. and EASTER, R.A. (1983). *Journal of Animal Science*, **57** (Suppl. 1), 94

CLEMENS, E.T., STEVENS, C.E. and SOUTHWORTH, M. (1975). *Journal of Nutrition*, 105

COLE, D.J.A., BEAL, R.M. and LUSCOMBE, J.R. (1968). *Veterinary Record*, **83,** 459–464

CORRING, T., AUMAITRE, A. and DURAND, G. (1978). *Nutrition and Metabolism*, **22,** 231–243

CRANWELL, P.D. (1985). *British Journal of Nutrition*, **54,** 305–320

CRANWELL, P.D., NOAKES, D.E. and HILL, K.J. (1976). *British Journal of Nutrition*, **36,** 71–86

CROMWELL, G.L. and BURNELL, T.W. (1987). *Animal Nutrition and Health*, **42,** (4) 14–16

EDMONDS, M.S., IZQUIERDO, O.A. and BAKER, D.H. (1985). *Journal of Animal Science*, **60,** 462–469

ETHERIDGE, R.D., SEERLEY, R.W. and HUBER, T.L. (1984). *Journal of Animal Science*, **58,** 1403–1410

FALKOWSKI, J.F. and AHERNE, F.X. (1984). *Journal of Animal Science*, **58,** 935–938

FORTE, J.G., FORTE, T.M. and MACHEN, T.W. (1975). *Journal of Physiology*, **244,** 15–31

FUNDERBURKE, D.W., KVERAGAS, C.L., VANDERGRIFT, W.L. and SEERLEY, R.W. (1982). *Journal of Animal Science*, **64,** 457–466

GIESTING, D.W. (1986). *Utilization of Soy Protein by the Young Pig*. PhD Thesis. University of Illinois, Urbana

GEISTING, D.W. and EASTER, R.A. (1985). *Journal of Animal Science*, **60,** 1288–1293
HENRY, R.W., PICKARD, D.W. and HUGHES, P.E. (1985). *Animal Production*, **40,** 505–509
KERSHAW, G.F., LUSCOMBE, J.R. and COLE, D.J.A. (1966). *Veterinary Record*, **79,** 296
KIDDER, D.E. (1982). *Pig News and Information*, **3,** 25–28
KIDDER, D.E. and MANNERS, M.J. (1978). *Digestion in the Pig*, Scientechnica, Bristol
KIRCHGESSNER, N. and ROTH, F.X. (1980). *Zeitschrift fur Tierphysiologie, Tierenahrung und Futtermittelkunde*, **44,** 239–246
KIRCHGESSNER, M. and ROTH, F.X. (1982). *Pig News and Information*, **3,** 259–264
LEWIS, A.J. (1981). *Nebraska Swine Research Report*, University of Nebraska, Lincoln
MANER, J.H., POND, W.G., LOOSLI, J.K. and LOWREY, R.S. (1962). *Journal of Animal Science*, **21,** 49–52
MANNERS, M.J. (1970). *Journal of the Science of Food and Agriculture*, **21**, 333–340
OKAI, D.B., AHERNE, F.X. and HARDIN, R.T. (1976). *Canadian Journal of Animal Science*, **56,** 573–587
PATIENCE, J.F., AUSTIC, R.E. and BOYD. R.D. (1987). *Journal of Animal Science*, **64,** 457–466
POLLMANN, D.S., DANIELSON, D.M. and PEO, E.R. JR (1980a). *Journal of Animal Science*, **51,** 577–581
POLLMANN, D.S., DANIELSON, D.M. and PEO, E.R. JR (1980b). *Journal of Animal Science*, **51,** 638–644
POND, W.G. and MANER, J.H. (1984). *Swine Production and Nutrition*, Avi Publishing Company, Westport, CT
RERAT, A.A. (1981). *World Review of Nutrition and Dietetics*, **37,** 229–287
ROOS, M.A., GIESTING, D.W. and EASTER, R.A. (1987). *Journal of Animal Science*, **65,** (Suppl. 1), 245
SCHUTTE, J.B. and VON WEERDEN, E.J. (1968). *CAFO as a Feed Additive in Diets for Young Pigs*. ILOB Report 569, Wageningen, The Netherlands
SLIVITSKII, M.G. (1975). *Vest. Sel'Khoz. Nauk.*, **7,** 75–80
STEVENS, C.E. (1977) In *Dukes' Physiology of Domestic Animals*, pp. 216–232. Ed. Swenson, M.J. Comstock, London
THOMLINSON, J.F. and LAWRENCE, T.L.J. (1981). *Veterinary Record*, **109,** 120–122
WHITE, F., WENHAM, G., SHARMAN, G.A.M., JONES, A.S., RATTRAY, E.A.S. and MCDONALD. I. (1969). *British Journal of Nutrition*, **23,** 847–857

18

# AETIOLOGY OF DIARRHOEA

J.W. SISSONS
*Nurish Products, Dairy Food Systems, Protein Technology International, Checkerboard Square, St Louis, MO 63164, USA*

## Introduction

Post-weaning diarrhoea amongst piglets and calves continues to be a serious and frustrating problem. The pathogenesis of the disease is complex and its aetiology is multi-factorial. Although severe and persistent diarrhoea undoubtedly involves intestinal microbial pathogens, it is clear that diet and the act of weaning itself are widely regarded as factors which predispose young animals to the condition. It is not known if the problem is increasing in prevalence, but in recent times certain limitations in feeding and husbandry practice may have made the disorder more difficult to control. These include an acute shortage of skim-milk, (the source of choice protein for weaner diets), the use of alternative proteins which may have antinutritional properties, production pressures to wean at younger ages and growing public concern over the use of oral antibiotics. It is therefore appropriate to review current understanding of diarrhoea from a dietary point of view, to examine the problems associated with using other feedstuffs as replacements for milk protein, to consider the effects of digestive disorders leading to diarrhoea on animal performance and to appraise possible remedies for overcoming or preventing nutritional scours amongst young farm animals.

## The pathogenesis of diarrhoea

As a symptom of a disorder in gastrointestinal function, diarrhoea is easily recognized as an increase in stool water excretion. Quantitatively it may be regarded as a decrease in faecal dry matter to below 20%. The problem is to identify the source and cause of this additional water. Physiologically it can arise through failings in fluid handling in either the small or large bowel or both. Fluxes of water and solute move through the intestinal mucosa in opposite directions. Normally these fluxes are large; in calves it has been estimated that intestinal bi-directional fluxes of fluid can be about 80 l/day. But in healthy animals, the net difference between them is very small. For example, as calves develop diarrhoea the net difference increases from about 50 ml/day to 1 l/day, and if this level reaches 2.5 l/day, dehydration and death will follow (Bywater, 1973; Bywater and Logan, 1974).

The process of fluid transport is governed by both active and passive driving forces. The former is a metabolic process which is governed by mucosal cyclic adenosine monophosphate (cAMP) and can result in electrolyte passage from the blood into the lumen. In contrast passive movements are driven by electrical and chemical potentials and hydrostatic pressure gradients across the mucosa (see review by Argenzio and Whipp, 1980). Clearly the aetiology of diarrhoea should be considered from the standpoint of processes which alter the bi-directional balance of fluid transport in favour of its accumulation in the lumen of the intestine. Mechanisms likely to bring this about include changes in the permeability of the gut wall, deranged contractile activity of the smooth muscle coat, altered osmolality of digesta, and mucosal inflammation. Physiological and immunological studies have shown that these dysfunctions can occur as a result of weaning calves and piglets onto whole milk substitutes or dry starter meals which contain large amounts of vegetable protein (Sissons, 1982; Newby *et al.*, 1985).

## WITHDRAWAL OF WHOLE MILK

Protective effects of milk against pathogenic *E. coli* have been demonstrated in piglets weaned onto a diet supplemented with sows' milk (Deprez *et al.*, 1986). The mechanism is not well characterized. Antibody and binding proteins of whole milk may prevent the attachment of microorganisms to enterocyte receptors and thereby minimize the stimulation of active secretion by bacterial exotoxins. Both *E. coli* and *Salmonella* have the ability to activate the enzyme adenylate cyclase located in the enterocyte membrane (Ooms and Degryse, 1986). The abundance of lactobacilli in milk may add further protection against coliforms by competitive growth or secretion of acid and bacteriocides (Fuller, 1986). These protective effects may account, in part, for the lower incidence of diarrhoea when skim-milk provides the major source of protein in the weaner diet. But these beneficial properties could be destroyed by overheating during processing of milk protein (see Roy, 1984).

Piglets which are abruptly weaned onto solid food usually starve for several hours before learning to eat dry meal or pellets. A subsequent tendency to over-eat so as to satisfy their extreme hunger may be the root cause of digestive upsets. So far there have been no satisfactory studies of digestive processes in piglets to test this notion. Electromyography in older pigs has shown that during fasting there is an increase in the numbers of migrating waves of pulsatile contractions passing along the small intestine (Laplace, 1984). If this happens in the early weaned and starving animal, then when food is consumed a combination of intestinal hypermotility and excessive feed intake could lead to an abnormal load of partially digested food filling the lumen of the small bowel. This digesta may provide substrates for an undesirable microflora whilst soluble components would increase osmotic pressure within the lumen. Thus, there may be value in developing feed additives which reduce motor activity of the distal stomach during the first few days after introducting solid food to piglets.

Post-weaning starvation is less of a problem for animals reared on liquid diets. But Roy (1984) has emphasized the importance of coagulation of casein in the true stomach of calves as a means of preventing the premature passage of undigested protein and fat to the duodenum. This property is impaired by severe heat treatment of milk protein. However, as a result or extensive studies by Roy and co-workers (see Roy, 1984) the need to control temperature during the processing of spray-dried milk powder for calf feeding in order to avoid diarrhoea is widely accepted.

## MATURATION OF DIGESTIVE FUNCTION

Secretion of digestive enzymes does not reach a maximum until piglets and calves reach about 5 weeks of age (Corring, Aumaitre and Durand, 1978; Kidder and Manners, 1980; Toullec, Guilloteau and Villette, 1984). It is also important to note that variation between animals in the quantities of specific enzymes secreted may be considerable. For example, Henschel (1973) showed that in some calves rennin, and in others pepsin, predominates during the first few weeks of life. For some animals, a slow maturation of enzyme synthesis and secretion probably limits their ability to digest early weaner diets especially when the composition deviates from that of whole milk. This is exemplified by data given in Table 18.1 showing that digestibility of nitrogen in milk substitutes containing non-milk protein increases with age and decreases with level of milk protein replacement. Part of the problem concerning level of substitution could be due to increased intake of antinutritional factors associated with the alternative protein source, or simply a lack of enzymes having the correct specificity for breaking down unusual structures. However, adaptation to diet is a phenomenon recognized in the digestion of some carbohydrates such as lactose (Huber, Rifkin and Keith, 1964). There is no evidence that this applies to protein.

Another weakness of early weaning is that development of the digestive processes may be halted or even depressed by radically changing the diet. Several workers have reported reductions in pancreatic enzyme activity in piglets weaned onto dry feeds at ages of less than 1 month (Hartmann *et al.*, 1961; Efird, Armstrong and Herman, 1982; Lindemann *et al.*, 1986). The extent of this depression may depend on the nature of the dietary protein since partial replacement of dried skim-milk with soyabean flour or fish protein concentrate in a milk substitute for calves led to reduced activity of pancreatic enzymes (Ternouth *et al.*, 1975). Likewise, soya or single cell protein was shown to reduce chymosin and pepsin activity in the calf abomasum, but this did not happen when fish protein was used in the calf diet (Ternouth *et al.*, 1975; Sedgman, 1980; Williams, Roy and Gillies, 1976). The exact nutritional implications arising from inhibited digestive enzyme activity by diet is not known, but these limitations would be expected to favour conditions leading to diarrhoea.

## ANTINUTRITIONAL FACTORS

Dried skim-milk, provided it has not been over-heated, has for long been regarded as a highly digestible protein source for both calves and piglets (Toullec, Mathieu and Pion, 1974; Roy, 1984; Armstrong and Clawson, 1980). But, because of restricted production of cows' milk within the European Economic Community, it has become both expensive and scarce. Considerable interest is now being given to cheaper alternatives such as protein from oil seeds (e.g. soyabeans and rapeseed) and arable legumes (e.g. peas and beans). Unfortunately the growth of piglets and calves reared on diets containing large amounts of vegetable protein is generally much lower than animals fed milk based diets, in part because of much lower digestibility (Meade, 1967; Combs *et al.*, 1963; McGilliard *et al.*, 1970; Bell, Royan and Young, 1974). Another reason is that these protein sources contain several biologically active substances which have antinutritional properties. Descriptions of these components and their effects on animal performance were given in a recent review by Wiseman and Cole (1988). Some of these substances, such as protease inhibitors, lectins and antigens, have been implicated in the onset of diarrhoea.

### *Enzyme inhibitors*

Protease inhibitors are widely distributed amongst plants used as protein sources for animal feed (Liener and Kakade, 1969). Their blockage of protein digestion may indirectly contribute to the bulkiness of digesta by increasing the load of undigested constituents in the gut lumen. They also possess the ability to stimulate hypersecretion of the pancreas in rats and chicks (Lepkovsky, Bingham and Pencharz, 1959; De Muelenaere, 1964), although this physiological effect has not been confirmed in calves (Gorrill *et al.*, 1967). Indeed, studies of relations between protease activity or trypsin inhibitor and growth are conflicting. For example, whilst Ducharme (1982) showed a positive correlation between activities of trypsin and chymotrypsin and growth in soya fed calves, others were unable to depress

**Table 18.1** APPARENT DIGESTIBILITY OF NITROGEN BY PIGLETS GIVEN WEANER DIETS CONTAINING DIFFERENT PROTEIN SOURCES

| *Protein source* | *Proportion of milk protein replaced* | *Age (days)* | *Digestibility of total nitrogen of the diet* | *Reference* |
|---|---|---|---|---|
| Skim-milk | 1.00 | 7–11 | 0.98 | Pettigrew *et al.*, 1977 |
| | | 11–15 | 0.97 | |
| | 1.00 | 14–25 | 0.96 | Seve *et al.*, 1983 |
| | | 26–32 | 0.95 | |
| | 0.76 | 28 | 0.80 | Christison and Parra de Solona, 1982 |
| | | 41 | 0.86 | |
| Whey | 0.24 | 7–11 | 0.99 | Pettigrew *et al.*, 1977 |
| | | 11–15 | 0.97 | |
| | 0.25 | 14–25 | 0.95 | Seve *et al.*, 1983 |
| | | 26–32 | 0.96 | |
| Soyabean | | | | |
| defatted | 1.00 | 21–28 | 0.74 | Combs *et al.*, 1963 |
| | | 29–35 | 0.76 | |
| | | 36–42 | 0.85 | |
| defatted | 0.77 | 35–36 | 0.87 | Neport and Keal, 1983 |
| | 1.00 | 35–56 | 0.86 | Neport and Keal, 1983 |
| defatted | 0.80 | 28 | 0.72 | Christison and Parra de Solano, 1982 |
| | | 41 | 0.78 | |
| Pea | | | | |
| ground, heated | 0.31 | 15–22 | 0.91 | Seve *et al.*, 1985 |
| | | 23–29 | 0.94 | |
| concentrate | 0.81 | 28 | 0.65 | Christison and Parra de Solano, 1982 |
| | | 41 | 0.75 | |

performance by adding trypsin inhibitor to calf milk substitutes (Kakade *et al.*, 1974).

Steam treatment of soyabean meal is claimed to substantially reduce trypsin and chymotrypsin activity (Circle and Smith, 1972). Even so, attempts to rear pre-ruminant calves on liquid diets containing more than 30–40% of the protein as heated soyabean flour have led to diarrhoea, poor growth and sometimes death (Gorrill and Thomas, 1967; Nitsan *et al.*, 1972; Kwiatkowska, 1973). This suggests that either the steam process does not destroy all forms of the inhibitor, or that heat stable factors are involved in the adverse response. Protease inhibitors exist in several forms and some of these are only unstable to heat in the presence of alkali (Obara and Watanabe, 1971). Alkali treatment has been reported to improve the utilization of heated soyabean flour when given in liquid feeds to young pigs (Lennon *et al.*, 1971). But this was not reproduced in all experiments in which fatal diarrhoea occurred. Some workers (De Groot and Slump, 1969; Gorrill, 1970) have warned that alkali treatment of soya protein may destroy certain amino acids and could yield lysinoalanine which has toxic effects in the kidneys of rats (Gorrill, 1970; Woodward, 1972). However, since there is much variation between soyabean varieties and content of protease inhibitor (Kakade *et al.*, 1972) in the longer term it may be possible to reduce enzyme inhibitors in legume seeds through plant breeding.

### *Lectins*

Lectins are characterized for their property of agglutinating red blood cells. There is doubt, however, as to whether this accounts for their fatal effects when injected into rodents, since non-toxic fractions exhibiting relatively high haemagglutinating activity have been isolated from different varieties of beans (Jaffe, 1969). Nevertheless, studies in rats have indicated that these glycoproteins strongly bind with membranes of enterocytes and could disrupt processes of intestinal absorption (Jaffe, 1969). Because the haemagglutinating property is possessed by many bacteria that show specific adherence to mucosal cells (Firon, Ofek and Sharon, 1984), it is feasible that plant lectins may compete with gut microorganisms for binding sites at the brush border. Thus, lectins could prevent beneficial organisms, such as lactobacilli, from colonizing the gastric and gut epithelium. This may lead to an imbalance in gut flora and allow the establishment of pathogenic strains of bacteria. However, the effects of lectins on intestinal function of young pigs and calves awaits investigation.

Studies *in vitro* have shown that soyabean lectins are readily inactivated by pepsin digestion and hydrochloric acid (Liener, 1955; Birk and Gertler, 1961). It is feasible, therefore, that lectins are rendered harmless by digestion. Many reports have also shown that haemagglutinating and toxic activities of lectins are also very susceptible to treatment with moist steam (Circle and Smith, 1972). Thus, at least for heated sources of protein, lectins may be of little antinutritional importance.

### *Dietary antigens*

Despite processing with heat to destroy enzyme inhibitors and lectins, diarrhoea still occurs amongst piglets fed non-milk protein. Recently much interest has focused on the idea that some animals may be intolerant of immunologically active globulins in legume protein and suffer from diarrhoea caused by inflammatory reactions in mucosal tissue.

Intestinal hypersensitivity to dietary antigens has been postulated to predispose diarrhoea which sometimes occurs when piglets are weaned onto soya diets (Miller *et al.*, 1983). However, in contrast to the persistence of diarrhoea in soya fed calves the intolerance to soyabean meal by piglets is transient. Diarrhoea occurs about 1 week after the introduction of the weaner diet and, in the absence of a secondary pathogen, faecal stools usually become firm again by 2–3 weeks post-weaning. During this transient period of hypersensitivity the gut mucosa of the piglet undergoes tissue inflammatory reactions. At the same time absorption is impaired and there is a reduction in the activity of brush-border disaccharides (Miller *et al.*, 1984; Miller *et al.*, 1986; Ratcliffe *et al.*, 1987). Studies of cutaneous reactions to intradermal injections of soya protein extract in piglets weaned onto a diet based on soyabean meal showed evidence of a delayed hypersensitivity response. The skin reactions were noted to coincide with observations of reduced capacity by the gut to absorb xylose (Newby *et al.*, 1985). The observed cutaneous response to soya implies that intestinal inflammation in weaned piglets could involve a cell mediated response to dietary antigens. Delayed hypersensitivity reactions begin with the primary exposure of a T-lymphocyte to a specific antigen resulting in T-cell activation. Upon secondary exposure to the same antigen the T-lymphocyte is induced to release lymphokines which have been linked with inflammation of mucosal tissue (Mowat and Ferguson, 1981).

According to Newby *et al.* (1985) the post-weaning immune response may be modified by primary exposure of the piglet to the same antigenic material during 'suckling', for example, through ingestion of creep feed. Some support for this view was obtained from experiments with cows' milk protein showing that the incidence and severity of the diarrhoea in piglets weaned onto native casein was increased if they ingested a small 'priming' dose of the cows' milk protein during suckling. Extending the pre-weaning exposure of the supposedly antigenic protein from 3 to 7 days reduced or prevented post-weaning scours (Miller *et al.*, 1984). Thus it was proposed that ingestion of a creep feed could either 'tolerize' or 'sensitize' the piglet, depending on the amount of antigen consumed prior to re-exposure of the same antigen at weaning.

This interesting hypothesis of pre-weaning sensitization to dietary antigens has serious implications for the value of creep feeding. But it must be noted that, so far, the idea has not been validated with protein sources other than that of cows' milk. When the oral tolerance experiments were repeated using soya, rather than casein, as the antigen source, Miller and co-workers were not able to sensitize the piglets by feeding a small amount of soya protein during the suckling phase (Miller, private communication). Although, feeding the animals relatively large amounts of soya antigen did prevent both xylose malabsorption and diarrhoea (Miller *et al.*, 1985). Since piglets are commonly born to sows which receive soyabean meal in their diet, small amounts of soya antigen passing to the suckling animal in colostrum and milk may have sensitized them in advance of a so-called 'priming' dose ingested as creep feed. But further doubt has been cast on the idea of sensitization during suckling in studies which showed that the architecture and function of mucosal tissue of piglets which had been forcibly 'primed', or offered creep feed, were not different from those in abruptly weaned animals (Hampson and Kidder, 1986; Hampson and

Smith, 1986; Hampson, Fu and Smith, 1988). Unfortunately there is a lack of information about the precise immune status of the animals in these studies. There is a need, for example, to understand whether factors other than diet could, through mechanisms of immune suppression, concurrently modulate the response of piglets to antigens in their feed.

## OSMOTIC COMPONENTS OF DIET

Raised osmotic pressure through the build up of non-absorbable solutes in the gut lumen can result in a considerable driving force for the movement of water. For instance, feeding young calves an excess of glucose or lactose (i.e. a daily intake of more than 9 g/kg live weight of hexose equivalent) will induce diarrhoea (Roy, 1969). A problem could also arise in circumstances where the activities of pancreatic and brush border enzymes are impaired or lack specificity. About 30% of soyabean meal comprises sucrose, stachyose, raffinose and complex polysaccharides (Rackis *et al.*, 1970). It is unlikely (though not proven) that any of these carbohydrates are digested by the calf since it does not possess suitable enzymes (Sissons, 1981). Except perhaps for the digestion of sucrose, the pig probably has the same digestive limitation (Kidder and Manners, 1978). However, studies of intestinal digesta flow in calves fed soya molasses (Sissons and Smith, 1976) have indicated that the induction of abnormal water movement by these carbohydrates is relatively small compared with the disturbed handling of fluid attributed to inflammation of the gut mucosa (see Table 18.2). Recent evidence (unpublished observations) supporting this view has been obtained from recordings of intestinal myoelectric activity. This work showed that the pattern of gut motility in calves suffering from diarrhoea which had been induced by feeding heated soyabean flour was distinctly different from motor activity in the same animals fed sufficient

**Table 18.2** EFFECT OF MILK OR SOYA PROTEIN ON THE MOVEMENT OF DIGESTA THROUGH, AND APPARENT NITROGEN ABSORPTION FROM, THE SMALL INTESTINE OF CALVES ORALLY SENSITIZED TO HEATED SOYA FLOUR

| *Order of giving feeds:* | *Small gut transit time* (h) | *Flow rate of digesta* (g/h) | *Net nitrogren absorption* (%) |
|---|---|---|---|
| HSF 1st feed (unsensitized) | 3.1 | 77 | 57 |
| HSF challenge feed (sensitized) | 1.4 | 165 | 25 |
| Casein | 3.5 | 48 | 85 |
| ETHSC | 3.1 | 69 | 74 |
| ETHSC + 30 g soya molasses | 2.2 | 82 | 77 |
| ETHSC + 30 g sucrose | 2.3 | 61 | 83 |

HSF = heated soyabean flour
ETHSC = ethanol treated and heated soya concentrate
After Sissons and Smith (1976)

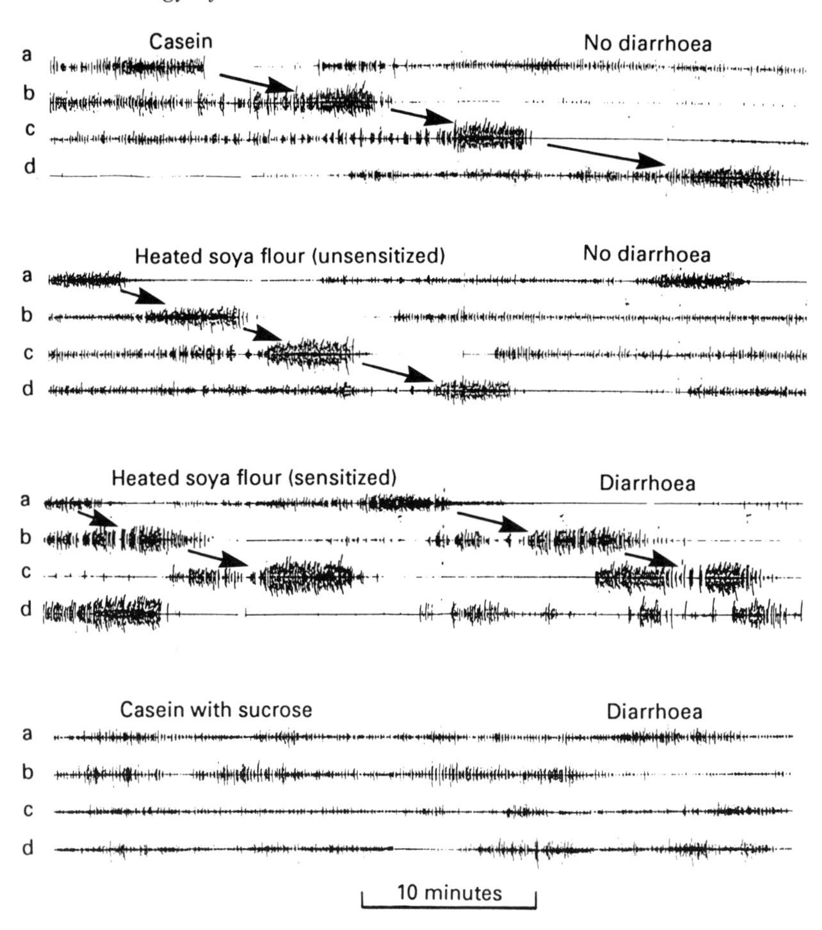


**Figure 18.1** Typical recordings of myoelectric activity made from the small intestine of a calf given feeds containing milk protein with and without a sucrose load (200 g) or heated soyabean flour on a first (unsensitized) or fifth (sensitized) occasion. Recording sites a–d were spaced about 2 m apart on the jejunum. Arrows denoted a propagating complex of contractions corresponding to peristalsis

sucrose to cause osmotic diarrhoea (see Figure 18.1). These recordings show that normal intestinal motility on the small intestine is characterized by cyclical phases of weak myoelectric activity (associated with motions which mix digesta) followed by stronger activity (associated with peristaltic contractions). Inducing a hyperosmotic condition with sucrose led to the replacement of the cyclic motor pattern with a mixing type of activity, whilst feeding antigenic soya protein had the effect of stimulating considerable peristalsis. Under these circumstances enhanced peristalsis might have a protective action by facilitating the rapid removal of offending antigens. Even though soluble molasses may not normally have a large effect on the osmotic pressure of digesta, they may be important in a state of malabsorption especially where there could be added contributions from soluble fatty acids and electrolytes.

## Effect of diarrhoea on performance

Despite concern to prevent diarrhoea there is little information on the effects of scouring on animal performance. Ball and Aherne (1982) showed that diarrhoea caused a significant slowing of growth of early weaned pigs between 21 and 35 days compared with animals not suffering from loose faeces. They also found that piglets with diarrhoea required 65–72% more feed per unit of gain than healthy animals. However, the degree of post-weaning diarrhoea was reported not to have significantly affected the growth rate to 90 kg live weight. In studies of pre-ruminant calves fed heated soyabean flour, flow rates of digesta at the distal ileum were negatively correlated with net nitrogen absorption (Sissons and Smith, 1976). On occasions when digesta flow was very high (about 4 litres/day) the amounts of nitrogen appearing at the ileum exceeded the amount given in a liquid feed. Much of this nitrogen loss probably arose from mucosal tissue and vascular exudation of fibrin as a result of inflammation.

## Dietary approaches to the control of diarrhoea

Methods of controlling diarrhoea should take account of the pathogenesis of the disorder, the species and whether the animal is being weaned onto a liquid or solid diet. Many studies indicate that provided animals are not severely infected with enteric microbial pathogens, dietary manipulation can be used to avoid or reduce diarrhoea. Some success has been achieved by altering intake or composition of feeds and through processing of dietary ingredients.

### RESTRICTION OF FEED INTAKE

Early weaned piglets show increased susceptibility to diarrhoea when they overeat. This is more likely to occur when solid food is given *ad libitum* than when intake is restricted during the first few weeks post-weaning (Palmer and Hulland, 1965; Smith and Halls, 1968; Hampson and Smith, 1986; Ball and Aherne, 1982). The precise physiological benefits of restricted feeding have, however, not been established. Reducing feed consumption may ensure that nutrient intake is consistent with a limited digestive capacity. At the same time, small meals are less likely to cause derangements in gastric and intestinal motor activity (Ruckebusch and Bueno, 1976) which might otherwise lead to the accumulation of digesta in the small intestine. Excessive food intake may also upset mechanisms which regulate gastric secretion. Failure of the calf abomasum to secrete sufficient acid and protease could be a reason why the drinking of large volumes of milk predisposes some calves to diarrhoea (Roy, 1984). Physiological studies in milk fed calves have shown that the enterogastric reflex control of abomasal emptying is impaired when liquid intake exceeds the normal capacity of the abomasum (Sissons and Smith, 1979; Sissons, 1983). This results in an abnormally high rate of abomasal digesta flow to the duodenum.

### ADDING FIBRE TO THE WEANER DIET

Several reports indicated that adding dietary fibre to a weaner diet for piglets had a beneficial effect by reducing the incidence and severity of diarrhoea (Palmer and

Hulland, 1965; Smith and Halls, 1968; Armstrong and Cline, 1976). The ameliorating effects of fibre on digestive function are unclear. But, dietary fibre stimulates the secretion of saliva, gastric juice, bile and pancreatic juice (Low, 1985). This, together with its effect of reducing nutrient density and increasing satiety may help to maintain a satisfactory balance between feed intake and the digestive capacity of the young pig. Another possible effect is that the yield of volatile fatty acids from fermentation of fibre in the colon alters the absorption of water (Crump, Argenzio and Whipp, 1980). Because fibre has a high water holding capacity it may increase the firmness of faecal stools. However, if there is a serious alteration in fluid handling by the small and, or large bowel, water retention by fibre could disguise critical losses of water and nutrients.

Studies in humans have investigated the relation between colonic motor processes and the passage of digesta during disease of the digestive tract (Connell, 1962). Curiously a diarrhoeal state has been linked with hypomotor activity of the colon whilst constipation is associated with hypermotility. It is well established that consumption of dietary fibre is an effective way of shortening transit time, and that fibre acts on colonic motility. Electromyographic studies in growing pigs showed that adding bran to a milk substitute diet stimulated increased numbers of propulsive contractions (Fioramonti and Bueno, 1980). It seems unlikely, however, that the beneficial effects of fibre in reducing diarrhoea are related to its effects on colonic motility, but rather its high water holding capacity.

### FAT LEVELS IN THE WEANER DIET

Milk substitutes with high levels of glucose or lactose have been shown to invoke diarrhoea in calves if the diet is low in fat (Blaxter and Wood, 1953; Mathieu and De Tugny, 1965). Fat is known to reduce the rate of abomasal emptying of digesta (Smith and Sissons, 1975) and this effect may slow the release of soluble carbohydrates so as to prevent a hyperosmotic condition developing in the lumen of the small intestine. Interactions between fat and carbohydrate on the digestive physiology of the young pig do not appear to have been investigated.

### PROCESSING TO REDUCE ANTINUTRITIONAL FACTORS

Recognition of a number of antinutritional factors in feedstuffs implicated in the genesis of diarrhoea and slowing of growth has stimulated a search for treatments for inactivation or removing offending constituents.

#### *Heat treatment*

As discussed earlier the need to treat protein sources containing lectins and protease inhibitors with heat is widely accepted. Heating brings about aggregate formation and loss of water solubility. Studies of soya protein indicate that molecules with few disulphide bridges are most susceptible to heat treatment. Thus, the Kunitz trypsin inhibitor (molecular weight 21 500) with two disulphide bridges in its polypeptide chain is less stable than the Bowman-Birk inhibitor (molecular weight almost 8000) which has seven disulphide linkages (Steiner, 1965; Bidlingmeyer, Leary and Laskowski, 1972). Even so, several studies have shown that steam treatment alone is not sufficient to prevent diarrhoea or overcome negative effects of soya flour on digestibility and growth of calves (Gorill and Thomas, 1967; Nitsan *et al.*, 1971; Sudweeks and Ramsey, 1972). Similarly, additional treatments appear to be necessary to improve the nutritive value of soyabean products for very young piglets. Newport (1980) reported severe scouring and high mortality amongst piglets weaned at 2 days of age onto a liquid diet containing 74% of the protein in the form of a soyabean isolate. This soya product was claimed to have undetectable levels of urease and trypsin inhibitor activity. However, older pigs may benefit from legume protein which has been steam treated. Recently Van der Poel and Huisman (1988) studied growing pigs given meals based on steamed beans (*Phaseolus vulgaris*) and showed that apparent digestibility of nitrogen, dry matter and fat measured at the ileum improved with duration of treatment.

Steam or toasting at 100°C has little effect on the antigenic activity of the major globulins of soya beans (Kilshaw and Sissons, 1979b). Inactivation of these immunological structures requires temperatures of at least 145°C (Srihara, 1984). Unfortunately, processing soyabean meal at such high temperatures may not yield a product suitable for feeding to calves (Srihara, 1984) or piglets, since the resulting polymerization of the protein molecules could impair digestibility (Hagar, 1984).

### *Alcohol treatment*

Some improvement in the utilization of soya protein by calves has been achieved when defatted soyabean flour was extracted with aqueous ethanol before toasting the product (Gorrill and Thomas, 1967; Gorrill and Nicholson, 1969; Nitsan *et al.*, 1971; Stobo, Ganderton and Connors, 1983). Studies of digestive processes in the small intestine suggest that the beneficial effects of processing soya flour with ethanol may be linked with alterations in protein structure rather than the removal of sucrose and oligosaccharides during extraction (Sissons and Smith, 1976; Sissons, Smith and Hewitt, 1979). Aqueous ethanol treatment appears to destroy the antigenic integrity of glycinin and β-conglycinin, the major globulins implicated in the gastrointestinal disorders of soya fed calves (Kilshaw and Sissons, 1979a). But the conditions of alcohol concentration and temperature required to prepare products with minimal levels of native antigenic activity are critical (see Figure 18.2); effective treatment was achieved using between 65–70% ethanol at 78°C (Sissons *et al.*, 1982a; Sissons *et al.*, 1982b). Aqueous organic mixtures are necessary to disrupt globular proteins as their structures are stabilized by hydrophobic side-chain residues located towards the centre of the molecule and hydrophilic groups positioned at the surface (Fukushima, 1969).

Although hot aqueous ethanol extraction of soya flour reduces antigenicity of native globulins and substantially improves its feeding value for calves, the product remains somewhat inferior to milk protein. For example, Stobo, Ganderton and Connors (1983) reported liveweight gains of 0.9, 0.7 and 0.2 kg/day during a 14-day period after weaning 2-day-old calves onto liquid diets in which all of the protein was derived from cows' milk or with 65% replaced by ethanol extracted soya concentrate or soyabean flour respectively. It is possible that new protein structures having immunological activity could be exposed through partial digestion. However, gut inflammatory responses to ethanol extracted soyabean concentrate were not seen in calves previously observed to react adversely to antigenic soya flour (Pedersen, 1986).

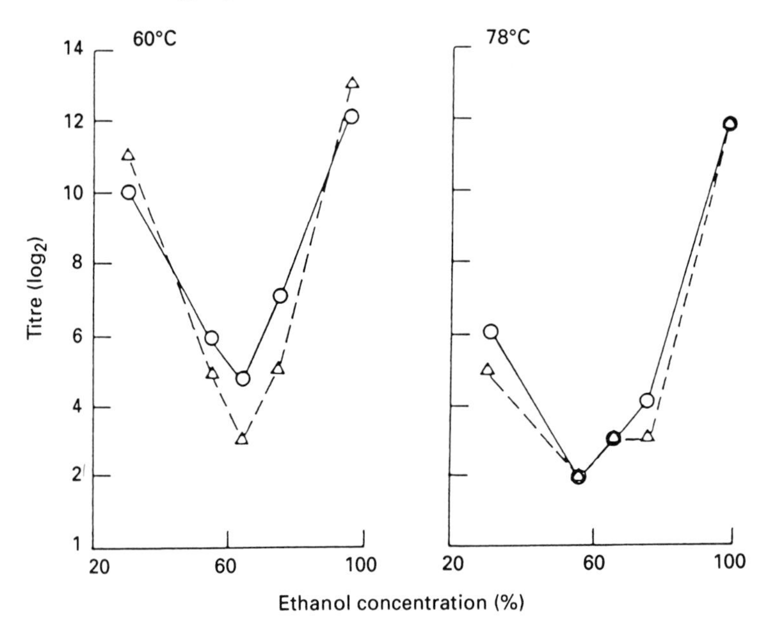


**Figure 18.2** Antigenically active glycinin (○) and β-conglycinin (△) contents of defatted soyabean flour extracted with different concentrations of ethanol at 60°C and 78°C (after Sissons *et al.*, 1982)

An attempt to wean 2-day-old piglets onto a liquid milk substitute containing large amounts of ethanol extracted soya concentrate led to severe scours and high mortality. The cause of death was not established. Decreasing the proportion of protein supplied by soya concentrate from 70 to 35% reduced deaths and scouring. Nevertheless, results for liveweight gain and feed conversion efficiency were much lower than performances observed in piglets fed a control diet based entirely on cows' milk protein (Newport and Keal, 1982). Measurements of digestive enzymes in these animals suggested that proteolysis of milk and soyabean proteins were equally efficient when the piglets reached 28 days of age.

## *Predigestion of protein*

Survival of antinutritional factors in the gastrointestinal tract of young animals could be influenced by maturity of the digestive function. It is also possible that endogenous enzymes may be ineffective in digesting some components of milk substitutes. For instance, proteases of the calf abomasum and small intestine have been reported to be less effective in their hydrolysis of soya protein compared with actions on milk protein (Jenkins, Mahadevan and Emmons, 1980). Ideally enzymes used in predigestion should be complementary to the actions of endogenous enzymes. Thus, there is a need to appreciate the character of structures of residual protein and carbohydrates of weaner feeds which resist digestion. *In vitro* investigations of the stability of soya antigens to actions by porcine and bovine proteases indicated that, under optimal conditions of pH for protease activity,

**Table 18.3** RELATIVE QUANTITIES OF IMMUNOLOGICALLY ACTIVE GLYCININ AND β-CONGLYCININ IN HEATED SOYA FLOUR AND SOYA ISOLATE PROTEIN

| *Enzyme treatment* | *Haemagglutination inhibition titre* (log 2) | | | |
|---|---|---|---|---|
| | *Anti-glycinin* | | *Anti-β-conglycinin* | |
| | *Soya flour* | *Soya isolate* | *Soya flour* | *Soya isolate* |
| No enzyme, pH 1.8 | 9 | 10 | 12 | 12 |
| Pepsin, pH 1.8 | >1 | >1 | 12 | 12 |
| No enzyme, pH 7.8 | 12 | 11 | 12 | 12 |
| Trypsin, pH 7.8 | 12 | 7 | 7 | 9 |
| Pepsin then trypsin | >1 | >1 | 7 | 5 |

After Sissons and Thurston (1984)

β-conglycinin but not glycinin was unaffected by pepsin, whilst both proteins retained their antigenic activity after treatment with trypsin (see Table 18.3).

The possibility that digestive difficulties might be overcome by predigestion of ingredients used in a weaner diet was demonstrated in experiments which showed that intestinal damage and diarrhoea amongst early weaned piglets fed casein did not occur when animals were given hydrolysed cows' milk protein (Miller *et al.*, 1983). The improvement was attributed to the absence of antigenic material in the hydrolysed product. In another study piglets weaned onto diets containing pre-digested milk or soya isolate protein absorbed more nitrogen compared with animals fed native protein (Leibholz, 1981). In contrast, supplementary enzyme treatments of soya protein did not result in improved growth performance of pre-ruminant calves (Fries, Lassiter and Huffman, 1958; Colvin and Ramsay, 1968). This disappointing result may have been related to the poor specification of the enzymes used in the pre-digestion treatment. As a note of caution, products arising from predigestion of non-milk sources should be assessed to ensure that the process does not increase thet activity of antinutritional components.

## PROBIOTICS

The normal microflora of the alimentary tract is thought to complement the digestive functions of the host and provide protection against invading pathogens. At times of stress, such as weaning, the 'balance' of gut bacteria may become disturbed in favour of enterotoxic bacteria (Barrow, Fuller and Newport, 1977). Recently there has been much interest in the possibility of using bacterial feed additives to stabilize the indigenous flora and prevent diarrhoea in young animals. Many claims for the efficacy of these preparations have been made by commercial companies. Whilst some published studies support these claims (Bechman, Chambers and Cunningham, 1977; Muralidhara *et al.*, 1977) others have failed to demonstrate a beneficial effect of probiotic organisms (Hatch, Thomas and Thayne, 1973; Elinger, Muller and Ganzt, 1978; Morrill, Dayton and Mickelsen, 1977). In a recent review of probiotic efficacy, Fuller (1986) proposed that organisms may fail to achieve a beneficial response because of non-adherence to gastric and gut epithelial tissue, inability to grow in the gut environment and a lack of specificity for the host. Clearly further work is needed on modes of action and desirable features of probiotic cultures before bacterial additives can replace antimicrobial agents as a means of controlling diarrhoea.

### PHARMACOLOGICAL TREATMENTS

Several pharmacological treatments have proved effective in preventing the adverse immune responses involved in certain human food allergies. The drug disodium cromoglycate has been used to prevent histamine release possibility by acting to stabilize mast cell membranes (Dahl and Zetterstrom, 1978). Other reports suggest that indomethacin can alleviate symptoms of food intolerance in humans (Buisserat *et al.*, 1978) and protect calves against cardiovascular anaphylactic shock (Burka and Eyre, 1974). However, neither drug showed any beneficial effects on intestinal reactions of calves which had been orally sensitized with heated soya flour (Kilshaw and Sissons, 1979a; Kilshaw and Slade, 1980). The absence of a response may have been due to a lack of specific action on mucosal mast cells by these treatments. But if suitable agents could be found for controlling intestinal allergic disorders in young farm animals they could be more appropriate for piglets where intolerance to dietary protein seems to be transient and therefore administration of a drug need not be continuous.

## Conclusions

Diarrhoea of nutritional origin amongst calves and piglets is fundamentally either a problem of mucosal inflammation and/or increased luminal osmolality, whilst deranged motility may well aggravate the primary cause. Tissue damage linked with adverse immune responses to dietary antigens appears to be the most serious problem and this can lead to considerable loss of nutrients. Much recent research has focused attention on immaturity and limitations of digestive and immune function.

Further studies are needed to clarify the effect of creep feed in predisposing piglets to dietary intolerance at weaning, together with maternal factors which may influence mechanisms regulating the balance between sensitivity and tolerance to immunologically active components. Also more knowledge is required of molecular aspects of digestion, especially of antinutritional structures which may be revealed by partial hydrolysis. Besides heat processing for eliminating lectins and protease inhibitors in legume protein, further treatment is required to destroy antigenic activity of globulins. This can be achieved with hot aqueous ethanol, but the exact conditions are critical. This treatment has so far only proved successful for calves.

At the present time, strategies for preventing diarrhoea in young farm animals continue to be based on formulating weaner diets with highly digestible ingredients. But the risk of digestive disorders will increase as nutrient sources of weaner diets deviate from those in whole milk. Possible control measures include restrictions in intake during the first week of weaning, lowering nutrient density and processing to eliminate antinutritional factors. Predigestion with novel enzymes and a putative protective and digestive role for probiotics need evaluation and therefore remain as future options.

## References

ARGENZIO, R.A. and WHIPP, S.C. (1980). In *Veterinary Gastroenterology – 1980*, pp. 220–232. Ed. Anderson, N.V. Lea and Febiger, Philadelphia

ARMSTRONG, W.D. and CLAWSON, A.J. (1980). *Journal of Animal Science,* **50**, 377–384

ARMSTRONG, W.D. and CLINE, T.R. (1976). *Journal of Animal Science,* **42**, 592–598

BALL, R.O. and AHERNE, F.X. (1982). *Canadian Journal of Animal Science,* **62**, 907–913

BARROW, P.A., FULLER, R. and NEWPORT, M.J. (1977). *Infection and Immunity,* **18**, 586–595

BECHMAN, T.J., CHAMBERS, J.V. and CUNNINGHAM, M.D. (1977). *Journal of Dairy Science,* **60**, 74

BELL, J.M., ROYAN, G.F. and YOUNG, C.E. (1974). *Canadian Journal of Animal Science,* **54**, 355–362

BIDLINGMEYER, U.D., LEARY, T.R. and LASKOWSKI, M. (1972). *Biochemistry,* **11**, 3303–3310

BIRK, Y. and GERTLER, A. (1961). *Journal of Nutrition,* **75**, 379–387

BLAXTER, K.L. and WOOD, W.A. (1953). *Veterinary Record,* **65**, 889–892

BUISSERAT, P.D., HEINZELMANN, D.I., YOULTEN, L.J.F. and LESSOF, M.H. (1978). *Lancet,* **i**, 906–908

BURKA, J.F. and EYRE, P. (1974). *Canadian Journal of Physiology and Pharmacology,* **52**, 942–951

BYWATER, R.J. (1973). *Research in Veterinary Science,* **14**, 35–41

BYWATER, R.J. and LOGAN, E.F. (1974). *Journal of Comparative Pathology,* **84**, 599–610

CHRISTISON, G.I. and PARRA DE SOLANO, N.M. (1982). *Canadian Journal of Animal Science,* **62**, 899–905

CIRCLE, S.J. and SMITH, A.K. (1972). In *Soybeans: Chemistry and Technology*, Vol. 1, *Proteins*, pp. 294–338. Ed. Smith, A.K. and Circle, S.J. The Avi Publishing Company, Connecticut, USA

COLVIN, B.M. and RAMSAY, H.A. (1968). *Journal of Dairy Science,* **51**, 898–904

COMBS, G.E., OSEGUEDA, F.L., WALLACE, H.D. and AMMERMAN, C.B. (1963). *Journal of Animal Science,* **22**, 396–398

CONNELL, A.M. (1962). *Gut,* **3**, 342–348

CORRING, T., AUMAITRE, A. and DURAND, G. (1978). *Nutrition and Metabolism,* **22**, 231–243

CRUMP, M.H., ARGENZIO, R.A. and WHIPP, S.C. (1980). *American Journal of Veterinary Research,* **41**, 1565–1568

DAHL, R. and ZETTERSTROM, O. (1978). *Clinical Allergy,* **8**, 419–422

DE GROOT, A.P. and SLUMP, P. (1969). *Journal of Nutrition,* **98**, 45–56

DE MEULENAERE, H.L.H. (1964). *Journal of Nutrition,* **82**, 197–205

DEPREZ, P., HENDE, C., VAN D., MUYLLE, E. and OYAERT, W. (1986). *Veterinary Research Communications,* **10**, 469–478

DUCHARME, G.A. (1982). In *Effect of Soybean Trypsin Inhibitor on Growth, Protein Digestibility, and Pancreatic Enzyme Activity in the Young Calf*, PhD Thesis, North Carolina State University

EFIRD, R.C., ARMSTRONG, W.D. and HERMAN, D.L. (1982). *Journal of Animal Science,* **55**, 1370–1379

ELLINGER, D.K., MULLER, L.D. and GANTZ, P.J. (1978). *Journal of Dairy Science,* **61**, 126

FIORAMONTI, J. and BUENO, L. (1980). *British Journal of Nutrition,* **43**, 155–162

FIRON, N., OFEK, I. and SHARON, N. (1984). *Infection and Immunity,* **43**, 1088–1090

FRIES, G.F., LASSITER, C.A. and HUFFMAN, C.F. (1958). *Journal of Dairy Science,* **41**, 1081–1087

FUKUSHIMA, D. (1969). *Cereal Chemistry,* **46**, 156–163

FULLER, R. (1986). *Journal of Applied Bacteriology*, 1S–7S

GORRILL, A.D.L. (1970). *Canadian Journal of Animal Science,* **50**, 745–747

GORRILL, A.D.L. and NICHOLSON, J.W.G. (1969). *Canadian Journal of Animal Science,* **49**, 315–321

GORRILL, A.D.L. and THOMAS, J.W. (1967). *Journal of Nutrition,* **92**, 215–223

GORRILL, A.D.L., THOMAS, J.W., STEWART, W.E. and MORRILL, J.L. (1967). *Journal of Nutrition,* **92**, 86–92

HAGAR, D.F. (1984). *Journal of Agriculture and Food Chemistry,* **32**, 293–296

HAMPSON, D.J. and KIDDER, D.E. (1986). *Research in Veterinary Science,* **40**, 24–31

HAMPSON, D.J. and SMITH, W.C. (1986). *Research in Veterinary Science,* **41**, 63–69

HAMPSON, D.J., FU, Z.F. and SMITH, W.C. (1988). *Research in Veterinary Science,* **44**, 309–314

HARTMANN, P.A., HAYS, V.W., BAKER, R.O., NEAGLE, L.H. and CATRON, D.V. (1961). *Journal of Animal Science,* **20**, 114–123

HATCH, R.C., THOMAS, R.O. and THAYNE, W.V. (1973). *Journal of Dairy Science,* **56**, 682

HENSCHEL, M.J. (1973). *British Journal of Nutrition,* **30**, 285–295

HUBER, J.T., RIFKIN, R.J. and KEITH, J.M. (1964). *Journal of Dairy Science,* **47**, 789–792

JAFFE, W.G. (1969). In *Toxic Constituents of Plant Foodstuffs – 1969.* Ed. Liener, I.E. Academic Press, New York and London

JENKINS, K.J., MAHADEVAN, S. and EMMONS, D.B. (1980). *Canadian Journal of Animal Science,* **60**, 907–914

KAKADE, M.L., SIMONS, N.R., LIENER, I.E. and LAMBERT, J.W. (1972). *Journal of Agriculture and Food Chemistry,* **20**, 87–90

KAKADE, M.L., THOMPSON, R.M., ENGELSTAD, W.E., BEHRENS, G.C. and YODER, R.D. (1974). *Journal of Dairy Science,* **57**, 650

KIDDER, D.E. and MANNERS, M.J. (1978). In *Digestion in the Pig – 1978.* Ed. Kidder, D.E. and Manners, M.J. Scientechnica, Bristol

KIDDER, D.E. and MANNERS, M.J. (1980). *British Journal of Nutrition,* **43**, 141–153

KILSHAW, P.K. and SISSONS, J.W. (1979a). *Research in Veterinary Science,* **27**, 361–365

KILSHAW, P.J. and SISSONS, J.W. (1979b). *Research in Veterinary Science,* **27**, 366–371

KILSHAW, P.K. and SLADE, H. (1980). *Clinical and Experimental Immunology,* **41**, 575–582

KWIATKOWSKA, A. (1973). *Prace i Materialy Zootech.,* **3**, 63–75

LAPLACE, J.P. (1984). In *Function and Dysfunction of the Small Intestine – 1984*, pp. 1–20. Ed. Batt, R.M. and Lawrence, T.L.J. Liverpool University Press, UK

LENNON, A.M., RAMSEY, H.A., ALSMEYER, W.L., CLAWSON, A.J. and BARRICK, E.R. (1971). *Journal of Animal Science,* **33**, 514–519

LEIBHOLZ, J. (1981). *British Journal of Nutrition,* **46**, 59–69

LEPKOVSKY, S., BINGHAM, E. and PENCHARZ, R. (1959). *Poultry Science,* **38**, 1289–1295

LIENER, I.E. (1955). *Archives of Biochemistry and Biophysics,* **54**, 223–231

LIENER, I.E. and KAKADE, M.L. (1969). In *Toxic Constituents of Plant Foodstuffs – 1969*, pp. 7–68. Ed. Liener, I.E. Academic Press, New York and London

LINDEMANN, M.D., CORNELIUS, S.G., EL KANDELGY, S.M., MOSER, R.L. and PETTIGREW, J.E. (1986). *Journal of Animal Science,* **62**, 1298–1307

LOW, A.G. (1985). In *Proceedings of the 3rd International Seminar on Digestive Physiology in the Pig – 1985*, pp. 157–179. Ed. Just, A., Jorgensen, H. and Fernandez, J.A. National Institute of Animal Science, Copenhagen, Denmark

McGILLIARD, A.D., BRYANT, J.M., BRYANT, A.B., JACOBSON, N.L. and FOREMAN, C.F. (1970). *Iowa State Journal of Science,* **45**, 185–195

MATHIEU, C.M. and DE TUGNEY, H. (1965). *Annales de Biologie Animale Biochimie Biophysique,* **5**, 21–39

MEADE, R.J. (1967). *Feedstuffs,* **39**, 18–21

MILLER, B.G., JAMES, P.S., SMITH, M.V. and BOURNE, F.J. (1986). *Journal of Agricultural Science,* **107**, 579–589

MILLER, B.G., NEWBY, T.J., STOKES, C.R., HAMPSON, D. and BOURNE, F.J. (1983). *Annales de Recherches Veterinarie,* **14**, 487–492

MILLER, B.G., NEWBY, T.J., STOKES, C.R., HAMPSON, D., BROWN, P.J. and BOURNE, F.J. (1984). *American Journal of Veterinary Research,* **45**, 1730–1733

MILLER, B.G., PHILLIPS, A., NEWBY, T.J., STOKES, C.R. and BOURNE, F.J. (1985). In *Proceedings of the 3rd International Seminar on Digestive Physiology in the Pig – 1985*, pp. 65–68. Ed. Just, A., Jorgensen, H. and Fernandez, J.A. National Institute of Animal Science, Copenhagen, Denmark

MORRILL, J.L., DAYTON, A.D. and MICKELSEN, R. (1977). *Journal of Dairy Science,* **60**, 1105

MOWAT, A.M. and FERGUSON, A. (1981). *Clinical and Experimental Immunology,* **43**, 574–582

MURALIDHARA, K.S., SHEGGEBY, G.G., ELLIKER, P.R., ENGLAND, D.C. and SANDINE, W.E. (1977). *Journal of Food Protection,* **40**, 288–295

NEWBY, T.J., MILLER, B.G., STOKES, C.R., HAMPSON, D. and BOURNE, F.J. (1985). In *Recent Developments in Pig Nutrition*, pp. 211–221. Ed. Cole, D.J.A. and Haresign, W. Butterworths, London

NEWPORT, M.J. (1980). *British Journal of Nutrition,* **44**, 171–178

NEWPORT, M.J. and KEAL, H.D. (1982). *British Journal of Nutrition,* **48**, 89–95

NEWPORT, M.J. and KEAL, H.D. (1983). *Animal Production,* **37**, 395–400

NITZAN, Z., VOLCANI, R., GORDIN, S. and HASDAI, A. (1971). *Journal of Dairy Science,* **54**, 1294–1299

NITZAN, Z., VOLCANI, R., HASDAI, A. and GORDIN, S. (1972). *Journal of Dairy Science,* **55**, 811–821

OBARA, T. and WATANABE, Y. (1971). *Cereal Chemistry,* **48**, 523–527

OOMS, L. and DEGRYSE, A. (1986). *Veterinary Research Communications,* **10**, 355–397

PALMER, N.C. and HULLAND, T.J. (1965). *Canadian Veterinary Journal,* **6**, 310–316

PEDERSEN, H.E. (1986). In *Studies of Soyabean Protein Intolerance in the Preruminant Calf*, PhD Thesis, University of Reading, UK

PETTIGREW, J.E., HARMON, B.G., CURTIS, S.E., CORNELIUS, S.G., NORTON, H.W. and JENSEN, A.H. (1977). *Journal of Animal Science,* **45**, 261–268

RACKIS, J.J., HONIG, D.H., SASAME, H.A. and STEGGARDCA, F.R. (1970). *Journal of Agriculture and Food Chemistry,* **18**, 977–982

RATCLIFFE, B., SMITH, M.W., MILLER, B.G. and BOURNE, F.J. (1987). *Proceedings of the Nutrition Society,* **46**, 101A

ROY, J.H.B. (1969). *Proceedings of the Nutrition Society,* **28**, 160–170

ROY, J.H.B. (1984). In *Function and Dysfunction of the Small Intestine – 1984*, pp. 95–132. Ed. Batt, R.M. and Lawrence, T.L.J. Liverpool University Press, UK

RUCKEBUSCH, Y. and BUENO, L. (1976). *British Journal of Nutrition,* **35**, 397–405

SEDGMAN, C.A. (1980). In *Studies of the Digestion, Absorption and Utilization of Single Cell Protein in the Preruminant Calf.* PhD Thesis, University of Reading, UK

SEVE, B., AUMAITRE, A., MOUNIER, A.M., LAPANOUSE, A., BRUNET, P. and PAUL-URBAIN, G. (1983). *Sciences des Aliments,* **3**, 53–67

SEVE, B., AUMAITRE, A., BOUCHEZ, P., MESSANGER, A., LEBRETON, Y. and LEVREL, R. (1985). *Sciences des Aliments,* **5**, 119–126

SISSONS, J.W. (1981). *Journal of Science Food and Agriculture,* **32**, 105–114

SISSONS, J.W. (1982). *Proceedings of the Nutrition Society,* **41**, 53–61

SISSONS, J.W. (1983). *Journal of Dairy Research,* **50**, 387–395

SISSONS, J.W. and SMITH, R.H. (1976). *British Journal of Nutrition,* **36**, 421–438

SISSONS, J.W. and SMITH, R.H. (1979). *Annales de Recherche Veterinaire,* **10**, 176–178

SISSONS, J.W. and THURSTON, S.M. (1984). *Research in Veterinary Science,* **37**, 242–246

SISSONS, J.W., SMITH, R.H. and HEWITT, D. (1979). *British Journal of Nutrition,* **42**, 477–485

SISSONS, J.W., NYRUP, A., KILSHAW, P.J. and SMITH, R.H. (1982a). *Journal of Science Food and Agriculture,* **33**, 706–710

SISSONS, J.W., SMITH, R.H., HEWITT, D. and NYRUP, A. (1982b). *British Journal of Nutrition,* **47**, 311–318

SMITH, H.W. and HALLS, S. (1968). *Journal of Medical Microbiology,* **1**, 45–49

SMITH, R.H. and SISSONS, J.W. (1975). *British Journal of Nutrition,* **33**, 329–349

SRIHARA, P. (1984). In *Processing to Reduce the Antigenicity of Soybean Products for Preruminant Calf Diets*, PhD Thesis, University of Guelph, Canada

STEINER, R.F. (1965). *Biochemistry, Biophysics Acta,* **100**, 111–121

STOBO, I.J.F., GANDERTON, P. and CONNORS, H. (1983). *Animal Production,* **36**, 512–513

STOBO, I.J.F. and ROY, J.H.B. (1977). *Animal Production,* **24**, 143

SUDWEEKS, E.M. and RAMSEY, H.A. (1972). *Journal of Dairy Science,* **55**, 705

TERNOUTH, J.H., ROY, J.H.B., THOMPSON, S.Y., TOOTHILL, J., GILLIES, C.M. and EDWARDS-WEBB, J.D. (1975). *British Journal of Nutrition,* **33**, 181–196

TOULLEC, R., GUILLOTEAU, P. and VILLETTE, Y. (1984). In *Physiologie et Pathologie Perinatales Chez Les Animaux de Ferme – 1984.* Ed. Jarrige, R. INRA, Paris

TOULLEC, R., MATHIEU, C.M. and PION, R. (1974). *Annales de Zootechnie,* **23**, 75–87

VAN DER POEL, A.F.B. and HUISMAN, J. (1988). *Proceedings of the 4th International Seminar on Digestive Physiology in the Pig*, p. 49. Jablonna, Poland

WILLIAMS, V.J., ROY, J.H.B. and GILLIES, G.M. (1976). *British Journal of Nutrition,* **36**, 317–335

WISEMAN, J. and COLE, D.J.A. (1988). In *Recent Advances in Animal Nutrition – 1988*, pp. 13–37. Ed. Haresign, W. and Cole, D.J.A. Butterworths, London

WOODWARD, J.C. (1972). *Federation Proceedings,* **31**, 695

19

# IMMUNITY, NUTRITION AND PERFORMANCE IN ANIMAL PRODUCTION

P. PORTER and M.E.J. BARRATT
*Unilever Colworth Laboratory, Bedford, UK*

## Historical perspective

Our understanding of the processes of disease and host resistance to infection developed over many centuries. It was characterized by a series of significant advances such as that by Jenner in 1796. He demonstrated the process of vaccination against smallpox by inoculating with (the closely related but non-pathogenic) cowpox.

Probably the most outstanding advances were made 100 years ago arising from the fortunate observation of Pasteur that disease causing organisms could be rendered harmless by various simple procedures and that inoculation of the host thereafter led to a process of resistance. Furthermore, this resistance could be transferred passively from one individual to another via cell free blood serum. These procedures led Pasteur, Koch, Roux, Von Behring and Erlich to develop a range of vaccines and antitoxic sera. The application of these, over a period of 20 years, practically doubled the life expectancy of man, and led dramatically to the improvement of health and hygiene in the twentieth century.

During these early days most of the attention was directed towards the blood-borne nature of resistance via 'antibodies'. Metchnikoff, from his vacation observations of cellular activities in a starfish larva discomforted by a thorn, had postulated cellular mechanisms of defence. These were not rationalized with the antibody theory until 1903 when the English scientist Amroth Wright surmised that antibodies coated bacteria thus 'opsonizing' them in preparation for destructive ingestion by phagocytic cells.

It is of further significance that a research student Bezredka, working under the tutelage of Metchnikoff at the Pasteur Institute around the period of the First World War, made the fundamental observation that resistance to infection need not correlate with the presence of antibodies in the blood. In particular, using his shigellosis model in the rabbit, he showed that resistance to infection of the gut was more related to the antibody detected locally from the intestine rather than antibody in the blood.

This observation of local intestinal immunity, which is highly pertinent to the context of immunology and its effect on animal nutrition, was almost totally disregarded until the 1960s when scientists in Europe and North America began to unravel and to characterize the function of the mucosal defence system of secretory immunity. Since the gut is the central organ of this immune system which mediates specific resistance at all mucosal surfaces, it is inevitable that it will feature strongly in this chapter. The immune system will be considered in the context of the macromolecular components 'antigens', which enter and challenge the alimentary tract. However, it must be remembered that they are of diverse origin, arising from maternal colostrum and milk, dietary proteins and polysaccharides as well as viruses, bacteria and parasites which may colonize the gut.

## Diversity of responses in the gut

In terms of antigenic challenge and immune function, the gut is the largest lymphoid organ of the body. The surface area of the intestinal tract can extend to many square metres and in its tissues are located millions of lymphoid cells. These carry out such diverse and complex functions as presenting antibody to the epithelium, regulating the production of antibody, destroying invading pathogens, destroying infected cells, tolerizing the host to harmless antigens and generally attending to a harmonious relationship between the host animal and its everyday environment.

These functions are not always beneficial to aspects of health and subsequent performance. Under some circumstances, frequently mediated by stress, the beneficial aspects are overtaken by deleterious aspects of immunity. Hypersensitivity mechanisms consequently arise which are damaging to the host tissues and can lead also to susceptibility to infectious opportunists which are always present in the intensive farm environment.

In modern agriculture the intensive rearing practices allow little time for the developing immune system of young animals to respond adequately to all the challenges of life represented by competition with fellow creatures, infectious organisms, toxic products and diet. Animal performance is certainly dependent upon the maintenance of integrity and function of the intestinal epithelium. Thus, it cannot escape recognition that resistance to *E. coli* infection in pigs is characterized by better food conversion and rate of growth post weaning (Porter *et al.*, 1973). Similarly, the removal of an allergenic component of soya in milk replacers has led to better health and performance in calves (Barratt, Strachan and Porter, 1978).

The fundamental questions are how can we enhance resistance to infection, regulate detrimental responses to nutrients, reduce the numbers and virulence of pathogens in the environment and exploit the dietary route as a means of beneficially modulating all essential resistance mechanisms of the host? With answers to these questions it would be possible to formulate feeds which would radically affect health, nutrition and performance. In dealing with these questions the ramifications of immunology are so vast that the subject would be difficult to comprehend as a whole. In the interests of brevity it is the intention here to limit concepts to the gut as an immunological organ and deal with factors of direct interest to nutritionists.

## Immune defence mechanisms for the gut

The non-immune factors which contribute to resistance in the gut include mucus, acidity, enzymes, symbiotic microflora and mechanical effects of peristalsis and evacuation. The non-specific immune mechanisms which reinforce this external protective barrier include the phagocytic processes of leucocytes which can emigrate into the lumen of the gut as well as operate in the intestinal tissue and the regional lymph nodes.

The specific immune defence of the gut is organized through the immunoglobulins, a family of structurally related glycoproteins equipped with receptors for macromolecular products recognized as foreign to the host (reviewed by Porter, 1979). The immunoglobulins can be produced systemically in organs such as the spleen or lymph nodes and be carried in the blood to extravasate into the intestinal tissues, only appearing in the lumen of the gut when damage and leakage occurs. The major blood-borne antibody class, IgG, follows this pattern. Alternatively, the immunoglobulins can be produced locally by cells which populate the intestinal tissues and take up strategic positions close to the intestinal epithelium where they present specialized immunoglobulins for active transport into the lumen. The major secretory antibody class, IgA, follows this pattern. Thus, in the mature state IgA dominates the external secretions for first line luminal defence. A second line defence is provided by the predominantly blood-borne IgG.

The two are not necessarily coexistent protective forces in that the host may be solidly immune to gut infection by virtue of IgA alone but unlikely to be so by virtue of IgG alone. To this end one must attend to the processes of immunization; IgG responses arise principally from the systemic application of antigen whereas IgA responses arise from the local application of antigens. The lymphocytes which give rise to the mucosal IgA response arise in the Peyer's patch lymphoid regions of the posterior small intestine. The Peyer's patches are covered by a thin membrane of cells capable of sampling the luminal contents and activating specific lymphocytes behind this membrane. These lymphocytes then begin the process of division, multiplication and differentiation into IgA secreting plasma cells while travelling through the mesenteric lymph, entering the blood stream via the thoracic duct and finally 'homing' to the tissues of the gut to present antibody for secretion into the lumen. This complex system of transport allows for the dissemination of such cells to other more remote external mucosal organs such as the urogenital tract, respiratory tract, salivary glands, ocular glands and mammary glands. Hence, the gut can act as a focus for a common mucosal immune system.

Dimeric IgA, together with the other polymeric immunoglobulin IgM, possesses a specialized peptide chain, 'j-chain', which binds the heavy chains of these immunoglobulins to produce polymers. The epithelial cells of mucosal surfaces synthesize a receptor for j-chain known as secretory component which binds the immunoglobulins for internalization in vesicles and transports them through the cytoplasm to the external brush border membrane. Here, the immunoglobulins are released with secretory component which then facilitates mucin binding and distribution over the external surface as a protective layer. IgG does not participate in this active process of local secretion and transport; it does not bind j-chain as a process of natural polymerization and therefore is not held on the epithelial membrane by secretory component.

On the other hand, IgG does participate in the process of natural transport from blood to colostrum in the mammary gland. Uptake of this immunoglobulin from the gut lumen then occurs in the neonate. This is because the mammary epithelial membranes immediately before birth are temporarily equipped with receptors for the constant region or 'tail' of the IgG molecule allowing for its short-term transport.

It is common knowledge that all farm species, unlike the primates, are born without intra-uterine transport of maternal immunoglobulins. Therefore, in order to meet the infectious environment of the postpartum world, mothers' antibodies must be rapidly transferred to the newborn and this is achieved by the colostrum. Adequate colostrum intake is essential for neonatal survival, but the quality of colostrum is dependent upon the immunological experience of the dam. The 'gut–mammary–gut' link of immunological experience from mother to suckling offspring is an important factor for postnatal health and appropriate preparation for weaning.

From the foregoing it is apparent that the dam, through her intestinal experience of her environment will, over a period of time, build up appropriate antibodies of the various protective classes, to pass on to the neonate through the colostrum and milk. This will provide passive protection for the offspring until it is competent to deal with infection pressures via its own active immune resources. Two things intervene to disturb this optimistic picture. First, under production pressures in modern intensive livestock farming many animals in first pregnancy are young and not extensively adapted to the pathogens they frequently harbour. They therefore pass on infection to their offspring during the stress of parturition and fail to provide appropriate colostral antibodies for protection. Second, the tendency to wean animals from the dam at an early age defaults passive antibody support before active competence to sustain protection has developed. Either circumstance will be detrimental to health and subsequent performance. The objective will be to plan for such contingencies with appropriate immunoprophylaxis and therapy.

## Immune destructive mechanisms in the gut

A range of glycoproteins in a complex enzyme cascade collectively known as 'complement' is involved with antibody in the destruction of invasive organisms. It is ironic that under certain circumstances the destructive properties of this cascade can be turned against the host.

Normally, for lytic function against invading bacteria and foreign cells, the complement cascade is 'fixed' on the membrane of the invader and destroys it. This activation is confined to a limited time and space, without harming the host. On the other hand if an animal absorbs dietary antigens from the gut which precipitate with antibody in the tissues to form immune complexes which activate complement, then inflammatory reactions take place within these tissues. The more the antigen is disseminated through the body via the blood circulation, the more the inflammatory reaction is generalized and can affect vital organs such as the lung, heart and kidneys as well as the gut.

This type of destructive immune reaction (Arthus type) was the major one induced by feeding soya protein in calf milk replacers (Barratt, Strachan and Porter, 1978).

Fortunately, the offending antigen was not characterized as the major storage protein of soya and it could be eliminated by appropriate processing of the feed product. The adverse pathophysiological effects which were mediated by IgG antibody reacting with the soya antigen in the tissues of the host were alarming to say the least. The following is the catalogue of disaster. First, at the level of intestinal physiology, there was widespread disturbance in gut function leading to vomiting, diarrhoea, rejection of the diet. Second, major morphological disturbances attributable to the inflammatory lesion in the gut tissues occurred, with partial villous atrophy, oedema and thickening of the lamina propria through mononuclear cellular infiltration.

This type of immune mediated damage is quite distinct from that of common allergic immediate type hypersensitivity. The immunoglobulin isotype responsible for allergy is almost exclusively IgE. Unfortunately, while this has been well characterized in man during the past 20 years, an analogue of this protein in farm animals has been difficult to identify. However, in a small percentage of calves (3–5%) fed antigenic soya there was also evidence of immediate hypersensitivity mediated by an IgE-like antibody (Barratt and Porter, 1979). IgE undertakes its damaging reaction in the tissues by clinging to mast cells and basophil granulocytes; crosslinking of the IgE molecules with allergen disturbs the membrane of these cells with consequent loss of granules into the tissues. The granules of these cells contain prostaglandins, chemotactic factors, histamine and enzymes such as bradykinin which induce the release of vasoactive amines such as 5-hydroxytryptamine from platelets. The reaction with allergen therefore is the harmful event leading to the release of a range of pharmacological mediators of discomfort, which initiates reactions broadly manifested in the tissues of the gut and respiratory tract, and more generally as anaphylactic shock.

## Immunoregulatory mechanisms of the gut

Various mechanisms of dealing with antigen by the gut have marked effects on the development or prevention of immune reactions. In most animals there will be detectable low levels of blood-borne antibody to food proteins; this is normally of the IgG class which cannot reach the gut lumen to block uptake of antigen. One small benefit of the allergic response initiated by immunoglobulin IgE is that pharmacological effects set in train a leakage of circulatory antibody through the epithelium, thereby facilitating the clearance of the offending antigen from the mucosa (Barratt and Porter, 1979). This 'gate keeper' effect of IgE is however the least comfortable mode of dealing with the problem and fortunately other more satisfactory approaches are available.

Most appropriate would be complete digestion in the lumen so that no intact antigen could challenge the intestinal mucosa, but in the event of incomplete digestion IgA may intervene to prevent absorption. As part of its function IgA binds antigen in the mucin on the epithelial surface. IgA is equipped to evade the activity of proteolytic enzymes, and indeed works in synergy with them, allowing for the complete digestion of bound antigen (Walker, 1982); thereby excluding any further access to the tissues and hence preventing the prospect of generating deleterious systemic reactions.

Naturally all biological systems have a threshold of capability and it is preferable to have a second string to one's bow. In the case of soya antigen in calves, the offending protein resists proteolysis, evades the IgA mucosal system and stimulates IgG complement mediated damage as well as IgE hypersensitivity (Barratt, Strachan and Porter, 1979). The normal response to uncleared circulating antigens derived from the diet is the induction of a state of specific immunological tolerance, i.e. a lowered systemic responsiveness. Ideally, tolerance and IgA antigen exclusion should both be established and operate simultaneously.

The ability to induce antigen tolerance in the antibody producing lymphocytes (B cells) is a property of a subpopulation of lymphocytes which derive from the thymus

(T cells). 'T suppressor' cell mechanisms operate both in the gut tissues and systemically to 'down regulate' the cells which would produce more antibody to each fresh ingress of dietary antigen (Ferguson, Mowat and Strobel, 1983). It is interesting that dietary protein antigens can induce this helpful phenomenon but bacterial polysaccharide antigens do not; these are T-cell independent antigens and hence no tolerance can result. This of course is necessary since no animal can afford to be unresponsive to infectious and toxic agents.

It is clear, therefore, that complex cellular interactions regulate the mucosal immune responses in the gut to orally presented antigens. Antibody producing B cells are directly influenced by thymus processed T lymphocytes. Other subsets of T cells induce the initiation of immune responses and differentiation of antibody producing cells. In addition the production of antibody to most antigens by B cells

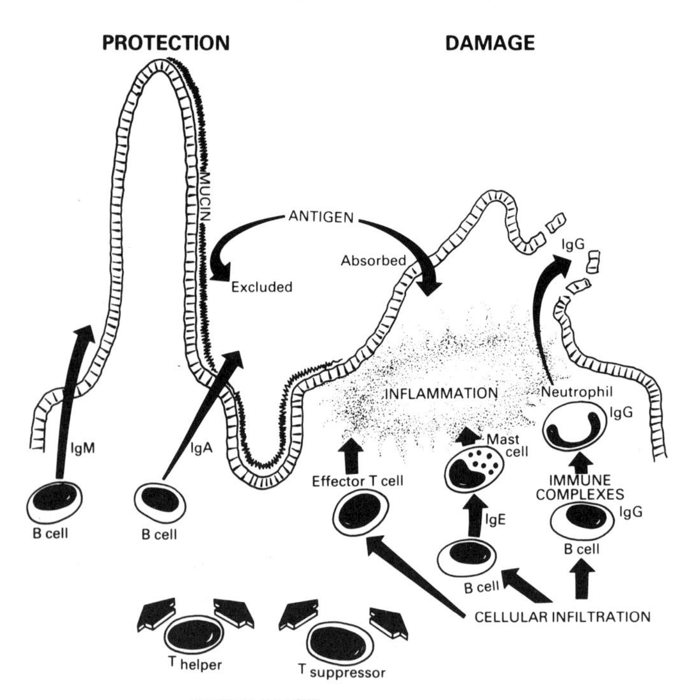


**Figure 19.1** The balance of immune activities involved in protection or destruction in the gut. Antibodies of the IgM and IgA isotypes are secreted into the mucin coat bathing the gut epithelial lining and here provide a protective barrier. Antigen that evades this barrier and is absorbed may trigger an inflammatory reaction that results in destruction of gut tissues. This damage can be mediated by a variety of immune mechanisms. Antibody of IgE isotype can activate mast cells to degranulate in the presence of antigen, releasing vasoactive amines. Immune complexes of antigen and antibody of IgG isotype can initiate complement mediated cell lysis and attract neutrophils into the area. Cytotoxic T lymphocytes kill cells directly without antibody involvement. Other populations of T lymphocytes regulate all of these immune activities by either helping or suppressing particular immune reactions

requires the assistance of soluble factors released by 'helper T cells'. On the other hand, 'T suppressor' cells restrain and regulate both the cytotoxic cells responsible for cell mediated immune reactions and also the helper T cells themselves.

There is consequently an intricate balance of helper:suppressor immunoregulatory T cells that are influenced by a wide variety of stimuli. Modulation of this balance by environmental influences can result in the enhancement of beneficial protective responses to disease causing organisms or alternatively, suppression of potentially damaging responses to harmless dietary antigens. This 'down regulation' of non-protective immune responses represents the normal tendency and can be brought about not only by active suppression by T cells but also by activation of macrophages, cells that first encounter antigens and then process and present them to the immune system.

So far we have dealt simply with each mechanism whereby antibody can act in the intestinal mucosa either protectively or harmfully. We have highlighted the processes of regulation via two subpopulations of T cells whose products either help B cells to form antibody or suppress B cells from forming antibody. In the harmful sensitizing responses of antibody either via IgE sensitizing mast cells or IgG immune complexes fixing complement, we have described the functional aspects of inflammation. An attempt to illustrate all these processes in the various effector and regulatory functions of the immune response is shown in *Figure 19.1*.

Finally in addition to all these actions there is one additional T cell effector function of importance in the gut, that of cell mediated immunity. Effector T cells exert their attack on antigens either directly, or by the release of soluble factors that stimulate other cell types to respond. Both the antibody and cell mediated immune responses, although initiated locally, prime systemic immune responses. Thus, localized protection may be provided at vulnerable mucosal surfaces other than that of the primary site of challenge. Typical of this is the tuberculin reaction, and a gut level of protection attributable to this mechanism has been demonstrated for salmonellosis in calves (Aiken, Hall and Jones, 1978).

## Feed strategies for health and performance

From the foregoing, the dietary ingredients, which in themselves provide the nutrients essential for the growth of the animal, are now evidently a potential source of antigenic challenge. They are also to be recognized as a substrate for microbial culture throughout the alimentary tract, and indeed may also be a source of ingested infectious agents.

Derangement of host metabolism by dietary hypersensitivity may lead to infection with opportunistic pathogens. Prevention of ingestion of infectious agents through the diet may be dealt with tactically by processing such as heat or acidification which has been undertaken with some calf milk replacers. The non-specific elements of host protection are difficult to regulate; mucus, acid and peristalsis are all normal gut functions. There is, however, some suggestion that acidification of diets may contribute to enhanced health and performance but authoritative data are not published in the literature.

Tactically there are few openings to deal with dietary hypersensitivity because insufficient is known about allergens and their damaging reactions in farm animals. The only well characterized dietary allergy has been associated with a soya antigen (Barratt, Strachan and Porter, 1978; Kilshaw and Sissons, 1979), and this was overcome by process removal of the immunologically characterized antigen with benefits in performance (Thomas, 1982).

Early weaning of piglets has been identified to induce sensitivity reactions to diet because of creep management (Miller *et al.*, 1984), and this leads to a malabsorption syndrome measured by xylose absorption. The malabsorption in turn provides substrate for microbial proliferation and gastrointestinal infection (Miller *et al.*, 1984). This problem can be obviated either by delaying weaning or abrupt early weaning with no access to creep diet and its sensitizing antigens. Alternatively and ideally, except for cost, the young animals should be fed a hydrolysed non-allergenic diet. The feeding of sufficient antigen preweaning in order to induce dietary tolerance is an improbable option for early weaning.

Feed strategies for inducing specific protection against microbial pathogens involve oral immunization. Enteric diseases associated with *E. coli* have been controlled via 'in-feed' vaccination of the sow prepartum and the piglets preweaning (Porter *et al.*, 1985). The antibody activities induced by heat inactivated *E. coli* vaccines are various, including antibacterial (Porter *et al.*, 1974) anti-enterotoxin (Linggood and Ingram, 1978) and anti-adhesion (Porter, Parry and Allen, 1977). Live organisms cultured in milk have been used to achieve the same ends (Kohler, Cross and Bohl, 1975) but the control of infection and the spread of virulence is substantially more doubtful than with the use of sterile non-replicating products. The environmental benefits of oral vaccination in minimizing the spread of virulence plasmids is now well proven (Porter and Linggood, 1983), and in the light of current concern over drug resistance the tactic of 'in-feed' vaccination provides important new controls in the host–pathogen relationship.

The process of oral vaccination to extend host resistance to *Salmonella* has been studied (Balger *et al.*, 1981) in the calf and for coccidiosis in poultry (Davis *et al.*, 1986). These approaches are not applicable to viruses which cannot be manufactured in quantities with sufficient economy to match the needs of the animal feeds industry.

While dealing with elements of specific antibody protection it is worth mentioning colostrum status in a performance context. We have adequate evidence in calves from a comparison of animals with low and high colostrum status, i.e. below or above a critical IgG blood serum level of 12 mg/ml respectively (Prior and Porter, 1980), that there are significant performance differences. The performance deficiencies of low colostrum status are substantially mitigated by oral vaccination with *E. coli* antigens.

Apart from *E. coli* products, the only other applications of bacterial products through the diet have been lactobacilli. These have not been used to induce changes in the immune system of the animal which ingests them but to produce changes in the character and metabolism of the intestinal flora. Thus, the beneficial effects may be due to colonization, competition for intestinal adhesion sites or changes in microbial metabolism in the gut. For example, the ability of *E. coli* to decarboxylate basic amino acids to form putrefactive amines is reduced by lactobacilli (Hill, Kenworthy and Porter, 1970) and the site of this activity is shifted from the small intestine to the colon, further benefiting the health and performance of the animal.

Microbial gut flora are known to influence immunity in a variety of ways, and bacterial lipopolysaccharide (LPS) in particular may have a profound effect not only on the immune system but also on macrophages of the reticuloendothelial system. Although LPS has traditionally been regarded as a B cell mitogen and macrophage activator, it is now apparent that an interaction between T cells and LPS occurs (Motta, Portnoi and Truffa-Bachi, 1986). In particular, bacterial endotoxin potentiates antigen specific proliferation of T helper cells. This effect is independent of interleukin I action, the T cell activation factor released by macrophages after LPS stimulation. The prospect thus emerges of multiple modulatory effects of bacterial endotoxin on gut immune responses. Not only could induction of tolerance be enhanced by selective increases of suppressor cell activity, but by polyclonal activation of B cell responses and increases in T cell help, both specific and non-specific protective immune responses could be improved.

Microbial endotoxins are only one example of non-specific enhancement of immune responses by orally administered material. Various micronutrients and other chemicals exert similar modulatory influences on protective immune responses to pathogens. Increased levels of dietary vitamin E for example, indirectly influence the proliferation of antigen specific B cells through enhanced cooperation with T helper cells (Tengerdy, Mathias and Nockles, 1980). Similar effects are obtained with vitamin A, achieved probably by increasing the lability of lysosomal membranes in macrophages. In consequence there is increased retention of ingested antigen within macrophages, an action which is more like that of a depot adjuvant. Levamisole, a chemical first used as an anthelmintic drug, has been shown to similarly, non-specifically, enhance B cell antibody responses by non-specific stimulation of macrophages and T cells (Kelly, 1978).

In conclusion a variety of non-specific and specific immune functions can be stimulated via the oral route. The cell mechanisms involved in gut immunology are complex but the objectives are quite simply to provide protective responses or prevent damaging responses. The guidelines for exploiting the immunobiology of the host intestinal system and the means of quantifying the effects are not well established. However, several interesting advances have been made employing bacterial products, chemicals and micronutrients. These have been reviewed in the light of current understanding of immune function and positioned in the context of feed strategies leading to improved health, which in turn can lead to improved nutritional performance.

## References


AITKEN, M.M., HALL, G.A. and JONES, P.W. (1978). *Research in Veterinary Sciences*, **24**, 370–374

BALGER, G., HOERSTKE, M., DIRKSEN, G., SEITZ, A., SAILER, J. and MAYR, A. (1981). *Zentralblatt für Veterinärmedizin B*, **28**, 759–766

BARRATT, M.E.J., STRACHAN, P.J. and PORTER, P. (1978). *Clinical and Experimental Immunology*, **31**, 305–312

BARRATT, M.E.J. and PORTER, P. (1979). *Journal of Immunology*, **123**, 676–680

BARRATT, M.E.J., STRACHAN, P.J. and PORTER, P. (1979). *Proceedings of the Nutrition Society*, **38**, 143–150

DAVIS, P.J., BARRATT, M.E.J., MORGAN, M. and PARRY, S.H. (1986). In *Proceedings of Georgia Coccidiosis Conference*, pp. 618–633. Eds. McDougald, L.R., Joyner, L.P. and Long, P.L., University of Georgia Press, USA

FERGUSON, A., MOWAT, A. MCI and STROBEL, S. (1983). *Annals of the New York Academy of Sciences*, **409**, 486–497

HILL, I.R., KENWORTHY, R. and PORTER, P. (1970). *Research in Veterinary Science*, **11**, 320–326

KELLY, M.T. (1978). *Journal of Reticuloendothelial Society*, **21**, 175–221

KILSHAW, P.J. and SISSONS, J.W. (1979). *Research in Veterinary Science*, **27**, 361–365

KOHLER, E.M., CROSS, R.D. and BOHL, E.H. (1975). *American Journal of Veterinary Research*, **36**, 757–764

LINGGOOD, M.A. and INGRAM, P.L. (1978). *Research in Veterinary Science*, **25**, 113–115

MILLER, B.G., NEWBY, T.J., STOKES, C.R. and BOURNE, F.J. (1984). *Research in Veterinary Science*, **36**, 187–193

MILLER, B.G., NEWBY, T.J., STOKES, C.R., HAMPSON, D.J., BROWN, P.J. and BOURNE, F.J. (1984). *American Journal of Veterinary Research*, **45**, 1730–1733

MOTTA, I., PORTNOI, D. and TRUFFA-BACHI, P. (1986). *Cellular Immunology*, **97**, 267–275

PORTER, P. (1979). *Advances in Veterinary Science and Comparative Medicine*, **23**, 1–21

PORTER, P. and LINGGOOD, M.A. (1983). *Journal of Infectious Diseases*, **6**, 111–121

PORTER, P., PARRY, S.H. and ALLEN, W.D. (1977). In *Immunology of the Gut: Ciba Foundation Symposium 46*, p. 55. Amsterdam, Elsevier

PORTER, P., KENWORTHY, R., HOLME, D.W. and HORSFIELD, S. (1973). *Veterinary Research*, **92**, 630–636

PORTER, P., KENWORTHY, R., NOAKES, D.E. and ALLEN, W.D. (1974). *Immunology*, **27**, 841–853

PORTER, P., POWELL, J.R., ALLEN, W.D. and LINGGOOD, M.A. (1985). In *The Virulence of Escherichia coli*, pp. 271–287. Ed. Sussman, M., Academic Press, London

PRIOR, M.E. and PORTER, P. (1980). *Veterinary Record*, **107**, 220–223

TENGERDY, R.P., MATHIAS, M.M. and NOCKELS, C.F. (1980). *Advances in Experimental Medicine and Biology*, **135**, 27–42

THOMAS, A. (1982). *Feed Compounder*, **2**, 30

WALKER, W.A. (1982). *Clinics in Immunology and Allergy*, **2**, 15–39

20

# NOVEL APPROACHES TO GROWTH PROMOTION IN THE PIG

P. A. THACKER
*Department of Animal Science, University of Saskatchewan, Saskatoon, Canada*

## Introduction

Antibiotics have played a major role in the growth and development of the pig industry for more than 30 years. Their efficiency in increasing growth rate, improving feed utilization and reducing mortality from clinical disease is well documented (Hays and Muir, 1979). However, consumers are becoming increasingly concerned about drug residues in meat products (Lindsay, 1984). In addition, it has been suggested that the continuous use of antibiotics may contribute to a reservoir of drug-resistant bacteria which may be capable of transferring their resistance to pathogenic bacteria in both animals and humans (Solomons, 1978). Thus it is possible that the future use of antibiotics in animal feeds may be restricted.

Alternative methods of growth promotion must be made available in order to allow the continued development of a viable pig industry. Recently, several new methods of growth promotion have been developed which may have potential for use with pigs. These include the use of growth hormone injections, somatostatin immunization, repartitioning agents, probiotics as well as various enzyme preparations.

## Growth hormone injection

The endocrine system plays an important role in the regulation of growth and in the partitioning of nutrients between muscle and adipose tissue (Schanbacher, 1984; Etherton and Kensinger, 1984). The hormones which are known to exert a significant effect on growth rate in pigs include insulin, growth hormone, thyroxine, glucocorticoids, oestrogen, testosterone and a variety of peptides loosley referred to as growth factors (Welsh, 1985). Manipulation of the endocrine system would therefore seem to have considerable potential as a method of increasing growth rate and improving the efficiency of feed utilization in commercial pig operations.

The principal hormone involved in stimulating growth is called somatotrophin or growth hormone (Spencer, 1985). Growth hormone is a protein, produced in the pituitary gland of the pig, which has been shown to stimulate hepatic synthesis of DNA, RNA and protein (Welsh, 1985). Growth hormone has also been shown to increase the concentration of free fatty acids in the blood, while decreasing amino acid breakdown (Welsh, 1985). These changes alter the way that the pig partitions the nutrients contained in the feed, with the end result being an increase in protein synthesis and a decrease in fat synthesis (Etherton *et al.*, 1987).

Growth hormone also stimulates the hepatic synthesis of a group of compounds called somatomedins which are a family of small peptides that mediate bone and muscle growth (Phillips and Vassilopoulou-Sellin, 1979). They exert their effect by causing an increase in the rate of multiplication of bone and muscle cells which is eventually translated into an increase in body size (Daughaday, 1982).

Since endogenously produced growth hormone has been shown to stimulate growth rate, a considerable amount of effort has been made to isolate growth hormone for use as a growth promoter (Turman and Andrews, 1955; Henricson and Ullberg, 1960; Lind *et al.*, 1968; Machlin, 1972). However, progress has been relatively slow due to problems in producing large quantities of growth hormone, the high cost of its production, as well as a lack of knowledge as to the most effective dosage, the correct time for treatment and its proper duration.

Recently, large amounts of pure, species-specific growth hormone have been produced by genetically altered bacteria, through recombinant-DNA technology (Goeddel *et al.*, 1979). The low cost, apparent abundance and purity of this biosynthetic growth hormone has rekindled interest in utilizing exogenous growth hormone as a growth promoter.

Daily injections of growth hormone (0–70 μg/kg body weight) have been shown to produce a 12.6% increase in average daily gain, a 17.5% improvement in feed conversion efficiency, a 24.7% decrease in fat content and a 14.7% increase in muscle mass (Etherton *et al.*, 1987; *Table 20.1*). Similar results have been reported by other workers (Baile, Della-fera and McLaughlin, 1983; Chung, Etherton and Wiggins, 1985; Etherton *et al.*, 1986; Boyd *et al.*, 1987).

In terms of growth rate, it would appear that the best response is obtained with a dosage of approximately 90 μg/kg body weight (Boyd *et al.*, 1987). At higher levels, feed intake is depressed with a concomitant decrease in growth rate. However, improvements in fat content and muscle mass continue up to the maximum levels of growth hormone tested.

The use of growth hormone as a growth promoter has many advantages. Unlike other growth promoters, such as antibiotics or anabolic steroids, growth hormone is not deposited in the body tissues. However, even if it was deposited in animal tissue, it would be broken down during cooking and digested in the gut just like any other protein. As a consequence, its use should find favour with consumers concerned with drug residues in meat products. In addition, since the half-life of growth hormone in plasma is estimated to be 8.9 min (Althen and Gerrits, 1976), a withdrawal period

**Table 20.1** EFFECT OF GROWTH HORMONE INJECTION ON PIG PERFORMANCE

| | *Dosage* (μg/kg bodyweight/day) | | | |
|---|---|---|---|---|
| | *0* | *10* | *30* | *70* |
| Liveweight gain (g/day) | 900 | 980 | 950 | 1030 |
| Feed conversion ratio (kg feed/kg liveweight gain) | 2.86 | 2.72 | 2.58 | 2.36 |
| Carcass fat (%) | 28.7 | 28.7 | 24.4 | 21.6 |
| Muscle mass (%) | 26.0 | 27.7 | 28.4 | 30.5 |

After Etherton *et al.* (1987)

should not be necessary, allowing producers to take advantage of its growth promoting effects right through to market weight.

Unfortunately, the use of growth hormone to increase growth rates in pigs has practical limitations which must be overcome before it can be utilized under commercial conditions. At present, daily injections must be given and fairly large amounts of the hormones must be injected. The increased labour required to inject pigs on a daily basis would probably prevent most producers from taking advantage of the benefits of growth hormone treatment. Further research is required to develop a delivery system which would allow pigs to be injected at less frequent intervals.

## Somatostatin immunization

The benefits of manipulation of the immune system to provide protection against disease are universally recognized (Quirke, 1985). The availability of effective vaccines against a wide variety of infectious organisms has facilitated the development of many modern intensive and highly efficient systems of animal production. However, it is only recently that attention has been focused on the possibility of utilizing the immune response to influence endocrine function and hence alter the growth rate of domestic animals (Spencer, 1986).

The secretion of growth hormone from the pituitary gland is controlled by hormones secreted from the hypothalamus. Growth hormone releasing factor (GH-RF), a 44 amino acid peptide, stimulates pituitary growth hormone secretion (Lance *et al.*, 1984), while somatostatin (SR-IF), a 14 amino acid peptide, inhibits growth hormone secretion (Brazeau *et al.*, 1973). The amount of growth hormone secreted depends on the balance between the degree of stimulation or inhibition by these two peptides (*Figure 20.1*). Therefore, a reduction in the levels of circulating somatostatin could lead to an increase in the secretion of growth hormone and thereby stimulate growth.

Inhibition of growth hormone secretion can be reduced by inducing an immune response in the animal against its own circulating somatostatin (Spencer, 1986). By coupling somatostatin to a foreign carrier protein and injecting it with a suitable adjuvant, it is possible to trick the animal's immune system into believing that the somatostatin is a foreign compound. Therefore, the body attempts to remove the somatostatin from the blood, with a concomitant increase in the levels of circulating growth hormone.

The preliminary results of an experiment conducted to determine the effects of somatostatin immunization on weaner pig performance are shown in *Table 20.2*. In this experiment, pigs were immunized (0–100 μg somatostatin) at five weeks of age and their performance was monitored for four weeks. During the four week experiment, pigs immunized against somatostatin gained approximately 12% faster than did control pigs while the efficiency of feed utilization was unaffected (Thacker and Laarveld, 1988).

The development of somatostatin immunization as a growth promoting technique is attractive from a practical standpoint. Unlike growth hormone injection which requires daily injections in order to obtain a response, somatostatin immunization requires only two injections spaced several weeks apart. Therefore, it may be possible to fit somatostatin immunization into a regular vaccination schedule.

The technique of somatostatin immunization appears to have a considerable amount of potential as a practical means of enhancing the growth rate of pigs. The most limiting aspect of the technique is the ability of inducing a consistent immune

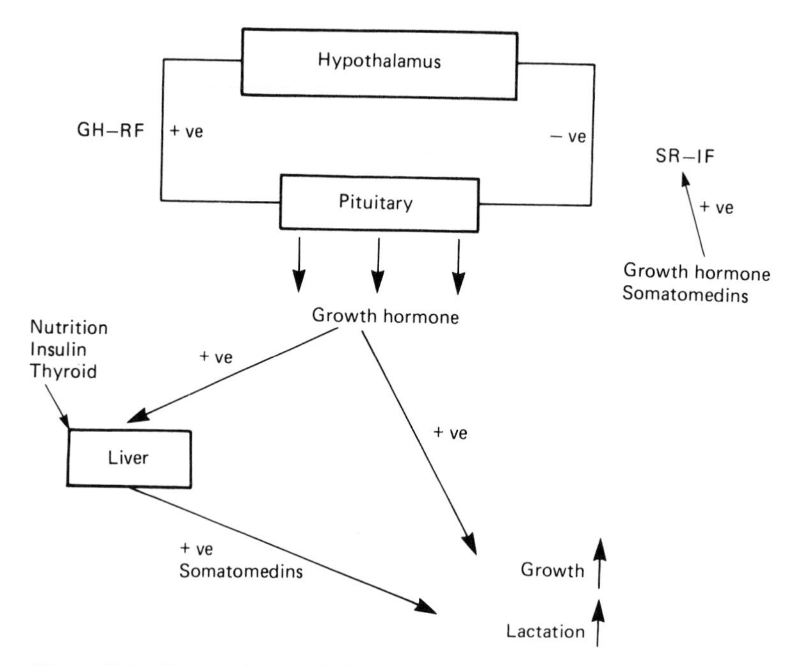


**Figure 20.1** Hormonal control of growth hormone secretion

response. Further research is required in order to determine the optimum immunization conditions required to elicit maximum growth stimulation. Areas currently being researched include the effect of different carrier proteins (ovalbumin, human alpha-globulin or bovine serum albumin) as well as various adjuvants (Havlogen, Freund's Complete Adjuvant, Freund's Incomplete Adjuvant, Ribi Adjuvant or Regressin).

## Repartitioning agents

The recent discovery of synthetic agents which have a repartioning effect on nutrient utilization in adipose tissue and skeletal muscle has caused a considerable amount of interest in the livestock industry (Asato *et al.*, 1984). These repartitioning agents, known as beta-agonists, have the ability to stimulate the production of lean muscle

**Table 20.2** EFFECT OF SOMATOSTATIN IMMUNIZATION ON THE PERFORMANCE OF WEANER PIGS (4–9 WEEKS)

| | *Dose of somatostatin injected* (μg) | | | |
|---|---|---|---|---|
| | 0 | 5 | 25 | 100 |
| Liveweight gain (g/day) | 488 | 500 | 558 | 546 |
| Feed intake (g/day) | 869 | 877 | 939 | 987 |
| Feed conversion ratio (kg feed/kg liveweight gain) | 1.8 | 1.7 | 1.7 | 1.8 |

Thacker and Laarveld (1988)

while limiting the synthesis and deposition of subcutaneous and internal fat (Jones *et al.*, 1985; Dalrymple and Ingle, 1986).

Beta-agonists derive their name from the way that they act on individual cells in the body. Cells have receptors on their outer surfaces that bind or latch onto blood-bound messengers (Stiles, Caron and Lefkowitz, 1984). These receptors are divided into alpha and beta types and are very specific as to what messengers they will accept. However, when the right one is present, its action on the receptor causes the whole cell to alter its metabolism.

A beta-agonist is a chemical messenger that activates a beta-adrenergic receptor. These beta-agonists stimulate the breakdown of fat in the cell and increase the rate at which the released fatty acids are oxidized (Mersmann, 1979). Under normal conditions, a large part of the energy obtained from the oxidation of these fatty acids would be lost as heat (Stock and Rothwell, 1981). However, under the influence of beta-agonists, more of the energy obtained from fatty acid oxidation is made available to the body for the protein synthesis (Baker *et al.*, 1983). The end result is a decrease in the amount of fat and an increase in the amount of protein in the body.

The beta-agonists which have been most widely tested are clenbuterol and cimaterol. Some typical research results (Jones *et al.*, 1985) are presented in *Table 20.3*. These results indicate that pigs fed diets containing cimaterol gain at approximately the same rate as control pigs but consume less feed. As a consequence, pigs treated with cimaterol exhibit a trend towards an improved feed conversion efficiency. In addition, there was a 13.8% reduction in carcass fat and a 10.6% increase in loin eye area as a result of dietary inclusion of the beta-agonist. Similar results have been reported by other workers (Moser *et al.*, 1986; Cromwell, Kemp and Stahly, 1987; Hanrahan *et al.*, 1986).

Unfortunately, the withdrawal of cimaterol from the diet for as short a period as seven days has been shown to result in compensatory accumulation of fat in subcutaneous and internal depots (Jones *et al.*, 1985). Therefore, if a withdrawal period is prescribed, it is unlikely that there would be any benefit from including beta-agonists in pig diets. In addition, several experiments have indicated a greater incidence of hoof lesions when beta-agonists were included in the diet (Jones *et al.*, 1985; Cromwell, Kemp and Stahly, 1987). The aetiology of these hoof lesions has not been determined. More research would appear to be warranted before repartitioning agents can be recommended for routine inclusion in commercial swine diets.

## Probiotics

Probiotics have been widely promoted as an alternative to the use of antibiotics in swine rations (Hale and Newton, 1979; Pollman, Danielson and Peo, 1980). Probiotics have the opposite effect to antibiotics on the micro-organisms in the digestive tract. Whereas antibiotics control the microbial population in the intestine by inhibiting or destroying micro-organisms, probiotics actually introduce live bacteria into the intestinal tract (Pollmann, 1986).

Both beneficial and potentially harmful bacteria can normally be found in the digestive tract of pigs. Examples of harmful bacteria are *Salmonella*, *Escherichia coli*, *Clostridium perfringens* and *Campylobacter sputorum*. Not only can these bacteria produce specific diseases known to be detrimental to the host but through competition for essential nutrients they can also decrease animal performance. In contrast to the effects of these disease causing micro-organisms, bacteria such as *Lactobacillus*

**Table 20.3** PERFORMANCE OF GROWING PIGS FED REPARTITIONING AGENTS (CIMATEROL)

| | *Dietary level* (ppm) | | | |
|---|---|---|---|---|
| | 0 | 0.25 | 0.50 | 1.00 |
| Liveweight gain (g/day) | 760 | 800 | 770 | 790 |
| Feed intake (kg/day) | 2.98 | 2.83 | 2.76 | 2.72 |
| Feed conversion ratio (kg feed/kg liveweight gain) | 3.92 | 3.54 | 3.58 | 3.44 |
| Leaf fat (kg) | 1.27 | 1.16 | 1.13 | 1.09 |
| Eye muscle area ($cm^2$) | 29.85 | 31.96 | 33.96 | 32.09 |

After Jones *et al.* (1985)

and the vitamin B-complex producing bacteria can be beneficial to the host. By encouraging the proliferation of these bacteria in the intestinal tract, it may be possible to improve animal performance.

The ideal situation would be always to have specific numbers of beneficial bacteria present in the intestinal tract. However, physiological and environmental stress can create an imbalance in the flora of the intestinal tract allowing pathogenic bacteria to multiply. When this occurs, disease and poor performance may result. Probiotics increase numbers of the desirable microflora in the gut thereby swinging the balance towards a more favourable microflora.

The mode of action of probiotics has not been clearly defined. It has been suggested that probiotics increase the synthesis of lactic acid in the gastrointestinal tract of the pig (White *et al.*, 1969 ; Thomlinson, 1981). This increased production of lactic acid is postulated to lower the pH in the intestine thereby preventing the proliferation of harmful bacteria such as *E. coli* (Mitchell and Kenworthy, 1976). The decrease in the number of *E. coli* may also reduce the amount of toxic amines and ammonia produced in the gastrointestinal tract (Hill, Kenworthy and Porter, 1970). In addition, there are reports which suggest that probiotics may produce an antibiotic-like substance (Shahani, Vakil and Kilara, 1976) and also stimulate the early development of the immune system of the pig.

The research conducted to determine the value of probiotics in pig diets has been inconclusive. The results of one experiment conducted to determine the effect of probiotics in starter diets are shown in *Table 20.4* (Fralick and Cline, 1982). The results of this experiment are typical of most of the research conducted with starter pigs with most workers reporting slight improvements in daily gain and feed conversion efficiency as a result of probiotic inclusion. However, this is not always the case and other workers have reported the opposite effect (Bebiak, 1979; Combs and Copelin, 1981; Pollmann *et al.*, 1982).

**Table 20.4** PERFORMANCE OF STARTER PIGS FED DIETS CONTAINING A PROBIOTIC[a]

| | *Control* | *Probiotic* |
|---|---|---|
| Liveweight gain (g/day) | 304 | 322 |
| Average daily intake (g) | 843 | 889 |
| Feed conversion ratio (kg feed/kg liveweight gain) | 2.8 | 2.8 |

After Fralick and Cline (1982)
[a]Bio-T (Ag-Mark, Inc., Frankfort, IN, USA)

Some of the reasons for the variability of results include the fact that the viability of microbial cultures may be dependent on storage method, strain differences, dose level, frequency of feeding, species specificity problems and drug interactions (Pollmann, 1985). The difficulty in maintaining a viable *Lactobacillus* culture in pig feeds may also partially explain the inconsistency in research results (Pollmann and Bandyk, 1984). It is well documented that temperature, change in pH and various antibiotics will decrease the viability of *Lactobacillus* cultures.

The value of adding probiotics to growing-finishing rations would appear to be questionable based on experimental data such as that shown in *Table 20.5* (Fralick and Cline, 1982). Several other experiments conducted using probiotics during the growing-finishing phase have also shown little benefit (e.g. Pollmann, Danielson and Peo, 1980). Several researchers have speculated that probiotics may actually have some negative effects on pig performance during the growing-finishing phase by competing for nutrients with indigenous organisms of the digestive tract, decreasing carbohydrate utilization and increasing the intestinal transit rate of digesta (Pollmann, 1985). Therefore, although the theoretical concept of probiotics appears promising, the documented evidence of their therapeutic value suggests that the search must continue for a workable alternative to antibiotics.

## Enzyme supplementation

Endogenous enzymes are required to break down the carbohydrates, proteins and fats in the diet into a form that can be utilized by the animal. Enzymes are also involved in activating and hastening the many chemical reactions which take place in the animal's body. Therefore, it is possible that pig performance could be improved by the addition of supplementary enzymes to the diet.

A considerable amount of research has been conducted in order to determine the value of supplementing pig feeds with some of the enzymes normally secreted in the digestive tract (Lewis *et al.*, 1955; Cunningham and Brisson, 1957; Combs *et al.*, 1960). Most of this work has been conducted with piglets weaned shortly after birth and has involved enzymes such as amylase, sucrase, pepsin, trypsin and pancreatin. Although of academic interest, this practice would appear to have limited practical application.

Recently, enzymes have been discovered which have the potential to break down deleterious compounds commonly found in pig rations such as the beta-glucans contained in barley and the soluble pentosans found in rye. This discovery has rekindled interest in the use of enzyme supplementation as a means of growth promotion.

Beta-glucans are water soluble polysaccharides found in the aleurone layer and

**Table 20.5** PERFORMANCE OF GROWING-FINISHING PIGS FED DIETS CONTAINING A PROBIOTIC[a]

| | *Control* | *Probiotic* |
|---|---|---|
| Liveweight gain (g/day) | 710 | 700 |
| Feed intake (kg/day) | 2.33 | 2.37 |
| Feed conversion ratio (kg feed/kg liveweight gain) | 3.28 | 3.38 |

After Fralick and Cline (1982)
[a]Bio-T (Ag-Mark, Inc., Frankfort, IN)

endosperm of barley kernels (Prentice and Faber, 1981). They consist of glucose units linked together by beta-1,4 and beta-1,3 linkages (Fleming and Kawakami, 1977; *Figure 20.2*). The beta-1,3 linkages confer upon the molecule a step-like structure that interferes with hydrogen bonding between adjacent chains resulting in increased water solubility.

Beta-glucans greatly lower the nutritional value of barley. They restrict weight gain through an increase in the viscosity of the intestinal fluid (Burnett, 1966) which interferes with the digestive process by impeding enzyme–substrate association as well as affecting the rate at which released nutrients approach the mucosal surface for absorption (Campbell *et al.*, 1986). It has also been suggested that beta-glucans allow microbial populations to assimilate a greater proportion of the nutrients contained in the feed into their own system thereby reducing the availability of these nutrients to the host.

The level of beta-glucan in a barley sample can vary from 1.5 to 8% depending on the cultivar of barley and the environmental conditions under which it was grown (Willingham *et al.*, 1960; Gohl and Thomke, 1976). Barley grown in areas with low rainfall will have higher levels of beta-glucans than that grown under conditions of adequate moisture (Aastrup, 1979). In addition, the hull-less varieties of barley contain higher levels of beta-glucans than the hulled varieties (Fox, 1981) while the malting varieties of barley may be lower in beta-glucans than the so-called feed varieties. Therefore, variations in beta-glucan content may explain some of the differences in feeding value often seen among barley cultivars.

Enzymes capable of breaking down beta-glucans are termed beta-glucanases. Treatment of barley with these enzymes has been shown to improve the nutritive

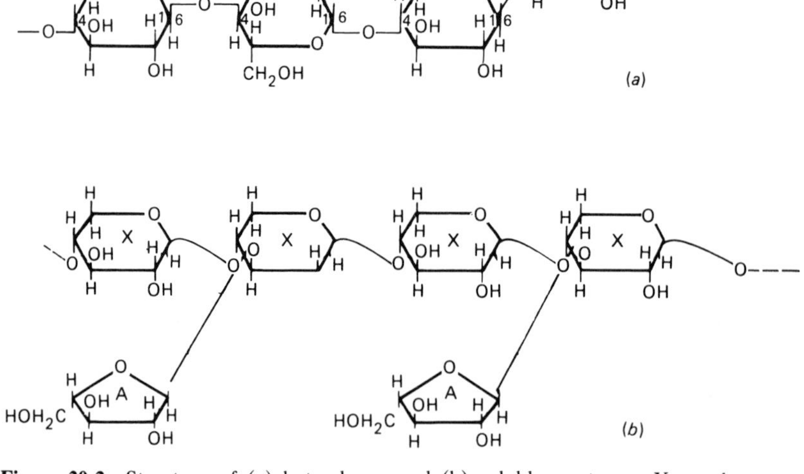

**Figure 20.2** Structure of (a) beta-glucan and (b) soluble pentosan. X = xylose; A = arabinose

value of barley for poultry (Burnett, 1966; White *et al.*, 1983). In one study, bodyweight gains were improved 9.4% and feed conversion efficiency 5.8% by the addition of a beta-glucanase enzyme to barley-based poultry rations (Campbell *et al.*, 1984; *Table 20.6*).

There has been little research conducted to determine the effects of beta-glucans on the performance of pigs. However, the few research reports available indicate that supplementation of barley-based pig diets with beta-glucanase may be beneficial (Newman *et al.*, 1980; Newman, Eslick and El-Negoumy, 1983).

The results of a recent trial conducted at the University of Saskatchewan are presented in *Table 20.7* (Thacker *et al.*, 1988). In this study, there were no significant differences in growth rate, feed intake or feed conversion efficiency between pigs fed hull-less barley diets supplemented or unsupplemented with beta-glucanase. However, digestibility coefficients for dry matter, crude protein and energy were marginally improved by beta-glucanase supplementation (*Table 20.8*).

Enzyme supplementation may also have potential as a means of improving the nutritive value of rye. At the present time, rye is not widely used in pig diets. The classic explanation for this relates to its high ergot content. However, recent research with poultry indicates that high levels of soluble pentosans in rye (*Figure 20.2*) may pose an even greater problem than does ergot. These pentosans are solubilized during digestion and result in a highly viscous intestinal fluid that interferes with digestion in a manner similar to the beta-glucans found in barley.

From an agronomic standpoint, rye is an attractive crop. It is high yielding, makes more effective use of water during spring run-off than cereals planted in the spring and allows for a more equitable distribution of a farmer's workload due to its early harvest. If the detrimental effects of the soluble pentosans could be overcome by enzyme treatment, then the market potential for rye would increase providing an alternative feed resource for use in pig feeding.

The results of an experiment conducted to determine the effects of enzyme supplementation on the performance of grower pigs fed diets containing rye are shown in *Table 20.9* (Thacker *et al.*, 1988). Although these are preliminary data, the results indicate that the performance of pigs fed rye-based diets supplemented with an

**Table 20.6** EFFECT OF BETA-GLUCANASE SUPPLEMENTATION OF BARLEY DIETS ON THE PERFORMANCE OF BROILER CHICKENS FROM 0–6 WEEKS OF AGE

| | *Feed intake* (g) | *Body weight* (g) | *Feed conversion efficiency* | *Cost of gain* (cents) |
|---|---|---|---|---|
| Wheat-corn | 3484 | 1996 | 1.75 | 0.81 |
| Barley | 3530 | 1718 | 2.08 | 0.73 |
| Barley + enzyme | 3738 | 1891 | 1.96 | 0.70 |

After Campbell *et al.* (1984)

**Table 20.7** EFFECT OF BETA-GLUCANASE ON THE PERFORMANCE OF GROWING PIGS

| | *Control* | *Enzyme* |
|---|---|---|
| Liveweight gain (kg/day) | 740 | 760 |
| Feed intake (kg/day) | 2.32 | 2.35 |
| Feed conversion ratio (kg feed/kg liveweight gain) | 3.13 | 3.11 |

Thacker *et al.* (1988)

**Table 20.8** EFFECT OF BETA-GLUCANASE SUPPLEMENTATION ON THE DIGESTIBILITY OF DRY MATTER, PROTEIN AND ENERGY BY PIGS (55 kg LIVEWEIGHT)

| | *Control* | *Beta-glucanase* |
|---|---|---|
| Dry matter digestibility (%) | 80.51 | 82.69 |
| Protein digestibility (%) | 75.06 | 77.72 |
| Energy digestibility (%) | 77.97 | 80.82 |

Thacker *et al.* (1988)

**Table 20.9** EFFECT OF ENZYME SUPPLEMENTATION OF RYE ON PIG PERFORMANCE

| | *Barley* | *Rye* | *Rye + enzyme* |
|---|---|---|---|
| Liveweight gain (g/day) | 750 | 700 | 750 |
| Feed intake (kg/day) | 2.20 | 1.96 | 1.88 |
| Feed conversion ratio (kg feed/kg liveweight gain) | 2.92 | 2.72 | 2.50 |

Thacker *et al.* (1988)

enzyme capable of breaking down soluble pentosans can equal that of pigs fed barley-based diets.

## Conclusion

Pig production continues to be an industry concerned with converting feed ingredients into meat with a degree of technical efficiency and economic management that allows a profit. New technology must be developed in order to allow the continued development of a viable pig industry. Several of the methods discussed in this chapter may find application as the growth promoters of the future.

## References

AASTRUP, S. (1979). *Carlsberg Research Communications*, **44**, 381–393

ALTHEN, T.B. and GERRITS, R.J. (1976). *Endocrinology*, **99**, 511–515

ASATO, G., BAKER, P.K., BASS, R.T., BENTLEY, J., CHARI, M., DALRYMPLE, R.H., FRANCE, R.J., GINGHER, P.E., LENCES, B.L., PASCAVAGE, J.J., PENSACK, M. and RICKS, C.A. (1984). *Agriculture and Biochemical Chemistry*, **48**, 2883–2888

BAILE, C.A., DELLA-FERA, M.A. and MCLAUGHLIN, C.L. (1983). *Growth*, **47**, 225–236

BAKER, P.K., AUST, T., DALRYMPLE, R.H., INGLE, D.L. and RICKS, C.A. (1983). *Symposium on Novel Approaches and Drugs for Obesity*. The 4th International Congress on Obesity, New York

BEBIAK, D.M. (1979). *Report of Swine Research*, Michigan State University, pp. 8–10.

BOYD, R.D., WRAY-CAHEN, D., BAUMAN, D., BEERMANN, D., DENEERGARD, A. and SOUZA, L. (1987). *Proceedings of the Maryland Nutrition Conference*, pp. 58–66.

BRAZEAU, P., VALE, W., BURGUS, R., LING, N., BUTCHEN, M., RIVIER, J. and GUILEMIN, R. (1973). *Science*, **179**, 77–79

BURNETT, G.S. (1966). *British Poultry Science*, **7,** 55–75
CAMPBELL, G.L., CLASSEN, H.L. and SALMON, R.E. (1984). *Feedstuffs*. 1984, May 7, 26–27
CAMPBELL, G.L., CLASSEN, H.L., THACKER, P.A., ROSSNAGEL, B.G., GROOTWASSNIK, J.W. and SALMON, R.E. (1986). *Proceedings of the 7th Western Nutrition Conference, Saskatoon*, pp. 277–250
CHUNG, C.S., ETHERTON, T.D. and WIGGINS, J.P. (1985). *Journal of Animal Science*, **60,** 118–129
COMBS, G.E. and COPELIN, J.L. (1981). Department of Animal Science Research Report AL-1981-6, Florida Agricultural Experimental Station, pp. 7–9.
COMBS, G.E., ALSMEYER, W.L., WALLACE, H.D. and KOGER, M. (1960). *Journal of Animal Science*, **19,** 932–937
CROMWELL, G.L., KEMP, J.D. and STAHLY, T.S. (1987). *Feed Management*, **38,** 8–11
CUNNINGHAM, H.M. and BRISSON, G.J. (1957). *Journal of Animal Science*, **16,** 370–376
DALRYMPLE, R.H. and INGLE, D.L. (1986). *Proceedings of the 47th Minnesota Nutrition Conference*, pp. 102–114
DAUGHADAY, W.H. (1982). *Proceedings of the Society of Experimental Biology and Medicine*, **170,** 257–263
ETHERTON, T.D., WIGGINS, J.P., CHUNG., C.S., EVOCK, C.M., REBHUN, J.F., WALTON, P.E. (1986). *Journal of Animal Science*, **63,** 1389–1399
ETHERTON, T.D. and KENSINGER, R.S. (1984). *Journal of Animal Science*, **59,** 511–528
ETHERTON, T.D., WIGGINS, J.P., CHUNG., C.S., EVOCK, C.M., REBHUN, J.F. and WALTON, P.E. and STEELE, N.C. (1987). *Journal of Animal Science*, **64,** 433–443
FLEMING, M. and KAWAKAMI, K. (1977). *Carbohydrate Research*, **57,** 15–23
FOX, G.J. (1981). PhD Dissertation. Montana State University
FRALICK, C. and CLINE, T.R. (1982). *Proceedings of the Purdue University Swine Day*, pp. 7–10
GOEDDEL, D.V., HEYNEKER, H.L., HOZUMI, T., ARENTZEN, R., ITAKURA, K., YANSURA, D.G., ROSS, M.J., MIOZZARI, G., CREA, A. and SEEBERG, R.H. (1979). *Nature*, **281,** 544–548
GOHL, B. and THOMKE, S. (1976). *Poultry Science*, **55,** 2369–2374
GUEST, G.B. (1976). *Journal of Animal Science*, **42,** 1052–1057
HALE, O.M. and NEWTON, G.L. (1979). *Journal of Animal Science*, **48,** 770–775
HANRAHAN, J.P., QUIRKE, J.F., BOMANN, W., ALLEN, P., MCEWAN, J.C., FITZSIMONS, J.M., KOTZIAN, J. and ROCHE, J.F. (1986). *Recent Advances in Animal Nutrition—1986*, pp. 125–138. Ed. Haresign, W. and Cole, D.J.A. Butterworths, London
HAYS, V.W. and MUIR, V.M. (1979). *Canadian Journal of Animal Science*, **59,** 447–456
HENRICSON, B. and ULLBERG, S. (1960). *Journal of Animal Science*, **19,** 1002–1008
HILL, I.R., KENWORTHY, R. and PORTER, P. (1970). *Research in Veterinary Science*, **11,** 320–326
JONES, R.W., EASTER, R.A., MCKEITH, R.K., DALRYMPLE, R.H., MADDOCK, H.M. and BECHTEL, P.J. (1985). *Journal of Animal Science*, **61,** 905–913
LANCE, V.A., MURPHY, W.A., SUEIRAS-DIAZ, J. and COY, B.H. (1984). *Biochemistry Biophysics Research Communication*, **119,** 265–272
LEWIS, C.J., CATRON, D.V., LIU, C.H., SPEER, V.C. and ASHTON, G.C. (1955). *Agriculture and Food Chemistry*, **12,** 1047–1050
LIND, K.D., HOWARD, R.D., KROPF, D.H. and KOCH, B.A. (1968). *Journal of Animal Science*, **27,** 1763 (Abstract)
LINDSAY, D.G. (1984). *Pig News and Information*, **5,** 219–222
MACHLIN, L.J. (1972). *Journal of Animal Science*, **35,** 794–800
MERSMANN, H.J. (1979). *Proceedings of the Reciprocal Meat Conference*, **32,** 93–107
MITCHELL, I. and KENWORTHY, R. (1976). *Journal of Applied Bacteriology*, **4,** 163–174

MOSER, R.L., DALRYMPLE, R.H., CORNELIUS, S.G., PETTIGREW, J.E. and ALLEN, C.E. (1986). *Journal of Animal Science*, **62,** 21–26

NEWMAN, C.W., ESLICK, R.F., PEPPER, J.W. and EL-NEGOUMY, A.M. (1980). *Nutrition Reports International*, **22,** 833–837

NEWMAN, C.W., ESLICK, R.F. and EL-NEGOUMY, A.M. (1983). *Nutrition Reports International*, **28,** 139–145

PHILLIPS, L.S. and VASSILOPOULOU-SELLIN, R. (1979). *American Journal of Clinical Nutrition*, **32,** 1082–1096

POLLMANN, D.S., DANIELSON, D.M. and PEO, E.R. (1980). *Journal of Animal Science*, **51,** 577–581

POLLMANN, D.S., KENNEDY, G.A., KOCH, B.A. and ALLEE, G.L. (1982). Agricultural Experimental Station Report of Progress No. 422, Kansas State University, pp. 86–91

POLLMANN, D.S. and BANDYK, C.A. (1984). *Animal Feed Science and Technology*, **11,** 261–267

POLLMANN, D.S. (1985). *Guelph Pork Symposium*, pp. 59–74

POLLMANN, D.S. (1986). *Recent Advances in Animal Nutrition—1986*, pp. 193–205. Ed. Haresign, W. and Cole, D.J.A. Butterworths, London

PRENTICE, N. and FABER, S. (1981). *Cereal Chemistry*, **58,** 77–79

QUIRKE, J.F. (1985). *Livestock Production Science*, **13,** 1–2

SCHANBACHER, B.D. (1984). *Journal of Animal Science*, **59,** 1621–1630

SHAHANI, K.M., UAKIL, J.R. and KILARA, A. (1976). *Journal of Cultured Dairy Produce*, **11,** 14–20

SOLOMONS, I.A. (1978). *Journal of Animal Science*, **46,** 1360–1368

SPENCER, G.S. (1985). *Livestock Production Science*, **12,** 31–46

SPENCER, G.S. (1986). *Domestic Animal Endocrinology*, **3,** 55–68

STILES, G.L., CARON, M.G. and LEFKOWITZ, R.J. (1984). *Physiological Reviews*, **64,** 661–743

STOCK, M.J. and ROTHWELL, N.J. (1981). *Biochemical Society Transactions*, **9,** 525–527

THACKER, P.A., CAMPBELL, G.L. and GROOTWASSINK, J.W. (1988). *Department of Animal and Poultry Science Research Reports*, University of Saskatchewan, Saskatoon, Saskatchewan, pp. 140–145

THACKER, P.A. and LAARVELD, B. (1988). Unpublished data

THACKER, P.A., CAMPBELL, G.L. and GROOTWASSINK, J.W. (1988). *Nutrition Reports International* (in press)

THOMLINSON, J.R. (1981). *Veterinary Record*, **109,** 120–122

TURMAN, E.J. and ANDREWS, F.N. (1955). *Journal of Animal Science*, **14,** 7–18

WELSH, T.H. (1985). *Animal Nutrition and Health*, **40,** 14–19

WHITE, F., WENHAM, G., SHARMAN, G.A., JONES, A.S., RATTRAY, E.A. and MCDONALD, I. (1969). *British Journal of Nutrition*, **23,** 847–858

WHITE, W.B., BIRD, H.R., SUNDE, M.L. and MARLETT, J.A. (1983). *Poultry Science*, **62,** 853–862

WILLINGHAM, H.E., LEONG, K.C., JENSEN, L.S. and MCGINNIS, J. (1960). *Poultry Science*, **39,** 103–108

21

# IMPACT OF SOMATOTROPIN AND BETA-ADRENERGIC AGONISTS ON GROWTH, CARCASS COMPOSITION AND NUTRIENT REQUIREMENTS OF PIGS

T. S. STAHLY
*Dept. Animal Science, Iowa State University, Ames, Iowa 50011, USA*

## Introduction

Recently, technologies have been developed that allow the economical synthesis of biologically-active compounds that dramatically alter the rate and efficiency of pork production as well as the composition of the products. In this chapter, two modifiers (somatotropin and beta-adrenergic agonists) of the neural-endocrine system in the pig will be characterized, their known biological effects identified and their potential impact on the optimum dietary regimen, specifically amino acid and energy needs, of the pig will be discussed.

## Somatotropin

Somatotropin is a natural protein of approximately 191 amino acid residues which is synthesized and secreted by the anterior pituitary gland. Natural and synthetic (recombinant) forms of porcine somatotropin (PST) are similar in nature but small differences exist such as in the formation of the disulphide bonds and the presence of an additional methionine at the terminal nitrogen. As a result, the bioactivity and potency of the various somatotropin sources may differ and must be individually defined (Boyd *et al.*, 1988; Boyd and Bauman, 1989).

In growing pigs, the administration of PST depresses the voluntary feed intake of pigs and alters the rate, efficiency and composition of growth (Table 21.1). Specifically, PST stimulates the growth of the major proteinaceous tissues such as skeletal muscles, internal organs (i.e. liver, kidneys, gastrointestinal tract), skin as well as bone. In contrast, the amount of fatty tissue in subcutaneous depots, the abdominal cavity as well as skeletal muscle is reduced by PST administration. These shifts in tissue accretion patterns result in significant changes in standard carcass measures of carcass yield (dressing percentage), leanness (i.e. loin eye area, % carcass muscle) and fat content (i.e. backfat thickness, leaf fat weight). These improvements in carcass leanness are associated with improved efficiencies of feed utilization since less feed (energy) is required for muscle *versus* fatty tissue accretion.

The biological impact of PST on the metabolic processes of the pig also alter dietary needs. Changes in voluntary feed intake, nutrient digestibility and relative

**Table 21.1** EFFECT OF PST ADMINISTRATION ON THE GROWTH RATE, EFFICIENCY OF FEED UTILIZATION AND CARCASS TRAITS OF PIGS

| *Criteria* | *PST dose* (μg/kg/day) | |
|---|---|---|
| | *0* | *60–130* |
| Growth and feed utilization[a] | | |
| Daily feed (kg) | 3.04 | 2.60 |
| Daily gain (kg) | 0.94 | 1.08 |
| Feed/gain ratio | 3.26 | 2.47 |
| Carcass traits[b] | | |
| Carcass yield (%) | 74.6 | 73.0 |
| Backfat – 10th rib (cm) | 2.34 | 2.03 |
| Carcass muscle (%) | 50.1 | 62.4 |

[a] Data adapted from Steele *et al.*, 1987; Machlin, 1972; Boyd *et al.*, 1986; Kraft *et al.*, 1986; Etherton *et al.*, 1987; and McLaren *et al.*, 1987. Pigs (barrows and gilts) administered PST from 54–96 kg body weight
[b] Data adapted from Bark *et al.*, 1989a unpublished data. Barrows administered PST from 23 to 115 kg body weight. Data corrected to 100 kg body weight

nutrient demands for maintenance functions and tissue accretion resulting from PST administration are likely to have the greatest impact on the optimum dietary regimen of PST treated pigs.

As PST administration results in a lower voluntary feed intake in pigs, higher concentrations of nutrients (i.e. minerals and vitamins) are needed to maintain a particular daily intake level. This lower level of feed consumption also influences the animals' ability to digest nutrients and accrue body protein. Lowering feed intake below *ad libitum* levels results in a slower rate of digesta passage (Kirchgessner and Roth, 1986) which is associated with small improvements (1–3%) in energy and protein digestibility. In addition, lowering the quantity of energy consumed restricts the potential accretion rate of body proteins in animals with high capacities for lean tissue growth due to genetic selection (Stahly, 1986; Reeds *et al.,* 1989) or PST administration (Campbell *et al.,* 1988; Nossaman *et al.,* 1989).

The maintenance requirements of the pig in terms of energy and probably protein also are altered by PST treatment. Energy maintenance requirements have been estimated to be increased up to 24% in pigs administered daily 50–100 μg PST/kg/day (Campbell *et al.,* 1988; Verstegen *et al.,* 1989). This is thought to be due at least partially to the greater metabolic activity of the internal organs (i.e. liver; Ferrell, 1988). Protein maintenance requirements also may be increased due to the greater protein mass in the PST-treated pigs. However, this relationship has not been well documented.

Based on the assumption that PST alters the rate but not the biological mechanisms (thus efficiency) of protein accretion, the amino acid and energy needs of PST-treated pigs can be estimated based on the nutrients needed to achieve specific accretion rates of major body tissues. Tissue accretion rates are altered by PST administration in a dose-dependent manner (Boyd *et al.,* 1986; Etherton *et al.,* 1987). Unfortunately, the dosage levels that (1) will be approved by regulatory agencies, and (2) will be economically priced for use in the field are not known.

**Table 21.2** POTENTIAL CHANGES IN RATE AND COMPOSITION OF GROWTH IN PIGS INDUCED BY PST ADMINISTRATION[a]

| *Criteria* | *PST dose* | | | |
|---|---|---|---|---|
| | *0* | *1X* | *2X* | *3X* |
| Body weight (g/day) | 900 | 945 | 990 | 1040 |
| Body composition[b] | | | | |
| Offal (%) | 22.3 | 22.8 | 23.3 | 23.8 |
| Carcass (%) | 72.6 | 72.1 | 71.7 | 71.1 |
| Carcass composition[c] | | | | |
| Muscle (%) | 57.80 | 61.80 | 65.80 | 69.80 |
| Fat (%) | 22.04 | 17.36 | 12.60 | 7.88 |
| Bone (%) | 13.47 | 13.72 | 13.97 | 14.22 |
| Skin (%) | 7.12 | 7.62 | 8.12 | 8.62 |
| Tissue accretion (g/day) | | | | |
| Muscle | 398 | 478 | 565 | 656 |
| Body nutrient accretion (g/day) | | | | |
| Protein | 139 | 168 | 200 | 235 |
| Fat | 279 | 213 | 136 | 57 |

[a] Assumptions: Pigs have a high genetic capacity for lean tissue growth, are housed in a thermoneutral environment and are self-fed a fortified maize–soyabean meal diet from 60–100 kg body weight
[b] Offal includes heads, lungs, heart, liver, kidneys, gastrointestinal tract and leaf fat, but not blood or hair
[c] Dissected muscle (corrected to 10% fat content), fat, bone and skin tissues expressed as a % of carcass weight

Thus, changes in growth rate, body composition and nutrient accretion rates potentially induced by four levels (0, 1X, 2X, 3X) of PST administration are shown in Table 21.2. The values of the 0 and 3X dose groups are those of pigs with a high genetic capacity for lean tissue growth (400 g/day) administered 0 and 70 μg, respectively, of PST/kg/day from 23 to 115 kg body weight (Bark *et al.*, 1989a, unpublished data). These body composition data were adjusted to the weight range of 60–100 kg. The values of the 1X and 2X dose group reflect the estimated response of pigs administered 20–30 and 40–50 μg PST/kg/day based on a linear response to PST administration.

The impact of these changes in tissue accretion rates on amino acid (lysine) and energy (ME) needs are estimated in Table 21.3. The estimates of the dietary amino acid needs are based on the following:

(1) the protein content and the amino acid composition of specific tissues [offal (gastrointestinal tract, organs, head and leaf fat), dissected carcass muscle, fat, skin, bone, blood and hair];
(2) the daily protein turnover and losses associated with maintenance of the body protein mass;
(3) the estimate of amino acid pattern of protein needed for body maintenance functions;
(4) the digestibilities of the amino acids in the feedstuffs used and the efficiency of incorporation of absorbed amino acids into tissue proteins.

The dietary energy (ME) needs are based on:

(1) the energy required for body fat and protein accretion;

**Table 21.3** NUTRIENT NEEDS OF PST-TREATED PIGS[a,b]

| *Criteria* | *PST dose* | | | |
|---|---|---|---|---|
| | *0* | *1X* | *2X* | *3X* |
| *Daily lysine needs* (g) | | | | |
| Biological need | | | | |
| Maintenance | 1.05 | 1.08 | 1.12 | 1.17 |
| Tissue accretion | 9.75 | 11.80 | 14.10 | 16.60 |
| Total | 10.80 | 12.88 | 15.22 | 17.77 |
| Digestive lysine need | | | | |
| Utilization of digested lysine (%) | 0.65 | 0.65 | 0.65 | 0.65 |
| Total | 16.61 | 19.82 | 23.42 | 27.34 |
| Dietary lysine need | | | | |
| Digestibility of lysine (%) | 0.83 | 0.835 | 0.84 | 0.845 |
| Total | 20.00 | 23.74 | 27.88 | 32.36 |
| *Daily ME needs* (MJ) | | | | |
| Biological need | | | | |
| Maintenance | 13.95 | 15.06 | 16.18 | 17.44 |
| Tissue accretion | | | | |
| Protein | 6.10 | 7.37 | 8.78 | 10.32 |
| Fat | 14.93 | 11.39 | 7.27 | 3.05 |
| Total | 34.98 | 33.82 | 32.23 | 30.81 |
| Estimated feed intake, (kg/day) | 2.67 | 2.58 | 2.46 | 2.35 |
| Dietary lysine (%) | 0.75 | 0.92 | 1.13 | 1.38 |
| Dietary ME, (MJ/kg) | 13.48 | 13.48 | 13.48 | 13.48 |

[a] Assumptions: Pigs have a high capacity for lean tissue growth, are housed in a thermoneutral environment and are self-fed a fortified, corn-soyabean meal diet from 60 to 100 kg body weight. Daily maintenance needs for ME and 0.75 protein assumed to be 522 kJ/day and 0.0036% of the body protein mass, respectively in the non-treated pigs. Energetic cost of body protein and fat accretion were assumed to be 43.9 and 53.5 kJ ME/g, respectively. Feed wastage assumed to be 3%.
[b] Assumptions: PST induced changes in tissue accretion patterns equivalent to those reported in Table 21.2

(2) the energy required for body maintenance in pigs maintained in a thermoneutral environment.

The protein contents of the various tissues were assumed to be 11–14.8% for offal, 20.2–21.8% for dissected muscle, 1% for dissected fat, 30% for dissected skin, 13.7% for dissected bone, 19% for blood and 94% for hair (Jorgensen *et al.*, 1988; Madsen *et al.*, 1965; unpublished data). The protein content of offal and muscle were increased as the animals' capacity for lean accretion increased due to genetic selection or PST administration (Bark *et al.*, 1989a,b, unpublished). The lysine contents of tissue proteins were assumed to be 6.1% in offal, 7.8% in muscle, 4.8% in fat and skin, 4.6% in bone, 8.5% in blood and 3.3% in hair (Jorgensen *et al.*, 1988; Madsen *et al.*, 1965). Daily maintenance needs for proteins were assumed to be 0.0036% of the body protein mass. Lysine content of proteins utilized in maintenance functions was assumed to be 2.3% (Fuller *et al.*, 1989).

The digestibility of lysine in the corn–soyabean diet was assumed to be 83% (Knabe and Tanksley, 1985; NRC, 1988) in the 0X dose group with small improvements in digestibility estimates as feed intake was reduced at higher PST levels. The efficiency of incorporating digested amino acids into maintenance as well as tissue proteins was assumed to be 65% (1.0 g protein retained/1.5 g synthesized; Reeds, 1989) in each of the four dosage groups. Initial observations in pigs indicated that PST stimulates protein synthesis and degradation to an equal degree (Campbell *et al.*, 1989). Furthermore, PST appears to stimulate the growth of the major proteinaceous tissues (i.e. muscle, bone, skin and offal – organs, gastrointestinal tract) which differ in their protein turnover rates and, potentially, efficiency of protein accretion by similar magnitudes (Table 21.2). Thus, the impact of PST on the efficiency of converting amino acids and energy consumed above maintenance into body tissue proteins may be minimal.

The optimum pattern of amino acids may be altered slightly by PST administration in that a relatively greater proportion of the biological demands for amino acids will be for tissue accretion (i.e. muscle, skin, bone) *versus* maintenance functions in the PST-treated pig. The optimum ratio of essential to non-essential amino acids and the ratio of lysine to methionine, cystine and possibly tryptophan have been estimated to be greater for tissue accretion than for maintenance functions (Fuller *et al.*, 1989).

Energy needs for maintenance were assumed to be 522 and 460 kJ ME/kg$^{0.75}$/day, respectively for pigs with high and medium genetic capacities for lean tissue accretion (Stahly, 1986). Energetic costs of protein and fat accretion were assumed to be 43.9 and 53.5 kJ ME/g, respectively, in both genotypes.

The magnitude of PST-induced changes in tissue accretion rates apparently is similar among genotypes, thus muscle accretion rates and nutrient demand will be lower in animals with reduced genetic capacities for lean tissue accretion. Genotypes of pigs with moderate and low capacities for lean tissue growth are estimated to have 5–6 and 10–12% units less carcass muscle than those with a high capacity (Stahly *et al.*, 1988, unpublished data). On this basis, the dietary lysine needs of pigs, in each PST treatment group, would be expected to be lowered by about 2 and 4 g/day for the medium and low genotypes, respectively.

PST administration is estimated to reduce the daily energy needs of the pigs even though 1X to 3X doses of PST are associated, respectively, with an 8–24% increase in the maintenance energy requirement. The net reduction in energy demand is due to the lower energy cost of muscle *versus* fatty tissue accretion. It is of interest that the estimated magnitude of reduction in energy intake needed to meet the tissue demands closely parallels that observed in growth studies (Table 21.1).

Estimates of the amino acid requirements derived from empirical experimentation is currently limited (Goodband *et al.*, 1988; Newcomb *et al.*, 1988; Campbell *et al.*, 1989; Fowler and Kanis, 1989). Furthermore, the dosage level used, the time period administered, as well as the dietary regimen (i.e. amino acid source and balance, feed intake) have varied substantially among studies. Estimates of the dietary lysine needs from the work of Goodband *et al.* (1988) in which approximately a 2X dose level of PST was administered to finishing pigs with a moderate to high capacity for lean tissue growth (Table 21.4) and that derived from factorial estimates (Tables 21.2 and 21.3) are quite similar. Rate and efficiency of growth in PST-treated pigs were maximized in pigs (60–100 kg) consuming 25–30 glysine/day *versus* the factorially-derived estimate of 26–28 g.

**Table 21.4** RESPONSES OF PST-TREATED (4 mg/day) PIGS TO VARYING DIETARY AMINO ACID LEVELS[a]

| Dietary lysine (%) | 0.6 | 0.8 | 1.0 | 1.2 | 1.4 |
|---|---|---|---|---|---|
| PST (mg/day) | 4 | 4 | 4 | 4 | 4 |
| Daily consumption | | | | | |
| Feed (kg) | 2.26 | 2.46 | 2.51 | 2.47 | 2.40 |
| Lysine (estimated) (g) | 13.6 | 19.7 | 25.1 | 29.6 | 33.6 |
| Daily gain (g) | 0.75 | 0.97 | 1.16 | 1.20 | 1.16 |
| Feed/gain ratio | 3.03 | 2.54 | 2.18 | 2.07 | 2.08 |
| Backfat (cm) | 2.16 | 2.13 | 2.29 | 2.16 | 2.26 |
| *L. dorsi* area ($cm^2$) | 31.4 | 39.1 | 40.5 | 42.3 | 42.5 |

[a] Adapted from Goodband *et al.* (1988)

## Beta-adrenergic agonists

Recently, several synthetic analogues of noradrenaline and adrenaline, called beta-adrenergic agonists, have been identified which, when fed to pigs, have the ability to accelerate muscle tissue accretion and reduce fat accretion (Mersmann, 1989).

In recent years, research with beta agonists has concentrated on the use of ractopamine with less extensive evaluations of cimeratol (Jones *et al.*, 1985), L-644, 969 (Wallace *et al.*, 1987) and GAH/034 (salbutamol) (Wood and Brown, 1987). The ingestion of beta agonists, such as ractopamine, have been shown to alter voluntary feed intake and rate, efficiency and composition of growth in pigs (Table 8.5). The magnitude of the responses is dependent on the dosage level and possibly the type of beta agonist used. In contrast to PST, beta agonists apparently increase the proportion of nutrients deposited in the carcass *versus* internal organs and gastrointestinal tract. This results in a greater carcass yield. Beta agonists also

**Table 21.5** IMPACT OF DIETARY RACTOPAMINE ADDITION ON GROWTH RATE, EFFICIENCY OF FEED UTILIZATION AND CARCASS TRAITS OF PIGS

| *Criteria* | *Dietary ractopamine (ppm)* | | | |
|---|---|---|---|---|
| | *0* | *5* | *10* | *20* |
| *Growth and feed utilization*[a] | | | | |
| Daily feed (kg) | 2.87 | 2.96 | 2.95 | 2.76 |
| Daily gain (kg) | 0.78 | 0.83 | 0.84 | 0.84 |
| Feed/gain ratio | 3.70 | 3.38 | 3.28 | 3.28 |
| *Carcass traits* | | | | |
| Carcass yield (%) | 71.4 | 72.1 | 72.1 | 72.8 |
| Backfat-10th rib (cm[a]) | 2.43 | 2.28 | 2.16 | 2.14 |
| Carcass muscle (%) | 53.1 | 55.5 | – | 62.0 |

[a] Data adapted from Anderson *et al.* (1987a); Cline and Forrest (1987); Crenshaw *et al.* (1987); Hancock *et al.* (1987); Prince *et al.* (1987); Veenhuizan *et al.* (1987). Pig weights averaged 64 kg and 105 kg at the initiation and completion of the studies

[b] Data adapted from Cline and Forrest, (1987); Crenshaw *et al.* (1987); Hancock *et al.* (1987); Prince *et al.* (1987); and Bark *et al.* (1989b), unpublished data for 20 ppm level

stimulate carcass muscle accretion and reduce carcass fat accretion. However, unlike PST, beta agonists appear to alter selectively the accretion patterns of proteinaceous tissues and fat deposits. Specifically, carcass muscle accretion is increased but that of skin and bone are not. Similarly, subcutaneous, abdominal and intermuscular fat deposits are reduced but apparently not intramuscular fat stores.

The magnitude of the response to a beta agonist, particularly in terms of carcass muscle accretion appears to be dependent on the genetic makeup of the animal (Table 21.6). Greater improvements in carcass muscle accretion are elicited by ractopamine in genotypes with high *versus* low genetic capacities for lean tissue growth (Bark *et al.*, 1989a, 1989b). Thus, the impact of a beta agonist on the dietary needs of the pig will be altered by the animal's genetic capacity for lean tissue accretion.

**Table 21.6** POTENTIAL CHANGES IN RATE AND COMPOSITION OF GROWTH IN PIGS FED RACTOPAMINE[a]

| *Criteria* | *Genotype*[b] | *High* | | *Medium* | |
|---|---|---|---|---|---|
| | *Ractopamine* | *0* | *2X* | *0* | *2X* |
| Body weight gain (g/day) | | 900 | 970 | 900 | 970 |
| Body composition[c] | | | | | |
| Offal (%) | | 22.3 | 20.3 | 22.6 | 20.6 |
| Carcass (%) | | 72.6 | 74.6 | 72.3 | 74.3 |
| Carcass composition[d] | | | | | |
| Muscle (%) | | 57.80 | 67.80 | 52.80 | 59.80 |
| Fat (%) | | 22.04 | 12.29 | 27.6 | 20.75 |
| Bone (%) | | 13.47 | 12.98 | 12.97 | 12.49 |
| Skin (%) | | 7.12 | 6.93 | 7.12 | 6.93 |
| Tissue accretion (g/day) | | | | | |
| Muscle | | 398 | 637 | 348 | 509 |
| Body nutrient accretion (g/day) | | | | | |
| Protein | | 139 | 202 | 126 | 167 |
| Fat | | 279 | 138 | 310 | 216 |

[a] Assumption: Pigs self-fed a fortified corn–soyabean meal diet and housed in a thermoneutral environment from 60 to 100 kg body weight
[b] Genotypes with high and medium capacities for lean tissue accretion
[c] Offal includes head, heart, liver, kidneys, gastrointestinal tract and leaf fat but not blood and hair
[d] Dissected muscle (corrected to 10% fat content), fat tissue, bone and skin tissues expressed as a % of carcass weight

Estimates of the dietary amino acid and energy needs of pigs fed a beta agonist (ractopamine) are shown in Table 21.7. Again, the nutrient needs are based on potential changes in carcass muscle and associated tissues and organs that occur in pigs fed varying levels of the beta agonist (i.e. 20 ppm of ractopamine) during the finishing stage of growth. As with PST, the lower feed intake of beta agonist fed pigs should be associated with a slight (1–3%) increase in nutrient digestibility. In contrast, beta agonists are assumed to have less effect on body maintenance requirements than PST. This assumption is based on the reduction in the relative weights of internal organs associated with beta agonist feeding. Furthermore, the selective enhancement of muscle protein stores which possess a slower turnover rate should minimize changes in body protein turnover and loss even though the

**Table 21.7** NUTRIENT NEEDS OF RACTOPAMINE-FED PIGS[a,b]

| *Criteria* | *Genotype*[a] | *High* | | *Medium* | |
|---|---|---|---|---|---|
| | *Ractopamine* | *0* | *2X* | *0* | *2X* |
| *Daily lysine needs* (g) | | | | | |
| Biological need | | | | | |
| Maintenance | | 1.05 | 1.13 | 0.97 | 1.03 |
| Tissue accretion | | 9.75 | 14.54 | 8.56 | 11.97 |
| Total | | 10.80 | 15.67 | 9.53 | 13.00 |
| Digestive lysine need | | | | | |
| Utilization of digested lysine (%) | | 0.65 | 0.68 | 0.65 | 0.665 |
| Total | | 16.61 | 23.04 | 14.66 | 19.55 |
| Dietary lysine need | | | | | |
| Digestibility of lysine (%) | | 0.83 | 0.84 | 0.83 | 0.835 |
| Total | | 20.00 | 27.43 | 17.66 | 23.41 |
| *Daily ME needs* (MJ) | | | | | |
| Biological need | | | | | |
| Maintenance | | 13.95 | 15.06 | 12.28 | 13.26 |
| Tissue accretion | | | | | |
| Protein | | 6.10 | 8.86 | 5.53 | 7.33 |
| Fat | | 14.93 | 7.38 | 16.59 | 11.56 |
| Total | | 34.98 | 31.30 | 34.40 | 32.15 |
| Estimated feed intake, (kg/day) | | 2.67 | 2.39 | 2.63 | 2.46 |
| Dietary lysine (%) | | 0.75 | 1.14 | 0.67 | 0.95 |
| Dietary ME, (MJ/kg) | | 13.48 | 13.48 | 13.48 | 13.48 |

[a] Assumptions: Refer to footnotes a and b of Table 21.5
[b] Assumptions: Ractopamine induced changes in tissue accretion patterns equivalent to those reported in Table 21.5

body protein mass is greater in ractopamine-fed pigs. The fact that beta agonists apparently depress protein degradation to a greater extent than they increase protein synthesis should improve the efficiency of converting dietary amino acids and possibly energy into body protein. Because the lysine content of muscle protein is relatively high compared with that of skin and bone, the optimum amino acid pattern for body protein accretion also may be altered in beta agonist fed pigs to more closely reflect that of muscle. Whether the slower turnover of muscle protein may cause the ideal pattern of amino acid needs for maintenance to more closely reflect those of endogenous protein secretions and skin and bone proteins has not been determined.

Specifically, the daily lysine needs of the two pig genotypes are estimated to increase by about 2.3 g as carcass leanness increased from 53 to 58% carcass muscle (standardized to 10% muscle fat). These data are based on pigs allowed to consume grain–soyabean meal diets *ad libitum* from 60 to 100 kg. Because ractopamine stimulates greater increases in carcass muscle in lean than fat genotypes, a larger increase in dietary lysine intake due to ractopamine feeding will be required in the lean genotypes to allow the maximum muscle accretion to occur. Because the ratio of lysine to other amino acids in muscle protein is higher than that in other major tissues, the lysine content of the ideal protein is calculated to rise by 0.2–0.4

percentage units, respectively, in medium and high lean growth genotypes with the dietary inclusion of 20 ppm ractopamine.

## Summary

Synthetic forms of porcine somatropin and beta adrenergic agonists have been shown to alter the rate and pattern of tissue growth in pigs dramatically. Incorporation of these modifiers of the neural-endocrine systems in pig production will place greater emphasis on the need to define the nutritional requirements of the pigs in light of the rate and patterns of potential tissue growth desired.

## References

Anderson, D. B., Veenhuizen, E. L., Waitt, W. P., Paxton, R. E. and Mowrey, D. H. (1987a). *Journal of Animal Science,* **65** (Suppl. 1), 130

Anderson, D. B., Veenhuizen, E. L., Waitt, W. P., Paxton, R. E. and Young, S. S. (1987b). *Federation Proceedings,* **46**, 102 (abstract)

Bark, L. J., Stahly, T. S. and Cromwell, G. L. (1989a). *Journal of Animal Science,* **67** (Suppl. 1), 212

Bark, L. J., Stahly, T. S. and Cromwell, G. L. (1989b). *Journal of Animal Science,* **67** (Suppl. 1)

Boyd, R. D. and Bauman, D. E. (1989). In *Animal Growth Regulation*, pp. 257–293. Ed. Campion, D. R., Hausman, G. J. and Morton, R. J. Plenum Press, New York

Boyd, R. D., Bauman, D. E., Beermann, D. H., DeNeergard, A. F., Souza, L. and Kuntz, H. T. (1986). *Proceedings of Cornell Nutrition Conference*, pp. 24–28

Boyd, R. D., Beermann, D. H., Roneker, K. R., Bartley, T. D. and Fagin, K. D. (1988). *Journal of Animal Science,* **66** (Suppl. 1), 236

Campbell, R. G., Johnson, R. J. and King, R. H. (1989). In *Biotechnology for Control of Growth and Product Quality in Swine Implications and Acceptability*, pp. 137–145. Ed. Van der Wal, P., Nieuwhol, G. J. and Politick, R. D. Pudoc, Wageningen, Netherlands

Campbell, R. G., Steele, N. C., Caperna, T. J., McMurty, J. P., Solomon, M. B. and Mitchell, A. D. (1988). *Journal of Animal Science,* **66**, 1643–1655

Cline, T. R. and Forrest, J. C. (1987). *Proceedings of the Purdue Swine Day*. West Lafayette, IN

Crenshaw, J. D., Swanteck, P. M., Marchello, M. J., Harrold, R. L., Zimprich, R. C. and Olsen, R. D. (1987). *Journal of Animal Science,* **65** (Suppl. 1), 308

Etherton, T. D., Wiggins, J. P., Evock, C. M., Chung, C. S., Rebhum, J. F., Walton, P. E. and Steele, N. C. (1987). *Journal of Animal Science,* **64**, 443

Ferrell, C. L. (1988). *Journal of Animal Science,* **66** (Suppl. 1), 23

Fowler, V. R. and Kanis, E. (1989). In *Biotechnology for Control of Growth and Product Quality in Swine Implications and Acceptability*. Ed. Van der Wal, P., Nieuwhol, G. J. and Politick, R. D. Pudoc, Wageningen, Netherlands

Fuller, M. F., Beckett, P. R. and Wang, T. C. (1989). *Proceedings of 10th Western Nutrition Conference.* Winnipeg, Canada

Goodband, R. D., Nelssen, J. L., Hines, R. H., Kropf, D. H., Thaler, R. C., Schricker, B. F. and Fitzner, G. E. (1988). *Journal of Animal Science,* **66** (Suppl. 1), 95

Hancock, J. D., Peo, E. R. and Lewis, A. J. (1987). *Journal of Animal Science,* **65** (Suppl. 11), 309

Jones, R. W., Easter, R. A., McKeith, F. K., Dalrymple, R. H., Maddock, H. M. and Bechtel, P. J. (1985). *Journal of Animal Science,* **61**, 905

Jorgensen, A., Fernandez, J. A. and Beck-Andersen, S. (1988). *Statens Husdyrbrugs forsog, Meddelelse*, NR 701

Kirchgessner, M. and Roth, F. X. (1986). *Proceedings of Tenth Symposium on Energy Metabolism*, EAAP. Publ. No. 32, Abstract 67

Knabe, D. and Tanksley, T. D. (1985). *Proceedings of Minnesota Nutrition Conference*, pp. 144–153

Kraft, L. A., Haines, D. R. and Delay, R. L. (1986). *Journal of Animal Science,* **63** (Suppl. 1), 218

Machlin, L. J. (1972). *Journal of Animal Science,* **35**, 794

Madsen, A., Mason, V. C. and Weidner, K. (1965). *Acta Agricultura Scandinavica,* **15**, 213

McLaren, D. G., Grebner, G. L., Bechtel, P. J., McKeith, F. K., Novakofski, J. E. and Easter, R. A. (1987). *Journal of Animal Science,* **65** (Suppl. 1), 245

Mersmann, H. J. (1989). In *Animal Growth Regulation*, pp. 337–357. Ed. Campion, D. R., Hausman, G. J. and Martin, R. J. Plenum Press, New York

Newcomb, M. D., Grebner, G. L., Bechtel, P. J., McKeith, F. K., Novakofski, J., McLaren, D. G. and Easter, R. A. (1988). *Journal of Animal Science,* **66** (Suppl. 1), 281

Nossaman, D. A., Schinckel, A. P., Miller, L. F. and Mills, S. E. (1989). *Journal of Animal Science,* **67** (Suppl. 1), 259

Novakofski, J. (1987). In *1987 Proceedings of the University of Pork Industry Conference*, pp. 84–92. Champaign, IL

NRC (1988). *Nutrient Requirements of Swine National Academy of Science National Research Council* (9th Edn), Washington, DC

Prince, T. J., Huffman, D. L. and Brown, P. M. (1987). *Journal of Animal Science,* **65** (Suppl. 1), 301

Reeds, P. J. (1989). In *Animal Growth Regulation*, pp. 183–210. Ed. Campion, D. R., Hausman, G. J. and Martin, A. J. Plenum Press, New York

Reeds, P. J., Cadenhead, A., Hay, S. and Fuller, M. (1989). *Journal of Animal Science,* **67** (Suppl. 1), 219

Stahly, T. S. (1986). *Proceedings of the Animal Nutrition Seminar.* AGRIFAIR, Leipzig, GDR

Steele, N. C., Campbell, R. G. and Caperna, T. J. (1987). In *Proceedings of the Cornell Nutrition Conference*, p. 15

Veenhuizen, E. L., Schmiegel, K. K., Waitt, W. P. and Anderson, D. B. (1987). *Journal of Animal Science,* **65** (Suppl. 1), 130

Verstegen, M. W. A., Van der Hel, W. and van Weerden, E. J. (1989). In *Biotechnology for Control of Growth and Product Quality in Swine Implications and Acceptability*, pp. 127–136. Ed. van der Wal, P., Nieunhof, G. J. and Politiek, R. D. Puduc Wageningen, Wageningen, Netherlands

Wallace, D. H., Hedrick, H. B., Seward, R. L., Daurio, C. P. and Convey, E. M. (1987). In *Beta Agonists and Their Effects on Animal Growth and Carcass Quality*, pp. 143–151. Ed. Hanrahan, J. P. Elsevier Applied Science, New York

Wood, J. D. and Brown, A. J. (1987). *Animal Production,* **44**, 477

22

# STRATEGIES FOR SOW NUTRITION: PREDICTING THE RESPONSE OF PREGNANT ANIMALS TO PROTEIN AND ENERGY INTAKE

I.H. WILLIAMS
*University of Western Australia, Nedlands, Perth*
W.H. CLOSE
*129 Barkham Rd, Wokingham, Berks, UK*
*and*
D.J.A. COLE
*University of Nottingham School of Agriculture, Sutton Bonington, UK*

## Introduction

Since nutrition is the primary factor influencing sow productivity it follows that the establishment of a successful feeding strategy to ensure optimum productivity must be based on a sound knowledge of the response of the animal to specified nutritional inputs. Productivity of the breeding herd is commonly measured as the number of pigs born, weaned or sold per unit time, for example, per year or throughout the breeding lifetime. It can be optimized by mating gilts at a young age to minimize the time between selection from the bacon pens and first conception, and then by keeping them in the breeding herd reproducing regularly for as long as possible.

Traditional feeding strategies have often utilized the body reserves of the gilt and sow to buffer short-term deficits in nutrient intake with minimum effect on the fetus or suckling piglets. For example, animals which are allowed to make large gains in body weight during pregnancy compensate by eating less food during lactation, often losing large amounts of body weight. Modern sows are managed differently from their counterparts of 20 years ago. They begin their reproductive life with fewer body reserves because they are mated younger and at a lighter body weight, and the amount of food is rationed during pregnancy so that they gain less body weight. As a consequence with modern sows it may be necessary to consider nutritional responses much more precisely. For example, weight loss during lactation may be associated with fewer ova at the next oestrus (Hardy and Lodge, 1969) and extended intervals between weaning and oestrus particularly in first-litter sows (Reese *et al.*, 1982; King and Williams, 1984). Cole (1982) has suggested that 'a strategy of maximum conservation' is more appropriate for current-day sows. The main aim of this strategy is to maintain body condition during lactation, and implies a high level of feeding during lactation (the period of maximum production) preceded by a carefully controlled and limited weight gain during pregnancy.

In devising any feeding strategy it is important that any allowances established are within the appetite limits of the animal, since low feed intake is frequently a problem with lactating sows. Although environment and genotype influence food intake, both previous and current nutritional regimens may have a profound effect. For example, it is well established that increasing the level of nutrition in pregnancy is associated with decreased appetite in lactation (Salmon-Legagneur and Rerat, 1961). In addition the quality of the diet is also of importance. Mahan and Mangan (1975) have shown that low levels of dietary protein during both pregnancy and lactation have been associated with low intake during the suckling period.

Responses to nutrients are not always obvious and often difficult to demonstrate, depending on the extent of body reserves of the sow and hence her ability to compensate during times of nutrient shortage when metabolic demand exceeds nutrient intake. It is evident that there is a whole complex of interrelationships which will influence the feeding strategy to be adopted. To understand these effects it is necessary to establish detailed relationships and requirements for both energy and protein to meet specified targets at various stages of the reproductive cycle.

Several models of growth based on empirical equations describing nutrient utilization have been developed and used successfully to predict the rate and composition of bodyweight gain in growing pigs (Whittemore and Fawcett, 1976; Fowler, 1979; Phillips and MacHardy, 1982). This chapter describes a simple model for the partition of nutrients during pregnancy as an aid to the establishment of feeding strategies for breeding animals. The model predicts maternal and conceptus weight gain and their composition from given amounts of dietary protein and energy fed to animals of specified body weight in a thermoneutral environment.

## Partition of nutrients

Dietary energy can be partitioned between the costs of maintenance, the energy contained in the products of conception, and the deposition of protein and fat in the maternal tissue, provided that relationships can be established between the retention and intake of nitrogen (N) and energy at various body weights.

### FACTORS INFLUENCING NITROGEN DEPOSITION DURING PREGNANCY

#### *Amount and quality of dietary nitrogen*

Relationships between the intake and retention of N of gilts fed different levels of energy intake during gestation, derived from the data of Kemm (1974), are shown in *Figure 22.1*. An approach similar to that of Black and Griffiths (1975) with lambs was used to evaluate the data in which the response was described in two phases. In the first phase, N retention was limited by N intake. When animals of a given body weight were fed increasing amounts of N under conditions of adequate energy supply, N retention increased linearly with N intake until it reached a maximum value. Diets in this phase will be referred to as diets which are deficient or limiting

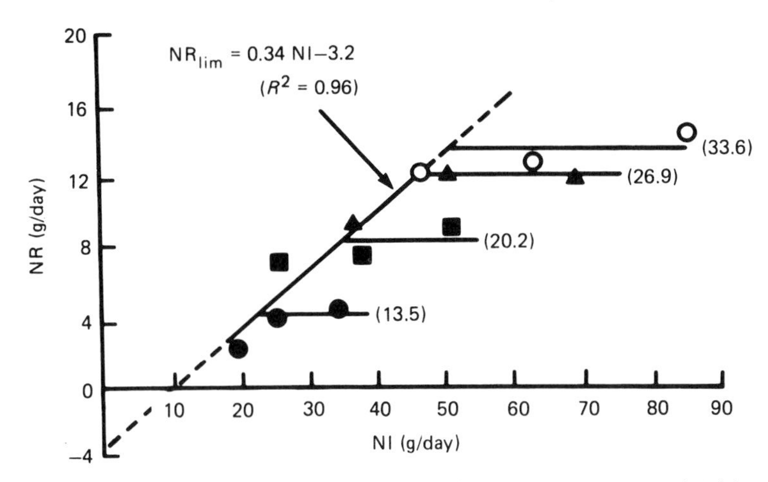


**Figure 22.1** The effect of nitrogen intake (NI) and digestible energy (DE) intake on the N retention (NR) of pregnant gilts (adapted from Kemm, 1974). Values in parentheses are the DE intakes (MJ/day) at which maximum NR occurs. The equation describing the N-limiting phase of the relationship is also given

in N or in crude protein (CP). N retention in this N-limiting phase ($NR_{lim}$) is described by equation (1).

$$NR_{lim} = 0.34\,NI - 3.2 \qquad (1)$$
$$(R^2 = 0.96)$$

where NI is the intake of dietary N. The slope of the line was 0.34 and is a direct measure of the biological value of dietary protein during pregnancy.

Additional data from several experiments where sows were fed diets limiting in N are given in *Figure 22.2*. From this it appears that sows retain N with a similar efficiency to gilts and equation (2) describes the pooled data for all animals.

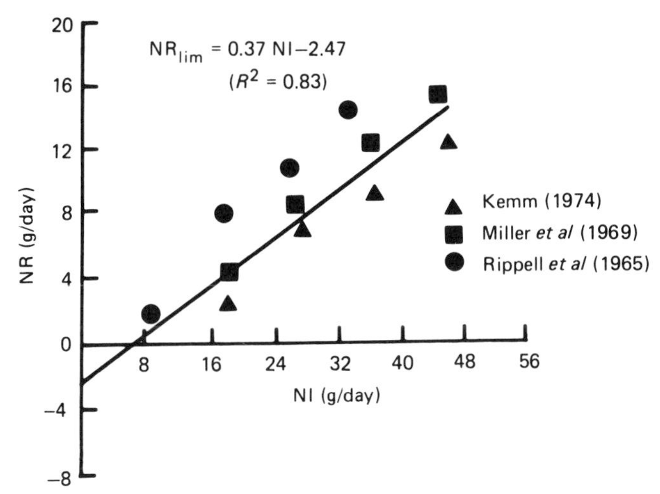


**Figure 22.2** The effect of nitrogen intake (NI) on the N retention (NR) of gilts and sows from a number of sources. The equation describing the relationship is given

$$NR_{lim} = 0.37\,NI - 2.47 \quad (2)$$

$$(R^2 = 0.83)$$

The relationship between the retention and intake of N given in equation (2) was used in all subsequent calculations and, although the estimate of total endogenous N was approximately half (68 mg/kg$^{0.75}$/day) that proposed by Carr, Boorman and Cole (1977), no adjustment was made.

*Energy intake*

In the second phase N retention was independent of N intake and, in animals of similar body weight, was determined by energy intake. This is clearly illustrated in *Figure 22.1* by the data of Kemm (1974). N retention was dependent on, and linearly

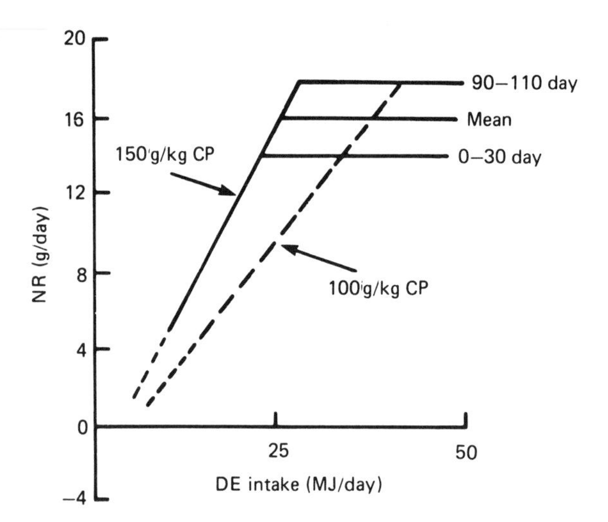


**Figure 22.3** Diagrammatic representation of the effects of digestible energy (DE) intake and crude protein content of the diet (100 and 150 g/kg) on the nitrogen retention (NR) of gilts (120 kg body weight at mating) at two stages of gestation (0–30 days and 90–110 days). The mean value throughout pregnancy is also indicated

related to, digestible energy (DE) intake between 13.5 and 27 MJ/day. Above this higher level of intake, which represents approximately 1.6 times the animal's maintenance energy requirement, there was no further increase in N retention. Hereafter diets in this phase will be termed sufficient or adequate in N or CP. *Figure 22.3* shows, diagrammatically, the relationships between N retention and DE intake used in the calculations for diets which contained adequate N (≥150g CP/kg) or were limiting in N (100g CP/kg).

In addition to the effect of N and energy intake, N retention is also influenced by the stage of pregnancy and body weight.

### *Stage of pregnancy*

W.H. Close (unpublished data) showed that maximum N retention in gilts increased from 14 g/day at day 30 of gestation up to 18 g/day at day 110. Two-thirds of this increase was accounted for by N deposition within the gravid uterus.

### *Body weight*

Body weight has a large effect on N retention as shown by the data of Carr, Boorman and Cole (1977). Their derived equation and some additional data, selected on the basis that neither DE nor N intake were limiting N deposition, are shown in *Figure 22.4*. Although there is considerable variation, the data indicate that

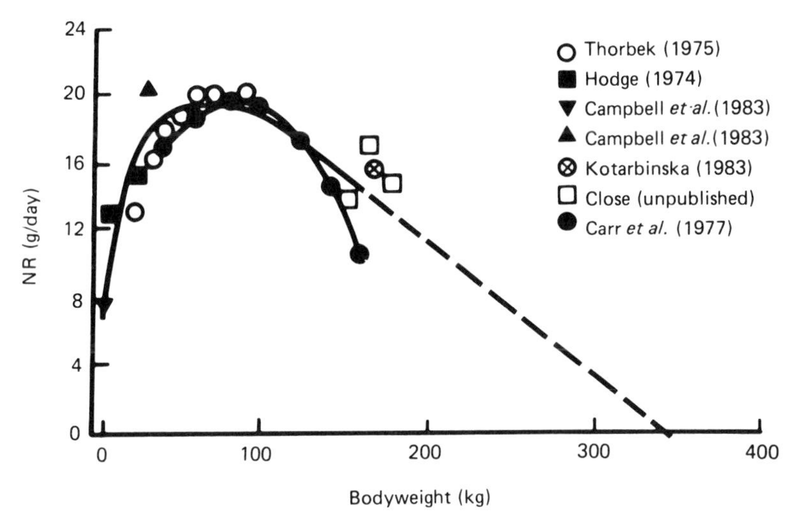


**Figure 22.4** The influence of body weight on the maximum rate of nitrogen retention (NR) in pigs. Only values for gilts and castrates have been included above 40 kg body weight. ---- is the line extrapolated to a mature body weight of 340 kg

N retention increases after birth, reaches a maximum value between 30 and 100 kg body weight and then begins to decline, approaching zero when mature body weight has been reached. There is no information on the mature body weights of modern sows and, from inspection of sow records at the National Institute for Research in Dairying at Shinfield, the value of 340 kg was chosen to represent the mature body weight of a modern sow at zero N retention.

## THE PARTITION OF ENERGY INTAKE

### *Deposition of fat and protein*

After making an appropriate allowance for the animal's maintenance energy requirement, the remaining DE was partitioned between protein and fat deposition according to the amount of N retained. The maintenance energy requirement was taken as 430 kJ metabolizable energy (ME)/kg$^{0.75}$/day (452 kJ DE/kg$^{0.75}$/day),

**Table 22.1** CALCULATION OF BODYWEIGHT GAINS TO DAY 110 OF PREGNANCY FOR GILTS WEIGHING 120 kg AT MATING AND FED A DIET CONTAINING 150 g CRUDE PROTEIN/kg AT VARIOUS DIGESTIBLE ENERGY (DE) INTAKES

| *DE* (MJ/day) | *ME for maintenance* ($ME_m$)[a] (MJ/day) | *Nitrogen retention* (g/day) | *Protein deposited* (P) (g/day) | *ME for protein* ($P_E$)[b] (MJ/day) | *ME for fat deposition* ($F_E$)[c] (MJ/day) | *Fat deposited* (F) (g/day) | *Lean tissue deposited* (L)[d] (g/day) | *Fat + lean deposited* (g/day) | *Daily gain* ($\Delta W$)[e] (g/day) | *Total gain* (kg/110 day) |
|---|---|---|---|---|---|---|---|---|---|---|
| 18 | 16.6 | 10.9 | 68 | 2.7 | –2.2 | –69 | 296 | 227 | 252 | 27.7 |
| 24 | 17.5 | 15.3 | 96 | 3.8 | 1.5 | 30 | 415 | 445 | 494 | 54.3 |
| 30 | 18.4 | 16.0 | 100 | 4.0 | 6.1 | 123 | 435 | 558 | 620 | 68.2 |
| 36 | 19.3 | 16.0 | 100 | 4.0 | 10.9 | 219 | 435 | 656 | 729 | 80.2 |
| 42 | 20.2 | 16.0 | 100 | 4.0 | 15.7 | 317 | 435 | 752 | 836 | 92.0 |
| 48 | 21.1 | 16.0 | 100 | 4.0 | 20.5 | 413 | 435 | 848 | 942 | 103.6 |

[a] $ME_m = 430\ kJ/kg^{0.75}/day$
[b] $P_E = P \times (0.0238 \div 0.60)$
[c] $F = F_E \div (0.0397 \times 0.80)$ when $(ME_M + P_E) > 0.95\ DE$ and [c] $F = F_E \div (0.0397 \div 0.8)$ when $(ME_M + P_E) < 0.95\ DE$
[d] $L = P \div 0.23$
[e] $\Delta W = (F + L) \div 0.9$

calculated from the mean of several recent estimates (Agricultural Research Council, 1981; Burlacu, Iliescu and Cărămidă, 1982; Close, Noblet and Heavens, 1985). Unlike estimates derived by the Agricultural Research Council (1981), no increase in the maintenance energy requirement with the progress of gestation was allowed, since recent experimentation (Burlacu, Iliescu and Cărămidă, 1982; W.H. Close, unpublished) indicated that it did not significantly increase during gestation. The efficiencies of utilization of ME for the deposition of protein and fat were taken as 0.6 and 0.8, respectively, with the corresponding energy values of 23.8 and 39.7 MJ/kg (Agricultural Research Council, 1981; Close, Noblet and Heavens, 1985).

As an example a calculation is described for a sow weighing 120 kg at mating and receiving 22.8 MJ ME (24 MJ DE) per day on a protein-adequate diet. Of the daily ME intake, 17.5 MJ ($0.43 \times 140^{0.75}$, where 140 kg represents the mean body weight of the animal throughout gestation) was required for maintenance. Protein deposition was 96 g/day (15.3 g N × 6.25), requiring 3.8 MJ ME [(0.096 × 23.8) ÷ 0.6], thus leaving 1.5 MJ ME/day for the synthesis and deposition of fat. Hence the daily retention of fat was 30 g/day [(1.5 × 0.8) ÷ 39.7] (*Table 22.1*).

### *Calculation of bodyweight gain*

Total bodyweight gain was calculated on the basis that lean and fat gain represents 90 per cent of the total gain; the remaining 10 per cent represents ash and gut fill. The lean tissue growth rate was calculated from the protein gain on the criterion that the protein gain represents 23 per cent of the combined protein plus water gain (Walach-Janiak *et al.*, 1983; Shields and Mahan, 1983). Thus in the example cited above, the daily lean tissue gain was 415 g (96 ÷ 0.23) and, since the gain in fat was 30 g/day, the total bodyweight gain was 494 g/day [(415 + 30) ÷ 0.90], that is a total

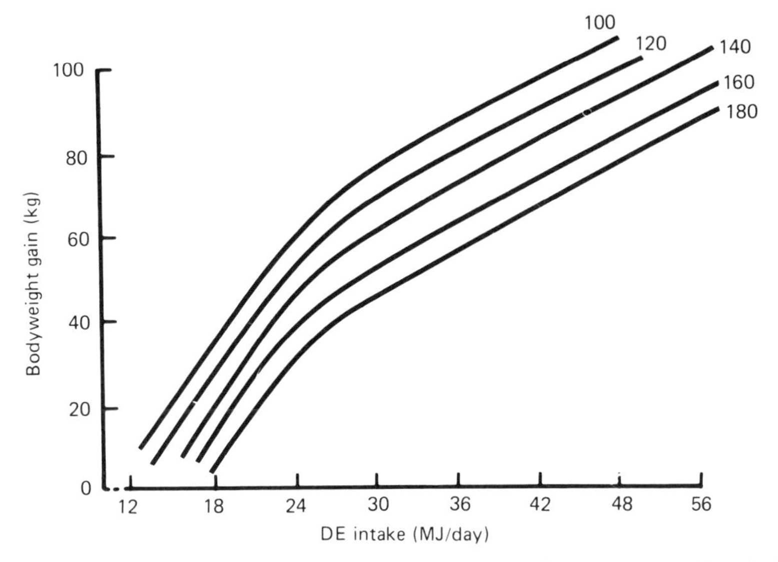


**Figure 22.5** Predicted effect of body weight at mating (100, 120, 140, 160 and 180 kg) and digestible energy (DE) intake on bodyweight gain of pregnant animals fed diets containing adequate nitrogen

of 54.3 kg throughout a 110-day gestation period. A more detailed account of the effect of energy intake on the rate and composition of the bodyweight gain of sows mated at 120 kg is given in *Table 22.1*, whereas *Figure 22.5* shows the effect of energy intake on the bodyweight gain of sows of different body weights at mating and fed protein-adequate diets.

### *Partition of nutrients between maternal and conceptus tissue*

The accretion of fat and protein in both the conceptus and maternal tissue depends upon nutrient supply and the priority for tissue deposition. In early gestation the needs of the conceptus tissue are small and nutrients are predominantly deposited in the maternal tissue. In late gestation, on the other hand, most nutrients are directed towards the conceptus tissue since its maximum rate of growth and development occurs at this stage. However, from the results of Kotarbinska (1983) and Noblet *et al.* (1985) it would appear that maximum N retention throughout gestation in both the conceptus and maternal tissue occurs at approximately similar energy intakes, that is, 25 MJ DE/day for a protein-sufficient diet (150 g CP/kg) or 37 MJ DE/day for a diet deficient in protein (100 g CP/kg). For a protein-sufficient diet, N retention in the conceptus tissue was calculated to be 1.7 g/day, at 18 MJ DE/day and increased linearly to a maximum of 2.7 g/day at 25 MJ DE/day.

Increases in both fresh weight and dry matter accompany the increase in N in the conceptus tissue, but the amount of fat is independent of energy intake and remains nearly constant at approximately 12 per cent of the dry matter content of the conceptus.

Since the rate of gain and composition of the total body and conceptus tissue is known, the gain in the maternal body can be calculated by difference. *Figure 22.6* therefore shows the effect of dietary energy and CP content of the diet on the partition of weight gain and its tissue composition into conceptus and maternal components for gilts and sows mated at 120 and 180 kg body weight, respectively (*Figure 22.6a* and *b*).

## Validation of the model

Predicted values of bodyweight gains during pregnancy are shown in *Figure 22.7* for gilts weighing 120 kg at mating and fed diets either adequate (150 g/kg) or deficient (100 g/kg) in CP at levels of DE between 18 and 48 MJ/day and compared with data from several experiments not used in deriving the basic relationships of the model. There was excellent agreement between the predicted and observed values. Both the magnitude and direction of the response to changes in DE intake were well predicted for all the available data. Prediction of the response to dietary CP was also in good agreement with the limited data available (Greenhalgh *et al.*, 1977, 1980). Thus at a constant DE intake of 25 MJ/day, increasing the dietary CP from 90 to 140 g/kg was associated with an 11 kg increase in body weight during pregnancy, and this compared favourably with the predicted change of 14 kg.

Multiparous animals of heavier body weight did not follow the predicted response nearly as closely as the gilts. For example, the data of Lee and Mitchell (1984) in *Figure 22.8* show that the predicted response of gilts is very close to experimental values but that the response of sows is underpredicted by nearly 30

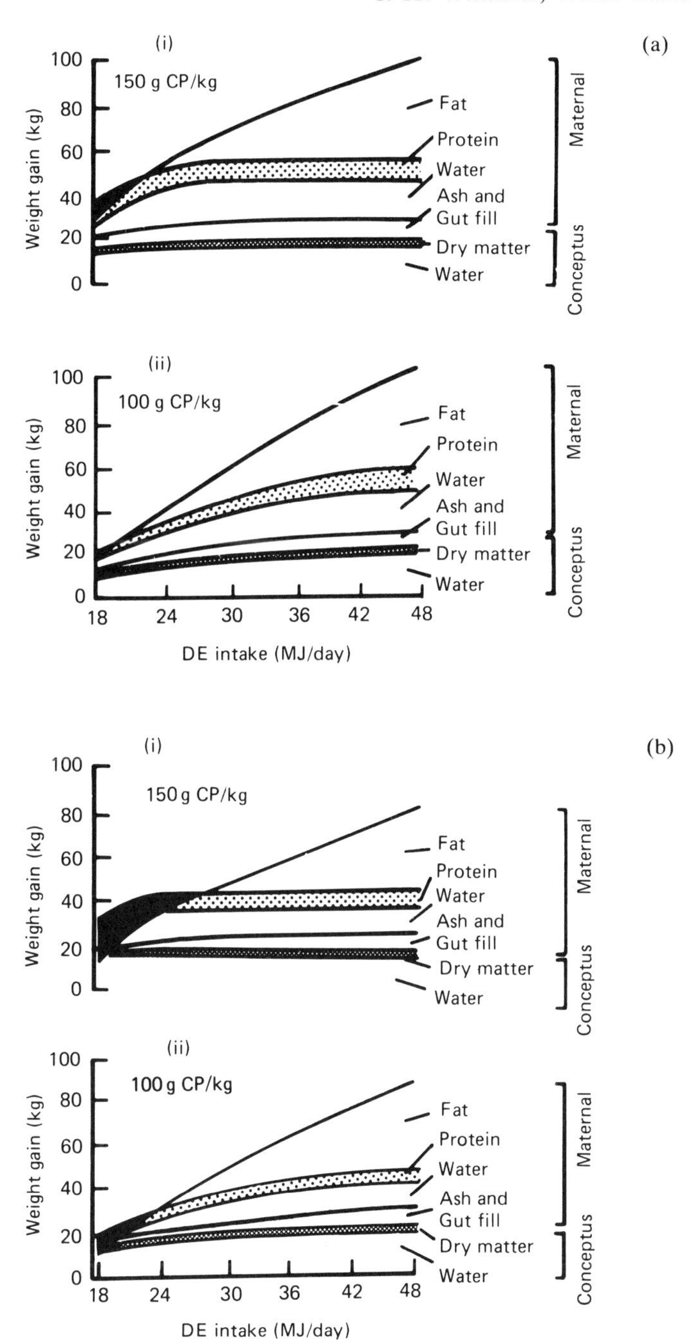


**Figure 22.6** The predicted effect of digestible energy (DE) intake on the partition of total bodyweight gains to 110 days of gestation, of gilts and sows, into maternal and conceptus components. The animals were mated at different body weights and fed diets containing either (i) 150 or (ii) 100 g crude protein/kg. ■, denotes loss of maternal body fat. (a) Gilts mated at 120 kg body weight; (b) Sows mated at 180 kg body weight

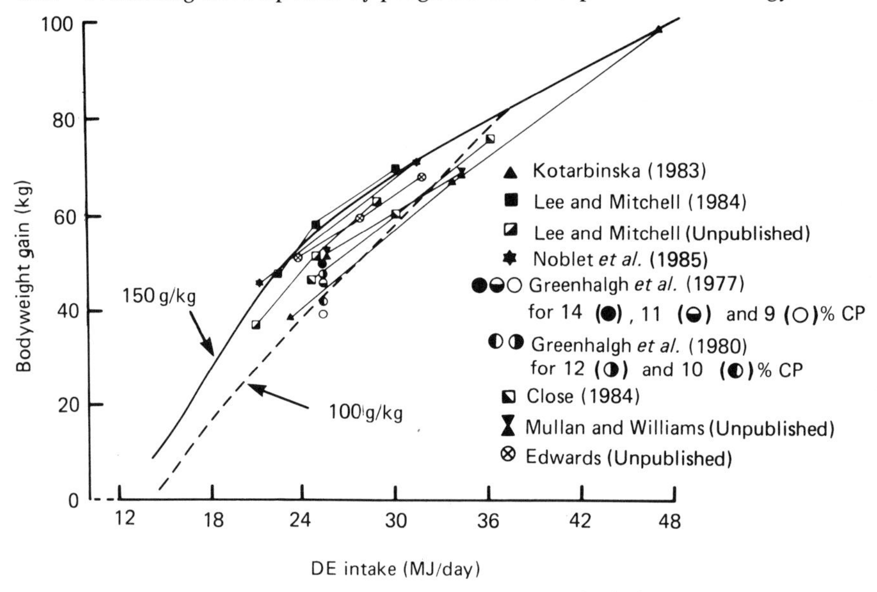


**Figure 22.7** Comparison of predicted and observed changes in the bodyweight gains of pregnant gilts (120 kg body weight at mating) in relation to digestible energy (DE) intake when fed diets supplying 100 g (----) and 150 g (——) of crude protein/kg

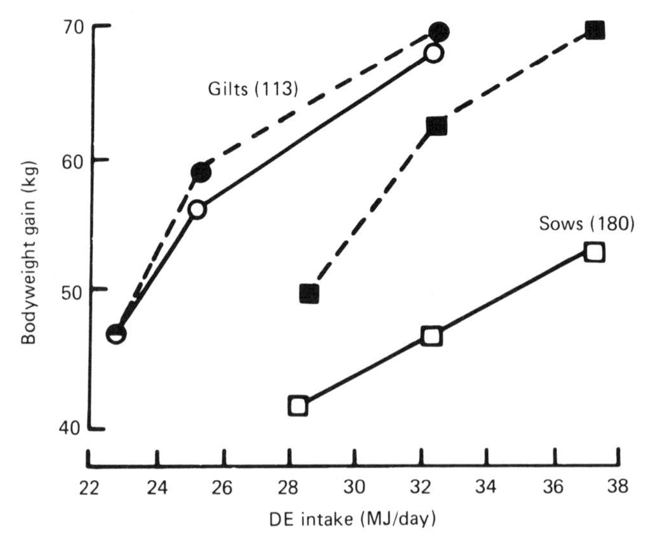


**Figure 22.8** Predicted (——) and observed (----) gains in body weight of gilts and sows in relation to digestible energy (DE) intake. (Values in parentheses are the body weights of the animals at mating) (Lee and Mitchell, 1984)

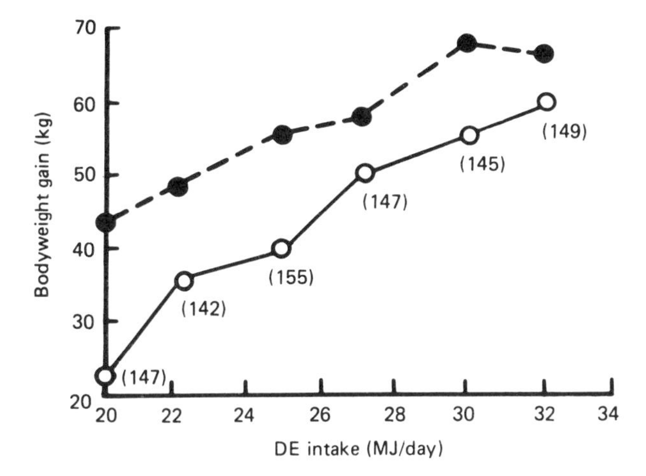


**Figure 22.9** Predicted (——) and observed (----) gains in body weight of sows during their second parity in relation to digestible energy (DE) intake. (Values in parentheses are the body weights of the sows at mating) (A.J. Harker and D.J.A. Cole, unpublished data)

per cent, especially at the higher energy intakes. Similarly, data from the University of Nottingham for second parity sows (A.J. Harker and D.J.A. Cole, unpublished) show a similar degree of underprediction (*Figure 22.9*). Despite these differences there was reasonable agreement in both the direction and magnitude of the response of sows to a change in feed intake.

## Differences between observed and predicted values

There are several possible reasons why the model is less successful in predicting the weight gains of heavy sows than gilts.

(1) In all calculations a constant value for the maintenance energy requirement of 430 kJ ME/kg$^{0.75}$/day has been assumed, regardless of body weight. If the maintenance energy requirement decreases with increase in body weight then more energy would become available for production giving a greater weight gain at any given energy intake. For example, from the experiments of Lee and Mitchell (1984) the weight gain of sows throughout pregnancy has been underestimated by 15 kg (*Figure 22.8*). This is equivalent to a growth rate throughout pregnancy of 144 g/day and, assuming that N retention is maximum and cannot increase, would represent a gain in fat deposition of 120 g/day. The ME required for this gain in fatty tissue is 6 MJ/day. For a sow weighing 180 kg, its maintenance energy would need to decrease by nearly 30 per cent to give a sufficient increase in productive energy.

On theoretical grounds there are reasons why maintenance may not be constant and might vary with body weight. Protein turnover per unit of metabolic body weight decreases as animals approach maturity and should reduce maintenance energy requirements. However, given that protein turnover accounts for a relatively small proportion of maintenance (Reeds *et al.*, 1980), large changes would be needed to reduce maintenance sufficiently to give the required increase in productive energy.

(2) The phenomenon of compensatory gain or pregnancy anabolism might explain the difference between the actual and predicted weight gains of sows. Such effects have been demonstrated in heavy sows (Salmon-Legagneur and Rerat, 1961) but not in gilts. Compensatory gain is well known in many species, particularly sheep, and generally occurs following a period of food restriction. The loss of body weight during lactation could be a predisposing factor for such compensatory gains in the subsequent pregnancy and might explain its occurrence in sows but not in gilts. Apart from increased food intake, several mechanisms have been suggested to explain compensatory growth including reduced basal metabolic rate, more efficient accretion of protein and/or fat, and increased amounts of water in the carcass. Since older, heavier sows are relatively lean compared with gilts, small changes in the ratio of water to protein in the body could have a large effect on weight gain.

(3) The most likely possibility to explain underprediction of weight gain in sows is that maximum N retention has been underestimated. There are very few published values of N retention for pigs above 100 kg body weight and, consequently, extrapolation was made from younger animals having set the mature body weight of modern sows at 340 kg. Since fat is energetically more expensive to deposit than lean (49.6 MJ/kg compared with 9.1 MJ/kg), it may be calculated for a sow weighing 180 kg at mating, that an extra 15 kg of body gain during pregnancy could be achieved by a 5–6 g/day increase in N retention.

## Predictions from the model and possible consequences

Energy is required during pregnancy by the developing conceptus and by the sow to maintain her own body tissues. It is common practice to ration sows to achieve set target gains of body weight on the basis that growth of the conceptus will not be retarded provided that some maternal gain is made. Suggested targets vary but one growth path which seems to be well accepted is that the sow should be given sufficient food to gain 15 kg in body weight per reproductive cycle for the first four cycles and then maintained at that body weight thereafter (Hillyer, 1980). Sows do not frequently consume sufficient food to gain weight during lactation and, therefore, most or all of the weight gain required in each cycle must be made during pregnancy.

The model described clearly demonstrates the consequences of such target changes in body weight on maternal body reserves. For example, a gilt beginning her reproductive life at a body weight of 120 kg and fed sufficient of a protein-adequate diet to give a maternal gain during pregnancy of 15 kg (total gain of maternal + conceptus tissue of 15 + 20 = 35 kg) would require a DE intake of approximately 18 MJ/day (see *Figure 22.7*). At this level of intake the net gain of 15 kg would be entirely lean tissue and would be accompanied by a loss in body fat of approximately 5 kg (see *Figure 22.6a(i)*, *Figure 22.10*). Bigger sows making similar gains in maternal weight would lose even more fat as a consequence of a lower energy intake relative to maintenance. Thus a gilt weighing 120 kg at mating and fed a protein-adequate diet in sufficient amounts to give a maternal gain of 15 kg in each pregnancy would lose approximately 5, 7, 8 and 8 kg of body fat in the first

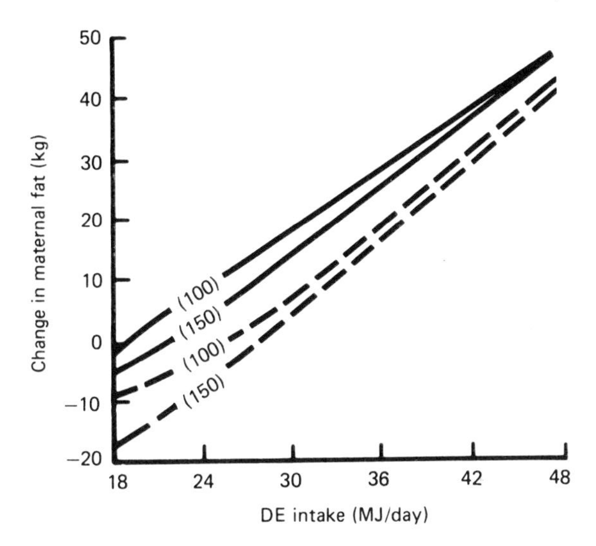


**Figure 22.10** The predicted changes in maternal fat in gilts (——, 120 kg body weight at mating) and sows (----, 180 kg body weight at mating) in relation to digestible energy (DE) intake, when fed diets supplying 100 or 150 g crude protein/kg

four reproductive cycles, that is, a total of 28 kg. Modern gilts weighing 120 kg at mating are unlikely to have sufficient body fat to cover losses of this magnitude. Furthermore, the predicted loss of 28 kg is likely to be an underestimate. Although some sows consume sufficient food to maintain body weight during lactation the majority lose body weight and most of this loss will be fat. In addition, it is likely that the model underestimates the amount of energy retained as protein in heavier sows and, therefore, overestimates the amount stored as fat.

Feeding a diet deficient in protein can result in the alternative outcome where body fat is gained rather than lost during pregnancy. For example, feeding a diet containing 100 g CP/kg in sufficient amounts to achieve a maternal gain of 15 kg or a total gain of 35 kg would require a DE intake of 22.5 MJ/day rather than the 18.0 MJ/day on a protein-adequate diet. Under this feeding regimen sows would be predicted to gain approximately 4 kg of fat per cycle or a total of 16 kg in the first four reproductive cycles.

The concentration of dietary protein has a very profound influence on body fat reserves of the sow since the animal is normally fed at a much lower level relative to its maintenance energy requirement when compared with the growing pig. As indicated above, feeding protein-adequate diets will reduce body fat to zero within four parities assuming that a gilt of 120 kg contains approximately 30 kg of body fat at mating and that there is no loss of fat during lactation. Depletion of body fat can however be prevented by increasing feed intake. For example, increasing DE intake of a protein-adequate diet to 22 MJ/day would maintain fat reserves during the first pregnancy (*Figure 22.10*). The gain in maternal body weight corresponding to this intake would be approximately 25 kg, that is 10 kg higher than the current target recommendations.

Clearly the components forming the basis of the model proposed here have not been the subject of sufficient research. However, in addition to providing a working hypothesis for nutrition in pregnancy, the model highlights the areas of inadequacy in our knowledge.

## Acknowledgements

We are most grateful to the following people for providing us with results of unpublished data: Sandra Edwards, MAFF, Terrington Experimental Husbandry Farm, Kings Lynn, Norfolk; A.J. Harker, University of Nottingham, School of Agriculture, Sutton Bonington, Loughborough, Leics; Pauline A. Lee and K.G. Mitchell, National Institute for Research in Dairying, Shinfield, Reading, Berks; B.P. Mullan, University of Western Australia, Nedlands, Western Australia.

## References

AGRICULTURAL RESEARCH COUNCIL (1981). *The Nutrient Requirements of Pigs*, p. 50. Commonwealth Agricultural Bureaux; Slough

BLACK, J.L. and GRIFFITHS, D.A. (1975). *British Journal of Nutrition*, **33**, 399–413

BURLACU, GH., ILIESCU, M. and CĂRĂMIDĂ, P. (1982). European Association for Animal Production Publication No. 29, 222–224

CAMPBELL, R.G. and DUNKIN, A.C. (1983). *British Journal of Nutrition*, **49**, 221–230

CAMPBELL, R.G., TAVERNER, M.R. and CURIC, D.M. (1983). *Animal Production*, **36**, 193–199

CARR, J.R., BOORMAN, K.N. and COLE, D.J.A. (1977). *British Journal of Nutrition*, **37**, 143–155

CLOSE, W.H., NOBLET, J. and HEAVENS, R.P. (1985). *British Journal of Nutrition*, **53**, 267–279

COLE, D.J.A. (1982). In *Control of Pig Reproduction*, p. 603. Ed. D.J.A. Cole and G.R. Foxcroft. Butterworths; London

FOWLER, V.R. (1979). In *Recent Advances in Animal Nutrition–1978*, p. 73. Ed. W. Haresign and D. Lewis. Butterworths; London

GREENHALGH, J.F.D., BAIRD, BARBARA, GRUBB, D.A., DONE, S., LIGHTFOOT, A.L., SMITH, P., TOPLIS, P., WALKER, N., WILLIAMS, D. and YEO, M.L. (1980). *Animal Production*, **30**, 395–406

GREENHALGH, J.F.D., ELSLEY, F.W.H., GRUBB, D.A., LIGHTFOOT, A.L., SAUL, D.W., SMITH, P., WALKER, N., WILLIAMS, D. and YEO, M.L. (1977). *Animal Production*, **24**, 307–321

HARDY, B. and LODGE, G.A. (1969). *Animal Production*, **11**, 505–510

HILLYER, G.M. (1980). In *Recent Advances in Animal Nutrition–1979*, p. 69. Ed. W. Haresign and D. Lewis. Butterworths; London

HODGE, R.W. (1974). *British Journal of Nutrition*, **32**, 113–126

KEMM, E.H. (1974). A study of the protein and energy requirements of the pregnant gilt (*Sus scrofa domesticus*). PhD Thesis, University of Stellenbosch

KING, R.H. and WILLIAMS, I.H. (1984). *Animal Production*, **38**, 241–247

KOTARBINSKA, MARIA (1983). *Pig News & Information*, **4**, 275–278

LEE, PAULINE, A. and MITCHELL, K.G. (1984). *Animal Production*, **38**, 528 (abstract)

MAHAN, D.C. and MANGAN, L.T. (1975). *Journal of Nutrition*, **105**, 1291–1298

MILLER, G.M., BECKER, D.E., JENSEN, A.H., HARMON, B.G. and NORTON, H.W. (1969). *Journal of Animal Science*, **28**, 204–207

NOBLET, J., CLOSE, W.H., HEAVENS, R.P. and BROWN, D. (1985). *British Journal of Nutrition*, **53**, 251–265

PHILLIPS, P.A. and MACHARDY, F.V. (1982). *Canadian Journal of Animal Science*, **62**, 109–121

REEDS, P.J., CADENHEAD, A., FULLER, M.F., LOBLEY, G.E. and MCDONALD, J.D. (1980). *British Journal of Nutrition*, **43**, 445–455

REESE, D.E., MOSER, B.D., PEO, E.R. JR., LEWIS, A.J., ZIMMERMAN, D.R., KINDER, J.E. and STROUP, W.W. (1982). *Journal of Animal Science*, **55**, 590–598

RIPPEL, R.H., RASMUSSEN, A.H., JENSEN, A.H., NORTON, H.W. and BECKER, D.E. (1965). *Journal of Animal Science*, **24**, 209–215

SALMON-LEGAGNEUR, E. and RERAT, A. (1961). In *Nutrition of Pigs and Poultry*, p. 207. Ed. J.T. Morgan and D. Lewis. Butterworths; London

SHIELDS, R.G. and MAHAN, D.C. (1983). *Journal of Animal Science*, **57**, 594–603

THORBEK, GRETE (1975). In *Studies on Energy Metabolism in Growing Pigs*, p. 100. Beretning fra Statens Husdyrbrugs forsøg no. 424

WALACH-JANIAK, M., RAJ, S., FANDREJEWSKI, H., KOTARBINSKA, MARIA and LASSOTA, M. (1983). Paper no. 5.31. 34th EAAP Annual Meeting, Madrid

WHITTEMORE, C.T. and FAWCETT, R.H. (1976). *Animal Production*, **22**, 87–96

23

# PREDICTING NUTRIENT RESPONSES OF THE LACTATING SOW

B. P. MULLAN
*Animal Industries, Department of Agriculture, Perth, Western Australia*

W. H. CLOSE
*129 Barkham Rd, Wokingham, Berks, UK*

*and*
D. J. A. COLE
*University of Nottingham School of Agriculture, Sutton Bonington, Loughborough, Leics, LE12 5RD, UK*

## Introduction

During the past two decades there have been numerous extensive reviews concerned with either establishing the nutritional requirements of the pregnant and lactating sow or assessing the consequences of nutrition on various aspects of sow performance (Lodge, 1962; Elsley, 1971; Elsley and MacPherson, 1972; Lodge, 1972; O'Grady, 1980; Agricultural Research Council, 1981; Cole, 1982; National Research Council, 1988). In general, these have shown that it is possible to make reasonable estimates of the short-term nutritional requirements of the sow during either a single pregnancy or lactation period, but only a few have considered the consequences of applying these short-term feeding recommendations to long-term nutritional or productive needs. It has been an explicit assumption that having established these requirements, it is possible to develop appropriate feeding strategies and that there are no restrictions to limit their application in practice. Whereas this may be true for the pregnant sow, which is generally fed on a restricted basis, it is not the case for the lactating animal which may not have the capacity to consume sufficient feed to meet its calculated daily needs. This, therefore, raises a number of questions. What are the consequences for the lactating sow if nutrient demand exceeds nutrient supply; how does the animal respond in both the short and long term? How may recent knowledge of nutritional, metabolic and physiological responses be interpreted and applied to provide a better understanding of the long-term nutritional needs and responses of the lactating sow?

The need to establish the response of the modern lactating sow to specified nutritional inputs is generally recognized and indeed, such reviews as ARC (1981) used information which was actually obtained in the 1960s and 1970s when the type of animal, its reproductive capacity, nutritional management and the husbandry and environmental conditions under which it was kept were markedly different from today. Modern sows are mated earlier and at a lighter body weight, they now begin their reproductive life with less body reserves and are expected to sustain a higher level of productivity for a longer period of time than hitherto. They are weaned earlier, kept individually in controlled environmental conditions and often creep feed is not provided with the consequence that growth and development of the suckling piglets are entirely dependent upon milk production and hence appetite of the sow. As a result, the nutritional needs and responses of the lactating sow must be considered more carefully and the consequences for not meeting these needs must be known. However, in assessing requirements it is recognized that responses during lactation cannot be treated in isolation from other events within the same or other parities. For example, both the level and composition of the diet fed during pregnancy influence appetite during lactation and this may have indirect consequences for subsequent reproductive performance, such as reduced ovulation rate and an extended weaning-to-oestrus interval. To take account of these assumptions Cole (1982) has suggested a strategy of maximum conservation which involves a low feeding level to control and limit weight gain during pregnancy followed by a high feeding level to minimize body weight loss and, as far as possible, maintain body condition during lactation.

## The factorial estimation of nutrient requirements

The objective of the sow during lactation is to produce sufficient milk to wean an adequate number of piglets of an acceptable body weight, with minimum variation, yet without utilizing excessive body reserves to prejudice subsequent reproductive performance. However, examination of the relationships between maternal feed intake and such aspects of productivity as piglet growth, change in body weight or backfat thickness of the sow during lactation, show the great variability that exists in the ability of the sow to nurture her young or maintain body condition (Figure 23.1). This suggests that many factors, which are not always included as experimental variables, have to be taken into account when assessing responses to nutrients and indicates the limited use that can be made of many empirical relationships to describe nutritional requirements and responses. The alternative approach described in this chapter is therefore based on the factorial procedure which allows the prediction of requirements and the determination and interpolation of changes in productivity according to metabolic and physiological principles. This complements a similar exercise carried out for the pregnant sow which has been reported at a previous University of Nottingham Feed Manufacturers Conference (Williams, Close and Cole, 1985) and which forms the basis of the new AFRC review of the nutrient requirements of sows (Agricultural and Food Research Council, unpublished). The overall objective is to provide a greater scientific basis to interpret changes in performance and to establish guidelines and indices which may then be used to develop feeding strategies to sustain optimum sow productivity.

The primary components necessary to establish the factorial assessment of the nutritional requirements of the lactating sow are:

(1) determination of the nutrient requirements of the suckling piglets at various growth rates, litter sizes and stages of lactation,
(2) calculation of the energy and nitrogen requirements for milk production necessary to sustain different rates of piglet gain,

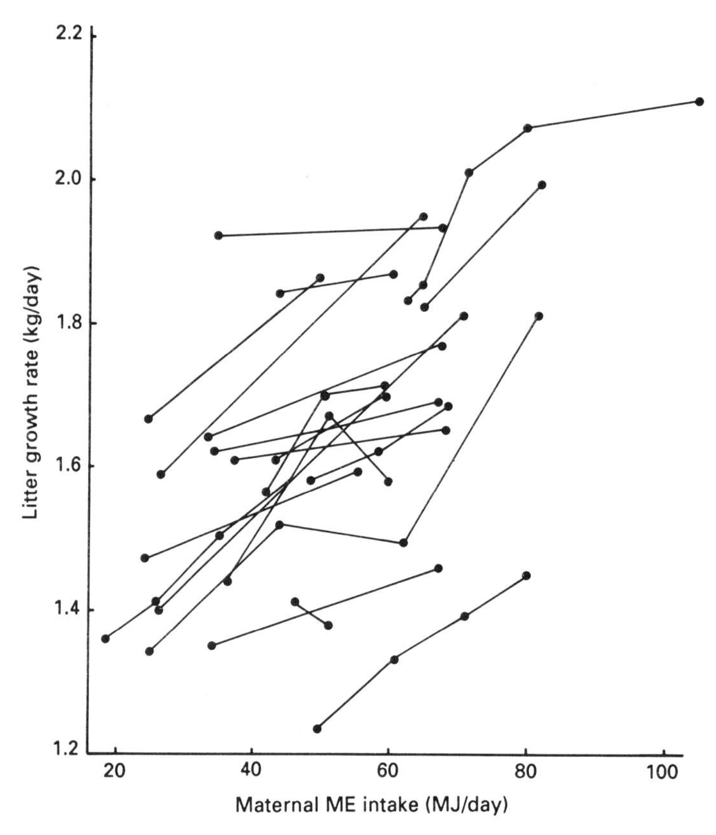


**Figure 23.1** The influence of maternal energy intake on growth rate of the litter, taken from various sources (O'Grady *et al.*, 1973; Reese *et al.*, 1982; Danielsen and Nielsen, 1984; King and Williams, 1984a and 1984b; King, Williams and Barker, 1984; Nelssen *et al.*, 1985; Harker, 1986; Johnston *et al.*, 1986; King and Dunkin, 1986a and 1986b; Noblet and Etienne, 1987; Eastham *et al.*, 1988; Yang *et al.*, 1988; Mullan and Close, 1989; Mullan and Williams, 1989)

(3) assessment of the maintenance energy and nitrogen needs of the lactating sow and hence the total nutrient requirements for both maintenance and milk production,
(4) estimation of the energy and nitrogen intake of the sow under *ad libitum* feeding conditions and those factors which influence it,
(5) calculation of the change in lean and fat and hence body weight of the sow together with corresponding changes in backfat thickness.

The factorial procedure assumes that it is possible to partition nutrient needs according to the requirements for maintenance, for tissue deposition within the body and for products formed but subsequently lost from the body. It assumes that sufficient experimentation exists to derive the relationship between nutrient intake and nutrient output according to the scheme presented in Figure 23.2. The maintenance requirement is therefore equivalent to that intake which maintains an animal in a state of equilibrium neither gaining nor losing nutrients. It is normally

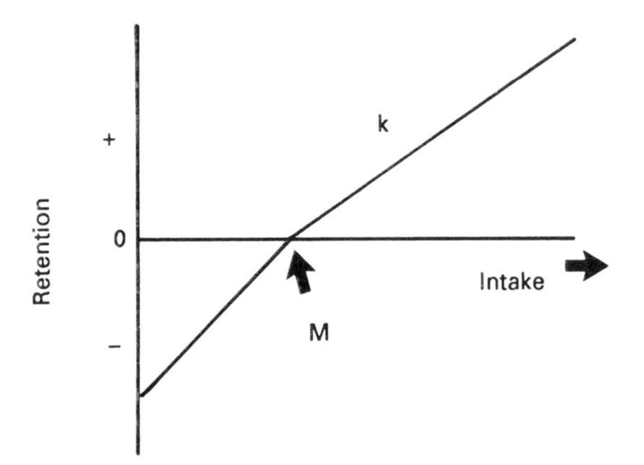


**Figure 23.2** Diagrammatic representation of the relationship between nutrient intake ($I$) and retention ($R$) in animals. $M$ = maintenance requirement, where R = 0; $k$ = net efficiency of nutrient utilization, that is $\Delta R/\Delta I$; total requirement ($T$) = $M + R/k$

expressed on a metabolic body size basis to take account of variations in body weight, age and maturity. Above maintenance intake the requirement for tissue deposition or product formation is a function of the rate of nutrient retention (or loss) and the net efficiency of nutrient utilization. This varies with the level of animal performance and increases as feeding level increases. Thus, in the diagrammatic representation in Figure 23.2, the total nutrient requirement ($T$) at any level of performance can be calculated as

$$T = M + R/k \qquad (23.1)$$

where $M$ is the requirement for maintenance, $R$ is the nutrient content of the product formed and $k$ is the net efficiency of nutrient utilization. This procedure has been used to calculate the energy and nitrogen needs and hence requires information on various aspects of the energy and nitrogen metabolism of both the suckling piglet and the lactating sow.

## THE NUTRITIONAL REQUIREMENTS OF THE SUCKLING PIGLET

A major prerequisite for the factorial estimation of the nutrient requirements of the suckling piglet is knowledge of the pattern and the rate of tissue accretion during lactation and those factors which influence it. If changes in growth rate are to be accommodated, then variation in the rate of protein and fat deposition must be known and these have been calculated from the experiments of Berge and Imdrebø (1954), Brooks *et al.* (1964) and Elliot and Lodge (1977). These show that the composition of the piglet tends to stabilize after the first week of life, but that the rate of fat accretion is much greater than that of protein (Figure 23.3). Thus the mean protein content was calculated to increase from 12% at birth to 14 and 15% at 1 and 2 weeks of age, respectively, and thereafter remained constant; the corresponding increases in mean fat content were from 1.3% at birth to 9, 13, 15 and 16% at 1, 2, 3 and 4 weeks, respectively. These values have therefore been

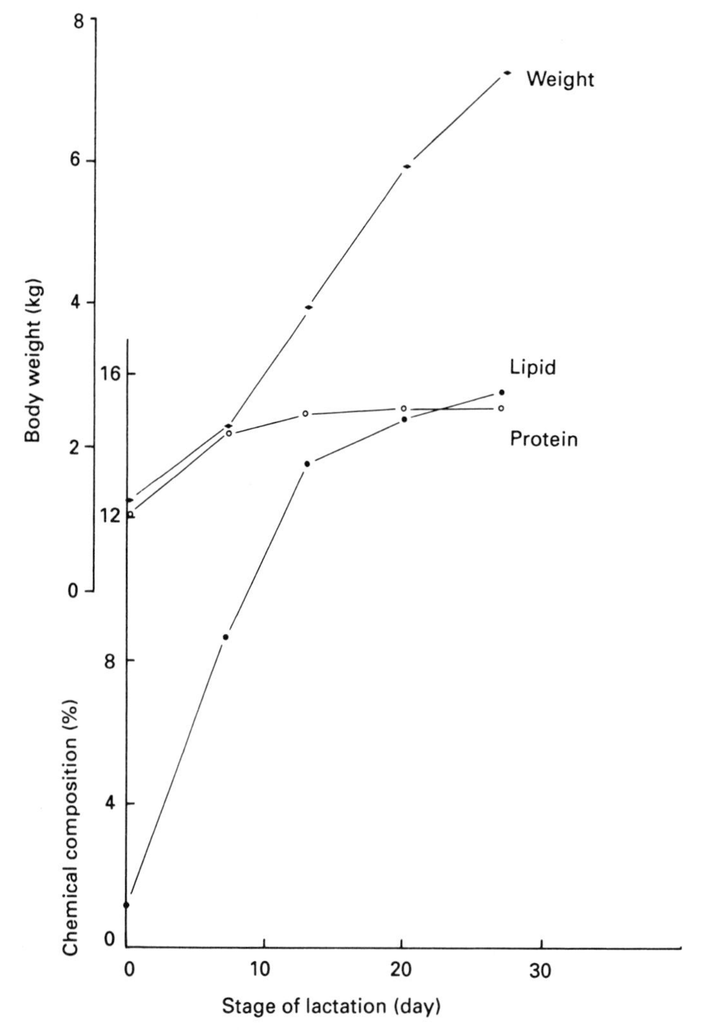


**Figure 23.3** The mean change in body weight and body composition of the suckling piglet, compiled from several sources (see text)

used to calculate the rates of protein and fat deposition and hence energy and nitrogen contents of suckling piglets at various stages of lactation and rates of growth.

From the above values of protein and fat content the daily rates of tissue accretion can be calculated and the energy requirements of the suckling pig determined from the knowledge that 1 g protein contains 23.8 kJ, 1 g fat contains 39.8 kJ, assuming that the net efficiency of energy utilization is 0.78 and that the maintenance energy requirement is 498 kJ/kg body weight$^{0.75}$ per day (Close, unpublished). Similarly, the nitrogen requirements have been calculated on the assumptions that 1 g protein contains 0.16 g nitrogen (N), the nitrogen requirement for maintenance is 0.34 g digestible N/kg body weight$^{0.75}$ per day and the efficiency of nitrogen accretion is 0.90 (Burlacu, Iliescu and Cárámida, 1986; Beyer, 1986).

**Table 23.1** THE ENERGY AND NITROGEN REQUIREMENTS OF SUCKLING PIGLETS (LITTER SIZE = 10; NO CREEP FEED PROVIDED)

| | *Stage of lactation* | | | |
|---|---|---|---|---|
| | *Week 1* | *Week 2* | *Week 3* | *Week 4* |
| Litter weight gain (kg/day) | 1.50 | 2.00 | 2.50 | 2.40 |
| Nitrogen deposition in gain (g/day) | 40.0 | 53.6 | 60.0 | 58.4 |
| Fat deposition in gain (g/day) | 280 | 390 | 480 | 470 |
| Maintenance | | | | |
| Energy (MJ ME/day) | 7.7 | 11.7 | 15.7 | 19.9 |
| Nitrogen (g N/day) | 5.3 | 8.0 | 10.7 | 13.6 |
| Tissue gain | | | | |
| Energy content (MJ ME/day) | 17.1 | 23.5 | 28.0 | 27.4 |
| Nitrogen content (g N/day) | 40.0 | 53.6 | 60.0 | 58.4 |
| Requirement for tissue gain | | | | |
| Energy (MJ ME/day) | 21.9 | 30.1 | 35.9 | 35.1 |
| Nitrogen (g digestible N/day) | 44.4 | 59.6 | 66.7 | 64.9 |
| Total requirement from milk | | | | |
| Energy (MJ ME/day) | 29.6 | 41.8 | 51.6 | 55.0 |
| Nitrogen (g digestible N/day) | 49.7 | 67.6 | 77.4 | 78.5 |

These values relate to the energy and nitrogen requirement for tissue gain in the piglets and can be converted into that required from milk on the basis that the digestibility of nutrients in milk is 0.95 (ARC, 1981) (Table 23.1).

## THE REQUIREMENTS FOR MAINTENANCE AND MILK PRODUCTION

If the maintenance requirement of the sow is known then it is possible to determine the total requirements for both milk production and maintenance. Similarly, if dietary nutrient intake is known the rate of gain or loss of energy and nitrogen during lactation can be determined by difference. From this the rate and composition of the change in body weight of the sow can be determined.

The best estimate of the maintenance energy requirement of the lactating sow has been calculated from several recent experiments to be 471 kJ ME/kg body weight$^{0.75}$ per day (Beyer, 1986; Burlacu, Iliescu and Cárámida, 1983; Kirchgessner, 1987; Noblet and Etienne, 1987; Verstegen *et al.*, 1985). The corresponding values for the net efficiency of energy utilization were 0.72 when dietary energy was in excess, or 0.87 when dietary energy intake was insufficient to meet nutrient needs and body tissue was mobilized to provide the deficit. For estimation of the dietary nitrogen requirements, the mean nitrogen for maintenance was calculated to be 0.38 g digestible N/kg body weight$^{0.75}$ per day, with a corresponding efficiency of utilization of 0.70 (Beyer, 1986; Burlacu, Iliescu and Cárámida, 1983, 1986). It has been further assumed that the digestibility of dietary nitrogen for the lactating sow is 0.88 (Noblet and Etienne, 1987). Details of these calculations are presented in Tables 23.1 and 23.2 and show the changes in the requirements of the sow during lactation at rates of piglet gain likely to be

**Table 23.2** NUTRIENT REQUIREMENTS AND BODY TISSUE CHANGE DURING LACTATION (LITTER SIZE = 10; NO CREEP FEED PROVIDED; 160 kg SOW POSTPARTUM)

| | *Stage of lactation* | | | |
|---|---|---|---|---|
| | *Week 1* | *Week 2* | *Week 3* | *Week 4* |
| Maintenance | | | | |
| Energy (MJ ME/day) | 21.2 | 20.9 | 20.6 | 20.2 |
| Nitrogen (g digestible N/day) | 17.1 | 16.9 | 16.6 | 16.3 |
| Milk production | | | | |
| Energy (MJ ME/day) | 41.6 | 58.9 | 72.4 | 77.2 |
| Nitrogen (g digestible N/day) | 74.7 | 101.7 | 116.4 | 118.0 |
| Total requirements | | | | |
| Energy (MJ ME/day) | 62.8 | 79.8 | 93.0 | 97.4 |
| Nitrogen (g digestible N/day) | 91.8 | 118.6 | 133.0 | 134.3 |
| Intake | | | | |
| Energy (MJ ME/day) | 51 | 67 | 74 | 78 |
| Nitrogen (g digestible N/day) | 87.0 | 113.2 | 125.8 | 132.7 |
| Deficit (from body reserves) | | | | |
| Energy (MJ ME/day) | −12 | −13 | −19 | −19 |
| Nitrogen (g digestible N/day) | −4.8 | −5.4 | −7.2 | −1.6 |
| Loss of body tissue | | | | |
| Lean (kg/day) | 0.16 | 0.18 | 0.23 | 0.06 |
| Fat (kg/day) | 0.24 | 0.26 | 0.36 | 0.41 |
| Body weight (kg/day) | 0.42 | 0.46 | 0.62 | 0.49 |

achieved in practice. However the question which arises is whether the sow has the capacity to consume sufficient nutrients in the food to meet total requirements and, if not, to what extent must body tissue, that is both lean and fat, be mobilized to meet the deficit. This requires knowledge of factors influencing the voluntary feed intake of the lactating sow.

## Voluntary feed intake

There are few estimates of the true voluntary feed intake of the lactating sow since by definition it refers to *ad libitum* feeding where the animal has continuous access to a supply of fresh food and water (Cole, 1984). However the National Research Council (1987) recently summarized the available data and reported that the average feed intake of gilts was 15% less than that for sows (two or more parities), 4.36 compared with 5.17 kg/day respectively, but commented on the great variability between estimates which reflects the multitude of factors that can influence feed intake. From limited experimentation they suggested that the change in energy intake (DE, MJ/day) with time after farrowing ($t$, days) was best estimated as

$$DE = 56.066 + 2.494t - 0.072t^2 \qquad (23.2)$$

with the relationship being valid for lactation periods of 28 days or less. However, this equation has been derived from experiments which did not necessarily practise

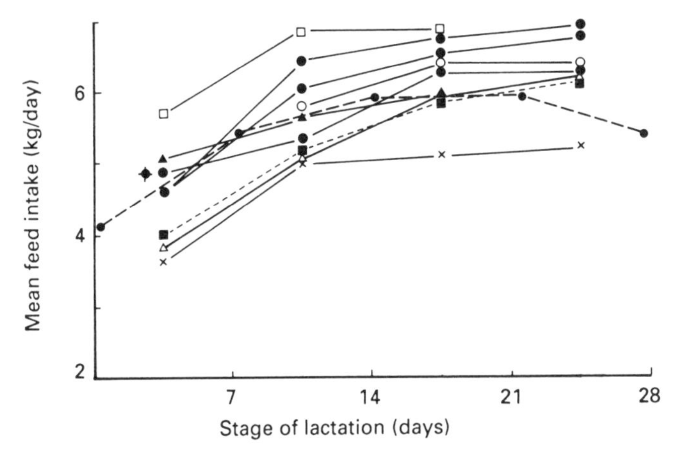


**Figure 23.4** The effect of stage of lactation on the voluntary feed intake of the sow taken from several sources (○—○, Mahan and Mangan (1975); x—x, O'Grady and Lynch (1978); ▲—▲, Stahly, Cromwell and Simpson (1979); △—△, Boyd *et al.* (1985); ✦ Moser *et al.* (1987); ●- - -●, NRC (1987); □—□, Mullan and Close (unpublished); ●—●, Zhu and Cole (unpublished); ■- - -■, predicted

true *ad libitum* feeding and does not take account of differences in body weight, level of performance or suckling intensity. On the basis of the data summarized in Figure 23.4, the approach adopted in this chapter has been to assume that a sow fed *ad libitum* will attain its maximum feed intake during the fourth week of lactation, and that mean feed intake in the preceding weeks is then a percentage of that maximum value, corresponding to 65, 85 and 95% for weeks 1, 2 and 3, respectively (Pettigrew *et al.*, 1986). In order to take account of the increase in maintenance energy requirement and hence increase in feed intake with increase in body weight, maximum intakes of 5.5, 6.1 and 6.7 kg/day were set for sows weighing 140, 160 and 180 kg postpartum, respectively, corresponding to sows in their first, second and subsequent parities and fed conventional cereal-based diets.

Although these values relate to sows suckling 10 piglets throughout lactation, it is recognized that milk yield varies with litter size (ARC, 1981). O'Grady, Lynch and Kearney (1985) indicated that the animals' voluntary feed intake increased by 0.2 kg per piglet per day during lactation, somewhat less than that of 0.5–0.6 kg per additional piglet per day calculated by Verstegen *et al.* (1985) from energy balance studies. The increase is, however, likely to be greater the lower the litter size so that at larger litter sizes the increased feed intake does not meet the additional nutrient demand and hence the growth rate of individual piglets will be reduced and/or the rate of body tissue mobilized from the sow increased.

There are a number of factors which may influence the voluntary feed intake of the lactating sow, the primary one being the environmental conditions to which the animals are exposed (O'Grady and Lynch, 1978). The environmental temperature is particularly important since both Lynch (1977) and Stansbury, McGlone and Tribble (1987) have shown that each 1°C increase in temperature above 21 and 18°C reduced feed intake by approximately 0.1 and 0.2 kg/day, respectively, with concomitant effects upon both the liveweight loss of the sow and the growth rate of the piglets. The reduced performance of the piglets resulted from the reduced milk yield of the sow and these effects illustrate the complexity of assessing and arranging the environment within the farrowing house to meet the needs of both the sow and her piglets.

There are also important associations both within and between parities which can affect feed intake of the lactating sow. It is well established that the more feed an animal consumes during pregnancy the lower its voluntary feed intake during lactation (Salmon-Legagneur and Rerat, 1962; Baker *et al.*, 1969; Harker, 1986; Mullan and Williams, 1989). Similarly, Mahan and Mangan (1975) demonstrated a close interaction between protein level fed during both pregnancy and lactation; the lower the crude protein content of the diet fed during both pregnancy and lactation the lower the voluntary feed intake during lactation. Maximum intake during lactation was achieved when a diet containing 120 g crude protein/kg was fed during pregnancy followed by one containing 180 g/kg during lactation. However, it may not necessarily be the food intake *per se* during pregnancy which influences subsequent voluntary feed intake but the body weight, rate of gain or level of fatness of the sow at farrowing (O'Grady, Lynch and Kearney, 1985; Mullan and Williams, 1989). If this is the case then feeding a diet to enhance the deposition of fat rather than body lean during pregnancy may have an adverse effect on feed intake during lactation.

Voluntary feed intake of sows may fluctuate considerably during lactation (McGrath, 1981) and there is considerable variation between animals of similar parity (Lynch and O'Grady, 1988). Data from the Moorepark herd indicate that 40% of sows during their first lactation failed to consume more than 4 kg/day whereas 45% consumed between 4.1 and 5.0 kg/day. Although the large range in individual intakes and concomitant changes in metabolism may provide an explanation of some of the variation in sow productivity in both the short and long term, there is need for an improved understanding of the control of feed intake in the lactating sow.

## Predicted changes in body weight, body composition and backfat thickness

Because nutrient intake, even under *ad libitum* feeding conditions, is often inadequate to meet metabolic needs, body tissue is mobilized to compensate for the nutrient deficit. For the practical levels of piglet performance indicated in Table 23.2, both energy and nitrogen intake were inadequate to meet requirements and hence both lean and fat reserves of the sow were mobilized throughout the 4-week lactation. The extent to which this reflects a reduction in body weight was calculated on the basis that each g of nitrogen is equivalent to 6.25 g protein and that protein gain represents 22% of the lean (Shields and Mahan, 1983). Since the energy deficit represents the loss of both protein and fat, the energy contained as fat can be calculated since each g of protein contains 23.8 kJ. The weight of fat mobilized is then calculated assuming an energetic efficiency of 0.85 and an energy content for fat of 39.7 kJ/g. Total body-weight change was calculated on the basis that lean and fat represent 95% of the total (Mullan, 1987). A more detailed account of the loss of body tissue which occurs during lactation is presented in Table 23.2. In this example, a sow weighing 160 kg postpartum and suckling 10 piglets, each gaining 200 g/day, would lose 13.8 kg body weight during a 4-week lactation, comprising a 4.4 kg reduction in the lean and an 8.9 kg reduction in the fat content of the body.

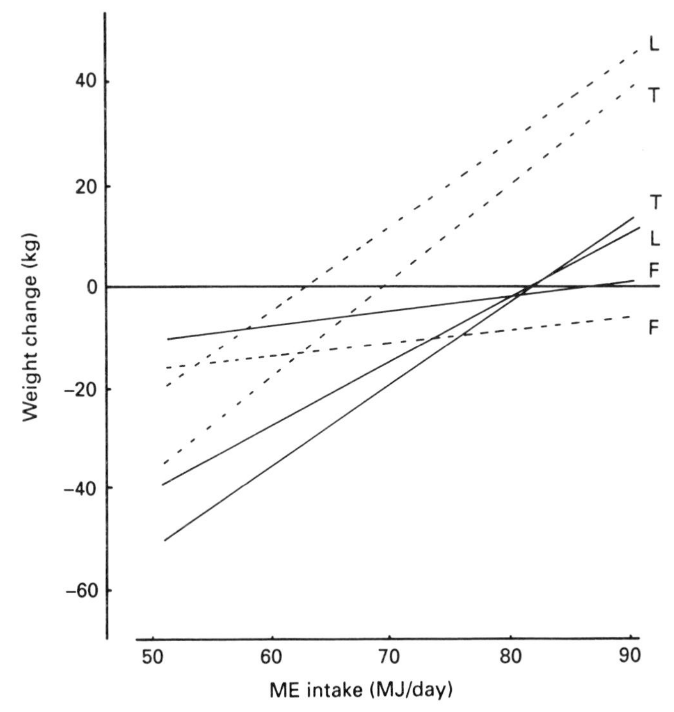


**Figure 23.5** Predicted changes in body weight and body composition of sows over a 28-day lactation when fed an isoenergetic diet containing either 140 (——) or 180 (– – –) g CP/kg at different levels of intake. T, L and F are the respective changes in body weight, body lean and body fat

Such procedures as those presented can be used to calculate the change in both body weight and body composition during lactation for a range of different nutritional circumstances and Figure 23.5 has been compiled to illustrate the effects of feeding diets of similar energy but varying protein content. Increasing the protein content of the diet leads to a higher rate of body gain at any given level of intake. At similar levels of bodyweight loss there were differences in the type and extent of body tissue mobilized so that more lean and less fat was catabolized at the lower dietary protein content. Such procedures therefore allow changes in both body weight, and body composition, to be predicted and interpolated for any given dietary circumstance, suckling intensity and body weight, and hence provide not only a greater appreciation of maternal responses but also of how changes in body composition and the dynamics of nutrient utilization during lactation may influence subsequent sow productivity.

While the model can be useful in assessing the likely consequences of a change in feeding practice, its application may be extended if the procedures can be used to predict the changes in body composition, and especially fat reserves, for additional production guidelines such as change in backfat thickness. This would provide a practical framework for testing the suggestion that it is the loss of body fat during lactation which may influence subsequent reproductive performance (Aherne and Kirkwood, 1985). However such procedures require knowledge of the relationship between body weight, backfat thickness and body fat content.

Quantitative data on the body composition of the lactating sow were not presented by ARC (1981). Indeed, in that review maternal weight loss during lactation was considered to be exclusively fat, although subsequent experimentation (King and Williams, 1984b; Etienne, Noblet and Desmoulin, 1985; Mullan and Williams, 1988), as well as the current model predictions, have shown that considerable quantities of maternal lean tissue are also mobilized during lactation. In addition, there have been a number of subsequent experiments which have examined the relationship between body lipid content and backfat thickness. Harker (1986) slaughtered gilts at mating, at days 90 and 110 of pregnancy and at weaning and found that generally there was a consistent relationship between total body lipid and backfat thickness at the different stages of the reproductive cycle. Consequently, it has been assumed in the present chapter that it is valid to pool data from sows slaughtered at different stages of reproduction (Harker, 1986; King, Spiers and Eckerman, 1986; King and Dove, unpublished; Mullan, 1987). The relationship between total chemical body lipid ($L$, kg), body weight ($W$, kg) and depth of backfat measured at the $P_2$ position ($P_2$, mm) derived from these data was:

$$L = W(0.128 + 0.0088\,P_2);\ \mathrm{RSD} = 3.46,\ R^2 = 0.76,\ n = 131 \qquad (23.3)$$

Equation 12.3 has been derived exclusively from first-parity gilts. Whittemore, Franklin and Pearce (1980) and Lee (1989) slaughtered sows following their second and third lactations, respectively, and at the same level of backfat the proportion of lipid in the body was less than that for the young sow. This may be due to a change in the chemical composition of the backfat with increased hydration of the fat depots during periods of fat mobilization (Lee, Close and Wood, 1989) or because backfat thickness and hence body reserves are lower in older sows.

From the above equation it may be calculated that a sow weighing 160 kg body weight postpartum and having a $P_2$ measurement of 20 mm would have a body fat content of 46.7 kg. On the basis of the results presented in Table 23.2 this would be reduced to 37.8 kg after a 4-week lactation, equivalent to a 5 mm decrease in $P_2$. This reduction in $P_2$ is consistent with the results of Eastham *et al.* (1988) who measured a 6.1 mm change in $P_2$ of sows which lost 15 kg body weight over a 28-day lactation, those of King and Dunkin (1986a) where a 19.6 kg loss in body weight was associated with a 5.7 mm change in $P_2$ and is also comparable with that reported by King (1987). It is important to note, however, that the above relationship has been derived from first-parity sows fed conventional diets and hence any predictions with older sows should be interpreted cautiously.

## Lactation and subsequent reproduction

Reproductive failure is the major reason for sows being culled prematurely from the breeding herd (Dagorn and Aumaitre, 1979). A major component of this can be attributed to a delayed resumption of oestrus activity after weaning particularly with young sows (King, 1987) and to the low levels of body reserves in sows at the start of lactation (King and Dunkin, 1986a; Mullan and Williams, 1989). The loss of body fat (Aherne and Kirkwood, 1985) and body protein (King, 1987) during lactation has been implicated as being responsible for the cessation of reproductive function, but it is not clear what signals (hormonal or metabolic) mediate the response to the nutritional state of the sow (Britt, Armstrong and Cox, 1988).

The metabolic state of the sow during lactation may also influence subsequent litter size. The failure of young sows to revert from a catabolic phase during lactation to an anabolic phase immediately after weaning has been suggested by Brooks (1982) to be the primary reason for some sows producing smaller litters in their second parity compared with the first. The beneficial effect on litter size of a delay in the age at which a gilt is first mated (Mercer and Francis, 1988) and a delay in the time of remating after the first lactation may indicate an association between the nutritional state of the animal and ovulation rate. Similarly the practice of split-weaning, that is weaning of the heaviest piglets some days before their smaller litter mates, has been shown to be beneficial in reducing the weaning to oestrus interval (Britt, 1988). This again demonstrates the possible association between the nutritional, metabolic and physiological state of the sow and its subsequent reproductive performance.

## Conclusions

The purpose of this chapter has been to describe a factorial procedure to predict the response of the lactating sow to nutritional manipulation. This objective has been achieved but there is little empirical information to validate the predictions and this must await further experimentation. However, while such exercises are useful it is

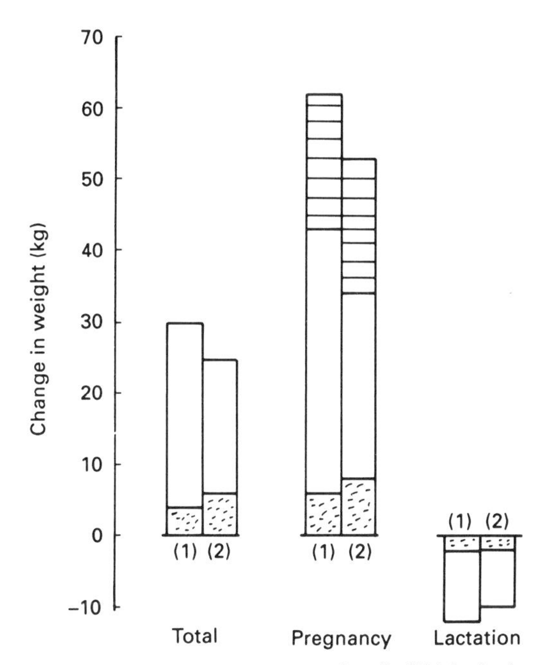


**Figure 23.6** Predicted response of a gilt (120 kg body weight at mating) during its first parity, and fed 2.0 kg/day throughout pregnancy and *ad libitum* during lactation: (1) a diet containing 160 g CP/kg during both pregnancy and lactation or (2) a diet containing 130 g CP/kg during pregnancy and 160 g CP/kg during lactation. All diets contain 13 MJ DE/kg. Total is the combined net response during both pregnancy and lactation. □, body lean; ⊟, body fat; ⊟, products of conception

recognized that the response of the sow during lactation cannot be considered in isolation from events within the same or other parities. It is therefore important to consider the reproductive cycle as a whole, that is pregnancy together with lactation, since O'Grady (1980) and Cole (1982) have suggested that it is the balance of nutrients throughout the whole reproductive period that is important. This has been attempted by extending the predictive model developed by Williams, Close and Cole (1985) for the pregnant sow to include that developed in this chapter for the lactating animal. A typical example of the response obtained is presented in Figure 23.6 and shows the extent to which body weight, and lean and fat reserves, change during both pregnancy and lactation under different dietary situations. In this particular example the changes which result when a single diet is fed during the whole reproductive cycle (160 g CP/kg) are compared with those which occur when a two-diet system in which separate diets are fed in both pregnancy (130 g CP/kg) and lactation (160 g CP/kg). Feeding the single diet system resulted in a higher gain in body lean and a lower gain in body fat, but a greater total bodyweight gain than the two-diet system. However when extended over four parities the predicted increases in body weights were similar, 111 and 107 kg, although lean gain was approximately 10 kg higher and fat gain 10 kg lower with the single compared with the two-diet system. While the consequences of such changes in body composition are not yet known, relating them to changes in reproductive performance should allow additional practical indices and guidelines to be developed which facilitate a better understanding of the associations between nutrition, metabolism and reproductive performance of sows.

## References

AGRICULTURAL RESEARCH COUNCIL (1981). *The Nutrient Requirements of Pigs.* Commonwealth Agricultural Bureaux, Slough

AHERNE, F.X. and KIRKWOOD, R.N. (1985). *Journal of Reproduction and Fertility,* (Supplement) **33**, 169–183

BAKER, D.H., BECKER, D.E., NORTON, H.W., SASSE, C.E., JENSEN, A.H. and HARMON, B.G. (1969). *Journal of Nutrition,* **97**, 489–495

BERGE, S. and INDREBØ, T. (1954). *Meldinger fra Norges Landbrukshøgskole,* **34**, 481–500

BEYER, M. (1986). *Untersuchungen zum Energie- und Stoffumsatz von graviden und laktierenden Sauen sowie Saugferkeln – ein Beitrag zur Prazisierung des Energie- und Proteinbedarfes.* Promotionsarbeit, aus dem Forschungszentrum fur Tierproduktion, Dummerstorf-Rostock

BOYD, R.D., HARKINS, M., BAUMAN, D.E. and BUTLER, W.R. (1985). *Proceedings of the 1985 Cornell Nutrition Conference*, pp. 10–19. Cornell University, Ithaca, New York

BROOKS, C.C., FONTENOT, J.P., VIPPERMAN, P.E., THOMAS, H.R. and GRAHAM, P.P. (1964). *Journal of Animal Science,* **23**, 1022–1026

BROOKS, P.H. (1982). In *Control of Pig Reproduction*, pp. 211–224. Ed. Cole, D.J.A. and Foxcroft, G.R. Butterworths, London

BRITT, J.H. (1988). *Journal of Animal Science,* **63**, 1288–1296

BRITT, J.H., ARMSTRONG, J.D. and COX, N.M. (1988). In *11th International Congress on Animal Reproduction and Artificial Insemination*, pp. 117–125. Dublin

BURLACU, G., ILIESCU, M. and CÁRÁMIDA, P. (1983). *Archiv für Tierernahrung,* **33**, 23–45

BURLACU, G., ILESCU, M. and CÁRÁMIDA, P. (1986). *Archiv for Animal Nutrition,* **36**, 803–825

COLE, D.J.A. (1982). In *Control of Pig Reproduction*, pp. 603–619. Ed. Cole, D.J.A. and Foxcroft, G.R. Butterworths, London

COLE, D.J.A. (1984). In *Fats in Animal Nutrition*, pp. 301–312. Ed. Wiseman, J. Butterworths, London

DAGORN, J. and AUMAITRE, A. (1979). *Livestock Production Science,* **6**, 167–177

DANIELSEN, V. and NIELSEN, H.E. (1984). *Paper presented at the 35th Annual Meeting of the European Association for Animal Production.* The Hague

EASTHAM, P.R., SMITH, W.C., WHITTEMORE, C.T. and PHILLIPS, P. (1988). *Animal Production,* **46**, 71–77

ELLIOT, J.I. and LODGE, G.A. (1977). *Canadian Journal of Animal Science,* **57**, 141–150

ELSLEY, F.W.H. (1971). In *Lactation*, pp. 393–411. Ed. Falconer, I.R. Butterworths, London

ELSLEY, F.W.H. and MacPHERSON, R.M. (1972). In *Pig Production*, pp. 417–434. Ed. Cole, D.J.A. Butterworths, London

ETIENNE, M., NOBLET, J. and DESMOULIN, B. (1985). *Reproduction Nutrition Development,* **25**, 341–344

HARKER, A.J. (1986). *Nutrition of the Sow*, PhD Thesis, University of Nottingham

JOHNSTON, L.J., ORR, D.E., TRIBBLE, L.F. and CLARK, J.R. (1986). *Journal of Animal Science,* **63**, 804–814

KING, R.H. (1987). *Pig News and Information,* **8**, 15–22

KING, R.H. and DUNKIN, A.C. (1986a). *Animal Production,* **42**, 119–125

KING, R.H. and DUNKIN, A.C. (1986b). *Animal Production,* **43**, 319–325

KING, R.H. and WILLIAMS, I.H. (1984a). *Animal Production,* **38**, 241–247

KING, R.H. and WILLIAMS, I.H. (1984b). *Animal Production,* **38**, 249–256

KING, R.H., SPIERS, E. and ECKERMAN, P. (1986). *Animal Production,* **43**, 167–170

KING, R.H., WILLIAMS, I.H. and BARKER, I. (1984). *Proceedings of the Australian Society of Animal Production,* **15**, 412–415

KIRCHGESSNER, M. (1987). In *European Association for Animal Production – Tenth Symposium on Energy Metabolism*, pp. 362–367. Ed. Moe, P.W., Tyrrell, H.F. and Reynolds, P.J. Rowman and Littlefield, Virginia

LEE, P.A. (1989). *Animal Production* (in press)

LEE, P.A., CLOSE, W.H. and WOOD, J.D. (1989). *Animal Production* (in press)

LODGE, G.A. (1962). In *Nutrition of Pigs and Poultry*, pp. 224–237. Ed. Morgan, J.T. and Lewis, D. Butterworths, London

LODGE, G.A. (1972). In *Pig Production*, pp. 399–416. Ed. Cole, D.J.A. Butterworths, London

LYNCH, P.B. (1977). *Irish Journal of Agricultural Research,* **16**, 123–130

LYNCH, P.B. and O'GRADY, J.F. (1988). *Proceedings of the International Conference – Improvement of Reproductive Efficiency in Pigs*, Italy (in press)

McGRATH, F. (1981). *Level of Feeding of Lactating Sows*, Thesis, University of Nottingham

MAHAN, D.C. and MANGAN, L.T. (1975). *Journal of Nutrition,* **105**, 1291–1298

MERCER, J.T. and FRANCIS, M.J.H. (1988). *Animal Production,* **46**, 493

MOSER, R.L., CORNELIUS, S.G., PETTIGREW, J.E., HANKE, H.E., HEEG, T.R. and MILLER, K.P. (1987). *Livestock Production Science,* **16**, 91–99

MULLAN, B.P. (1987). *The Effect of Body Reserves on the Reproductive Performance of First-Litter Sows*, PhD Thesis, University of Western Australia

MULLAN, B.P. and CLOSE, W.H. (1989). *Animal Production* (in press)

MULLAN, B.P. and WILLIAMS, I.H. (1988). *Animal Production,* **46**, 495

MULLAN, B.P. and WILLIAMS, I.H. (1989). *Animal Production,* **48**, 449–457

NATIONAL RESEARCH COUNCIL (1987). *Predicting Feed Intake of Food-Producing Animals.* National Academy Press, Washington, DC

NATIONAL RESEARCH COUNCIL (1988). *Nutrient Requirements of Swine – Ninth Revised Edition.* National Academy Press, Washington, DC

NELSSEN, J.L., LEWIS, A.J., PEO, E.R. and CRENSHAW, J.D. (1985). *Journal of Animal Science,* **61**, 1164–1171

NOBLET, J. and ETIENNE, M. (1987). *Journal of Animal Science,* **64**, 774–781

O'GRADY, J.F. (1980). In *Recent Advances in Animal Nutrition – 1980*, pp. 121–131. Ed. Haresign, W. Butterworths, London

O'GRADY, J.F., ELSLEY, F.W.H., MacPHERSON, R.M. and McDONALD, I. (1973). *Animal Production,* **17**, 65–74

O'GRADY, J.F. and LYNCH, P.B. (1978). *Irish Journal of Agricultural Research,* **17**, 1–5

O'GRADY, J.F., LYNCH, P.B. and KEARNEY, P.A. (1985). *Livestock Production Science,* **12**, 355–365

PETTIGREW, J.E., CORNELIUS, S.G., EIDMAN, V.R. and MOSER, R.L. (1986). *Journal of Animal Science,* **63**, 1314–1321

REESE, D.E., MOSER, B.D., PEO, E.R., LEWIS, A.J., ZIMMERMAN, D.R., KINDER, J.E. and STROUP, W.W. (1982). *Journal of Animal Science,* **55**, 590–598

REESE, D.E., PEO, E.R. and LEWIS, A.J. (1984). *Journal of Animal Science,* **58**, 1236–1244

SALMON-LEGAGNEUR, E. and RERAT, A. (1962). In *Nutrition of Pigs and Poultry*, pp. 207–223. Ed. Morgan, J.T. and Lewis, D. Butterworths, London

SHIELDS, R.G. and MAHAN, D.C. (1983). *Journal of Animal Science,* **57**, 594–603

STAHLY, T.S., CROMWELL, G.L. and SIMPSON, W.S. (1979). *Journal of Animal Science,* **49**, 50–54

STANSBURY, W.F., McGLONE, J.J. and TRIBBLE, L.F. (1987). *Journal of Animal Science,* **65**, 1507–1513

VERSTEGEN, M.W.A., MESU, J., VAN KEMPEN, G.J.M. and GEERSE, C. (1985). *Journal of Animal Science,* **60**, 731–740

WHITTEMORE, C.T., FRANKLIN, M.F. and PEARCE, B.S. (1980). *Animal Production,* **31**, 25–31

WILLIAMS, I.H., CLOSE, W.H. and COLE, D.J.A. (1985). In *Recent Advances in Animal Nutrition – 1985*, pp. 133–147. Ed. Haresign, W. and Cole, D.J.A. Butterworths, London

YANG, H., PHILLIPS, P., WHITTEMORE, C.T. and EASTHAM, P.R. (1988). *Animal Production,* **46**, 494

24

# NUTRITION OF THE WORKING BOAR

W.H. CLOSE
*129 Barkham Rd, Wokingham, Berks, UK*
*and*
F.G. ROBERTS†

## Introduction

Despite the importance of the boar to herd fertility, it has received little attention and is the most neglected animal in the pig unit. Boars are often kept in pens that are too small or badly designed and in poor climatic and social environments. Hygiene can be poor and the nutrition of the boar has received scant attention. Information relating to the nutritional requirements and responses of the breeding boar is therefore limited and this topic received little mention in the reviews of ARC (1981) and NRC (1988). Many recommendations have therefore been based on the breeding sow.

If boars are to be reared for breeding purposes, then physical soundness and future reproductive performance are as important as good growth rate. Young boars are normally selected according to an index which includes such characteristics as growth rate, appetite, feed conversion efficiency, lean tissue growth rate, carcase quality and breeding potential. They are normally fed to appetite and it is assumed that this does not prejudice subsequent reproductive capacity. It may however affect their physical ability to perform since the tendency to leg weakness may be exacerbated by high growth rates and by feeding to appetite (Grondalen, 1974; Hanssen and Grondalen, 1979; Kesel *et al.*, 1983). This suggests that the priorities for nutrients and the nutritional requirements of the breeding boar may differ markedly from those bred for meat production or from the breeding sow. The purpose of this chapter is therefore to review the nutritional requirements and responses of the breeding boar at all stages of development, to assess whether nutrition influences sexual development and reproductive capacity, and to make recommendations on appropriate feeding strategies to ensure good reproductive performance.

## Nutrition during rearing

Under normal conditions boars attain puberty between 5 and 8 months of age, when they weigh 80–120 kg body weight. Age is more important than body weight in determining the onset of puberty (Einarsson, 1975). Spermatogonia first appear in the testes of the boar at 2 months of age, spermatocytes at 3 months and spermatozoa at 5 months (Westendorf and Richter, 1977; Hughes and Varley,

†Present address: Department of Food and Nutritional Sciences, Kings College, Kensington, London W8 7AH.

1980). Poppe *et al.* (1974b) trained 122-day-old boars to mount a 'dummy' sow and noted that the first ejaculate appeared at 140 days of age when the animals weighed 55–65 kg body weight. Fertility is low immediately following puberty but increases to a maximum at 15–18 months of age when the ejaculate contains 20–100 × $10^9$ sperm in some 200–400 ml of semen. There may be differences in the age of attainment of puberty, and Sellier, Dufour and Rousseau (1973) and Kim *et al.* (1976) have shown that crossbred boars reach puberty earlier than their purebred counterparts, by as much as 40 days. When rearing boars, the objective is, therefore, to allow them to attain puberty at a normal rate rather than to stimulate the precocious attainment of sexual development.

Nutrition during the rearing period may influence the attainment of puberty. Dutt and Barnhart (1959) fed young growing boars at 100, 70 and 50% of the NRC requirements (Beeson *et al.*, 1953) and observed that the age of puberty ranged between 203 and 212 days. The body weights at the time of puberty were significantly different between the groups, being 101, 78 and 61 kg, respectively. Kim *et al.* (1977) also observed that the attainment of puberty was delayed when the feed intake of the young boar was reduced. A 30% reduction in feed intake resulted in a 42 and 30 day delay in puberty for purebred and crossbred boars, respectively. Althen, Gerrits and Young (1974) fed diets containing 100, 150 and 200 g crude protein/kg and concluded that reproductive development in young boars was completed by 230 days of age. After this age the diet containing 100 g protein/kg was utilized without detrimental effect on either the content of pituitary gonadotrophins or development of the reproductive organs. Uzu (1979) concluded that boars fed a low-protein diet (120 g protein/kg diet) had significantly lower growth rates than those receiving diets containing 180 or 230 g crude protein/kg. Semen could be collected from boars in all groups after a body weight of 88 kg was reached. However the age of first semen collection varied between treatment groups, with boars fed the 120 g protein/kg diet being 193 days of age compared with 182 and 177 days for those fed the 180 and 230 g crude protein/kg diets, respectively. There was no long-term effect of dietary treatment on sexual development, suggesting that the onset of reproductive activity does not require a higher supply of protein than that necessary for the optimization of growth rate.

This evidence suggests that nutrition during the rearing period can influence both the age of puberty and the rate of sexual development in young boars. However, unless severely undernourished this does not appear to impose any lasting damaging effect upon reproductive capacity other than the obvious effects of the growth and mature body size of the animal. In practice, most young boars are offered feed close to their appetite potential during rearing and at this level there is unlikely to be any effect upon sexual development or subsequent reproductive capacity.

## The effects of nutrition on reproductive performance

The reproductive performance of a working boar can be described by three separate characteristics: (i) libido, (ii) sperm production, and (iii) viability and fertilizing capacity of the sperm cells.

Although it is acknowledged that nutrition has a profound effect on the sexual and reproductive development in the pubescent animal (Phillips and Andrews, 1936; Niwa, 1954) current comments will be confined to the nutrition of the mature animal.

LIBIDO

Libido is usually assessed in terms of the proportion of successful mountings by a boar, where a successful mounting results in the production of an ejaculate into a sow or dummy. Libido of boars has been scored empirically (Berger *et al.*, 1981) as follows:

0 = no sexual interest
1 = some sexual interest
2 = great deal of interest
3 = one or more false mounts
4 = one correct mount
5 = repeated correct mounts
6 = penis extension
7 = intromission
8 = ejaculation

Most libido scores are within the range 3.5 to 5.5 (Berger *et al.*, 1981; Ju, Cheng and Yen, 1985).

The few studies reporting the role of nutrition on libido in boars fail, in general, to show any appreciable effects with varying energy and protein intakes. Stevermer *et al.* (1961) subjected boars to three nutritional regimes over a 15-month period and measured several reproductive characteristics. During the course of the experiment the boars gained 136 and 39 kg on the high and medium levels of feeding, but lost 60 kg on the low level of feeding, respectively. They concluded that libido was not consistently affected by treatment, except in the last month of the experiment when restrictedly-fed boars refused to serve the artificial vagina during a 2-week period. Dutt and Barnhart (1959) had previously reported no effects of level of feeding on libido although these experiments were of shorter duration than those of Stevermer *et al.* (1961). Ju, Cheng and Yen (1985) found no significant difference on libido of boars fed varying levels of lysine or methionine. Similarly Kemp *et al.* (1988) reported no effects of different levels of dietary lysine, although other reproductive characteristics were affected. These results therefore suggest that it is only in prolonged conditions of under- or over-feeding, or in extremes of body condition, that the libido of boars is impaired.

A phenomenon sometimes overlooked in pig production systems is that a reduction in libido, or rather the actual physical inability to mount a sow, is more likely to result from sustained high energy intakes. This results in an increase in growth rate, larger body size, higher body fatness and in lethargy. It is, therefore, preferable to keep boars relatively light so that they can serve young gilts and to minimize leg weakness (Westendorf and Richter, 1977).

SPERM PRODUCTION

The process of sperm production (spermatogenesis), including the process of the transformation of a stem cell to a spermatozoon, takes between 25 (Singh, 1962) and 34 (Swierstra, 1968) days. In addition, the passage of sperm cells through the epididymis requires a further 10–14 days (Singh, 1962). Thus, feeding studies reporting a preliminary period of less than 6 weeks will not reflect the true response of the experimental treatment *per se* but some residual 'pre-trial' effects.

The nutrition of the animal can influence semen quantity and this needs to be considered in conjunction with aspects of the environment since it is known that

**Table 24.1** THE INFLUENCE OF PLANE OF NUTRITION ON SEMEN CHARACTERISTICS OF BOARS

| *Semen characteristics (average per ejaculate)* | *Plane of nutrition*[a] | | |
|---|---|---|---|
| | *100%* | *70%* | *50%* |
| Volume (ml) | | | |
| Total | 173±24.2[b] | 137±21.1 | 121±9.5[c] |
| Minus gelatinous material | 145±19.4 | 119±17.0 | 102±7.9 |
| Motile cells (%) | 84±1.7 | 82±2.2 | 84±1.7 |
| Sperm cells concentration (1000/mm$^3$) | 378±26 | 417±40 | 398±21 |
| Abnormal cells (%) | 6.5±0.4 | 8.6±0.80 | 7.5±0.6 |

(From Dutt and Barnhart, 1959)
[a]Based on NRC requirements (Beeson *et al.*, 1953)
[b]Each value is the average of nine ejaculates (three weekly collections from three boars on each plane of nutrition)
[c]Significantly different from the 100% plane of nutrition at the 5% level

ambient temperature can affect semen production (see later). In the study of Dutt and Barnhart (1959) plane of nutrition was found to significantly affect semen production in young boars. There was a reduction in overall semen production in animals fed a medium (70%) or low (50%) plane of nutrition compared with those on a high plane of nutrition (100% of NRC requirements, Beeson *et al.*, 1953). Boars on the higher plane of nutrition produced larger semen volumes, but motility, sperm cell concentration and percentage of morphologically abnormal cells were not significantly affected (Table 24.1). The proportion of total semen volume composed of gelatinous material was 0.153, 0.154 and 0.157 for the three groups, respectively, and was unaffected by nutritional treatment. Stevermer *et al.* (1961) also concluded that widely varying planes of nutrition can be tolerated by boars without detrimental effect on spermatozoa, since neither motility nor fertility was affected over a prolonged period. Yen and Yu (1985) also obtained no significant response to variations in energy intake *per se* other than to an increase in the level of feeding.

The number of ejaculated sperm cells per week from boars on three different feeding levels is illustrated in Figure 24.1 (Kemp, den Hartog and Grooten, 1989). It is interesting to note that the differences between treatments only became significant after 8 weeks of the experiment, reflecting the 6-week adaptation period required to remove effects associated with the animals' previous nutritional state and resulting from the length of time required for the maturation and transportation of spermatozoa. Those animals initially receiving the high and low levels of feed had their feeding levels reversed during a second treatment period whereas those initially receiving the medium feeding level continued on this regime. Eight weeks after this change, the differences in the number of sperm ejaculated in each of the treatment groups were no longer significant, indicating the direct effects of nutrition on semen production. This study has particular relevance to the keeping of pigs in AI centres since many are fed at a level below the medium level used in these experiments. Thus, in terms of achieving maximum sperm output, these feeding regimes may be too low.

Several studies have indicated the importance of protein, and more specifically the amino acids lysine, methionine and cystine, on the number of sperm cells

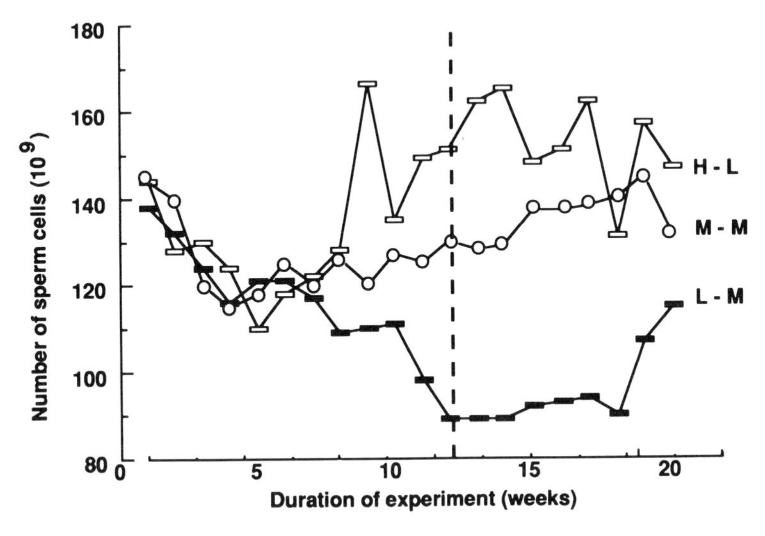


**Figure 24.1** The influence of feeding level on sperm cell production of working boars. L = low, M = medium and H = high feeding level. The broken line represents the time at which feeding level was changed. (Adapted from Kemp, den Hartog and Grooten, 1989)

produced. There appears to be a minimum level of protein necessary to ensure maximal sperm production since low protein levels have been shown to reduce the naumber of sperm cells ejaculated (Kolenko, 1977; Yurin, 1981; Ovchinnikov, 1984). For example, Hühn (1969) fed diets containing 121, 161 and 222 g digestible crude protein/kg dry matter and observed a significant effect of dietary protein level on total sperm production at intakes below 161 g/kg. There was a small but non-significant effect at intakes above this level. However, in subsequent studies opinion is divided as to the effects of specific amino acids. Poppe *et al.* (1974a,b) carried out an often-cited experiment in which AI boars received either a low-protein diet (507 g/d) or a high protein diet (744 g/d), unsupplemented or supplemented with synthetic lysine or methionine. The animals were then kept on a low mating frequency for 6–8 weeks followed by a period of high mating frequency for a similar period. At the low mating frequency, sperm production between the various experimental groups was not significantly different. However, in the intensively-used boars, sperm production was improved by the addition of extra protein and/or amino acids, particularly the addition of methionine. This suggested an increased requirement for sulphur-containing amino acids during periods of extreme reproductive activity. Van der Kerk and Willems (1985), on the other hand, found no beneficial effects of protein, lysine or methionine supplementation in AI boars fed a standard basal diet. The quality and number of sperm cells produced was not improved and supplementation with synthetic methionine was actually found to have some negative effects, particularly on the morphology, and therefore the viability, of the sperm cells. Similarly, Medding and Nielsen (1977) and Ju, Cheng and Yen (1985) found no effect of protein or amino acid supplementation of the diet on semen quality. However, in both of these later studies the mating frequency approximated to the low intensity usage in the studies of Poppe *et al.* (1974a,b).

**Table 24.2** THE INFLUENCE OF AMINO ACID INTAKE ON SEMEN PRODUCTION IN BOARS, TAKEN FROM VARIOUS SOURCES

| *Source*[a] | *Age/body weight* | *Lysine intake* (g/d) | *Methionine + cystine intake* (g/d) | *Sperm production* (×10⁹) | | |
|---|---|---|---|---|---|---|
| | | | | (1) | | (2) |
| 1. | 8–9 months | 42.1 | 20.7 | 74.2 | | 49.1 |
| | (1 = low usage) | 54.1 | 20.7 | 84.0 | | 59.7 |
| | (2 = high usage) | 42.1 | 30.7 | 90.3 | | 64.4 |
| | 15–16 months | 50.7 | 22.6 | 84.0 | | 40.5 |
| | | 62.7 | 34.6 | 75.6 | | 50.4 |
| | | | | (3) | | (4) |
| 2. | 120–170 kg (3) | 42.1 | 20.7 | 71.6 | | 48.5 |
| | | 54.1 | 20.7 | 78.2 | | 52.1 |
| | 170–250 kg (4) | 42.1 | 36.7 | 81.4 | | 61.9 |
| | | 29.7 | 15.0 | 86.1 | | 48.4 |
| | | | | (5) | (6) | (7) |
| 3. | 4–5 months (5) | 18.9–29.2 | 8.6–15.9 | 4.3 | 15.5 | 20.3 |
| | 5–7 months (6) | 18.9–29.2 | 11.6–27.9 | 6.3 | 17.0 | 47.5 |
| | 7–8 months (7) | 24.9–37.4 | 11.6–27.9 | 5.3 | 18.3 | 46.2 |
| 4. | 6–7 months | 11.0 | 11.8 | | 29 | |
| | | 18.9 | 11.8 | | 24 | |
| | | 11.0 | 17.7 | | 23 | |
| | | 18.9 | 17.9 | | 25 | |
| 5. | Adult boars | 18.5 | 15. | | 53.9 | |
| | | 18.5 | 17.5 | | 54.5 | |
| | | 22.0 | 19.5 | | 51.5 | |
| 6. | Adult boars | 18.4 | 11.9 | | 38.0 | |
| | 210–270 kg | 32.4 | 21.9 | | 37.3 | |

[a]1. Hühn *et al.* (1973)
2. Poppe *et al.* (1974a)
3. Poppe *et al.* (1974b)
4. Ju *et al.* (1985)
5. Van der Kerk and Willems (1985)
6. Kemp *et al.* (1988)

It is difficult to reconcile the results of many of these experiments since they were conducted with animals of different age and body weight, of different mating frequencies, fed diets containing different levels of protein and amino acids which were supplemented with varying levels of lysine and methionine (Table 24.2). However, with the exception of the work of Hühn *et al.* (1973) and Poppe *et al.* (1974a,b) the evidence suggests that supplementation with lysine or methionine has little effect on sperm production in breeding boars, unless severely over-worked, and it is reasonable to conclude that current levels of dietary protein and amino acids are sufficient to ensure good semen characteristics.

VIABILITY AND FERTILIZING CAPACITY OF SPERM CELLS

Arguably, the only true method of assessing sperm viability, and therefore fertilizing capacity, is to record the number of successful pregnancies in sows inseminated with a single ejaculate from a given boar. However, this becomes

**Table 24.3** FERTILITY RESULTS IN GILTS BRED BY BOARS ON THREE PLANES OF NUTRITION

| *Item* | *Plane of nutrition*[a] | | |
|---|---|---|---|
| | *100%* | *70%* | *50%* |
| No. gilts bred | 9 | 9 | 9 |
| No. pregnant 25 to 28 days | 7 | 8 | 7 |
| Total no. corpora lutea (all gilts) | 102 | 103 | 106 |
| No. of normal embryos | 56 | 65 | 59 |
| No. of degenerate embryos | 3 | 4 | 4 |
| Per cent of corpora lutea represented by normal embryos: | | | |
| All gilts bred | 54.9 | 63.1 | 55.7 |
| Pregnant gilts only | 75.5 | 74.1 | 70.2 |

(From Dutt and Barnhart, 1959)
[a]Based on NRC requirements (Beeson *et al.*, 1953)

extremely difficult to implement in practice and it is more common to assess qualitative parameters of semen production in terms of the percentage of motile cells, the vitality of the movement and abnormalities arising in acrosomal morphology.

Dutt and Barnhart (1959) found that, on average, 70% of the sperm were motile in a single ejaculate, whilst the occurrence of morphological abnormalities varied between 5 and 13%. However, plane of nutrition had no effect on sperm viability or on subsequent sow fertility (Table 24.3). Variation between boars within a given plane of nutrition was greater than that between planes of nutrition, and fertility of the boars did not differ significantly between groups.

Stevermer *et al.* (1961) and Kemp *et al.* (1989) also failed to demonstrate any effects of the level of nutrition on semen quality and viability of boars. In the latter study, the percentage of motile cells, non-return percentage and vitality of moving sperm (e.g. vigorous, straightforward movement or slow, spasmodic movement) were assessed over two separate periods during which each boar was on a different plane of nutrition.

The effects of protein and amino acids in the diet on semen quality is uncertain and in some cases conflicting. For example, Zaripova and Shakirov (1978) found a reduction in the percentage of abnormal sperm when extra lysine or lysine plus methionine was fed to working boars. However, Poppe *et al.* (1974a,b), Ju, Cheng and Yen (1985) and Van der Kerk and Willems (1985) reported no positive effect on any reproductive characteristic following amino acid supplementation in the diet. In fact the latter authors reported an increased incidence of abnormalities in acrosomal morphology and a higher percentage of returns in boars fed additional methionine.

Most studies have investigated the response of semen characteristics to protein and amino acid levels in the diet under a single set of production conditions. Little attention has been given to different environmental factors and production variables such as the intensity of boar use. For example, Hühn *et al.* (1973) investigated the effect of amino acid supplementation on semen properties of both young and old boars, specifically within the context of frequency of use and the interval between services. They found that in both young and old boars an increase in daily lysine and methionine supply improved various semen characteristics, including semen production capacity, when they were used infrequently and assessed by 'daily semen harvests'. This improvement compensated in part for the linear decrease in semen yield observed as the interval between matings extended beyond 7 days. However, amino acid supplementation had little effect on the qualitative parameters of the semen. Realistically, the observation that supplementation with sulphur-containing amino acids improves semen yield in animals of infrequent sexual activity is of limited practical application since boars would generally be used at least twice per week within a conventional production system.

All the above studies therefore indicate that energy and protein intake *per se* have a limited effect on semen quality, and therefore fertilizing capacity, in the working boar. This does not necessarily mean that nutrition has no role in determining semen quality. Certain anti-nutritional properties of feeds may influence semen production. For example, zearalenone, which binds with oestrogens evoking a similar physiological response to that elicited by other oestrogens, has been reported to decrease libido in boars fed mouldy corn (Bristol and Djurickovic, 1971; Berger *et al.*, 1981). Similarly, the intake of high levels of zearalenone (>6000 ppm) by prepubertal boars has been shown to reduce the onset of puberty and cause a reduction in testes size with corresponding reductions in semen quality (Christensen *et al.*, 1972). Aflatoxin $B_1$ has also been shown to affect adversely semen characteristics and fertility in breeding boars (Picha, Cerovsky and Pichova, 1986). Those animals with the highest levels of aflatoxin $B_1$ residues in their sperm experienced more fertility disorders, lower sperm concentration, lower survival of sperm and a larger proportion of sperm with morphological abnormalities.

## Requirements for energy

The daily energy requirements for boars can be assessed as the summation of the following components:

Requirement for maintenance
Requirement for body gain
Requirement for semen production
Additional requirement associated with mating activity
Requirement for extra heat production, when kept below their lower critical temperature

### MAINTENANCE

Although it is generally recognized that the maintenance energy requirement of growing boars is greater than that for females or castrated males (Fuller, 1980; Ludvigsen, 1980), estimates for breeding boars are limited and have normally been based on values obtained from the mature sow (for example, Kemp, 1989). Several recent estimates have, however, been determined for growing entire male animals and these have been extended to accommodate the breeding boar (Table 24.4). The mean of these values has been taken as the best estimate of the energy requirement for maintenance of animals of different body weights. The overall relationship between the energy requirements for maintenance (M, MJ/day) and body weight (W, kg) was calculated as:

$$M = 0.763\, W^{0.665} \qquad (24.1)$$

**Table 24.4** THE MAINTENANCE ENERGY REQUIREMENTS (M) OF BREEDING BOARS (MJ ME/d)

| *Body weight* (W, kg) | *Source*[a] | | | | | *Mean* |
|---|---|---|---|---|---|---|
| | *1* | *2* | *3* | *4* | *5* | |
| 50 | 11.5 | 10.4 | 11.0 | 9.3 | 9.3 | 10.3 |
| 100 | 17.7 | 15.9 | 16.6 | 15.6 | 15.7 | 16.3 |
| 150 | 22.7 | 20.4 | 21.2 | 21.1 | 21.2 | 21.3 |
| 200 | 27.2 | 24.3 | 25.2 | 26.3 | 26.3 | 25.9 |
| 250 | 31.2 | 27.8 | 28.8 | 31.1 | 31.1 | 30.0 |
| 300 | 34.9 | 31.1 | 32.2 | 35.5 | 35.7 | 33.9 |
| 350 | 38.4 | 36.2 | 35.3 | 39.9 | 40.1 | 38.0 |

[a]1. Jentsch *et al.* (1989) $M = 1.017\ W^{0.62}$
2. Laswai *et al.* (1991) $M = 0.959\ W^{0.61}$
3. Noblet, Karege and Dubois (1989) $M = 1.050\ W^{0.60}$
4. Hofstetter and Wenk (1987) $M = 0.493\ W^{0.75}$
5. AFRC (1990) $M = 0.495\ W^{0.75}$

and values derived from this equation have been used in the factorial estimation of energy requirements.

## GROWTH

The rate of growth desirable for a breeding boar is a controversial subject. On the one hand, it is desirable to keep the boar relatively light so that he can serve young gilts without causing injury either to the gilts or himself. On the other hand, an

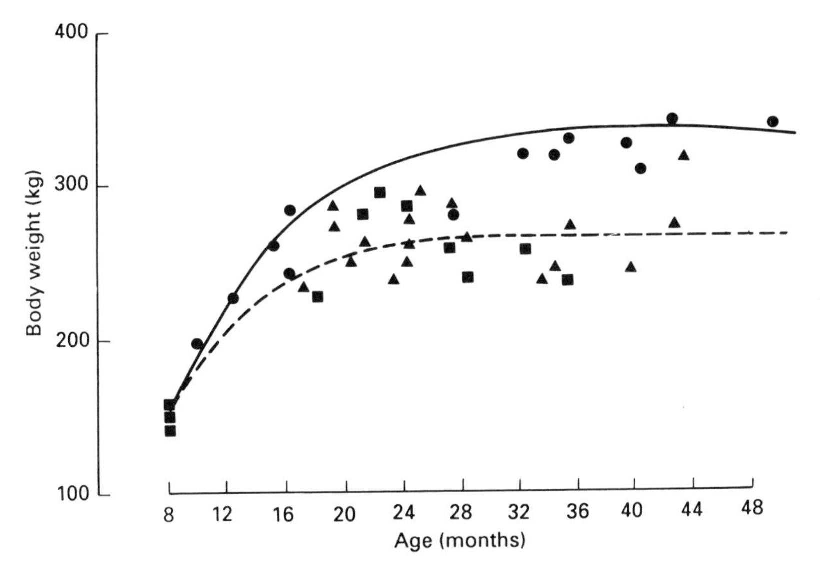


**Figure 24.2** The relationship between age, body weight and feed intake of working boars. --- animals fed 2.7 kg/d (▲, values from the Rowett Research Institute, courtesy of Dr V.R. Fowler, and ■ IGER Shinfield). —— animals fed 3.5 kg/d (●, values from the AI Centre, Shinfield, courtesy of Dr C. Glossop)

emaciated boar may be liable to injury from the environment and severe undernutrition may eventually reduce reproductive capacity.

On selection into the breeding herd, young boars should therefore be fed so that they maintain an adequate rate of growth. There is little information available on what this should be, although estimates can be made if the relationship between the body weight and age of the animal is known. Figure 24.2 has therefore been compiled from a number of sources, from which the growth rates of boars at various body weights can be calculated. Whether these represent the ideal rates of gain is, however, open to question.

For the purposes of the present calculations, values of 0.50, 0.40, 0.30, 0.20 and 0.10 kg/d have been taken as acceptable growth rates for animals of 100, 150, 200, 250 and 300 kg body weight, respectively. It is further assumed that the mature body size of a modern working boar is 350 kg. Since each 1 kg gain in body weight contains 160 g protein and 250 g fat (ARC, 1981), the energy requirements for growth can be calculated. The energy values for protein and fat are 23.8 and 39.7 MJ/kg, respectively, with corresponding efficiencies of utilization of ME of 0.54 and 0.74 (ARC, 1981). Thus each g of protein and fat deposited requires 44.1 and 53.6 kJ ME, respectively. Using these values, the energy requirement for growth decreases from 10.2 to 2.0 MJ/d as growth rate is reduced from 0.5 to 0.1 kg/d, with increase in body weight from 100 to 300 kg.

### MATING ACTIVITY

An inevitable and additional cost associated with reproduction is the activity associated with mating. This should increase with increase in body weight and with the frequency of boar use. It is, however, difficult to measure this additional energy expenditure precisely but Kemp *et al.* (1990) have calculated that the heat production of boars associated with the collection of semen when mounting a 'dummy' sow was 18 kJ/kg $W^{0.75}$/d. This value was therefore taken to represent the additional energy requirement associated with the mating activity of the boar.

### SEMEN PRODUCTION

The frequency with which boars are used for service is variable, but for the purpose of this calculation a rate of one service per day has been assumed. The volume of semen produced is approximately 250 ml per ejaculate with an energy and protein content of 1.04 MJ and 37 g/kg, respectively (Hafez, 1974). The energy contained in each ejaculation therefore contains 0.26 MJ. Since the energy yielding substrates are mostly sugars, especially fructose and citric acid, and since the efficiency of energy utilization of carbohydrates is 0.6 (ARC, 1981), the daily energy requirement for semen production is 0.43 MJ (0.26÷0.6).

### TOTAL REQUIREMENTS

The summation of the requirements, presented in Table 24.5, indicates that the energy requirements of the working boar increase from 27.5 to 37.4 MJ ME/d as body weight increases from 100 to 350 kg. These values compare with estimates of 34.2 and 36.9 MJ ME/d for animals of 150 and 350 kg body weight, respectively, calculated by Kemp (1989). The major energy cost is that associated with the maintenance of the animal, representing between 0.60 and 0.90 of the total ME intake, whereas that associated with mating activity and semen production is small and represents no more than 0.05 of the total requirement. Assuming a dietary DE content of 13.0 MJ/kg, then the present calculations suggest that the feeding level of the breeding boar should increase from 2.2 kg/d at 100 kg body weight to 3.2 kg/d at 350 kg body weight, and these compare with estimates of between 2.7 and 3.1 kg/d determined by Kemp (1989).

An alternative procedure adopted in practice is to feed young boars to achieve a limited body weight gain and then to maintain their body weight at a certain level. For example, if the objective is to maintain a body weight of 200 or 250 kg, then from Table 24.5 it may be calculated that the boar requires 28.7 and 33.2 MJ DE/d, respectively, corresponding to feed intakes of 2.2 and 2.6 kg/d. Thus the information provided in Table 24.5 may be used to determine the feeding requirements of boars depending upon the strategy developed in practice.

### CHANGES IN THE CLIMATIC ENVIRONMENT

The values presented in Table 24.5 refer to animals living in optimal environmental conditions. However, boars are often kept in pens in parts of the building with low stocking density and in poor environmental conditions. In such circumstances the environment may be below the animal's lower critical temperature ($T_c$) and feeding requirements must be increased if the boar is not to lose body condition. The first step is, therefore, calculation of the $T_c$ of boars in relation to the variation in both body weight and feeding level, presented in Table 24.5.

Kemp *et al.* (1989) have established that the $T_c$ of a 250 kg boar at its maintenance energy intake is approximately 20°C in draught-free conditions and on a well-insulated floor. It is further known that $T_c$ decreases by 1°C for each 60 kg increase in body weight (Holmes and Close, 1977; Geuyen, Verhagen and Verstegen, 1984) and for each 80 kJ/kg $W^{0.75}$/d increase in ME intake. Using this information it may be calculated that for the ranges of body weights and ME

**Table 24.5** FACTORIAL ESTIMATES OF THE DAILY ENERGY REQUIREMENTS OF BREEDING BOARS

| Body weight (kg) | 100 | 150 | 200 | 250 | 300 | 350 |
|---|---|---|---|---|---|---|
| Growth rate (kg/d) | 0.50 | 0.40 | 0.30 | 0.20 | 0.10 | – |
| Protein gain (g/d) | 80 | 64 | 48 | 32 | 16 | – |
| Fat gain (g/d) | 125 | 100 | 75 | 50 | 25 | – |
| | | | *Requirements* (MJ/d) | | | |
| Maintenance[a] | 16.31 | 21.36 | 25.86 | 30.00 | 33.87 | 37.53 |
| Protein deposition[b] | 3.53 | 2.82 | 2.12 | 1.41 | 0.71 | – |
| Fat deposition[c] | 6.71 | 5.36 | 4.02 | 2.68 | 1.34 | – |
| Mating activity[d] | 0.57 | 0.77 | 0.96 | 1.13 | 1.30 | 1.46 |
| Semen production[e] | 0.43 | 0.43 | 0.43 | 0.43 | 0.43 | 0.43 |
| Total ME (MJ/d) | 27.55 | 30.74 | 33.39 | 35.65 | 37.65 | 39.42 |
| Total DE (MJ/d)[f] | 29.0 | 32.3 | 35.2 | 37.5 | 39.6 | 41.5 |

[a] $0.763\,W^{0.665}$, where W is the body weight (kg)
[b] (Protein gain × 0.0238) ÷ 0.54
[c] (Fat gain × 0.0397) ÷ 0.74
[d] $0.018\,W^{0.75}$, where W is the body weight (kg)
[e] 0.26 ÷ 0.6
[f] ME ÷ 0.95

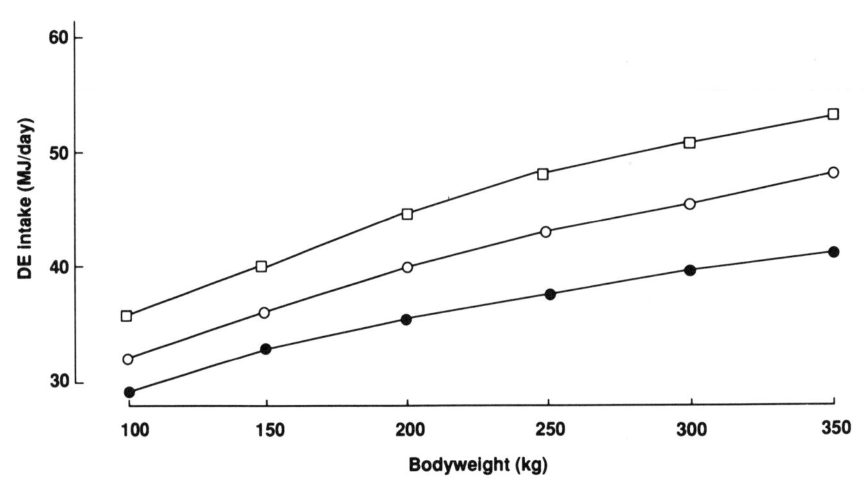


**Figure 24.3** The influence of environmental temperature on the energy requirements of boars of various body weights: □-□, 10°C, ○-○, 15°C, ●-●, 20°C

intakes (890–484 kJ ME/kg $W^{0.75}$/d) presented in Table 24.5 the $T_c$ remains constant at 20°C. When the environmental temperature falls below this level, there is an increase in heat loss of 16 kJ/kg $W^{0.75}$/d (Kemp *et al.*, 1989). Thus the extra energy intake required to meet this thermal need can be calculated to increase from 0.50 to 1.29 MJ DE/d per 1°C as body weight increases from 100 to 350 kg. Figure 24.3 illustrates the additional energy requirement of boars kept at environmental temperatures of 10, 15 and 20°C, that is at their respective critical temperatures. However, temperature is not the only component of the environment which can affect critical temperature and hence feed requirements. Components such as air movement, radiation, floor type and bedding also influence heat exchange (Close, 1981) but there is little available information on these appropriate to boars.

The increased thermal demand associated with poor environmental conditions occurs at the expense of both protein and fat deposition. Kemp *et al.* (1989) have shown that for each 1°C decrease in temperature below $T_c$, protein deposition was reduced by 10 g/d and fat deposition by about 20 g/d. Thus, fat is more depressed than protein deposition at lower ambient temperatures, indicating the importance of ensuring proper nutrition when boars are kept in poor housing conditions in order to maintain body condition and to ensure good reproductive performance.

## Requirements for proteins and amino acids

There have been few experiments to assess the requirements for protein or specific amino acids on the growth and development of the mature breeding boar. Most experiments have been designed to investigate aspects of semen characteristics and reproductive performance (see earlier). However, the intake of protein and amino acids can influence the growth and development of the young growing boar and these have been well documented. For example, ARC (1981) reviewed protein and amino acid requirements of growing pigs and used the concept of 'ideal protein', that is, the protein which supplies the animal with amino acids in exactly the proportion in which they are required and which is therefore potentially fully utilizable. More recently Stahly *et al.* (1983) observed that lower lysine levels (6.5 and 5.0 g/kg) within the body weight ranges of 34 to 63 and 63 to 100 kg, respectively, not only had an adverse effect upon the growth rate and feed:gain ratio of boars but also delayed the development of sexual behaviour. Those boars fed the low-lysine diet produced their first ejaculate later and at a higher body weight than those receiving the high-lysine diet. However, spermatogenic function was not impaired since the low-lysine group had more mature spermatozoa in their first ejaculate.

Yen, Cole and Lewis (1986a,b) have determined the lysine requirements of young growing boars to be 21.6 and 25.3 g/d within the body weight range 20–55 kg and 50–90 kg, respectively. These intakes promoted maximum growth rate, highest body lean content and optimal feed:gain ratio. Fuller and Wang (1987) and Wang and Fuller (1989) have further refined the ideal protein needs of pigs and have calculated that the sum of the essential amino acids for growth is 48 g/100 g protein,

**Table 24.6** PROTEIN AND AMINO ACID REQUIREMENTS OF ADULT BREEDING BOARS (BASED ON REQUIREMENTS OF PREGNANT SOWS)

| | *ARC (1981)* | *NRC (1988)* |
|---|---|---|
| Protein (g/d) | – | 228 |
| Amino acids (g/d) | | |
| Arginine | 0.0 | 0.0 |
| Histidine | 2.6 | 2.8 |
| Isoleucine | 7.4 | 5.7 |
| Leucine | 6.4 | 5.7 |
| Lysine | 8.6 | 8.2 |
| Methionine + cystine | 5.8 | 4.4 |
| Phenylalanine + tyrosine | 6.6 | 8.6 |
| Threonine | 7.2 | 5.7 |
| Tryptophan | 1.4 | 1.7 |
| Valine | 9.2 | 6.1 |

whereas for maintenance it is only 20 g/100 g protein. The requirement for the mature boar will be considerably less than that. Speer (1990) has calculated that the nitrogen requirement for the maintenance of a 120 kg breeding gilt is only 6.5 g/d. The requirement for semen production is also small, and as it contains only 3.4 g protein/100 g, then intakes of 5–10 g protein/d would suffice to meet this dietary need (Westendorf and Richter, 1977). Similarly, the amino acid requirements will also be small since semen contains only 40–90 mg/100 g, depending upon the frequency of boar usage (Johnson *et al.*, 1972).

Thus, these results suggest that the protein and amino acid requirements of the working boar are small in relation to those of the young growing animal. A protein intake of 230 g/d and an amino acid pattern similar to that of the pregnant sow (ARC, 1981) will meet the daily requirements of the working boar (Table 24.6).

## Requirements for minerals

Calcium (Ca) and phosphorus (P) are the most important to consider in the context of mineral nutrition of the working boar as they are crucial not only to optimal growth rate but also to bone mineralization, and hence the overall soundness of the limbs. There has been an increase in leg abnormalities in boars over the last 25 years, including joint abnormalities and lesions on the heel and toes, partly attributable to nutrient inadequacies (Gronalden, 1974). However, the incidence of leg weakness (osteochondrosis) varies between breeds (van der Wal *et al.*, 1987) making precise recommendations for mineral requirements more difficult.

It is generally accepted that higher dietary levels of Ca and P are required for optimum bone mineralization than for maximum performance during the growth processes of the boar (Hickman and Mahan, 1980). Hines, Greer and Allee (1979) recommended concentrations of 9.3 and 7.5 g/kg of Ca and P, respectively, in the diet for maximum bone development but only 5.5 and 4.5 g/kg for growth. Similarly, Hickman, Mahan and Cline (1983) found that in young boars the requirements for Ca and P were higher than those of 7.5 and 6.0 g/kg, respectively, recommended by NRC (1979). Kornegay and Thomas (1981) and Kornegay, Thomas and Baker (1981) also found that maximum bone development occurred at Ca and P levels of 1.25 times the NRC (1973) recommendations for mineral requirements of 6.5 and 5.0 g/kg respectively, during the growing phase, and 5.0 and 4.0 g/kg respectively, during the finishing phase. However, in the former study little effect of dietary calcium and phosphorus on foot soundness and toe lesions was observed, although the number of toe cracks was reduced in animals fed 1.25 and 1.5 times the NRC recommendations for Ca and P levels in the diet.

**Table 24.7** REQUIREMENTS FOR MINERALS FOR GROWING AND BREEDING BOARS

| *Minerals* | *Growing boars* | *Breeding boars* | *Adult boars*[a] |
|---|---|---|---|
| | (amounts/kg dry matter) *(AFRC, 1990)* | | (amount/kg) *(NRC, 1988)* |
| Calcium[b] | 9.5 g | 7.0 g | 7.5 g |
| Phosphorus[b] | 7.5 g | 5.5 g | 6.0 g |
| Chlorine | 1.5 g | 1.5 g | 1.2 g |
| Sodium | 1.3 g | 1.3 g | 1.5 g |
| Potassium | 2.5 g | 2.5 g | 2.0 g |
| Magnesium | 400 mg | 400 mg | 420 mg |
| Iron | 50 mg | 50 mg | 80 mg |
| Zinc | 100 mg | 100 mg | 50 mg |
| Manganese | 15 mg | 10 mg | 10 mg |
| Copper[c] | 4.0 mg | 4.0 mg | 5.0 mg |
| Cobalt[d] | — | — | — |
| Iodine | 0.5 mg | 0.5 mg | 0.2 mg |
| Selenium | 0.2 mg | 0.2 mg | 0.1 mg |

[a]Calculated assuming a daily feed intake of 1.9 kg
[b]Requirements above and below 50 kg, respectively
[c]Copper may be increased in circumstances where it is a permitted growth promoter
[d]Cobalt has not been shown to be required other than as a component of cyanocobalamin ($B_{12}$)

Soundness of limbs is an important index to consider as foot-related problems are a major contributory factor to loss of libido and the inability of the boar to mount a sow or dummy. Kesel *et al.* (1983) reported that boars which were fed 150% of the NRC (1973) recommendations for dietary Ca and P had heavier and thicker-walled metacarpals which had greater bone strength than boars fed 100% of recommendations. The weight, size, thickness, strength and ash content increased with age, with a small decline in the rate of increase, suggesting a reduction in the Ca and P requirements with age (Kornegay, Thomas and Baker, 1981). This reduction with age may be less than that for breeding sows and Liptrap *et al.* (1970) have shown that boars have significantly heavier bones with higher moment of inertia but with lower breaking stress values than gilts. Thus, for boars up to 50 kg, the dietary requirements for Ca and P should be set close to those suggested by Hines, Greer and Allee (1979) of 9.5 and 7.5 g/kg, respectively. For boars above 50 kg body weight and during their breeding life the requirements should be set at 7.0 and 5.5 g/kg, respectively. Thus, as the animal grows the Ca:P ratio should be within the range 1.3 to 1.5:1.

There is no evidence to suggest that the Ca:P ratio is important in spermatogenesis or semen quality *per se*. In contrast, zinc does have an established role in spermatogenesis since zinc deficiencies are implicated in the retardation of the development of the leydig cells, a reduced response to luteinizing hormone and reduced testicular steroidogenesis (Hesketh, 1982). A level of 100 mg/kg is therefore recommended.

Boars show no additional requirements for the other major minerals in relation to sows and the recommended mineral requirements for growing or breeding boars are shown in Table 24.7.

## Requirements for vitamins

It has been suggested that the working boar has no additional vitamin requirements above those of the breeding sow (adjusted according to maintenance requirements). However, the role of biotin in the diet of boars is becoming increasingly important as a result of its association with foot lesions and the attendant implications this has for reproductive performance in the boar. There is also the suggestion that vitamins E and C may be of special importance in the diet of stress-susceptible breeds.

### BIOTIN

Biotin is an essential water-soluble, sulphur-containing vitamin which although widely distributed in nature is present in relatively small concentrations and has a limited bioavailability. The symptoms associated with deficiency of biotin in the diet of pigs include reduced growth rate, hair loss, dry scaly skin, spasticity of the hind limbs, erosion of the soft toe heel and extensive cracking of the toe heel and horn (Kornegay, 1986). Foot and toe lesions are generally associated with lameness and there has been a reported increase in foot abnormalities in pigs over the last 25 years (van der Wal *et al.*, 1987) due in part to the poor conditions of housing and poor floor surfaces (Penny, Osborne and Wright, 1963; Lepine *et al.*, 1985).

Although Ca and P have been shown to be extremely important in the general development of limb soundness in boars, supplementation in the diet has met with only limited success in actually reducing the incidence of foot lesions. Supplementation of the diet with biotin has, on the other hand, been shown to significantly strengthen the hoof (Webb, Penny and Johnston, 1984). It was shown initially that experimentally-induced biotin deficiencies in both young pigs and sows produce hoof cracks and lesions that are responsive to biotin supplementation. Similarly, foot lesions in sows housed in confinement can be reduced, although not completely prevented, by adding biotin to diets containing commonly-used feed ingredients (Penny *et al.*, 1980; Bryant *et al.*, 1985a,b).

Webb *et al.* (1984) showed that supplementation of the sow's diet with 1 mg of D-biotin per kg of diet significantly strengthened the hoof. Brooks, Smith and Irwin (1977) showed that supplementing the diet of pregnant and lactating sows with 150 and 250 μg D-biotin per kg, respectively, over a 6-month period, reduced the number of foot lesions by 28%. Penny *et al.* (1980) examined the effect of supplementary biotin on a herd of sows affected with a high incidence of foot lesions. The treated group received 1160 μg/d during gestation and 2320 μg/d during lactation, additional to the 200 μg/kg estimated to be present in the basal or control diet. The results showed no significant improvement in established lesions, but gilts entering the herd, and which had received supplemental biotin, developed significantly fewer lesions on the weight-bearing soft surfaces of their feet than did the controls and were similar in the lateral digits of their hind feet.

The precise mode of action of biotin in the prevention of foot lesions is still uncertain. It is known that biotin increases the compressive strength and hardness of the hoof wall whilst decreasing the hardness of the heel bulb tissue (Webb *et al.*, 1984). A soft heel bulb presumably acts as a cushion, minimizing stresses and absorbing strain energy. As leg weaknesses and foot lesions are a major cause of the inability of the boar to mount a sow and/or a dummy, and a general loss of libido, the supplementation of biotin to the diet is a potential route whereby overall reproductive performance could be increased. However, under current production systems, where boars are frequently housed on poorly-designed floors, the animals are unlikely to recover since the extent of the injury may exceed the capacity of the hoof for growth and repair. This indicates that certain problems experienced by the working boar need to be addressed at a more basic level than that of nutritional status or dietary supplementation. Despite this, it is suggested that diets for working boars contain at least 300 μg biotin/kg, but if foot problems exist then this should be increased to 1 mg/kg diet.

### VITAMIN E

The continuous selection of animals for faster growth may also have resulted in an increase in the incidence of porcine stress syndrome. Stress-susceptible pigs can undergo a fatal malignant hypothermia when exposed to strenuous activities such as exercise and transportation. This frequently leads to mortality and a carcase condition referred to as PSE (pale, soft and exudative) meat which represents both a considerable financial and management problem. In terms of their metabolic and biochemical status, stress-susceptible pigs under resting conditions are characterized by increased plasma activities of creatine kinase, pyruvate kinase and conjugated dienes and thiobarbituric acid reactive substances (Duthie and Arthur, 1987). These components are indices of cell membrane damage and lipid peroxidation and can be decreased by the addition of large quantities of vitamin E (a potent antioxidant) to the diet.

Hoppe *et al.* (1989) and Duthie *et al.* (1989) have shown that the conditions responsible for stress-susceptibility are decreased when diets are supplemented with vitamin E. In addition, when fed in conjunction with increased dietary levels of vitamin C (an additional antioxidant) there may be an additional protective effect on cell membrane integrity. These antioxidants appear to confer protection to stress-susceptible pigs at the biochemical level, but they do not always improve the carcase characteristics of PSE meat even though the number of pigs producing a fatal stress response during transportation may be reduced. This indicates that a genetically-based syndrome can be modified by dietary intervention. The specific importance of reducing stress-susceptibility in the working boar is related to its ability to perform. The excitement associated with mounting a sow and/or dummy may be sufficient to induce a fatal stress-response in susceptible individuals and so where a particular stress-sensitive boar is used supplementation of the diet with antioxidants could prove helpful. However, it may be better not to use such animals in order to prevent the inheritance of this trait throughout the herd.

## VITAMIN C (ASCORBIC ACID)

It is generally assumed that pigs can synthesize adequate amounts of vitamin C to meet their metabolic needs (NRC, 1979). However, there is increasing evidence that the pig's requirement for vitamin C is greater under stress, such as changes in environmental conditions and when excessive handling occurs. Thus, working

**Table 24.8** REQUIREMENTS FOR VITAMINS AND ESSENTIAL FATTY ACIDS FOR GROWING AND BREEDING BOARS

| *Vitamins* | *Growing boars* (amounts/kg dry matter) *(AFRC, 1990)* | *Breeding boars* | *Adult boars*[a] (amount/kg) *(NRC, 1988)* |
|---|---|---|---|
| Thiamin | 1.5 mg | [b] | 1.0 mg |
| Riboflavin | 2.5 mg | 3.0 mg | 3.7 mg |
| Nicotinic acid | 14.0 mg | [b] | 10.0 mg |
| Pantothenic acid | 10.0 mg | 10.0 mg | 12.0 mg |
| Pyridoxine | 2.5 mg | 1.5 mg | 1.0 mg |
| Cyanocobalamin | 10.0 μg | 15.0 μg | 15.0 μg |
| Biotin | 300.0 μg | 300.0 μg | 210.0 μg |
| Folic acid (*p*-aminobenzoic acid) | [b] | [b] | 0.3 mg |
| Myoinositol | [b] | [b] | [b] |
| Choline | [b] | [b] | 1.3 g |
| Ascorbic acid | [b] | [b] | [b] |
| Vitamin A | 1200 i.u. | 2310 i.u. | 4000 i.u. |
| Retinol | 400.0 μg | 700.0 μg | |
| Vitamin D | 200 i.u. | 200 i.u. | 200 i.u. |
| Cholecalciferol | 5.0 μg | 5.0 μg | |
| Vitamin E | | | |
| DL-α-tocopherol acetate | 15.0 mg | 15.0 mg | 22.0 mg |
| Vitamin K | | | |
| Menaphthone | [b] | [b] | 0.5 mg |
| Essential fatty acids | | | |
| Linolenic acid | 7 g | 7 g | [b] |
| Arachidonic acid | 5 g | 5 g | [b] |

[a]Calculated assuming a daily feed intake of 1.9 kg
[b]Insufficient data exist to make an estimate

boars may benefit from the addition of supplemental vitamin C in their diet on the assumption that it improves metabolism resulting in better growth, improved feed conversion and better foot and leg structure (Riker *et al.*, 1967). However, the results from studies on the merits of vitamin C supplementation are equivocal. The studies by Cleveland, Bondari and Newton (1987) failed to show any effects of supplemental vitamin C on performance traits or composite soundness scores of the foot and leg structure, although the straightness of the front leg was improved. Neilsen and Vinther (1984) reported more positive results since animals fed an additional 1 g vitamin C per day exhibited a lower incidence of bent and crooked front legs. Pigs given 700 ppm of vitamin C in their diet were found to have better locomotion scores compared with animals receiving 0 or 350 ppm (Nakano, Aherne and Thompson, 1983) but Strittmatter *et al.* (1977) had previously failed to detect any effects on overall joint or visual soundness in pigs receiving an additional supplement of 780 ppm vitamin C. It is, therefore, not possible to make a definitive statement regarding supplemental requirements for vitamin C in the diet of the working boar, although a potential role in ameliorating leg abnormalities cannot be ignored.

The vitamin requirements for both the growing and breeding boar are presented in Table 24.8.

## Dietary fibre in the nutrition of the working boar

During rearing, boars are fed at a level of 0.75–1.00 of appetite, but this is reduced later in life so that feed intake may be no higher than that required to maintain the animal in energy equilibrium. However, restriction of feed intake may result in hunger and reduction in feeding behaviour and this may enhance frustrated feeding motivation and sterotypic behaviour (Lawrence, Appleby and Macleod, 1988). Increasing the fibre content or bulkiness of the ration has been shown to improve welfare and reduce hunger on restricted feeding regimes (Appleby and Lawrence, 1987; Dantzer and Morméde, 1983; Lawrence, Appleby and MacLeod, 1988) and may therefore have application in the feeding of boars.

The precise role that dietary fibre may play in reducing hunger and increasing satiety over a prolonged period of time may be two-fold. Firstly the considerable 'bulking' effect that plant fibre produces will increase overall gut fill producing a feeling of 'fullness' and satiation. Secondly, there is evidence to suggest that gastric emptying and intestinal transit times will also be prolonged (Ray *et al.*, 1983; Rainbird, 1986) so that the duration of this increased fullness will be extended (Di Lorenzo *et al.*, 1988). Increasing the soluble fibre content of a meal may also attenuate the absorption of nutrients such as glucose from the small intestine and this offers the potential for greater overall absorption of food and hence increase in the efficiency of feed utilization.

There may also be positive health and welfare advantages associated with the feeding of fibrous foods. Working boars are kept for a longer period of time than animals reared for meat production and they are more likely to develop health problems associated with chronic nutritional inadequacies. For example, lesions of the *pars oesophagea* are common in cereal-fed animals, resulting in reduced feed intake and poorer overall performance. Increasing the fibre content of the diet has been shown to reduce the severity and incidence of such lesions (Potkins, Lawrence and Thomlinson, 1984). Thus, increased levels of fibre in the diet of the working boar can be recommended on the basis of (i) reduction in overall feed and production costs, (ii) increase in satiety due to prolonged transit times, increased gut fill and reduction of postprandial 'rebound hypoglycaemia', (iii) increased feed efficiency due to attenuated absorption, and (iv) overall improvement in the welfare, well-being and long-term health of the animal. There is no evidence to suggest that dietary fibre influences the reproductive performance of boars, but it may improve their ability to thermoregulate under cold conditions because of the high heat increment associated with fermentation in the large intestine (Close, 1987).

The only area in which to exercise caution when increasing the fibre content of the diet is with regard to mineral requirements, as a variety of insoluble fibre sources have been shown to reduce their bioavailability (Kelsey, 1986). Thus, under certain circumstances Ca and P requirements may need to be reviewed in the light of other dietary modifications.

## Environmental effects on reproductive performance

The environmental conditions within which boars are kept not only influences their feeding requirements (see earlier) but also their reproductive performance. Low environmental temperatures *per se* do not influence sperm production, provided the dietary energy intake is sufficient to compensate for any additional extra thermoregulatory heat produced by the animal (Swierstra, 1970; Kemp *et al.*, 1988). High environmental temperatures, on the other hand, decrease both sperm motility and sperm production and Stone (1982) has established that the critical air temperature at which this occurs is 29°C. The boar seems to be more sensitive to higher environmental temperature than the males of other species and this is probably related to its limited ability to sweat and hence to lose heat by evaporation (McNitt and First, 1970; McNitt, Tanner and First, 1972). Thus, exposure of boars to high environmental temperatures results in rapid increases in respiration rate and rectal temperature and reduced reproductive performance. However, the effects of heat stress take 2–6 weeks to become manifest and a period of at least 5 weeks after the end of exposure of boars to heat stress is needed for the quality and quantity of semen to be returned to normal levels.

The fertility of sows mated with heat-stressed boars is reduced (Table 24.9)

**Table 24.9** FERTILITY OF GILTS ARTIFICIALLY-INSEMINATED OR NATURALLY MATED WITH CONTROL OR HEAT-STRESSED BOARS

| *Treatment* | *No. of boars* | *No. of gilts* | *Gilts pregnant at 30±3 days after breeding* | |
|---|---|---|---|---|
| | | | *%* | *% embryo survival* |
| Artificially inseminated | | | | |
| Control | 6 | 88 | 41 | 71±4 |
| Heat-stressed | 6 | 77 | 29 | 48±5 |
| Naturally mated | | | | |
| Control | 6 | 37 | 82 | 82±2 |
| Heat-stressed | 6 | 60 | 59 | 78±5 |

Wettermann *et al.* (1976, 1979)

(Wettermann *et al.*, 1976; Wettermann, Wells and Johnson, 1979). The conception rate of gilts either artificially inseminated or naturally mated with heat-stressed boars was also reduced. However, embryo survival was only influenced when gilts were artificially inseminated. This suggests that the spermatozoa from heat-stressed boars may be less suitable for preservation.

## Summary and conclusions

This chapter indicates that there is a paucity of information on the nutritional requirements and responses of the working boar, despite its importance to herd fertility. In the past it has been the practice to feed boars a diet similar to that of the pregnant sow, but this may not always meet their specific nutrient needs and there are distinct differences between sexes and also between animals reared for the production of meat and those for breeding.

The evidence suggests that nutrition during rearing can influence both the age of puberty and the rate of sexual development, but these are unlikely to be impaired under current feeding regimes and practices. However, nutrition may improve reproductive performance and there are specific dietary needs for energy, protein and amino acids to ensure good libido, semen production and fertility. Over-feeding should be avoided since this will increase body size and may reduce libido.

Requirements for energy, protein and amino acids must take account of the animal's body weight, rate of gain, frequency of use and environmental conditions within which it lives. For energy, estimates have been derived according to the factorial principle of nutrient partitioning and range between 29.0 and 41.5 MJ DE/d, with an additional 3% for each 1°C decrease in temperature below an environmental temperature of 20°C. Lysine levels below 5 g/kg reduce growth rate and delay sexual development and a protein intake of 230 g/d with an amino acid pattern similar to that of the pregnant sow is recommended.

There is no clear evidence that the mineral and vitamin requirements of boars are greater than those of the breeding sow but calcium and phosphorus levels may need to be increased to ensure adequacy of bone development. Similarly, biotin has been implicated in the soundness of limbs and requirements may need to be increased up to 1000 μg/kg diet if leg problems occur.

A major practical problem is that the levels of feeding for boars are generally low and this may result in hunger, frustration, abnormal behaviour and generally poor welfare. The use of bulky or fibrous diets may overcome these problems and improve health, provided they supply sufficient nutrients to meet the dietary needs of the animal.

In the future, there is a need to provide a better understanding of the nutritional requirements and responses of boars so that feeding and management strategies can be based on sounder scientific principles. It is also important to understand the basic metabolic and physiological responses of boars and how these may be manipulated by nutrition to ensure and sustain good reproductive performance. The welfare needs of the boar must also be better understood and there may be considerable potential for the development of high fibre diets for the working boar; this will also ensure good health, behaviour, welfare and reproductive performance.

**References**

Agricultural and Food Research Council (1990). Technical Committee on Responses to Nutrients, Report No. 4, Nutrient Requirements of Sows and Boars. *Nutrition Abstracts and Reviews* Series B: Livestock Feeds and Feeding, **60**, 383–406

Agricultural Research Council (1981). *The Nutrient Requirements of Pigs*. Slough: Commonwealth Agricultural Bureaux

Althen, T.G., Gerrits, R.J. and Young, E.P. (1974). *Journal of Animal Science*, **39**, 601–605

Appleby, M.C. and Lawrence, A.B. (1987). *Animal Production*, **45**, 103–110

Beeson, W.M., Crampton, E.W., Cunha, T.J., Ellis, N.R. and Leucke, R.W. (1953). *Nutrient Requirements of Swine*. National Research Council Publication, 295, 2

Berger, T., Esbenshade, K.L., Diekman, M.A., Hoagland, T. and Tuite, J. (1981). *Journal of Animal Science*, **33**, 1559–1564

Bristol, F.M. and Djurickovic, S.C. (1971). *Canadian Veterinary Journal*, **12**, 132

Brooks, P.H., Smith, D.A. and Irwin, V.C.R. (1977). *Veterinary Record*, **101**, 46–50

Bryant, K.C., Kornegay, E.T., Knight, J.W., Webb, J.E., Jr, and Notter, D.R. (1985a). *Journal of Animal Science*, **60**, 136–144

Bryant, K.C., Kornegay, E.T., Knight, J.W., Webb, J.E., Jr, and Notter, D.R. (1985b). *Journal of Animal Science*, **60**, 145–153

Christensen, C.M., Mirocha, C.J., Nelson, G.H. and Quast, J.F. (1972). *Applied Microbiology*, **23**, 202

Cleveland, E.R., Bondari, K. and Newton, G.L. (1987). *Livestock Production Science*, **17**, 277–283

Close, W.H. (1981). In *Environmental Aspects of Housing for Animal Production*, pp. 149–166, Ed. Clarke, J.A. Butterworths, London

Close, W.H. (1987). In *Pig Housing and the Environment*, pp. 9–24. Eds Smith, A.T. and Lawrence, T.L.J. Occasional Publication of the British Society of Animal Production, No. 11. British Society of Animal Production

Dantzer, R. and Mormède, P. (1983). *Applied Animal Ethology*, **10**, 233–244

Di Lorenzo, C., Cardiff, M.W., Hajnal, F. and Valinzuela, J.E. (1988). *Gastroenterology*, **95**, 1211–1215

Duthie, G.G. and Arthur, J.R. (1987). *American Journal of Veterinary Research*, **48**, 309–310

Duthie, G.G., Arthur, J.R., Nicol, F. and Walker, M.J. (1989). *Research in Veterinary Science*, **46**, 226–233

Dutt, R.H. and Barnhart, C.F. (1959). *Journal of Animal Science*, **18**, 3–13

Einarsson, S. (1975). Cited in Leman and Rodeffer (1976)

Fuller, M.F. (1980). In *Recent Advances in Animal Nutrition—1980*, pp. 157–169. Ed. Haresign, W. Butterworths, London

Fuller, M.F. and Wang, T.C. (1987). In *Manipulating Pig Production*, pp. 97–111. Ed. APSA Committee, Werribee, Australasian Pig Science Association

Grondalen, F. (1974). *Acta Veterinaria Scandinavica*, **15**, 555–573

Geuyen, T.P.A., Verhagen, J.M.F. and Verstegen, M.W.A. (1984). *Animal Production*, **38**, 477–485

Hafez, E.S.E. (1974). *Reproduction in Farm Animals*, 3rd edition, 497 pp. Lea and Febiger, Philadelphia

Hanssen, J.T. and Grondalen, T. (1979). *Zeitschrift für Tierphysiologie, Tierernährung und Futtermittelkunde,* **42**, 65–83
Hesketh, J.E. (1982). *Journal of Comparative Pathology*, **92**, 239–247
Hickman, D.S. and Mahan, D.C. (1980). *Journal of Animal Science*, **51** (Supplement 1), 201–202
Hickman, D.S., Mahan, D.C. and Cline, J.H. (1983). *Journal of Animal Science*, **56**, 431–437
Hines, R.H., Greer, J.C. and Allee, G.L. (1979). *Journal of Animal Science*, **49** (Supplement 1), 101
Hofstetter, P. and Wenk, C. (1987). In *Energy Metabolism of Farm Animals*, pp. 122–125. Ed. Moe, P.W., Tyrrell, H.F. and Reynolds, P.J. EAAP Publication No. 32. Rowman and Littlefield, New Jersey
Holmes, C.W. and Close, W.H. (1977). In *Nutrition and the Climatic Environment*, pp. 51–73. Ed. Haresign, W., Swan, H. and Lewis. D. Butterworths, London
Hoppe, P.P., Duthie, G.G., Arthur, J.R., Schoner, F.J. and Wiesche, H. (1989). *Livestock Production Science*, **22**, 341–350
Hughes, P.E. and Varley, M.A. (1980). *Reproduction in the Pig*, 241 pp. Butterworths, London
Hühn, U. (1969). Cited in Westendorf and Richter (1977)
Hühn, U., Kleemann, F., König, I. and Poppe, S. (1973). *Archiv für Tierzucht*, **16**, 347–358
Jentsch, W., Hoffmann, R., Schiemann, R. and Wittenberg, H. (1989). *Archives of Animal Nutrition*, **3**, 279–297
Johnson, L.A., Russel, V.G., Gerrits, R.J. and Thomas, C.H. (1972). *Journal of Animal Science*, **34**, 340–434
Ju, J.C., Cheng, S.P. and Yen, H.T. (1985). *Journal of Chinese Society of Animal Production*, **14**, 27–35
Kelsey, J. (1986). In *Dietary Fibre: Basic and Clinical Aspects*, pp. 361–372. Ed. Vahauncy, G.V. and Kritchevsky, D. Plenum Press, New York
Kemp, B. (1989). Investigations on breeding boars to contribute to a functional feeding strategy. PhD Thesis, University of Wageningen
Kemp, B., den Hartog, L.A. and Grooten, H.J.G. (1989). *Animal Reproduction Science*, **20**, 245–254
Kemp, B., Grooten, H.J.G., den Hartog, L.A., Luiting, P. and Verstegen, M.W.A. (1988). *Animal Reproduction Science*, **17**, 103–113
Kemp, B., Verstegen, M.W.A., den Hartog, L.A. and Grooten, H.J.G. (1989). *Livestock Production Science*, **23**, 329–340
Kemp, B., Vervoort, F.P., Bikker, P., Janmaart, J., Verstegen, M.W.A. and Grooten, H.J.G. (1990). *Animal Reproduction Science*, **22**, 87–98
Kesel, G.A., Knight, J.W., Kornegay, E.T., Veit, H.P. and Notter, D.R. (1983). *Journal of Animal Science*, **57**, 82–98
Kim, J.K., Suh, G.S., Park, C.S. and Sul, D.S. (1976). Research Report Office Rural Development, Livestock Series, 18, 29–42
Kim, J.K., Suh, G.S., Sul, D.S., Kim, V.B. and Lee, Y.B. (1977). In Hughes and Varley (1980)
Kolenko, V. (1977). *Zhivotnovodsto*, **5**, 70–71
Kornegay, E.T. (1986). *Livestock Production Science*, **14**, 65–89
Kornegay, E.T. and Thomas, H.R. (1981). *Journal of Animal Science*, **52**, 1049–1059
Kornegay, E.T., Thomas, H.R. and Baker, J.L. (1981). *Journal of Animal Science*, **52**, 1070–1084

Laswai, G.H., Close, W.H., Sharpe, C.E. and Keal, H.D. (1991). *Animal Production*, **52**, 570

Lawrence, A.B., Appleby, M.C. and MacLeod, H.A. (1988). *Animal Production*, **47**, 131–137

Leman, A.D. and Rodeffer, H.E. (1976). *Veterinary Record*, **98**, 457–459

Lepine, A.J., Kornegay, E.T., Notter, D.R., Veit, H.P. and Knight, J.W. (1985). *Canadian Journal of Animal Science*, **65**, 459–472

Liptrap, D.O., Miller, E.L., Ullrey, D.E., Keahey, K.K. and Hoefer, J.A. (1970). *Journal of Animal Science*, **31**, 540–549

Ludvigsen, J.B. (1980). In *Energy Metabolism*, pp. 115–118. Ed. Mount, L.E. Butterworths, London

Medding, A.J.H. and Nielsen, H.E. (1977). *Statens Husdyrbrugsforsog*, **175**, 2

McNitt, J.I. and First, N.L. (1970). *International Journal of Biometeorology*, **14**, 373–380

McNitt, J.I., Tanner, C.B. and First, N.L. (1972). *Journal of Animal Science*, **34**, 117–121

Nakano, T., Aherne, F.X. and Thompson, J.R. (1983). *Canadian Journal of Animal Science*, **63**, 421–428

National Research Council (1973). *Nutrient Requirements of Domestic Animals, No. 2 Nutrient Requirements of Swine*, 7th revised edition. National Academy of Sciences—National Research Council, Washington

National Research Council (1979). *Nutrient Requirements of Swine*, 9th revised edition. National Academy Press, Washington

National Research Council (1988) *Nutrient Requirements of Pigs*, National Academy Press, Washington

Neilsen, N.C. and Vinther, K. (1984). In *Ascorbic Acids in Domestic Animals*, pp. 39–41. Ed. Wigger, I., Tagwerker, F.J. and Moustgaard, J. The Royal Danish Agricultural Society, Copenhagen

Niwa, T. (1954). *Bulletin National Institute for Agricultural Science*, **8**, 17 (Animal Breeding Abstract 1955, 23, 403)

Noblet, J., Karege, C. and Dubois, S. (1989). In *Energy Metabolism of Farm Animals*, pp. 57–60. Ed. van der Honing, Y. and Close, W.H. EAAP Publication No. 43. Pudoc, Wageningen

Ovchinnikov, A.A. (1984). *Zhivotnovodstvo*, **7**, 14–36 (cited from *Nutrition Abstracts and Reviews*, **55**, 843)

Penny, R.H.C., Cameron, R.D.A., Johnson, S., Kenyon, P.J., Smith, H.A., Bell, A.W.P., Cole, J.P.L. and Taylor, J. (1980). *Veterinary Record*, **107**, 350–351

Penny, R.H.C., Osborne, A.D. and Wright, A.I. (1963). *Veterinary Record*, **75**, 1225–1240

Picha, J., Cerovsky, J. and Pichova, D. (1986). *Veterinarini Medicini*, **31**, 347–357

Phillips, R.W. and Andrews, F.N. (1936). *Missouri Agricultural Experimental Station Bulletin* No. 331

Poppe, S., Hühn, U., Kleemann, F. and König, I. (1974a). *Archiv für Tierernährung*, **24**, 499–512

Poppe, S., Hühn, U., Kleemann, F. and König, I. (1974b). *Archiv für Tierernährung*, **24**, 551–565

Potkins, Z.V., Lawrence, T.L.J. and Thomlinson, J.R. (1984). *Animal Production*, **38**, 534

Rainbird, A.L. (1986). *British Journal of Nutrition*, **35**, 99–109

Ray, T.K., Mansell, K.M., Knight, L.C., Malmud, L.S., Owen, O.E. and Boden, G. (1983). *American Journal of Clinical Nutrition*, **37**, 376–381

Riker, J.T., Perry, T.W., Pickett, R.A. and Heidenrich, C.J. (1967). *Journal of Nutrition*, **92**, 99–103

Sellier, R., Dufour, L. and Rousseau, G. (1973). *Animal Breeding Abstracts*, **41**, 1191

Singh, G. (1962). *Annals of Biology, Animal Biochemistry and Biophysiology*, **1**, 403–406

Speer, V.C. (1990). *Journal of Animal Science*, **68**, 553–561

Stahly, T.S., Zavos, P.M., Edgerton, L.A. and Cromwell, G.L. (1983). *Journal of Animal Science*, **57** (Supplement 1), 81

Stevermer, E.J., Kovacs, M.F., Jr, Hoekstra, W.C. and Self, H.L. (1961). *Journal of Animal Science*, **20**, 858–865

Stone, B.A. (1982). *Animal Reproduction Science*, **4**, 283–299

Strittmatter, J.E., Ellis, D.J., Hogberg, M.G., Miller, E.R., Parsons, M.J. and Trapp, A.L. (1977). Report of Swine Research. Michigan Agricultural Station and Michigan State University, 111–115

Swierstra, E.C. (1968). *Anatomical Record*, **61**, 171–186

Swierstra, E.C. (1970). *Biology of Reproduction*, **2**, 23–28

Uzu, G. (1979). *Annales de Zootechnie*, **28**, 431–441

Van der Kerk, P. and Willems, C.M.T. (1985). *Zeitschrift für Tierphysiologie, Tierernährung und Futtermittelkunde*, **53**, 43–49

Van der Wal, P.G., Goldegebuure, S.A., van der Valk, P.C., Engel, B. and van Essen, G. (1987). *Livestock Production Science*, **16**, 65–74

Wang, T.C. and Fuller, M.F. (1989). *British Journal of Nutrition*, **62**, 77–89

Webb, N.G., Penny, R.H.C. and Johnston, A.M. (1984). *Veterinary Record*, **114**, 185–189

Westendorf, P. and Richter, L. (1977). *Ubersicht für Tierernährung*, **5**, 161–184

Wettermann, R.P., Wells, M.E. and Johnson, R.K. (1979). *Journal of Animal Science*, **49**, 1501–1505

Wettermann, R.P., Wells, M.E., Omtvedt, I.T., Pope, C.E. and Turman, E.J. (1976). *Journal of Animal Science*, **42**, 664–669

Yen, H.T., Cole, D.J.A. and Lewis, D. (1986a). *Animal Production*, **43**, 141–154

Yen, H.T., Cole, D.J.A. and Lewis, D. (1986b). *Animal Production*, **43**, 155–165

Yen, H.T. and Yu, I.T. (1985). *Proceedings of the 3rd AAAP Animal Science Congress*, **2**, 610–612

Yurin, M. (1981). *Svinovodstvo*, **1**, 31–32 (cited from *Nutrition Abstracts and Reviews*, **51**, 5097)

Zaripov, L. and Shakirov, Sh. (1978). *Svinovodstvo,* **15**, 14–15 (cited from *Nutrition Abstracts and Reviews, Series B*, **48**, 5432)

# INDEX

Absorption
  of synthetic amino acids, 106–108
Acid detergent fibre (ADF)
  analysis of, 139
Acidification of diets
  by lactic acid producing microbes, 263
  interaction between diet type and acid response, 260
  response to inorganic acids, 261
  response to organic acids, 258
Additives to diets, 248
ADF (see acid detergent fibre)
Aflatoxin
  semen quality in boars, 354
Allergenic responses to diet, 291
Amadori compound, 119
Amino acids
  availability
    assays to determine, 131
    in feeds, 85
    protein deposition, 41
  bioavailability, 65
  body protein maintenance, 120
  catabolism, 121
  determining requirements, 76–84
  digestibility
    assays, 125
    ileal digestibility, 128
    *in-vitro* estimation, 130
  digestion and absorption of synthetic amino acids, 106–108
  efficiency of utilisation, 124
  estimates of piglet requirements, 81
  excesses, 72
  free, in piglet diets, 76–77
  ileal digestibility, 85
  inevitable catabolism, 124
  limitations, assessment of, 71
  metabolism (protein), 108–111
  nutrition, pigs and poultry, 60
  oxidation, in pigs, 78–79
  protein deposition, 39
  requirements, 53, 56, 64
    boars, 358
Amino acids (*cont.*)
    body composition, factors affecting, 64
    environmental temperature, 68
    food composition and food intake factors, 65
    frequency of feeding, 65
  spermatogenesis in boars, 350
  unabsorbed, 118, 124
Anabolic drive, 123
Anastomosis
  ileal rectal, 127
Anti-diuretic hormone, 228
Antibiotics
  interaction with dietary fibre, 157
  use in the pig industry, 295
Anti-nutritional factors, 269
  dietary antigens, 271
  enzyme inhibitors, 270
  lectins, 271
Apparent digestibility
  effect of cellulose, 144
  effect of fibre, 142
Arginine, requirement of piglet, 83
*Aspergillus ficuum*
  phytase production, 171
AUSPIG model, 42

Bacterial infection of digestive tract, 243, 250
Barley
  phytase, 165, 167
Beta-adrenergic agonists, 312–314
  effect on body composition, 313
  effect on growth rate, 313
  energy requirements, 314
  food intake, 314
  lysine requirements, 314
Beta-glucanase
  addition to barley based diets, 302
Bicarbonate, 225
Bile secretion in pigs, 244
Biotin
  boar requirements, 361
Boars
  amino acid requirements, 358

Boars (*cont.*)
heat stress, 365
libido, 349
lower critical temperature, 357
mineral requirements, 360
protein requirements, 358
taint, 5
tryptophan requirement, 83
vitamin requirements, 361
Body weight gain
calculation of in pregnant pigs, 323
predicted effect of DE intake in pregnant sows, 325
Breed effects
dietary fibre, 157

Calcium
boar requirements, 360
pig feeds, 165
Calves, pre-ruminant diarrhoea, 267
Cannulation of pig digestive tract, 86–90
Carcass quality in pigs
effect of age and appetite, 16
effect of energy value of feeds, 18
effect of level of feed, 16
effect of protein, 15
Cereals, ileal digestibility, 86
Carcass composition, 21
beta-adrenergic agonists, 314
effect of dietary fibre, 156
porcine somatotropin, 307–309
Chloride, 231, 236
Cholesterol, 5
Cottonseed meal, ileal digestibility, 98, 101

Dehydration, 217, 220
Deoxyketosyl derivative, 119
Diarrhoea
aetiology of, in pigs and pre-ruminant calves, 267
dietary approaches to control, 275
effect on performance, 275
electrolyte disturbance, 231
*Escherichia coli*, 268
pathogenesis, 267
post-weaning, 275
probiotics, 279
Diet, and piglet health, 291
Dietary undetermined anion, 235
Dietary fibre, 244, 252
analytical methods, 138
breed effects on use, 157
definition, 137
effect on
absorption from the large intestine, 145
absorption from the small intestine, 141
body composition, 156
digestive secretions, 140
growth and feed conversion efficiency, 154
the whole animal, 153
voluntary feed intake, 153

Dietary fibre (*cont.*)
enzymic assay, 139
for sows, 156
influence on
energy digestibility, 151
mineral digestibility, 152
nitrogen digestibility, 150
nutrient absorption in the large intestine, 149
total nutrient digestibility, 150
total transit time, 152
interaction with antibiotics, 157
processing, 158
properties of, 138
sows, 156
volatile fatty acids, 147
Dietary electrolyte balance, 235
Digestible energy, 48
prediction equations, 18
Digestion in pigs, 243
Digestion, of synthetic amino acids, 106–108
Disease, effects on animal production, 285–293

Early weaned piglet
and health, 292
physiological difficulties for, 256–258
Electrolyte disturbance, 231
diagnosis of, 234
diarrhoea, 231
starving and re-feeding, 233
stress, 233, 238
vomiting, 231
Electrolyte balance
tibial dischondroplasia, 238
Electrolytes, 225
Endogenous protein, 129
Energy
and growth in boars, 355
intake in pregnant pigs, 320
maintenance of boars, 354
mating, 356
metabolism of synthetic amino acids, 111–113
protein deposition, 30
requirements, 47
semen quality, 356
Enzyme supplementation
in piglet diets, 301

Fat
deposition in sows, 321
effect of sex and diet, 27
factors controlling, 25
intramuscular fat, 22, 23
role in meat quality, 20
Fat content in carcass and feeding level, 16
FDNB, 130
Feed additives, 248
Feeding method for pigs, 245
Fibre (see also dietary fibre)
apparent digestibility, 142

Fibre (*cont.*)
  effect of digestive processes on, 140
  in diets, 244, 252
Fishmeal
  supplements, 167
Food intake
  phytase, 171
Foot lesions
  biotin, 361

Gastric pH
  inability of young pig to maintain correctly, 257
  problems associated with increases in, 257
Genotype
  growth in pigs, 42
  protein deposition, 36, 123
Grading standards of pig carcasses, 13
Growth promoters
  potential types of for pigs, 295–304
Growth hormone (see also porcine somatotropin)
  as a growth promoter, 295
  control of secretion of, 297
  involvement in growth regulation, 295
Growth rate
  beta-adrenergic agonists, 312–314
  porcine somatotropin, 309
Gut
  immune defence mechanisms, 286
  immune destructive mechanisms, 288
  immunoregulatory mechanisms, 289

Halothane gene, 23, 24
Health
  animal, effect of diet, 291
Heat stress, 365
Histidine, 71
  requirement of piglet, 79–81

Ideal protein for pigs, 40
Ileal digestibility, 86
  amino acids, 118, 125
  cereals, 86
  cottonseed meal, 98, 101
  meat and bone meal, 90, 100
  soyabean meal, 90
  use in ration formulation, 96
Ileal
  anastomosis, 127
  cannulation, 126
Immune defence mechanisms, in the gut, 286
Immune destructive mechanisms, in the gut, 288
Immunisation, oral, 292
Immunoglobulins, in the gut, 287
Immunoregulatory mechanisms, in the gut, 289
Inorganic acids
  as dietary acidifiers, 258

Intramuscular fat
  effect on eating quality, 22, 23
Ionic composition of body fluids, 226

Lactating sow
  factorial estimation of nutrient requirements, 333
  nutrient partition, 338
  prediction of
    body composition, 340–342
    nutrient responses, 332
  reproduction, 342
  voluntary food intake, 338, 339
Lactic acid
  producing microbes
    as dietary acidifiers, 258
  water intake, 193
Lean growth rates, 15–18
Lean tissue, composition and quality, 21
Level of feeding
  carcass fat, 21
Libido in boars, 349
Liquid diets
  water to meal ratio
    effect on digestibility, 196
    effect on intake, 196
Litter size, 339
Lower critical temperature of boars, 357
Lysine, 60, 61, 64
  availability, 119, 120
  beta-adrenergic agonists, 314
  catabolism, 122
  efficiency of utilisation, 117
  microbial growth in the hind gut, 126
  libido in boars, 349
  porcine somatotropin, 310–312
Lysine-arginine antagonism, 237

Magnesium, 225
Maillard reaction
  amino acid availability, 119, 132
Meat quality
  boar taint, 5
  breed effects, 6
  chilling effects, 7
  consumer perceptions, 1
  definition of, 20
  dietary effects, 5
  marbling fat, 5
  slaughter effects, 6
Meat and bone meal, ileal digestibility, 90, 100
Metabolism
  amino acids for,
    energy yielding purposes, 111
    protein deposition, 108–111
Methionine, 60, 64, 66, 128
Micro-organisms
  hind gut, 126
  in gut, effect on immunity, 292
Mineral requirements
  boars, 360

Models
amino acids, 120
AUSPIG, 42

Net energy for growth, 41
Neutral detergent fibre (NDF)
analysis of, 138
Nitrogen
deposition in sows during pregnancy, 318
intake, effects of nitrogen retention, 319
retention
effect of body weight on, 321
effect of DE intake on, 319
effect of stage of pregnancy, 321
Non-starch polysaccharides
analysis of, 139
Nutrient requirements
porcine somatotropin, 310
Nutrient absorption
effect of dietary fibre, 149
Nutrition, and fatness, 15
Nutrition
rearing of boars, 347
reproduction in boars, 348
Nylon bag, amino acid digestibility, 129

Pale soft exudative (PSE) meat, 6, 8, 23, 362
Particle size in pig diets, 248
Partition of energy intake, in pregnant sows, 321
Partition of nutrients
between maternal and conceptus tissue, 324
in the pregnant sow, 323
Phosphate, 225
Phosphorus
availability
balance technique, 164
slope ratio, 164
boars, 360
digestibility
animal products, 167
plants, 165
supplements, 168
fetal growth, 174
growth, 173, 175
maintenance, 172
pollution, 163, 176
sows, 174
Phytase
barley, 165–167
microbial, 171
plant, 170
wheat, 166
Phytates
plants, 165
Pig meat consumption, 1–4
Piglet, suckling
nutritional requirements, 335
Porcine somatotropin (PST), 38, 307–312
dosage levels, 308

Porcine somatotropin (*cont.*)
effect on body composition, 307–309
effect on food intake, 307–309
effect on live weight gain, 307–309
Pork
fat content, 2
processing, 3
Post-weaning diarrhoea
effect of dietary acidification on, 258
Potassium, 225, 229
Poultry
amino acid nutrition, 245
digestion and absorption of synthetic amino acids, 106–108
Probiotics, 249
as porcine growth promoters, 299
mode of action of, 300
Proline, requirement of piglet, 83
Protein requirements, 51, 54, 56
of boars, 358
Protein intake
and energy intake, 30
and genotype, 36
efficiency of utilisation, 117
genotype, 123
ideal protein, 39
lean deposition, 30
sex, 35
spermatogenesis in boars, 351
Protein
deposition in sows, 321
effect on carcass quality, 15
metabolism of synthetic amino acids, 108–111
PSE, see pale soft exudative muscle
PST, see porcine somatotropin

Ractopamine, 313–314
Repartitioning agents
as growth promoters, 298
Requirements
energy, 48
protein, 51, 54, 56

Semen quality
aflatoxin, 354
amino acids, 353
energy, 356
zearalenone, 354
Sex
carcass quality, 27
growth rates, 18
Skatole, 5
Skimmed milk, in piglet diets, 77,, 81–83
Sodium, 225, 229
Somatostatin immunisation
as a growth promoting technique, 297
Sows (see also lactating sow)
dietary fibre, 156
Soyabean meal, ileal digestibility, 90
Sperm cells
fertilising capacity, 352

Spermatogenesis
  plane of nutrition, 349
  protein intake, 349

Tibial dischondroplasia
  electrolyte balance, 238
Toxic substances
  recommended limits in water, 193
Tryptophan, requirement of growing boar, 83

Ulceration of digestive tract, 249

Vaccination, oral, 292
VFA, see volatile fatty acids
Vitamin C
  boars, 363
Vitamin requirements
  boars, 361
Vitamin E
  pale, soft exudative (PSE) meat, 362
Volatile fatty acids
  effect of dietary fibre, 153
  metabolism, 148
  production from dietary fibre, 147

Water
  balance, 181
  demand and ambient temperature, 189
  factors affecting, 181–190
Water (*cont.*)
  lost through skin, faeces, urine and respiration, 185–186
  produced from oxidation, 182
  quality, 192–193
  requirements, 179–199
    for metabolism, 183
    lactating sows, 209–213
    newly weaned piglets, 218–220
    nutritional factors, 187, 188
    unweaned piglets, 213–218
  salt poisoning, 188
  supply from food, 182
Water supply, 191
  delivery rate, 194–195, 212
  delivery systems, 196–198
  drinkers, 192, 193
Water quality, 202–208
  chemical standards, 204–208
  newly weaned piglets, 218
  physical standards, 204
Water quantity
  lactating sows, 208–213
  newly weaned piglets, 218–220
  unweaned piglets, 213–218

Zearalenone, 354